"A masterful overview of China's environmental processes from the twentieth century to the present." —Peter C. Perdue, author of *China Marches West: The Qing Conquest of Central Eurasia*

"A major study with statistics and details on every environmental issue, from locations of plastic disposal to debates over big dams." —Eugene N. Anderson, author of *The East Asian World-System: Climate and Dynastic Change*

"Essential reading for anyone seeking to better understand China's environmental predicament." —Micah S. Muscolino, author of *The Ecology of War in China: Henan Province, the Yellow River, and Beyond, 1938–1950*

"Should be read by everyone who is worried about the ecological future of China and beyond." —Ling Zhang, author of *The River, the Plain, and the State: An Environmental Drama in Northern Song China, 1048–1128*

射小利，害大謀，急近功，遺袁患

Aim for small advantage and wreck a great plan,

fret over short-term success and leave behind long-term suffering.

LIN YUAN 林元 (Yuan dynasty), *Shishou chong kai gu xue ji* 石首重開古穴記
(Notes on reopening the old gap at the head of the stone)

CONTENTS

Lists of Illustrations, Maps, and Tables   *ix*
Preface and Acknowledgments   *xiii*
Note on Conventions   *xix*
Introduction: Understanding China through Its Ecological History   *1*

PART I. SETTING THE SCENE

1. A Tour of China's Social-Ecological Systems   *25*
2. Development, Revolution, and Science   *63*

PART II. LAND, WATER, AND FOOD

3. Feeding a Starving Nation, 1949–1957   *93*
4. Three Years of ~~Natural~~ Disasters, 1958–1961   *118*
5. Normal Socialist Agriculture, 1962–1978   *139*
6. Solving the Warm and Full Problem, 1978–1998   *157*
7. Every Last Drop, 1998–2022   *176*
8. Dammed If You Do, 1993–2021   *201*
9. A Toxic Cornucopia, 2000–2022   *230*
10. Big Ag and Its Ecosystem Effects, 2002–2022   *256*

PART III. CITIES AND INDUSTRY

11. A Revolution of Steel, 1949–1961   *287*
12. Normal Socialist Industry, 1962–1980   *309*
13. Factory to the World, 1984–2015   *327*
14. Building an Urban Continent, 1980–2022   *351*
15. Sludge Abides, 1990–2022   *378*
16. Paradoxes of Eco-Development, 1998–2022   *401*
Conclusion: Chinese History as Ecological History   *423*

Glossary of Chinese Terms, Sayings, and Slogans   *441*
Notes   *447*
References   *455*
Index   *527*

LISTS OF ILLUSTRATIONS, MAPS, AND TABLES

ILLUSTRATIONS

I.1 Temporal and spatial scales of a boreal forest and
related disturbances 9

I.2 Temporal and spatial scales of cycles in an industrial
social system 9

I.3 The Malthus-and-Boserup ratchet 11

I.4 Graphic illustration of social-ecological system resilience 13

I.5 Schematic of the relationship between productivity
and resilience 18

I.6 The adaptive cycle 18

1.1 Farming on the North China Plain, 1930s 27

1.2 Transplanting rice seedlings, Shisantian, Taiwan, 1973 28

1.3 Flows of goods and services in a precollectivization household
in China Proper 29

1.4 Lattice window with decorative carvings, Manshuiwan,
Sichuan, 1994 31

1.5 House in Manshuiwan, Sichuan, built of local
materials, 1994 31

1.6 Extended family compound in Xiyuan, Taiwan, 1970 32

1.7 Yishala, a tightly packed village in Panzhihua,
Sichuan, 1988 33

1.8 Rice terraces with courtyard house and forests
in the background, Datiekeng, Taiwan, 1973 40

1.9 Nuosu men in front of a traditional house in Ebian
Subprefecture, Sichuan, 1913 44

1.10 Family sitting around the hearth in a traditional
Nuosu house, 1993 44

1.11 Gathering firewood, Apiladda Valley, Sichuan, 2009 47

1.12 Walking sticks left by pilgrims on a circuit of Zhayizhaga
Mountain in Jiuzhaigou, Sichuan, 2008 50

1.13 Ecological flows in a pastoral nomadic household 59

2.1 Two models of development compared schematically 67

2.2   Marxist-Leninist version of flows of goods and services
      in a pre-Founding Han village   68

2.3   Conventional and ecosystemic periodization of PRC history   79

2.4   PRC history conceived as the adaptive cycle   80

2.5   Revisionist Marxist-Leninist schematic
      of historical teleology   87

3.1   Intensification, buffer removal, and the adaptive cycle
      in lowland communities   94

3.2   Intensification, buffer removal, and the adaptive cycle
      in mountain migrant communities   95

3.3   Staple production and population, 1949–2021   99

3.4   New Year's poster from 1949, with the song "Tractor" celebrating
      prospective agricultural mechanization   99

3.5   Inputs into agriculture under Stalinist industrialization
      and the Great Leap Forward   101

3.6   Flows of goods and services in a collectivized household
      in China Proper   110

3.7   Profile of the Yellow River on the North China Plain   113

4.1   Peasant teams constructing the Ming Tombs Reservoir
      near Beijing, 1958   121

4.2   Unidirectional causality in Great Leap Forward thinking   132

4.3   A more realistic model of Great Leap Forward policy   132

5.1   Capital inputs added to the strategies for increasing grain
      production beginning in 1962   143

5.2   A sign for US agrochemicals used on dragon fruit plantations
      in Guangxi, 2018   150

5.3   Nuosu Yi farmers spreading *dimo* on Green Revolution corn
      seedbeds, 2012   151

7.1   Changes in sediment regime after water sediment regulation
      began in 2002   187

8.1   Trackers or haulers in the Chang River gorges, 1913   202

8.2   Cleaning up garbage from the Sanxia Reservoir, 2013, and
      from the Three Gorges Dam, 2014   217

9.1   Flows of goods and services in the postpeasant Chinese
      household   254

10.1 Relationships between productivity and resilience with
      increasing inputs of fertilizer and pesticides and antibiotics   261

10.2 Inputs and resilience in CCCM agriculture *266*

11.1 Relations between the physical elements of heavy industry
      built in China under the Stalinist model *288*

11.2 Masses digging and gathering iron ore, 1958 *303*

11.3 Masses operating small blast furnaces, 1958 *303*

11.4 An illicit charcoal kiln in the hills of Liangshan,
      Sichuan, 2007 *304*

12.1 *Gandalei* settlement in Daqing, 1977 *315*

12.2 Oil workers building *gandalei* houses, 1970s *315*

12.3 *Flowers of Vanadium and Titanium* statue in downtown
      Panzhihua, 1988 *319*

12.4 Mountaintop mining of magnetite ore, Panzhihua, 1988 *320*

14.1 A typical urban scene in Beijing, 1992 and 2000 *352*

14.2 Baishiqiao Road in Beijing, 2000 *361*

14.3 Calligraphy teacher Liu Lanbo, Tuanjiehu Park, Beijing, 2007 *361*

14.4 Waste separation bins at a residential complex in Chengdu *363*

14.5 Ruins of old Beichuan and new "minority-themed"
      Beichuan, 2011 *369*

14.6 Streets in Kashgar's old town, before and after urban renewal,
      2013 and 2016 *372*

14.7 A trestle between tunnels on the Chengdu-Xichang Expressway,
      2012 *374*

15.1 Water use in China, 1980 and 2008 *387*

16.1 Energy consumption and energy intensity of China's
      economy, 1980–2020 *402*

16.2 China's total energy consumption, 1996 and 2019 *403*

16.3 China's installed electrical generation capacity by
      source, 2019 *405*

MAPS

I.1   The three macroecological zones of China *7*

1.1   Agricultural regions of China Proper, mid-twentieth century *26*

1.2   Regions of Zomia now in the People's Republic *42*

1.3   Schematic of ideal Akha village layout in Sipsongpanna,
      Yunnan *47*

1.4   Chinese Central Asia, showing Xinjiang oases *53*

3.1   Heilongjiang Reclamation District *103*

3.2   Reclamation in the Huai River basin, 1950s   *106*

3.3   Yellow River basin   *112*

3.4   Expansion of the Dujiangyan Irrigation Area, 1953–2021   *115*

7.1   The shrinking of Dongting Lake, 1650s to 1970s   *181*

7.2   South-to-North Water Transfer Project   *192*

7.3   Water transfer projects between the Chang, Han, and Wei River basins   *196*

8.1   High-capacity dams   *202*

8.2   Hydrological effects of Sanxia Dam   *209*

8.3   Hydrologic basins of Dongting and Poyang Lakes   *212*

8.4   Lancang-Mekong River dams   *221*

8.5   Proposed dams on the Nu-Salween River   *224*

11.1 Anshan, showing urban area, Angang steel mill, and associated mines   *291*

12.1 Sawmill complex in Shaowu, Fujian   *312*

14.1 Beijing subway system   *367*

14.2 High-speed rail network, 2019   *375*

15.1 Waste-processing sites in and around Beijing, 2010s   *395*

TABLES

2.1   Two philosophies of development, revolution, and science   *78*

6.1   Increases in production of representative foods in China, 1980–1998   *159*

6.2   Animal agriculture and animal agricultural products, 1979–2008   *165*

7.1   Changes in surface area of major lakes in the middle Chang River basin, 1949–1980   *180*

9.1   Annual amount of animal-derived food and protein availability per day per capita in China and the United States, 2010s   *234*

13.1 China's consumer goods industry, 1980–2010s   *342*

13.2 China's heavy industry, 1980, 2000, and 2018   *343*

15.1 Increases in production of materials needed for cities and vehicles, 1979–2018   *388*

16.1 Monthly air quality index (AQI) averages and concentrations of selected air pollutants in 2014 and 2021–2022 for selected cities in China   *414*

PREFACE AND ACKNOWLEDGMENTS

On 25 August 2022 Chinese media reported that authorities in Ruzhou, Henan, had drained a twelve-hectare artificial lake in search of an errant alligator gar, a medium-sized "weird fish" (*guai yu*) imported from subtropical North America. The fish, said to be "endangered," was a bit less than a meter long and was probably originally purchased as a pet by a local family and then abandoned. It was labeled a threat to local ecosystems and native species. After the lake was "basically drained dry" with no gar to be found, the environmental authorities figured it had escaped to hide in a culvert, so they began to inspect the intake pipes. Weibo and TikTok influencers "swarmed to the site for a glimpse of the fish" (BJNews 2022; Lyric Li 2022). A few days later, "Chinese authorities . . . listed it as one of the 10 major invasive organisms threatening China's ecosystem" (Zuo 2022).

When I started the work that led to this book, artificial lakes were rare in Chinese cities, TikTok hadn't been invented, and exotic fish imported as pets were not yet a thing. My book project began when Joe Esherick and Yingjin Zhang invited me to a conference at the University of California, San Diego, in April 2007: "Paradigms in Flux: New Perspectives on Shifting Grounds in Contemporary China and Chinese Studies." I wrote a long and somewhat tedious paper entitled "Ecosystem Cycling and the Periodization of Late Twentieth-Century Chinese History." When the organizers decided not to assemble a volume from the conference papers, I asked several friends how I might condense the paper for publication in a journal. Gene Anderson told me, "Don't condense it, expand it. Into a book."

Now, in late summer 2022, in addition to chasing after weird fish, China is experiencing a record heat wave, almost certainly influenced by global climate warming. The Chang (Yangtze) River has slowed to a trickle, only two years after floods brought immense devastation, and Poyang Lake, once China's largest, is at record low levels, exposing long-submerged historical sites. Sichuan and Chongqing are recording historic high temperatures, ranging up to a startling 45 degrees C (Shepherd and Livingston 2022). It would be impossible to keep up with the ever-accelerating environmental changes and still bring a book to press, especially a generalist one like this. I'm a recovering anthropologist trying to write history without any archival sources, a white guy pretentiously thinking he knows a lot about China, an amateur ecologist attempting to integrate many kinds of technical science into a general volume. A work by such a dilettante will inevitably contain mistakes and omissions, but without the kindness of colleagues and friends, it would have contained a lot more.

Although they have not commented directly on the book manuscript or its contents, Vaclav Smil, Judith Shapiro, and Robert Marks have written pioneering works that inspired me to write about ecological history, and Jim Scott provided important if only partially heeded advice not to make the thing any longer than it already is.

Several people gave me extensive comments over and over on much or all of the manuscript, and their overall advice has been crucial to developing my overall arguments. Gene Anderson, in addition to suggesting the project in the first place, has been an encyclopedia of important facts and concepts, as well as a cautionary voice when I started to let my ideas run away with me. Eddie Schmitt and Ross Doll read drafts of many chapters and used both their wide empirical knowledge and their gifts for conceptualization to help me sharpen my arguments. Ed Grumbine has given advice both technical and general, particularly reminding me that sometimes I tend to say too much and admonishing me to keep the numbers up to date, especially since the numbers change so fast in China. Jamie Banfill has also provided useful general comments, while David Pietz and Peter Perdue were both encouraging and helpfully cautionary in their official comments.

Many people shared the excitement and tribulation of interdisciplinary field research over the years. Most prominently, in Liangshan, Mgebbu Lunzy, Mgebbu Ashy, Mgebbu Jyjy, Ma Fagen, Ma Vuga, Hxisse Vuga, Lyrgu Gogo, Lurlur Adda, and the late Mgebbu Vihly, and, in Jiuzhaigou, Yang Qingxia and her family shared invaluable indigenous knowledge and opened local doors. Tom Hinckley, Keala Hagmann, Zeng Zongyong, and Lauren Urgenson taught me what I know about forest ecology, and Amanda Schmidt and Brian Collins continually and sometimes successfully tried to educate me in basic geomorphology. Li Xingxing has been a treasured and loyal friend and research companion since the late 1980s, and Eddie Schmitt was both challenging and helpful all along. Field research in the Chengdu Plain by Dan Abramson, Jennifer Tippins, Matt Hale, Jenna Pang, and Wu Shuang was essential in supplementing my otherwise very superficial knowledge of the area. Others who enriched the field experience and the ecological and environmental knowledge I gained from it are Li Yongxian, Rachel Meyer, Dick Olmstead, Victoria Poling, Barbara Grub, Tang Ya, Joanne Ho, Kayanna Warren, Alex Kyllo, Steve Rigdon, Heather Simmons, Haldre Rogers, Sara Jo Viraldo, Tom DeLuca, Christine Trac, Mark Ingman, Geoff Morgan, Abby Lunstrum, and Andrew Scanlon.

In such a general book as this, I have had to depend heavily on constructive criticism of individual sections from people who know much more than I about specific topics. I've learned about Central Asia from Chris Atwood, Jonathan Schlesinger, and Darren Byler. Gonçalo Santos provided experiential knowledge of rice farming, as did Astrid Cerny on Central Asian migratory pastoralism. Janet Sturgeon clarified some important points about Zomian livelihoods other than the ones I know firsthand. Micah Muscolino, Chris Courtney, Eddie Schmitt, and Jing Xu made important suggestions about my treatment of developmental ideology in chapter 2. Susan Whiting has provided good advice on writing about politics for someone who, like me, understands it only a little, while Clair Yang has done the same for someone who understands economics even less. Ian Thomson's patient instructions on using the rich online resources of the University of Washington's Tateuchi East Asian Library allowed me to write the book during the COVID pandemic when I could not travel.

Ross Doll gave me more than one round of important suggestions about all the agricultural chapters, and Huang Yu generously provided unpublished information about shrimp farmers and their diets. Jack Hayes, Denise Glover, Luke Habberstad, Zhiguo Ye, and members of the Northwest China Forum and anonymous reviewers for *Human Ecology* helped sharpen the arguments about the Great Leap Forward in chapter 4, while He Wenhai contributed valuable empirical material.

Chapters 7 and 8 (on water) benefited greatly from careful readings and advice from Bryan Tilt, Darrin Magee, Chris Courtney, Arunabh Ghosh, David Shankman, Ding Xingli, Gao Yan, and Clark Alejandrino.

I never could have written even semicoherently about steelmaking without generous advice and important corrections from Don Wagner, and I want to thank Francesca Bray for introducing me to Don. About industry in general, Covell Meyskens and Victor Seow have been very generous with advice and unpublished material. In addition, Covell provided important clarification about Mao Zedong's anatomical metaphors.

Dan Abramson's careful, critical reading of the chapters on urbanization was crucial to correcting empirical errors as well as aligning my argumentation with important work in the area of urban planning, and Kam Wing Chan selflessly shoveled me more material than I could use about migration and urban population growth. Mark Selden and T. J. Cheng importantly clarified the ways rural residents could or could not travel in earlier periods. Joshua Goldstein read my urban ecology material very carefully and provided important corrections. Anna Lisa Ahlers and Mette Halskov Hansen generously allowed me to see their then-unpublished manuscript on air pollution and gave me important critical readings of my own treatment of that topic.

Many people have contributed photographs that I hope liven up my accounts. They include Angela Zito, Wang Yongchen, Bryan Tilt, Dan Abramson, Eddie Schmitt, Alessandro Rippa, and the Joseph Needham Institute at Cambridge, headed by its ever-helpful director, John Moffett. Madeleine Yue Dong, in addition to being a great colleague all these years, put in a lot of time helping me track down pictures and references. Hou Li allowed me to use two photos from her book on Daqing, while bpk Bildagnetur, CAB International, the Harvard-Yenching Institute, and the Burke Museum at the University of Washington all allowed me to use photos from their collections free of charge. Janet Sturgeon, Joshua Goldstein, and Joanna Lewis were equally generous with maps and figures from their published works.

I am very proud of the original maps in this volume, drawn by talented cartographer Lily Demet Crandall-Oral. I thank Lily for a major contribution. I am also grateful to Kaitlin Banfill for rescuing me from my total inability to draw and producing figure 1.13 on short notice. Also, special thanks to Zhou Shuxuan, who shared the map of the sawmill complex in her hometown, put together from memory with her father, Zhou Shaoping. Bo Zhao also helped me with some outline maps.

Books are known by their titles, of course, but readers are usually unaware of the long and agonized discussions leading to the words that appear on the cover and in the catalogs. For this one, suggestions from Gene Anderson, Chris Courtney, David Pietz, Eddie Schmitt, Jonathan Lipman, Victor Seow, Ross Doll, and of course Lorri Hagman led the way to a final decision.

Finally, posting queries to the Sinologists group on Facebook has always yielded fascinating results, whether it is advice on tricky translations or access to obscure sources. Particular thanks go to Eli Alberts, Eric Schluessel, and James Meador for clarifying the nature of Harbin bread, as well as to Amy Gordanier for resolving a particularly thorny translation dilemma.

Jing Xu very kindly proofread the glossary and the Chinese-language sources in the references list and found a few typos that my non-native eyes somehow missed.

Needless to say, all errors of commission or omission, committed out of haste, personal quirks, or sheer stubbornness, are mine alone and no fault of the people I have mentioned here or of anyone I've inadvertently omitted.

I have been working with University of Washington Press for more than forty years now, and for more than twenty-five of those years Lorri Hagman has been my patient, encouraging, sometimes appropriately stern editor. Her support and guidance have contributed materially to making this book a reality. Press editors Marcella Landri and Joeth Zucco, along with copy editor Maureen Bemko, assisted greatly in bringing the book to its final form.

So many authors have rightly pointed out that spouses bear an unfair burden of book writing. That has certainly been true for Barbara, my wife and companion of more than half a century. Her caring, patience, indulgence, constructive critique, and shining, no-nonsense intellect have all been lifelong inspirations for me. I can only thank her for all these and more.

NOTE ON
CONVENTIONS

Writing about China in English always involves disputable choices in transcription, formatting, and translation. The following are the choices I have made:

Chinese terms are transcribed in the standard Hanyu Pinyin system. Chinese characters can be found in the section Glossary of Chinese Terms, Sayings, and Slogans.

Chinese personal names are written in the Chinese order, with surname first, except when they refer to authors who have listed their names with the given name first.

The river known as Yangtze in most English-language writing but as Changjiang (Long River) in Chinese is referred to here as the Chang River.

The last three paramount leaders of the Chinese regime (i.e., Jiang Zemin, Hu Jintao, and Xi Jinping) are known by their most important title, General Secretary (of the Chinese Communist Party), rather than the ceremonial title of President (of the People's Republic).

The time of the founding of the People's Republic in 1949 is referred to as the "Founding," a more ideologically neutral term used in translating the Chinese term "Jianguo," rather than "Liberation" (Jiefang), or "Revolution" (Geming), which in Chinese refers to the whole process of social change that followed the Founding.

The period from 1953 to 1979 is referred to as the "high socialist" period, to reflect the state socialist economy created and maintained during that time. "Maoist" refers here only to ideology, not to the time period.

I have geekily and crankily chosen to translate various slogans and sayings in ways that I think hew more closely to the meaning of the original than do the customary English renderings. Where there might be confusion, I also give the customary forms.

I use the metric system exclusively, since it is standard in almost all countries. The only exceptions are Chinese *derivatives* of metric units not used internationally: one *jin* is 500 grams, and one *mu* is a fifteenth of a hectare, or 666.7 square meters.

Finally, some lowland parts of Taiwan have been part of the China Proper social-ecological system since the early seventeenth century, so I have used some examples from there in the text and illustrations of chapter 1. This implies nothing about the political status of Taiwan.

AN ECOLOGICAL HISTORY
OF MODERN CHINA

INTRODUCTION Understanding China through Its Ecological History

When we try to pick out anything by itself, we find it hitched
to everything else in the Universe.

JOHN MUIR, *My First Summer in the Sierra*

Green waters and blue mountains are gold and silver mountains.

XI JINPING, 2013

China's 1.4 billion citizens now live lives of material plenty that would
have been unimaginable in October 1949, when Mao Zedong stood atop
Tian'anmen and proclaimed the founding of the People's Republic. At
that time all but a small urban elite and perhaps a few rural landlords
lived in danger of starvation, fatal infection, or death by marauding
army. For the 80 percent or more of China's people who were subsistence farmers, harvests were unpredictable; food was monotonous,
mostly starchy grains, and often insufficient; clothes were patched and
repatched; houses were built of mud; roads, where they existed at all,
were rutted and muddy. Most farmers could only dream of such amenities as electricity, running water, indoor plumbing, schools, or clinics,
let alone cars, household appliances, or televisions. Agricultural infrastructure was in disrepair after twelve years of war; forests continued
to be cut down for firewood; memories of famine were recent in many
areas. Non-Han populations in the western parts of the country lived
even simpler lives, mostly with no access to roads, education, or public
health services. China was, in simple terms, a poor country.

As I write in 2022, China is an upper-middle-income country. About
65 percent of its citizens live in cities, mostly in high-rises amid wide
boulevards and tastefully designed, crowded parks. The educated make
their living as professionals and business employees, while the less
fortunate are factory, construction, and service workers. Most of the
35 percent who remain in rural areas no longer farm; those who do use
machinery and chemicals to produce yields several times those of their

grandparents' day, marketing widely varied foods that give ordinary people access to the rich culinary tradition once a monopoly of the wealthy elite. Like their urban compatriots, rural people connect to the world through the (censored) internet and ubiquitous mobile phones. China has the world's longest high-speed rail network, produces more than half of the world's steel, and is a leader in emerging electronic and energy technologies. Though social inequality is severe, all but the poorest live lives that would have seemed wildly luxurious to their mid-century ancestors. China is, in simple terms, a development success story.

Also as I write in 2022, China has become a villain for world environmentalists. In growing from impoverished giant to wealthy superpower in seventy years, China has sacrificed whatever resilience its ecosystems once possessed. It has polluted and poisoned its air, water, and soil; it has paved over large parts of its territory; it has turned forests into plantations and seen deserts expand, mangroves disappear, and lakes come to resemble green paint. China also contributes to global environmental degradation. It emits far more greenhouse gases than any other nation, and where richer countries once exported pollution to China along with manufacturing jobs, China now reaches outward in search of raw materials, animal feed, agricultural land, and even fish.

As a result, in the past three decades Chinese people have become increasingly concerned about the environment. National leaders worry about retaining legitimacy and sustaining high rates of industrial and agricultural growth. Scientists worry about chemicals in the air, water, and soil, as well as about hydrological changes caused by dams, desertification caused by inappropriate land use, declines in forest quality caused by hasty logging and hastier reforestation, and coastal hazards resulting from wetland reclamation. Urban dwellers worry about dirty air and polluted water. Urban planners worry about flood vulnerability, heat islands, and sprawl. Consumers worry about pesticide-laden food. Farmers worry about eroded, contaminated, or compacted soil and about diminishing groundwater supplies and vulnerable irrigation works. Environmentalists worry about biodiversity loss and children's lack of exposure to nature. Many of these worries have prompted action in the last three decades, with varying degrees of success.

This book is the story of China's "miraculous" development and of the losses that development has brought. It shows how a country can become both modern and vulnerable, giving its people both prosperity and pollution, a conjunction that raises several important questions. Just as Freud's *Civilization and Its Discontents* posited that repression of desire was an inevitable result of the social contract, is environmental degradation an inevitable result of economic development? Could growth have happened in a "greener" way? How reversible is environmental degradation, now that both officials and the public are committed to reversing it? If early stages of economic growth inevitably bring environmental degradation, do later stages inevitably reverse it, as the so-called environmental Kuznets curve (EKC) would predict?

As humans draw more and more energy and materials from ecosystems, these systems appear to become less and less resilient—less able to absorb disturbances or shocks and to continue functioning. Having become dependent on all the material inputs that have brought about economic growth, are ecosystems destined to be forever fragile and to become vulnerable to ever-smaller disturbances? If so, is China doomed to a fragile, polluted, vulnerable environment for the foreseeable future? To answer questions about possible futures, we need to understand the process by which China's environment has gotten to its current parlous state. Pursuit of this understanding demands a history of China as a system of social-ecological systems, in other words, an ecological history.

WRITING ECOLOGICAL HISTORY

Ecological history is more than another name for environmental history, more than just a focus on human-environment interactions. Ecological history deals with how "[a thing] is hitched to everything else in the universe" (Muir 1911, 211). This perspective sees China and its constituent parts as *systems* of interconnected elements—land, water, air, trees, crops, animals, bacteria, chemicals, people, policies, organizations, laws, literature, film, art, values, and beliefs—and shows how these elements have interacted over the course of time. An ecological history, as Bao Maohong (2004, 478) says, must center our history on the "*interaction of humans, society, and the rest of nature.*"

Although environmental history has begun to take its rightful place in the historiography of the People's Republic of China (see, e.g., Bao Maohong 2004; Pietz 2015; Courtney 2018b; J. Goldstein 2021), the standard developmental narrative of China's history focuses on politics, economics, and policy. According to the official version, the Revolution following the Liberation of 1949 stabilized a long-chaotic society and rescued the nation from a century of imperialistic exploitation and humiliation.[1] Under the Communist Party's leadership, the 1950s were a time of socialist construction, expanding industry, building agricultural infrastructure, and solidifying educational and public health systems. The Great Leap Forward between 1958 and 1961 was a tragic mistake, a detour from this constructive road. However, official histories have yet to portray it in the stark terms employed by unofficial and foreign historians.[2] But after a brief period of recovery, the official narrative tells us, politics and ideology got the upper hand over rationality, and the Cultural Revolution of 1966–76 was a disaster for China's development. Economic growth stalled, millions suffered from unjust political persecution, and popular support for the regime was severely undermined. Only with the overthrow of the radicals whom Mao Zedong sponsored in his final years was China again launched upon its forward trajectory, which has led through the so-called Reform and Opening process (often referred to simply as the Reform) to the dynamic, prosperous, and powerful nation that is China today.

In this standard story, ecology and environment play very small roles until around the turn of the twenty-first century, when leaders realized that they could no longer avoid environmental problems. Only when Hu Jintao's regime, which began in 2002, formulated the narrative that "ecological civilization" was replacing the "industrial civilization" of the previous era and its associated environmental costs, does environment take a prominent place in the official story.

Putting environment and ecology at the center of PRC history from the beginning leads to a different periodization of this history, consigning the Cultural Revolution to a minor role. Instead, I see four major periods: reconstruction and building socialism from the Founding to the Great Leap, 1949 to 1957; the Great Leap Forward and its immediate aftermath, from 1957 to 1962; environmentally unconscious development, from 1963 to 1998; and eco-developmentalism (Haddad and Harrell 2020), from 1998 to 2022 and beyond.

Ecological history begins with social-ecological systems—multi-stranded webs of humans, other living species, and components of the physical environment. In social-ecological systems there are no independent variables. History proceeds not by cause and effect but instead by growth, shrinkage, shifts, flips, and feedback loops at the nexus of humans and other parts of nature. Social-ecological systems are "neither humans embedded in an ecological system nor ecosystems embedded in human systems, but rather a different thing altogether" (B. Walker et al. 2006). They "include flows of energy and materials, but also flows of ideas, power, and social relations" (Abel 2007, 56–57). Social-ecological systems have both spatial and temporal components; the relations between spatial parts change over time, and the temporal changes over time work differently in different spaces.

SPATIAL STRUCTURES—HIERARCHIES, CORE-PERIPHERY STRUCTURES, AND MISMATCHES

Three aspects of systems in general are important for analyzing a social-ecological system. First, the component parts of any system are themselves systems at a smaller scale, forming hierarchies of nested systems at different scales. The multilevel hierarchy of market systems in late imperial China (Skinner 1964–65) is a social example; an ecological example is the system of watersheds of a main river, its tributaries, the tributaries of those tributaries, and so forth. Second, any system at any level has a core and a periphery, which exchange people, goods, and information asymmetrically. In marketing or urban systems, more goods, people, and services usually flow from the rural periphery to the urban core; in a watershed system, water, energy, nutrients, and organisms run downhill from the tributaries to the main river and out to the ocean. Third, flows of different things—energy, water, people, animals, power, ideas—operate at different spatial and temporal scales. Economic and administrative systems overlap but do not coincide; they also operate at different scales, as do weather systems, vegetation patches, and watersheds. Just as important, ecological systems and the social systems they overlap are rarely congruent, leading to misunderstandings and difficulties in management (Cumming, Cumming, and Redman 2006).

What we see when we look at either social or ecological systems, then, depends on the magnification of our scope. Using only the microscope of the ethnographer studying the household or the macroscope of the historian studying provinces or macroregions, we will miss both the phenomena that are clear only at a different magnification and the cross-scale interactions that affect phenomena at any scale. Also, human actors (B. Walker et al. 2006) have their own scales at which they observe, think, analyze, and act. We thus need to use both an etic, or arbitrary, scale, selected according to what we want to see, and an emic scale that reflects what actors in the system see and act upon. Otherwise, we are in danger of creating our own scale mismatch between the scale of our analysis and the scale of the thing we are analyzing.

Some properties of any system at a higher scale are emergent; they are not the simple sum of and cannot be explained totally by processes at a lower level. Thus, a cell has emergent properties that cannot be precisely predicted from the physical properties of the molecules that constitute it, and an ecosystem is not just the sum of the properties of its species. These properties of living systems emerged historically in the evolution of life, of species, and of ecosystems involving the interactions of multiple species over multiple scales (Levin 1992, 1956). States are a social parallel; they emerged *de novo* in the historical process of pristine state formation (Fried 1967), and we could not have predicted their form or functioning from the psychological or physiological characteristics of the humans who inhabit the states or from the social organization of the villages, tribes, and chiefdoms out of which the states emerged. In densely populated rural China, for example, a marketing community consisting of several villages contains flows of rents and sales, and usually also of marriage relations, that do not exist at the village level and could not be predicted from the nature of flows within a village. Similarly, in pastoral areas in Chinese Central Asia, larger-scale political-military confederations are not predictable from the lineage and clan structures of migrant communities.

The People's Republic of China is continental in scale, but it is not a continent, nor is it a natural entity. China's boundaries are the outcome of the political history of the Qing empire (1644–1911) and are based only very partially on topographic barriers such as mountain ranges, rivers, or deserts and even less on divisions between ecological zones. The economic and ecological core of the Qing empire and its PRC successors lies in what, using an old term, I call China Proper. The empire's far peripheries reached southwestward from this core into the complex, vertical topography of what James Scott, following Willem van Schen-

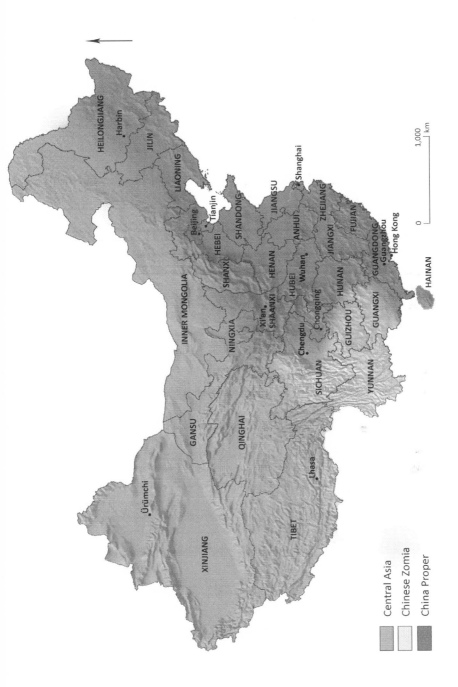

Map I.1. The three macroecological zones of China. Map by Lily Demet Crandall-Oral.

Central Asia

Chinese Zomia

China Proper

del, has called Zomia (Scott 2009; van Schendel 2002). The Qing empire also stretched northward and northwestward into the pastoral zones of Central Asia (Lary 2007, 2; Perdue 2005, 13–50). As the Qing entered the world order organized by nations in the nineteenth century, its borders were systematized and ratified through military and diplomatic confrontations with neighboring powers. When the Republic succeeded the Qing in 1911–12 and the People's Republic succeeded the Republic in 1949, the borders remained largely where they had stood at the end of the Qing, leaving parts of Zomia and Central Asia within PRC territory.[3] Because the Communist Party regime has had the power to open and close its borders, the flows of people, goods, and ideas have been different within the borders of China and across those borders (Sturgeon 2005; Turner, Bonnin, and Michaud 2015). In this sense, China as a whole is a social-ecological system.[4]

TEMPORAL STRUCTURES—CYCLICAL AND SECULAR CHANGE

A social-ecological system is not just about space but also about time; ecological history involves tracing both long-term secular changes and cyclical oscillations. Any system at any spatial scale changes in different ways over different time scales in response both to its own inner dynamics and to disturbances that affect the system from outside. Lance Gunderson and C. S. Holling's (2002) diagram (figure 1.1) illustrates the relationships between spatial and temporal scales in a boreal forest. The components of socioeconomic systems (figure 1.2) show a similar array of scales in their temporal and spatial variation.

In an ecosystem, solar energy, nutrients, water, and other things flow faster and slower in cyclical patterns over the course of a day and of a year. These cycles are driven by changes in the amount of solar energy reaching the system both through direct insolation and through conduction by air and water, as well as by the genetic predisposition of the various species to grow, mature, reproduce, and decay. Similarly, any government's cycles of budgeting, tax collection, and expenditure work themselves out over a yearly fiscal cycle. But in any system, stocks and flows are not the same from the peak or trough of one cycle to the peak or trough of the next. Spring and summer precipitation in

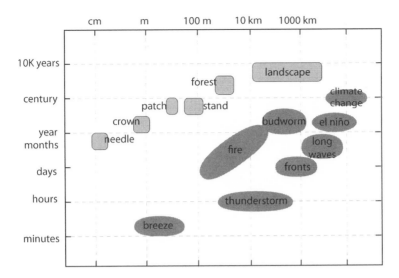

Figure I.1. Temporal and spatial scales of a boreal forest and related disturbances. From *Panarchy*, edited by Lance H. Gunderson and C. S. Holling. Copyright © 2002 Island Press. Reproduced by permission of Island Press, Washington, DC.

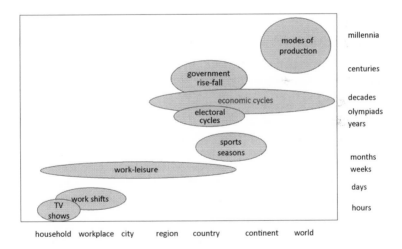

Figure I.2. Temporal and spatial scales of cycles in an industrial social system. Log time and space scales are not precise, so no numbers are shown.

different parts of China Proper varies according to El Niño–Southern Oscillation cycles, as well as longer cycles of ocean temperatures in the Pacific (F. Yang and Lau 2004). Similarly, China's yearly rounds of congresses, plenums, budgets, and reports are distributed over longer cycles of quinquennial party congresses and less predictable changes of paramount leadership. At even longer scales, there are climate cycles on the order of the Medieval Optimum of approximately 800–1300 and the Little Ice Age of approximately 1450–1900 (B. Yang et al. 2002) and political cycles such as the dynastic cycle of imperial Chinese history.

Both the Qing empire and the People's Republic, in their early decades, achieved remarkable gains in population and production over about a century and a half in the Qing case and over just half a century in the People's Republic case. The Qing and PRC cycles are two tightenings of what biological anthropologist James Wood (1998) called the MaB (Malthus-and-Boserup) ratchet (figure 1.3). Thomas Malthus (1798) pointed out that population could grow faster than the ability to produce subsistence resources, primarily food. When production could not keep up with population, a society would move from a condition of plenty to one of misery, or insufficiency. Ester Boserup (1965) pointed out, however, that population increase or political change might force people to adopt innovative technology or work harder, thus allowing a larger population or a higher level of consumption. Wood demonstrated that this meant the size of a human population is never in stable equilibrium. In conditions of plenty, people will have more children or consume more resources per capita, and after a while the population will overshoot the carrying capacity of its ecosystem. If technology provides a way to increase per capita production, then plenty will be restored under the new technological regime, and people will once again be motivated to overshoot. With each episode of major technological change, the system gets ratcheted up a notch. In the Qing, the most important technological innovation was the spread of the New World crops of corn, potatoes, and sweet potatoes; in the early People's Republic, it was fossil fuels. In both cases, the new technology allowed people to thrive at first, but in applying the new technology, people consumed resources that previously buffered the system from disturbances, thereby decreasing the system's resilience. In the late Qing, few buffers against ecological disasters were left, compounding misery. Nowadays, the People's Republic experiences

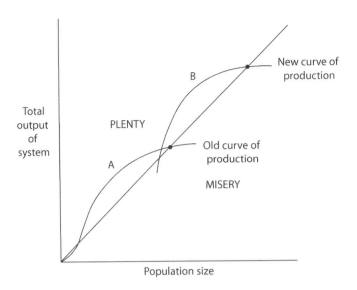

Figure 1.3. The Malthus-and-Boserup ratchet. Technological improvements in cycle A and cycle B allow the ratio of output to population to grow, putting the people into a state of plenty (to the upper left of the straight line), but this allows population to increase and overshoot carrying capacity at that level of technology, crossing the line into misery (to the lower right of the line) and necessitating (if possible) new technological improvements. Redrawn from James W. Wood, "A Theory of Preindustrial Population Dynamics: Demography, Economy, and Well-being in Malthusian Systems." *Current Anthropology* 39 (1): 99–135, 1998. Copyright 1998 by the Wenner-Gren Foundation for Anthropological Research. All rights reserved 0011-3204/98/3801-0005$3.00.

plenty, but this book asks the question of how much resilience the system has lost in the process.

Because no cycle will be just like the previous one, cycling systems do not tend to self-correct toward a stable equilibrium. This process is partly stochastic—for example, after a fire burns a forest, something as random as the amount of rainfall after the fire or the direction of the wind that blows seeds to the burned-over area may determine which opportunistic species reestablish themselves (B. Walker et al. 2006). Similarly, the "return to work" after the COVID pandemic promises to involve a lot more online conferencing than before. Slow variables, or things that change incrementally, can change the properties of the system from within, often in ways that are not noticeable until they have already altered the system's capacity to respond to shocks or disturbances

(Bennett, Cumming, and Peterson et al. 2005, 946; B. Walker et al. 2006). Shifting agriculture in Chinese Zomia, for example, follows a cycle of forest cutting, cultivation, abandonment, and regrowth. This succession may be stable across a few cycles, but it may give rise to change in slow variables such as growth of human population or gradual erosion of topsoil, rendering it much more vulnerable to disturbances (Harrell et al. 2022), so that a major disturbance such as a clear-cut may destroy the ability of the system to restore itself. When these slow variable changes combine with major disturbances, the system may enter a new state, including one of reduced productivity.

DISTURBANCE AND RESILIENCE

Resilience is the ability of a system to absorb disturbances and return to a functioning state. This does not mean, however, that social-ecological systems are like Chinese *budao weng* dolls, weighted at the bottom so they bounce back when you push them over. Resilience in an ecosystem is instead defined as the magnitude of disturbance that a system can absorb without "flipping" into an alternative stable state in which the variables that control the system are fundamentally altered (Gunderson and Holling 2002, 27–28).

The ball-in-basin metaphor illustrates the relationship between slow variable change, disturbance, and resilience. In figure I.4, the basin represents all possible values of the variables that control the system and allow it to maintain its basic properties. The depth of the basin is the range of variability that a system can tolerate and still retain its basic functioning—in other words, its resilience. The ball represents the values of those variables in the absence of disturbance. The arrow is the disturbance—it moves the ball, changing the state of one or more variables in the system.

For example, in a well-managed system of rice agriculture, local residents participate in an equitable regime governing maintenance of terraces and waterworks, and wetlands have not been converted to paddies. The system will be highly resilient, as represented by the deep basin in figure I.4A. A major disturbance such as a flood or drought might cause short-term damage; it might even wipe out one season's crop, but strong institutions could restore the system's functioning the following season.

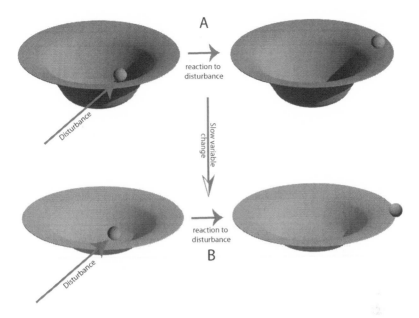

Figure I.4. Graphic illustration of social-ecological system resilience. In A (*above*), the basin is deep, representing a resilient system. Even a major disturbance does not move the system into an alternative stable state in which it is controlled by different variables. Over time, slow variables may shift the system to state B, where it is less resilient. The same magnitude of disturbance sends the system into an alternative state—the ball falls out of the basin and cannot return.

Changes in slow variables could, however, lower the resilience of the system, making the basin shallower. If, for example, population in the local area grew, more wetlands or lake margins might be converted to paddy land, reservoir storage capacity might be reduced, and the local irrigation association might have difficulty enforcing rules for water allocation. The basin would become shallower (figure I.4B). The same disturbance or even a weaker one could damage the system to the point that it could not easily be repaired. Much land might be lost, people would become refugees, they would no longer have the institutional means to maintain the necessary infrastructure, and the system would go into an alternative state—it would no longer be a rice-growing community.

Two resilience-promoting features are especially important for PRC ecological history. One is functional redundancy, when one actor or a set of actors eliminated from the system (such as a species that becomes extinct locally or a community leader who dies suddenly) can be replaced by another actor or set of actors (Low et al. 2003, 87). A second is diversity—genetic diversity within a species, species diversity within a patch, or patch diversity within an ecosystem (Levin 1992, 1959; Kremen, Iles, and Bacon 2012; Kremen and Miles 2012; Cabell and Oelofse 2012). Diversity reduces the chance that a disturbance will wipe out an irreplaceable component and knock the whole system out of functioning.

No system has infinite resilience; there is always some disturbance big enough to fundamentally alter its nature and functioning. A natural disaster or an invading army may cause the collapse of a civilization or a major polity, especially when human activity has decreased the resilience of the system. For example, deforestation apparently decreased the resilience of many classic Maya political structures, rendering them vulnerable to periodic droughts (Diamond 2005; Griffin et al. 2014). Similarly, World War I collapsed three major empires, and out of the rubble emerged many nation-states. Or a disaster may collapse only a local social-ecological system, as Hurricane Katrina did in New Orleans, which has reemerged in attenuated form (Hobor 2015). When European imperialist powers confronted the Qing empire in the late 1830s, it was in a very vulnerable state, due to the slow variables of population increase, resource degradation, and institutional decay. A big enough disturbance can collapse a system even in a state of rapid growth, as was the case in many places in China during the Great Leap Forward (Harrell 2021).

Complete lack of disturbance, however, also has its dangers. A certain amount of disturbance may promote long-term resilience; the "intermediate disturbance hypothesis" (Connell 1978; Grime 1973) explains why. With no disturbance at all, diversity diminishes and, along with it, resilience. Intermittent fires in western North American forest ecosystems actually promote diversity and thus long-term resilience (Hessburg and Agee 2003, 45). Total fire suppression, as practiced by the US Forest Service from the late nineteenth to the late twentieth centuries, eliminated the small fires that acted as intermediate disturbances, resulting in fuel buildup and making the forest ecosystems less resilient to the high temperatures and low humidities that caused big fires such as those in Yellowstone in 1988.[5] Applied more broadly, the intermediate disturbance hypothesis explains how human-managed ecosystems can be more biodiverse (and thus more resilient) than systems in the same location unmodified by humans (Peña 2005, 90; B. Walker and Salt 2006,

125–38; d'Alpoim Guedes et al. 2020). It also calls into question the post-modern "deep ecology" or "eco-fascist" notion that humans are always destructive to ecosystems (see Ellis 1996).

PREDICTABILITY AND UNCERTAINTY

Precisely because they are complex, social-ecological systems are unpredictable. They involve such a huge variety of inputs (energy, water, nutrients, policies, prices) and actors (individuals, species, patches, subsystems, institutions) that they are impossible to model mathematically (Levins 1966; Pilkey and Pilkey-Jarvis 2007). This has been demonstrated repeatedly in attempts to predict and manage such ecosystems as fisheries (Francis n.d.; Holling and Meffe 1996), especially when managers try to maximize a single variable.

Because social-ecological systems are so complex as to be unpredictable, we will never be able to model all possible contingencies. There will almost always be others—what Lance Gunderson (2003, 36–37) calls "ecological surprises." Hurricane Katrina was one such; despite the best models of climate, levee strength, and storm surges, no one predicted the convergence of climatic, social, and political factors that led to that disastrous event. The outbreak of World War I, just when Europeans were feeling a new era of peace, prosperity, and harmony, is another example of a surprise.

It has been difficult for environmental managers to accept the unpredictability of ecosystems. This has been particularly true for scientific modernists (Scott 1998) who hold faith in the power of reason to engineer the world. Marxism-Leninism, a species of scientific modernism built out of faith in both science and human agency (Engels [1876] 1939), was particularly vulnerable to hubristic fantasies that nature could be engineered into maximum material productivity. The hyperrationalism (Glover, Hayes, and Harrell 2021) born of this double faith has been a constant feature of PRC approaches to social-ecological systems at scales from the continental disaster of the Great Leap Forward to more recent local attempts at rational management, such as breeding shrimp scientifically in the Leizhou Peninsula (Y. Huang 2012) or restoring grasslands in northwest China by planting trees (Shixiong Cao 2008).

There are alternatives to this kind of rigid, hyperrational scientism.

C. S. Holling and Gary Meffe (1996) proposed "golden-rule management"—modeling and prediction not of specific outcomes but of plausible ranges of key system variables or, in another version, of plausible scenarios for the future state of the system. In China as of now, however, planning for resilience has continued to take a back seat to planning for productivity and to further attempts to bring socio-ecosystems under simplified human control.

BUFFERS

Ecosystem resilience depends heavily on buffers or guarantors that can absorb disturbances, so that a system does not easily flip into an alternative state (Gunderson 2003, 45; Ludwig, Walker, and Holling 2002). These buffers can be of several kinds:

- Ecological buffers, such as wetlands, which absorb overflow from lakes or rivers, or forests, which stabilize soil and decrease runoff
- Infrastructural buffers, such as terraces, which allow intensive agriculture but retard erosion, or irrigation water storage systems, which protect against both too much rain and too little
- Institutional buffers, such as kin networks, which share resources in times of scarcity, or governing organizations, which regulate the use of resources
- Cultural buffers, such as moral disapproval of greed, which restrains overexploitation of resources, or beliefs that wasting resources will bring supernatural punishment.

None of these buffers is cost-free to maintain, which is one reason why resilience in social-ecological systems tends to decline over time unless addressed directly. Ecological buffers have an opportunity cost in short-term productivity of a system; people sometimes recognize this and will often forgo short-term gains in favor of long-term stability. People in a condition of misery, however, will usually shift resources from ecosystem services to human consumption (Solow 1993). "Desperate ecocide"—destroying the buffers in a local ecosystem in order to survive—is an extreme example (Blaikie and Brookfield 1987, 240). Infrastructural buffers may increase both productivity and resilience up to a certain inflection point, but they are very costly to maintain. Institutional buffers depend on a social contract that is increasingly strained when participants in the institutions view the self-sacrifice of participation as larger than the personal gains. In a common property system, such perceptions may lead to what Garrett Hardin (1968) called the "tragedy of the commons," something that can be prevented only by strong common institutions (Ostrom 1990). In a hierarchical system of governance, such perceptions result in the governors' no longer governing in the interest of the governed or in the perception that governors are violating the rules of a moral economy (Scott 1976). Finally, cultural buffers may be either ignored in times of desperation or changed by exposure to different sets of values, including capitalist accumulation. In all of these cases, the weakening of buffers can lead to ever more temptation (or desperation) to act in short-term self-interest and thus weaken the buffers further, driving the social-ecological system into collapse and reorganization.

Alternate strengthening and weakening of buffers, or gains and losses in resilience, has happened differently in China's three great ecological zones. In Chinese Central Asia, herders living a traditional life did not try to construct infrastructural buffers; they realized that in that zone of low productivity, any increase in efficiency almost immediately led to decreases in resilience. They did, however, benefit from protection by extensive institutional and cultural buffers. In Zomia, infrastructural buffers were also difficult to construct and maintain, but some did exist, along with institutional and cultural buffers. In China Proper, by contrast, the very existence of intensive agriculture was completely dependent on infrastructural buffers to replace many (but not all) of the ecological buffers that would have served the same function in a less intensive system. Even with these buffers, however, the extreme intensification of Chinese agrarian life both in the Qing empire and in the People's Republic meant that the resilience of the system could diminish rapidly if any of the remaining buffers were neglected or destroyed.

RESILIENCE AND PRODUCTIVITY

Many processes of development in recent times have enabled production to increase at the expense of ecosystem resilience. Monocrop industrial

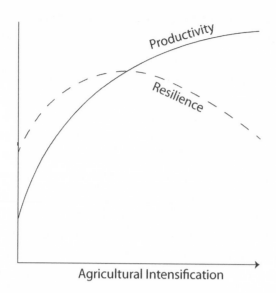

Figure 1.5. Schematic of the relationship between productivity and resilience.

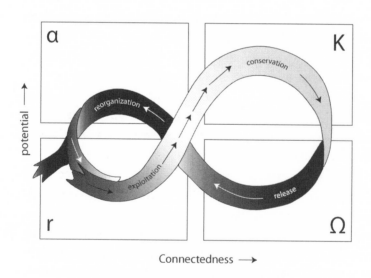

Figure 1.6. The adaptive cycle. From *Panarchy*, edited by Lance H. Gunderson and C. S. Holling. Copyright © 2002 Island Press. Reproduced by permission of Island Press, Washington, DC.

agriculture and industrial strip mining are prominent specific examples, while the entire industrial revolution based on fossil fuel energy is a more general one. The relationship between productivity and resilience, or efficiency and resilience, is, however, not simple or linear. Central and southern China's historic civilization expanded for more than a millennium on the basis of wet-rice agriculture, one of the most intensive systems of food energy production ever developed. At a certain level of productivity, a community growing wet rice could enjoy a more stable and resilient livelihood than it would have had before their ancestors installed the paddy system, with its soil conservation and annual recycling of nutrients. It was only at a further degree of intensification, when tying up too much land in paddies and diverting too much water for irrigation removed the remaining ecological buffers from the system, that productivity sacrificed resilience and led to increased frequency of disasters (Jiayan Zhang 2014). Similarly, many of the projects that increased productivity in the early Qing and again at the beginning of the People's Republic increased resilience at the same time. Lower frequency of disasters accompanied continent-wide moves into material plenty, but further development made the system more vulnerable once again. The relationship between productivity and resilience in an agricultural ecosystem is illustrated in figure 1.5.

THE ADAPTIVE CYCLE

When increases in production lead to decreases in resilience, major disturbances can cause a social-ecological system to collapse. Energy, resources, species, institutions, and individuals remain, however, and they usually reorganize themselves into a new system, which then tends to grow or develop anew. The model of the adaptive cycle (figure 1.6) illustrates this recurring temporal pattern.

This model consists of two loops, each with two phases. The "front" loop is the growth part of the cycle. It begins with the r phase of exploitation, as a system is organizing its resources and building more complex structures and subsystems within it. Eventually, in a kind of gradual process characteristic of logistic growth in systems from forests to bureaucracies, the curve inflects. The organization gets more complex, the system locks its resource flows into ever more rigid infrastructures and institutions, and the system moves into a K, or conservation, phase, using the great majority of its internal and external energy inputs just to maintain itself, causing it to become rigid and lose resilience. More and more vulnerable to major disturbances, the system loses its ability to restore itself, and when a disturbance is big enough, the system crosses a threshold. It enters the "back" loop and transforms or collapses into the rapid and destructive omega, or release, phase, leaving the system in a state of chaos, when most of the energy is dissipated and not harnessed by humans. But the system is then freed to reorganize itself during the alpha phase, usually in a somewhat different form. The next round of the cycle rarely duplicates the previous round, allowing secular change.

As C. S. Holling, Lance Gunderson, and Garry Peterson (2002, 95–98) point out, a system in a K phase is not always vulnerable to immediate collapse. It can instead enter a "rigidity trap" where the system is so well buffered by institutions and infrastructure that it is quite resilient against disturbances up to a certain size. The trade-off, however, is that it becomes impossible to alter the system's condition, so that its functioning becomes more and more dependent on particular buffers. If a greater disturbance strikes the system, then it will enter an even more devastating back loop. This happened all across China with the Great Leap Forward of the late 1950s; more recent local examples have included disease outbreaks in massive pig-farming operations and re-current floods in the middle Chang River region.

Intermediate disturbance can stabilize the adaptive cycle. A disturbance of intermediate size during the front loop of a system's cycle can retard the otherwise inevitable inflection and movement toward the brittle conservation phase. Small disturbances and the adjustments they facilitate function as a sort of safety valve or vaccination against large disturbances and catastrophic effects (Scheffer et al. 2002, 206–7). At the same time, bureaucracies in particular are averse to allowing small disturbances to happen (Holling and Meffe 1996, 329). In a highly institutionalized or bureaucratic system, "the priority of resource management becomes stability and not resilience" (Davidson-Hunt 2003, 68), and there develops a "pathology of natural resource management" (Holling and Meffe 1996; Gunderson 2003, 33) in which bureaucratic managers preserve themselves and the systems they manage at all costs. This can end in system collapse, as happened with the Soviet Union in the 1980s.

Contemporary Chinese social philosopher Yu Jianrong has warned against the twenty-first-century Chinese regime's bureaucratic tendency to seek stability at all costs. He claims that this tendency has led to a kind of "rigid stability" (*gangxing wending*) characteristic of the K phase of the adaptive cycle, making the current social order dangerously vulnerable to disturbance (Yu Jianrong 2012). In the ecological realm, this attitude has often led to a tendency to implement a "fix to fix the fix." Managers become leery of removing infrastructure that has enabled increases in production, even when it has created pollution, degradation, or diminished resilience. Their solution is to create more infrastructure, which further hardens the existing rigidity trap. Pollution control equipment in heavy industry is a rather benign example, but some water control and diversion projects directed at problems created by previous water control and diversion projects are examples of fixes that only lead to further decreases in resilience.

....

Are China's hyperproductive local, regional, and national social-ecological systems now in a K phase, dangerously lacking resilience to disturbances that will inevitably happen? We know that their continued functioning is increasingly dependent on infrastructural and institutional buffers. Ecological buffers have been greatly diminished by more than seventy years of rapid development, and the environmental consciousness that could become a cultural buffer is still in its infancy. There is a problem with predicting resilience, however. Despite some research on such systemic problems as "critical slowing down" or increases in "flickering" (Scheffer et al. 2012; Barnosky et al. 2012), in the end we can only know a system's true resilience in retrospect. If it survived a shock, it was resilient; if not, it was not. Even in a detailed account such as this one, we can only describe how the systems, apparently but not provably vulnerable, got to their present state. Before we do so, however, we must look in more detail at the spatial ecology and temporal cycles of China's three major ecological zones.

PART I

Setting the Scene

1. A Tour of China's Social-Ecological Systems

Much can be understood about a civilization from its landscapes

RICHARD STRASSBERG, *Inscribed Landscapes*, 1994

Traveling from the dry, flat monotony of the North China Plain to the subtropical lushness of the Guangdong coast to the misty rurality of the Sichuan Basin, it is hard to conceive of them as parts of a single ecosystem. Nevertheless, as parts of China Proper, they have been shaped for millennia by the forces of agrarian empire, and their differences pale in comparison to the way they contrast with the rugged vertical topography of Zomia or the expansive steppes and deserts of Central Asia, where the forces of China Proper–based empire have encroached more slowly. The natural environments of these three ecological zones of the People's Republic provide the background for the way the development projects of the Chinese Communist Party (CCP) have shaped their different ecological histories.

CHINA PROPER

China Proper historically supported one of the world's densest agricultural populations. With intensive land use, most natural ecological buffers against disturbance were sacrificed over the course of history in favor of gains in productivity, and people had to compensate by strengthening infrastructural and institutional buffers. These infrastructures and institutions, in turn, required intensive upkeep, and if the upkeep failed, there was not much resilience left in the system, leading to ecological crises.

Two requirements roughly determine China Proper's ecological boundaries.[1] There must be enough rainfall to support agriculture, and there must be patches of topography large and flat enough to generate an agricultural surplus to support ruling classes of nonfarmers (Whitney 1980, 114; Perkins 1969, 212). Within these boundaries, a northwest-southeast gradient from dry to wet divides the area into two

Map 1.1. Agricultural regions of China Proper, mid-twentieth century, after Buck 1937. Map by Lily Demet Crandall-Oral.

primary subzones (map 1.1). In the north, the climate is marginal for agriculture. Yearly rainfall averages approximately 350 to 600 millimeters but varies widely from year to year, and only a few areas have reliable surface water supplies that allow irrigation without fossil-fuel energy inputs (Lohmar et al. n.d., 278; Jinxia Wang et al. 2006, 276). Staple crops historically depended on summer precipitation. The growing season is short—there are only 190 to 210 days with a mean temperature above 50 degrees C (B. Liu et al. 2010), and 60 to 75 percent of yearly rain falls in June, July, and August (Smil 1993, 39; F. Yang and Lau 2004, 1626). In the loess lands of northwestern China, mountain areas are punctuated by intensively cultivated alluvial plains and valleys, but on the vast, flat alluvial North China Plain, crops could be grown almost everywhere, and land transport was easy and convenient (figure 1.1). Millets were the original staple grown here, supplemented by barley and wheat about

Figure 1.1. Farming on the North China Plain, 1930s. Photo by Hedda Morrison, used courtesy of the President and Fellows of Harvard College.

4,500 years ago (Yong Zhou et al. 2018), by sorghum in the Song period (Perkins 1969, 50), and by corn and sweet potatoes in the Columbian Exchange (Crosby 1972). Farmers on the Shandong coast in the 1920s, for example, grew wheat, millet, barley, soybeans, corn, sweet potatoes, and peanuts. In gardens near their houses, they grew vegetables, including "cabbage, turnips, onions, garlic, *chiu-tsai* (*jiucai*), garlic chives, *yuan-sui*, radishes, cucumbers, spinach, several kinds of string beans, squashes, peas, and melons" (Martin Yang 1945, 16).

The south of China Proper is the inverse of the north: better climate but worse topography.[2] Here rainfall is greater, from six hundred to as much as two thousand millimeters a year, with lesser amounts concentrated in the summer (F. Yang and Lau 2004, 1626). Warmer temperatures provide a longer growing season. This climate allows irrigated paddy rice agriculture, which gives higher yields per unit area than the rainfall agriculture of the north. However, the topography of the south is much more mountainous, and rice paddies could be built only in alluvial plains or deltas or on hillsides gentle enough for terraces. Dry fields supported

Figure 1.2. Transplanting rice seedlings, Shisantian, Taiwan, 1973.
Photo by the author.

less desirable staples such as corn and tubers or sometimes orchards or
tea gardens. In the northern and western parts of the rice zone, where
spring was cooler and drier, rice grew only in the summer, so farmers
grew dry winter-spring crops such as wheat, corn, or rapeseed. In the
southernmost areas, particularly the southeast, there was enough heat
and water for both spring and summer crops of rice (Buck 1937, 83–86).

Wet-rice agriculture is far more productive per unit of land than any
other widespread form of farming, but it requires more labor: plowing
and harrowing fields, transplanting shoots by hand, weeding two or
three times, maintaining appropriate water levels, harvesting, threshing,
winnowing, and drying (figure. 1.2). By the late Qing, rice cultivation
typically required about three times the labor per unit of land to produce
twice the yield of dry-grain agriculture (Whitney 1980, 111–15).

Rice fields were intimately connected to other parts of the ecosystem. They depended on fertilizer, mostly manure, for soil building and
enrichment, as well as on streams, lakes, reservoirs, and ditches for
water supply and drainage. They also hosted large numbers of beneficial animals (Bambaradeniya et al. 2004). Frogs ate harmful insects and
provided animal protein; fish could be raised in polyculture with rice

(Fernando 1993); ducks and geese fed on the gleanings after the harvest and provided eggs and meat; dragonflies ate mosquitoes; wild birds ate insects and could be hunted for food.

Since rice agriculture could support a denser population, in rice-farming areas more people lived within a day's walk of local markets, and more surplus was extracted as sales, rents, and taxes, thus supporting a larger landowning leisure class (Skinner 1964–65, 32–34).

The Chinese Household in Its Social-Ecological Systems

The household was the basic unit of production and consumption; both its ecological and its social dimensions were embodied in the word *jia*, meaning both house and family. An ideal household contained one or more patrilineally related men, with their wives, widowed mothers, and unmarried daughters. Sons took in wives, daughters married out, and, usually after their father's death, sons divided into separate households. A household had no fixed size or composition but rather experienced a temporal cycle of growth and division in each generation.

Households held both private and commons rights (Ostrom 1990) to resources at spatial scales from the house to the village, the village landscape, and the standard marketing community. Figure 1.3 shows the flows of goods and services at these several scales.

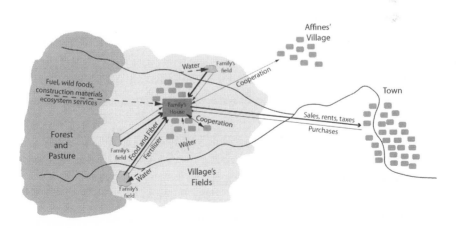

Figure 1.3. Flows of goods and services in a precollectivization household in China Proper. Solid arrows are private goods; dashed arrows are common-pool goods.

Houses everywhere were built of local materials, usually with walls of mud, brick, stone, wood, or some combination. By the late Qing, wood was scarce in most places and used only for pillars, posts, beams, and door or window frames (see J. Jing 2000, 7; Mobo Gao 1999 43–45; Osgood 1963, 97). Where trees were abundant, however (Leonard n.d.), houses could be built of wood. Stone, where available, might be the primary building material; otherwise it might be used for foundations only (Osgood 1963, 97) or not at all. Roofs were made of tile from local kilns or thatched with crop straw (see Osgood 1963, 99–100; Martin Yang 1945, 38–41; Skinner 2017, 53–55). In parts of the loess plateau in the northwest, people lived in cave houses, wholly or partially dug into the hillsides (see Golany 1992; Myrdal 1965, 44–48; Xin Liu 2000, 43–50).

Houses provided maximum protection from the elements with minimum energy use. Thick mud walls insulated residents from both summer heat and winter cold (Mobo Gao 1999, 43–44). In very cold places, there were no windows (Y. Yan 2003), since any opening would let in cold air. In the north, heat from the kitchen stove was recirculated through a series of flues under a brick platform, or *kang* (see Gamble 1963, 18; Xin Liu 2000, 45; Hinton 1966), where people slept, socialized, did crafts, and even entertained guests (Y. Yan 2003, 116; Meyer 2015). In hot areas, verandas covered by wide, awning-like eaves provided a well-ventilated place to sit or squat and work or chat out of the rain or the direct sunlight, while open-grilled windows provided indoor ventilation (figures 1.4, 1.5).

The ideal house was usually built around an open courtyard (figure 1.6). A main room, or *ting*, contained an altar to the ancestors, seating for guests, and often storage space for miscellaneous items. Bedrooms belonging to nuclear families within the *jia* opened into the *ting* or the courtyard or both. A kitchen often sat to the side or back of the *ting*. Poorer people might have no courtyard, or no *ting*, only bedrooms and a kitchen. In the caves of Shaanxi, there might be one large room, with the kitchen and guest area toward the outside and private family quarters toward the back (Xin Liu 2000). Very poor people might live in a single room.

Domestic animals were kept in rooms opening onto an outer part of

Figure 1.4. Lattice window with decorative carvings, Manshuiwan, Sichuan, 1994. Photo by the author.

Figure 1.5. House in Manshuiwan, Sichuan, built of local materials, including river cobble, cut stone, and wood, 1994. Photo by the author.

Figure 1.6. Extended family compound in Xiyuan, Taiwan, built around an open courtyard with the *ting* in the middle and related households' rooms in the wings, 1970. Photo by the author.

the courtyard, or in a separate building, enclosed by a fence that might also enclose a vegetable garden or an orchard. Pigs provided occasional nourishment, but a pig was also, as Mao Zedong (1959) famously said, "a small-scale organic fertilizer factory." In many areas the toilet adjoined the pigpen, thus collecting both human and pig excrement in a single pit, whence it could be scooped out and mixed with straw or grass to ferment and make *nongjia fei*, farmhouse fertilizer. Pigs could also be sold for income. Poultry—chickens everywhere, and ducks where water was abundant—also provided eggs and occasionally meat, contributed small amounts of manure to the fertilizer, helped with pest control, and brought in a little cash when sold at market.

Draft animals contributed manure and could be sold on occasion, but they mainly provided labor for farming and hauling. For this reason, beef was rarely a favored food (Martin Yang 1945, 47). In the north and the southwest, horses, mules, or donkeys might pull carts, while oxen pulled plows. In the south, people kept fewer animals, usually a water buffalo or ox to pull plows and harrows (C. K. Yang 1959, 38). In a few places, people kept no draft animals, and they hoed rather than plowed their fields (Fei 1939, 159). Sheep and goats, needing pasture, were marginal to China Proper and raised primarily in hilly or desert regions with lots of land that was too poor or dry or steep to farm. People in China Proper rarely or never consumed any kind of milk products.

THE VILLAGE

Most rural Chinese lived in villages—dense clusters or tightly packed rows of houses (figure 1.7). Closely related households, formed by family division, often shared courtyards or built adjacent houses; in some places descendants of a common patrilineal ancestor lived near each other and formed a corporate lineage that often held resources in common (see Freedman 1958, 1966; Baker 1968; Potter 1968; J. Watson 1974). A single lineage might contain wealthy landlords or even officials, as well

Figure 1.7. Yishala, a tightly packed village in Panzhihua, Sichuan, 1988. The mountain in the background was historically a forest commons. Photo by the author.

as their poor relations who were tenants of either the landlord families or the lineage itself. Not everyone belonged to a lineage, however, and in some places lineages were unimportant or nonexistent (Hinton 1966, 21; Harrell 1982, 117–20).

Households within a village exchanged goods and services, the most important being human and animal labor (P. Huang 1985, 151–52), and in irrigated regions they coordinated their agricultural schedules (Santos 2004, 153). People also socialized regularly with members of other households. Although village boundaries were clearly marked, villagers also had close relations with households outside their own village, usually either distant agnates or affines. Village households also held common rights to a variety of resources within the village, most commonly roads, lanes, and wells (Gamble 1963, 17; Smith [1899] 1970, 20–23).

THE VILLAGE LANDSCAPE

Outside the village itself, households had private rights to agricultural fields and orchards—through either ownership (all the product minus taxes), tenancy (part of the product), or both. They also held common rights to waterways, waterworks, forests, and pastures. In North China, many poor households worked for wages on land owned by wealthy managerial landlords (P. Huang 1985).

People managed their fields to provision the members of the household (including, in most cases, the animals) as efficiently as possible. With limited resources, this meant balancing needs for food, clothing, fuel, construction materials, and ceremonial expenses. Almost no household grew everything it ate or used or ate and used everything it grew, instead selling agricultural or handicraft products in markets and buying things it did not grow or make.

Farmers ate a mostly plant-based diet, consuming mainly staple grains or tubers, plus vegetables—fresh in season or preserved out of season. Since animal foods require much more energy to produce, most people ate just enough (or sometimes not enough) animal foods to provide sufficient protein (Smil 2002, 129–30; Martin Yang 1945, 32–33). Wealthier people enjoyed "better" diets, including more protein and fat, a greater variety of vegetables and other flavoring foods, and even "higher-quality" staples, such as wheat or rice, instead of poor people's foods such as corn or sweet potatoes (Martin Yang 1945, 24; Skinner 2017, 43–45).

Most households also shared common pool resources, most often forests, water, and fisheries. Not all villages had access to forests, especially after the large-scale deforestation of the Qing (see Elvin 2004, 84–85; Marks 1998, 318–27; Marks 2012, chaps. 4–6), but villages in areas as widespread as Kao Yao in Yunnan (Osgood 1963, 118–19), Weihai in Shandong (R. F. Johnston 1910, 167), Gao Village in Jiangxi (Mobo Gao 1999, 8), and Zhaojiahe in northern Shaanxi (Xin Liu 2000, 84) all had forest commons.

THE TOWN AND THE STANDARD MARKETING COMMUNITY

Above the levels of the village and its landscape was a hierarchy of spatial systems structured by the flow of goods, people, and ideas between rural households and markets, as well as between adjacent levels of a hierarchy of markets, culminating at the largest scale in "physiographic macroregions" (Skinner 1977a, b; Skinner, Henderson, and Yuan 2000). Each system at each level had a more urban and wealthy core—a city or market town—and a more rural, poor periphery, and each system contained all or part of these systems at the next smaller level. Marketing systems reflected the structure of ecosystems, whose peripheries had more forests and grasslands, while intermediate areas had more grain farms and core areas had more vegetable farms. The flows of economic goods and social interactions were thickest at the smallest scale, the standard marketing area, which in Qing times typically contained twelve to forty villages (Skinner 1964–65).[3] Household products flowed into the markets and into the coffers of the state and the landlords, and the products sold in the markets flowed to the households for consumption. As water and forest products traveled more from the core to the periphery than vice versa, rural households, through rent, taxes, and differential prices, contributed more to the town than they received in return.

These four concentric zones—the house, the village, the village landscape, and the marketing community—constituted the ecological and social world of the Chinese farm household until the advent of fossil fuel energy and modern economic change. Resources within this social-ecological system were vulnerable to both slow variable change within the system and disturbances from outside, but the degree of vulnerability or resilience to these disturbances depended on the effectiveness of buffers.

Buffers, Guarantors, and Everyday Sustainability
in the Traditional Chinese Household

In order to maximize both productivity and resilience, Chinese households needed to protect the house, maintain social reciprocity within the village, conserve the private and common resources of the landscape, and keep relationships with the town at a level acceptable in the moral economy (Scott 1976). But trade-offs between productivity and resilience were complicated and difficult to manage. Some actions boosted both productivity and resilience, but others increased short-term productivity while decreasing resilience (see figure I.5). In the highly modified ecosystem of intensive agriculture in China Proper, ecological buffers were weakened through intensification, thus making infrastructural, institutional, and cultural buffers even more central to resilience.

ECOLOGICAL BUFFERS

Ecological buffers protected a household's private and common resources against natural disturbances, mainly weather and, secondarily, insect plagues. Several landscape elements served as ecological buffers:

- Wetlands absorbed surplus rainfall, as around China Proper's two largest lakes, Dongting in Hunan and Poyang in Jiangxi, where rich, low-lying farmlands were developed early for rice cultivation.
- Forests deflected hard rainfall before it hit the ground and loosened the soil, took up water and kept it out of streams, prevented runoff, and decreased sediment flow that could disrupt downstream hydrology.
- Fallow land in dry-farming zones preserved soil fertility without using up valuable fertilizer that could enrich actively farmed fields.
- Crop diversity buffered against storms, droughts, diseases, and pests that affected one crop but not others.
- Surplus one year buffered temporally against disturbances in subsequent years.

As agriculture intensified, lands and waters that had previously served as ecological buffers were turned to production and had to be replaced by infrastructural, institutional, and cultural buffers.

With intensification, infrastructural buffers partly compensated for the loss of ecological buffers. Many waterworks and other agricultural infrastructure could increase both resilience and productivity at low intensity and replace ecological buffers at moderate levels of intensification but lost their buffering capacity and diminished resilience with further intensification (see figure 1.5):

- Reservoirs, such as Xiang Lake in northern Zhejiang (Schoppa 1989, 6), can store excess rainfall and replace wetlands that previously buffered against floods. Creating new paddy land at lake margins, however, reduced the volume of water that a lake could contain, decreasing resilience.
- River dikes, such as those built along rivers in the North China Plain over the centuries (Ling Zhang 2016; Pietz 2015), could prevent flooding in the short run, but as sediment built up in the riverbed, the dikes had to be raised higher and higher, increasing the severity of floods if the dikes were breached.
- Polder dikes, such as those built between and around Dongting and Poyang Lakes and in the Jianghan Plain in Hubei and Jiangxi (Perdue 1982; Jiayan Zhang 2014) and to form the Pearl River delta (C. K. Yang 1959), could increase farmed area, but if they occupied too much ecological buffer land, they reduced water storage capacity and rendered farm livelihoods dependent on the dikes.
- Terraces, for rice in the south or for dryland crops in the northwest, could buffer against erosion, preserve topsoil, and permit farming to continue. Built on already-farmed sloping land, they could boost both resilience and productivity, but built on forested land, they eliminated ecological buffers.

INSTITUTIONAL BUFFERS

Like infrastructure, institutions can replace ecological buffers lost with intensification. Some institutions were based in the village and its household networks; others operated at wider scales.

- Everyday labor exchanges in water regulation, planting, and harvest kept labor demands manageable.

- Relatives, including affines, provided credit and gave money to help meet the major expenses of events such as weddings and funerals, as well as to help after disasters.
- Lineages were both egalitarian and hierarchical (R. Watson [1981] 2004, 79) and could either raise or lower resilience. All households in a lineage, rich or poor, had equal rights to education, ritual meals, and emergency relief, but lineages also extracted wealth by renting collectively owned land to their poorer members.
- Properly managed commons, including local irrigation systems, protected vulnerable lands. At Kaixiangong in Jiangsu, every household with rice fields within a polder contributed labor to pump water in and out; elected managers enforced penalties against slackers and those who built illegal dikes or took water out of turn (Fei 1939, 166–77). Elected villagers managed a similar system in the Daba Mountains of northeastern Sichuan (Parker 2013, 36).
- At a wider scale, government-managed grain storage and distribution systems buffered the differential effects of weather. The Qing maintained supplies of surplus grain and distributed it to areas hit by natural disturbances, preventing many famines in times of droughts or floods (Lillian Li 2007, 221–49; Will 1990, 182–208).
- Markets could either increase or decrease resilience. Without markets, when ecological and institutional buffers are insufficient, there can be food shortages. A nearby market can offer food from other areas not hit by the disturbance, if households can afford to buy it. A market thus increases resilience but exacts a price. But if households switch from subsistence to cash crops (P. Huang 1990, 22–24), they can become dependent on the market and vulnerable to market disturbances.

CULTURAL BUFFERS

None of these ecological, infrastructural, or institutional buffers could work without households' willingness to act so that the buffers could do their work, and this depended on culturally transmitted values and practices.

- Generational continuity. Households depended partially on resources inherited from past generations, which would flow to future generations. This meant taking care of resources so that they would grow rather than shrink.

- Reciprocity with other households. Household reciprocity could act as an institutional buffer because reciprocity itself was a cultural buffer. Gestures of symbolic reciprocity—gifts, courtesies, invitations, and common participation in ritual— demonstrated sincerity and reliability, restraining mistrust and enabling households to call on trusted relatives and neighbors for help in times of trouble (Y. Yan 1996).
- Frugality. Frugality or thrift was probably the most important cultural value contributing to household resilience. According to anthropologist Fei Hsiao-tung,

> To be content with simple living is part of early education. Extravagance is prevented by sanctions. . . . Thrift is encouraged. Throwing away anything which has not been properly used will offend heaven. . . . For instance, no rice should be wasted. . . . Clothes are used by generations, until they are worn out [and then] used for making the bottom of the shoes or exchanged for sweets and porcelain.

> In a rural community where production may be threatened by natural disasters, content and thrift have practical value. If a man spends all his income, when he fails to have a good harvest, he will be forced to raise loans which may cause him to lose a part of his right over his land. (Fei 1939, 119–20)

Anthropologist Martin Yang (1945) recorded almost precisely the same sentiments from Taitou, a village in coastal Shandong.

Many other practices testify to this ethic of frugality. Living in caves or windowless, semisubterranean houses saved fuel. Everything organic that did not have a better use went into the fertilizer pit. Morality stories of ruin brought on by profligacy were well known from operas and puppet shows.

- Aesthetic appreciation of landscape. The landscape of rural Taiwan in 1972–73 was still recognizably agrarian. Around 2000, I showed a picture of it (figure 1.8) to a grandmother. She remarked, "Look how beautiful it was then, and now it has turned so ugly," expressing grief for a lost landscape aesthetic.

- Both colonial administrator R. F. Johnston and Martin Yang record similar sentiments and similar regrets in Shandong:

  > From the present denuded condition of the hills one would hardly suppose that the people of Weihaiwei cared much for trees: yet as a matter of fact they value them highly for their shade and for their beauty. (R. F. Johnston 1910, 168)

  > Not long ago . . . the village was admired by travelers who approached it from the south. Before one reached the edge of the river one could hardly see the village because of a thick green wall of trees. . . . Unfortunately, a great part of that is gone. (Martin Yang 1945, 5)

- Johnston and Yang, as well as Pamela Leonard (n.d.), who talks about moral landscapes influenced by *fengshui*, are all describing resilient landscapes, including trees, clear-flowing streams, ordered fields, and houses built of local materials.

Figure 1.8. Rice terraces with courtyard house and forests in the background, Datiekeng, Taiwan, 1973. Photo by the author.

Buffers in place kept local ecosystems resilient. But slow variables, cross-scale interactions, and the seemingly inevitable turning of the Malthus-and-Boserup ratchet could erode the buffers and decrease the system's resilience. Households trying to maintain their resources and consumption levels in the short run optimized fertility at far beyond the replacement level, contributing to population growth that itself encroached on all the buffers enumerated above. Protecting the house might thus contradict the goal of maintaining the fields and commons. Conflicting interests of different households could fray the bonds of balanced reciprocity. Powerful individuals or groups in the town might increase their extractive demands so the household could not protect itself without degrading the fields and commons or straining its bonds with other households, causing further encroachment on buffers and further decreases in the household's systemic resilience.

ZOMIA

Between China Proper and the intensive agricultural zones of India and Southeast Asia, Zomia (Scott 2009; van Schendel 2002) is a land of high mountain ranges and deep river valleys (Henck et al. 2011). Within very short distances, climate varies from semiarid subtropical through moist temperate to alpine. In alluvial basins, people have long practiced intensive agriculture, and small states have developed over the past two millennia (see, e.g., Leach 1954; Wiens 1954; Backus 1981; Wang N. 1985; Hsieh 1995; Whittaker 2008). In wide stretches of mountain slopes between these small, relatively flat areas, people's livelihoods until very recently depended on shifting agriculture, small-scale herding, and forestry (see, e.g., Altieri 1995, 130–36; Berkes 1999, 60–68; Jianchu Xu et al. 2005; Sturgeon 2005, 120–22; Urgenson et al. 2010; Trac et al. 2013). Population densities were much lower than in China Proper but greater than in Central Asia (map 1.2).

With no large states, Zomia had no borders, and current borders between China, Vietnam, Laos, Myanmar, Thailand, and northeastern India have little to do with the ethnic or even ecological divisions in the area. Similarly, the boundaries between China Proper and Zomia are neither clear nor fixed. In most Zomian landscapes, different ethnic and linguistic groups occupied ecological niches at different altitudes

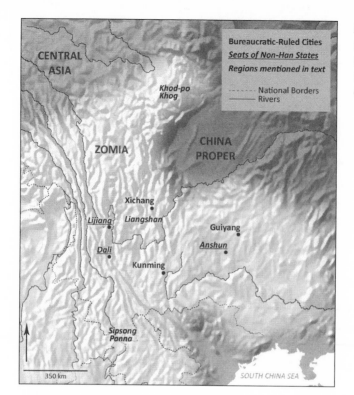

Map 1.2. Regions of Zomia now in the People's Republic. Map by Lily Demet Crandall-Oral.

and ordinarily interacted less with large states or empires the higher up they lived (see, e.g., Cheung 1995; Hsieh 1995; Hansen 1999; Tapp 2001).

For example, near Lugu Lake on the Sichuan-Yunnan border, Na and Prmi people occupied the basins near the lake, growing dry grains and tubers on permanent fields and keeping cattle, sheep, goats, and pigs, while Nuosu swiddened the surrounding mountains (see, e.g., Shih 2010; Weng 1993; Blumenfield 2010; H. Cai 2001). In Sipsongpanna on the border between Yunnan and Laos, Tai-speaking peoples ruled a small feudal state in the lowlands and farmed wet rice, while several "hill peoples"—Akha, Jinuo, Lahu, and Bulang—practiced shifting cultivation, along with herding and forestry, and some of them paid feudal dues to the Tai overlords.[4] Around Anshun in Guizhou, Han lived in walled cities, towns, marketplaces, and villages along transport routes and were directly governed by the Qing state. Nasu Yi lived in the middle altitudes, tributary to but not governed by the Qing. Miao lived in the high mountains, trying—not always successfully—to avoid paying dues to the Yi (Cheung 1995).

Zomian Household Ecologies

Zomian households, like those in China Proper, did not migrate with the seasons, but they used more diverse resources, depended less on market exchange, and covered a wider territory in their daily and yearly rounds. Two detailed cases illustrate Zomian ecologies.

THE NUOSU OF SOUTHERN SICHUAN

Nuosu people have lived for about eighteen hundred years in Liangshan, now in southwestern Sichuan. In their core areas, Nuosu occupy all elevation niches; where they have migrated in the last few centuries, Nuosu are part of a vertical mosaic consisting of Han people in the river valleys, Qiangic peoples in the higher basins, and Nuosu on mountain slopes and in high valleys.

Nuosu built houses of local materials—mud walls in many areas and stone in a few places where it was abundant; in forested areas they used wood, often elaborately decorated (figure 1.9). Each nuclear family had its own house, centered on a circular hearth where people sat on felt capes or mats spread out on the packed earth (figure 1.10) and slept there around the fire on cold nights. Using a wok or kettle placed on hearth-stones over the fire or hung from the wooden rafters, people could boil or steam potatoes or buckwheat cakes or fry buckwheat pancakes. On special occasions, meat, chopped into large chunks, would boil for hours while hosts and guests drank liquor or locally brewed beer and recited songs and poetry. Domestic animals ran freely in and out, sharing the space with their human owners.

Nuosu ordinarily did not live in compact villages. A few closely related households might build their houses inside a single walled compound, but most houses were scattered about the landscape. Households in a typical community were related by clanship, marriage, or ties of lord and serf or slaveholder and slave. Kinship ties were more important to Nuosu than ties of locality; people related more to the aesthetics of certain landscapes than to particular places on a map.

Nuosu drew their livelihood from three ecological zones: field, pasture, and forest. They grew a mix of grain crops at different elevations. In the rare, low-lying alluvial plains, those crops might include rice. More frequently, people grew buckwheat, oats, wheat, and barley wher-

Figure 1.9. Nuosu men in front of a traditional house in Ebian Subprefecture, Sichuan, 1913. Photo by Fritz Weiss, copyright and courtesy bpk Berlin.

Figure 1.10. Family sitting around the hearth in a traditional Nuosu house, 1993. Photo by the author.

ever possible in household-owned permanent fields in river valleys or as short-fallow crops on relatively level benchlands. Swiddens on ridgetops or south-facing slopes belonged to the household that first cleared them, but when abandoned fields reverted to forest, anyone belonging to a local clan segment could make a clearing. By the early Qing, corn and potatoes had replaced buckwheat as the most important sources of calories. A variety of turnips called *vama*, grown on high ground, could be eaten fresh to quench the thirst or dried and preserved to make sour soup.

Pastures supported livestock: horses, mules, donkeys, cattle, sheep, goats, and pigs. Pigs provided meat and fertilizer; cattle provided traction, leather, and meat; horses, transportation and prestige; sheep and goats, meat and wool for cloth made by women or felt made by men. Nuosu ate more meat than did poor farmers of China Proper—not every day, but households would slaughter animals for weddings, funerals, priestly rituals, or visits from distant relatives, and larger or smaller circles of relatives and neighbors would partake of the meat (see figure 1.10). Nuosu rarely consumed milk products and did not milk their cattle, sheep, or horses.

Forests completed the resource triangle. Households needed firewood constantly, especially in the wintertime (figure 1.11), and used timber from straight-trunked pines or firs to build their houses. Specialists used wood to make plows, wagons, and other agricultural implements. People gathered a large variety of wild plants and fungi for food and for human and animal medicine, as well as pine branches to protect mud walls against rain, lichen as bedding for baby animals, and bamboo to weave baskets and trays. They hunted deer, bear, red pandas, upland game birds, and other sources of rare and prized animal foods and skins. Most important, Nuosu valued forests for their ecosystem services, particularly clean water and protection against soil erosion and runoff.

There were very few markets, very little surplus to sell, and very little incentive to produce surplus, but specialists, including priests, blacksmiths, silversmiths, and lacquerware makers, traveled widely, exchanging their wares or services for grain, silver, or animals. Cookware, silk, and in later times commercial cloth were purchased from itinerant Han traders. However, until the rise of opium cultivation and the entry of firearms and ammunition into Nuosu territory in the nineteenth century (Hill 2001, 1036–37), households and watersheds were basically independent, both economically and ecologically.

THE AKHA OF SIPSONGPANNA

The Akha, known as Hani in Chinese, live in warm hill regions of China, Laos, Myanmar, and Thailand, almost always in a higher-elevation niche in a system of vertical stratification.

Akha in Sipsongpanna, not needing much insulation in their warm forest homes, built their small houses using wood from the local forest for stilts and frames, bamboo for walls, and tough *Imperata* grass thatch for roofs. They penned livestock under the floor at night (Sturgeon 2005, 124).

Akha clustered their houses in small, linear hamlets *within* the forest (map 1.3). They forbade cutting in either protected forests surrounding the houses (Sturgeon 2005, 124–25) or in hillside forests preserved to protect the watershed. In more distant mountain forest commons, people cut wood for fuel and construction, gathered food and medicinal plants, and hunted abundant wild game. Even there, someone who cut a small amount of primary forest paid a fine in liquor; a person who cut a larger amount would have to slaughter a pig and feed every family in the hamlet (Sturgeon 2005, 126).

Any member of the hamlet could make a swidden, anywhere in the secondary forest downhill from the hamlet, to grow upland dry rice, their primary grain crop, as well as "a rich array of vegetables." An actively farmed swidden belonged to the family that cleared it, but once abandoned and reverted to forest, it also reverted to commons, and any village family could make its swidden there as long as the forest was again mature and ready to be cut (Sturgeon 2005, 8). In the moist primary forests called *sanpabawa*, villagers could cut rattan at specific times and use it to make headgear, knife handles, and the edges of baskets (Sturgeon 2005, 124–25; Jianchu Xu et al. 2005).

Buffers and Guarantors in a Varied Environment

Zomian environments, lacking the productive potential of China Proper, could support only moderate population density. Some hillsides could be terraced, and swidden rotations could be shortened somewhat. But it

Figure 1.11. Gathering firewood, Apiladda Valley, Sichuan, 2009. Photo by the author.

Map 1.3. Schematic of the ideal layout of an Akha village in Sipsongpanna, Yunnan. Map by John Ng, from Janet C. Sturgeon, *Border Landscapes*, © 2005, University of Washington Press.

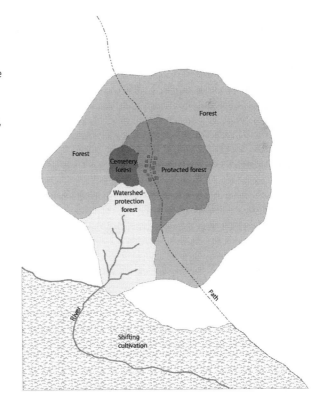

was impossible to push this intensification too far, to create a landscape as anthropogenic as that of China Proper. Infrastructural buffers were mostly infeasible, so Zomians rarely created rigidity traps that diminished their landscapes' resilience. Rather, they maintained important ecological, institutional, and cultural buffers.

ECOLOGICAL BUFFERS

- Patch diversity. Akha farmers had rules for who could use what resources from swiddens, pastures, infrequent wet-rice fields, and many different kinds of forests (Sturgeon 2005, 120). Nuosu also modified and used different patches in different ways; a proverb states that "sunny slopes are good for grain; shady slopes are good for trees." Tibetans living at Jiuzhaigou in the Khod-po Khog region of Amdo built villages in a rich layer of loess at middle elevations, where they could grow diverse grains and vegetables (Henck et al. 2010; Urgenson et al. 2014). Lower down, they managed coniferous forests for timber and deciduous patches for firewood, clearing a few small patches for cultivation, later left as winter pastures. At the higher elevations, they pastured their yaks and sheep.
- Temporal diversity or "landscape plasticity" (Sturgeon 2005, 25). In the swidden cycle, land rotated from forest to newly cleared fields planted with grain, to more mature fields with tree crops or perennials, to immature forest, and back to mature forest again (Berkes 1999, 60–64). This cycle itself could change at a longer temporal scale, however, as a larger area once used for swiddens might be allowed to revert to forest for the long term and not be cut for centuries (Berkes 1999, 121). Conversely, a field could be kept clear for pasture after its agricultural fertility was gone, as in Jiuzhaigou.

INSTITUTIONAL BUFFERS

- Mixed common and private property rights. For Akha, Lahu (Shanshan Du 2002, 13–17; Jianxiong Ma 2013), and Lisu (Harwood 2013), almost all rights to *productive* property were held in common. The specific designs of their commons conformed with almost all of Elinor Ostrom's (1990, 90) famous "design principles" for successful common-property regimes, including clearly defined and delineated rights to withdraw resources, rules for changing the rules, effective

monitoring, appropriate penalties for violators, respect for con-
flict-resolution mechanisms, and lack of interference from more
powerful outside forces (Sturgeon 2005, 121–28, and pers. comm.).

Nuosu aristocratic clans controlled territories where they, their
commoner retainers, and the serfs bound to either all had rights to
cultivate swiddens (Li S. 2000; Mgebbu 2003). Pastures and water
resources were also held in common. Individual households or
groups of closely related households had exclusive rights to products of forests extending from their houses up to the nearest ridgetops; others needed owners' permission to cut here. More distant
forests were open to anyone but so far away that overexploitation
was impractical. Valley agricultural fields that could be farmed
every year or in short-fallow rotations were individually owned as
private property. This system both promoted careful stewardship of
intensively farmed lands and buffered against random disturbances
to resources that were subject to stochastic variation.

- Generalized reciprocity—the obligation to give freely to members of
  one's close network and expect the same from them, without making any exact accounting of the amount of the gift (Sahlins 1972).
  This contrasts to the Han norm of "balanced reciprocity," in which
  a favor establishes a calculable debt. For example, leaders of a Laluo
  community growing silkworms told me that if one household ran
  short of mulberry leaves, a relative or neighbor would contribute
  extra leaves free of charge, *not like Han people, who would keep track
  of who owed whom how much.* Similarly, when a hailstorm obliterated the corn crop in a Nuosu community in 2004, people told me
  they were unconcerned. Not only did they have potatoes and buckwheat that were less affected than the corn (patch diversity), but relatives from villages where it had not hailed would help out.

CULTURAL BUFFERS

As in China Proper, in Zomia people developed cultural-ethical norms
that reinforced the action of ecological and institutional buffers:

- Supernatural protection for natural features. For Naxi people, who
  live at mid-high elevations in northwestern Yunnan, uncultivated
  parts of the landscape have a supernatural patron called Shu, who

protects against resource destruction caused by human temptation and greed (Jianchu Xu et al. 2005). Akha ancestors, as guardians of the forest, also enforce of the rules of their common-property regime (Sturgeon 2005, 124–25).

- Sacred precincts. These are either off-limits to humans altogether or have severe strictures against appropriating resources. The cemetery forest of the Akha is one example; others are the sacred mountains of many Buddhist peoples, including the lowland Tai of Sipsongpanna (Jianchu Xu et al. 2005; Liu Hongmao et al. 2002), and the middle- to high-elevation Amdo Tibetans of Jiuzhaigou, where pilgrims circumambulating a sacred mountain must take in everything they use, take out everything they don't, and not leave anything inside the sacred precinct (figure 1.12). Watercourses are similarly sacralized—they turn wheels containing Buddhist scriptures and may not be polluted, literally or symbolically.

Figure 1.12. Walking sticks left by pilgrims on a circuit of Zhayizhaga Mountain in Jiuzhaigou, Sichuan, 2008. Later pilgrims reuse them, so as not to waste resources on the sacred mountain. Photo by the author.

Sacred precincts are often centers of biodiversity, harboring species that are rare or extirpated elsewhere, as on Tai "holy hills." In the Nuosu community I know well, a rare conifer, *Keteleeria davidiana*, grows in a sacred grove, where community members swear in a yearly ritual not to violate environmental or other ethical principles. During the Cultural Revolution, a young man who violated a taboo on cutting the trees was soon struck dead by a mysterious disease.

- Temporal prohibitions. Beliefs about resources mark off not only space but time. The Nuosu year is divided into a growing season, from the first rhododendron bloom in spring until the last harvest in the fall, when neither hunting nor felling trees is allowed, and a complementary killing season, from fall to spring, when people can both hunt wild game and cut down trees. Violations can bring hailstorms or other supernatural retribution.

- Social-ecological parallels. Like the Akha, who connect the ancestors and the natural world, Nuosu draw parallels between the social and natural worlds in couplets, or *lurby*:

Sy zzu i pamu, yy zzu i pamu: "Trees are senior relatives; water is a senior relative."

Bbo ggut mu a nde, pu nyo jjy wep a zze: "Don't neglect thanks for a gift given to you; don't allow the fertility of land to decline."

Pu nyo mu su vi; vi ke she su vi: "Land belongs to those who work it; disputes belong to those who get into them." In other words, you are responsible for your actions, social or ecological.

- Aesthetic appreciation of landscape. A passage from *Kepu Jjylur Shy-wa-te* (Kepu n.d.), a Nuosu ritual book summoning spirits, expresses the beauty of nature and its connection to leaders who enforce social and environmental rules:

White grain hangs above the water
Out of the water rice grows
Rice grows luxuriantly
Spirit of the heavenly ruler

Beautiful like a wild goose
Like a goose with a golden bill
Its golden bill glittering and scintillating
Spirit of the heavenly judge

The wings of the white kite
Flapping its wings
Banking and soaring in the four directions
Spirit of the heavenly priest

The peoples of Zomia thus see themselves as part of the natural world, rather than apart from it, and they have seen that while humans can and sometimes must modify the landscape, there are limits. There is no such thing as mastery over nature, and if people were to eliminate ecological buffers, infrastructure could not easily replace them.

CHINESE CENTRAL ASIA

Inner Mongolia, Xinjiang, and the northern parts of Tibet are ecologically part of the wider Central Asian dry zone that stretches from Mongolia to the Caspian Sea (map 1.4). The People's Republic claimed parts of this zone purely because the Qing held onto them toward the end of its rule, when it was joining the international order of nations with definite borders.

Outside of some mountain areas, rainfall is sparse in Central Asia, ranging from thirty-five to three hundred millimeters per year. This dry climate makes agriculture difficult or impossible in most places, although some primarily pastoralist societies were able to practice limited agriculture in favored local environments in times of favorable climate (Di Cosmo 1994; Di Cosmo 1999, 13). Only in oases watered by runoff from mountain ranges has farming prevailed consistently throughout history. Outside these isolated agricultural areas, humans must acquire food calories by raising ruminant animals, which convert the cellulose and other complex molecules found in grasses to nutrients that humans can digest as milk or meat. Each of what Mongols call the "five snouts"— horses, cattle, camels, sheep, and goats (Hanson 2004, 82–83)—serves a different set of functions. Horses provide traction, milk, prestige, and,

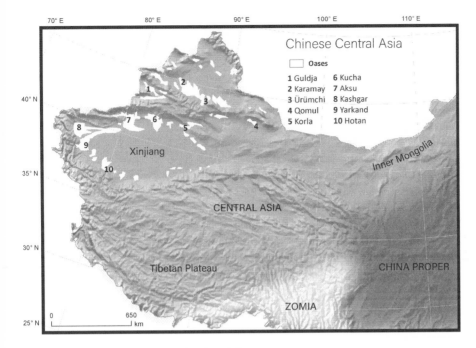

Map 1.4. Chinese Central Asia, showing Xinjiang oases.
Map by Lily Demet Crandall-Oral.

among the Turkic peoples at least, occasionally meat (Cerny 2008, 248). Cattle (including yaks in Tibet and higher-elevation parts of Mongolia) give milk, pull carts, and can be eaten and their hides used for clothing, ropes, and other leather goods. Yak wool is also important for making felt. Camels, found primarily in the driest regions, are the ideal transport animal—large and strong—and can also be milked and eaten. Sheep are economically and dietarily the most important animal, being a source of meat, milk, wool, and skins. Goats can graze steeper terrain than sheep and are somewhat hardier; they provide milk, meat, and hides, and some varieties provide cashmere.

The primary productivity of grasslands (the amount of solar energy converted to grass) is very small to begin with, limited by low rainfall and in many places by cold temperatures (Webb et al. 1978). Ruminant animals are very inefficient at converting vegetable material to humanly digestible meat: it takes about thirty kilograms of forage to produce one kilo of beef or mutton (Pimentel and Pimentel 2003, 662S). Milk products are much more efficient converters; about eight kilos of forage can produce one kilo of nutritionally valuable components of milk, namely fat, sugars, and protein. Because of the small volume of grass and the inefficiency of conversion, the pastoral economy supports only a tiny fraction of the population density that China Proper can support, perhaps one to three persons per square kilometer (Whitney 1980, 109–11). These ecological limits influence not only material culture and migration patterns but also property rights, community and political structure, and ecological thinking.

Animals need pasturage at almost every season—hay can substitute for only a short time—but grass grows in different places at different times. People who rely on animals must be able to move with their herds through the cycle of the seasons (Ekvall 1968, 34–36; M. Goldstein and Beall 1990, 72–79) and take nearly everything with them as they move, even their housing. The Turkic *yurt* or Mongolian *ger* is light enough that a camel or a couple of packhorses can carry it, compact enough that a stove in the middle can warm a lot of people, and portable enough that it can be taken down or put up again in a few hours (M. Goldstein and Beall 1994, 57–63). On the Tibetan plateau, the sturdy and portable yak-wool tent serves the same purpose (M. Goldstein and Beall 1990, 58–64; G. Tan 2016).

Marginal returns to labor in a pastoral economy quickly go to zero, and there is no way to increase productivity beyond its original, very low level. The condition of pastures, unlike that of agricultural lands, depends almost entirely on the weather and past use—shortening the grazing rotation or increasing the density of animals degrades the pasture (Silveira et al. 2013). There is no equivalent to irrigation; the only remedy for a dry or cold pasture is to go somewhere else, and the only way to recondition a pasture after it has been grazed down is to go somewhere else for a while. Even while herders are camped in one place, they need to take their animals to different pastures on different days. And since the weather varies from year to year, they need to be able to migrate to different areas or to the same areas at different times, from one year to another.

Within these ecological constraints (Perdue 2005, 15–19), peoples in different parts of Central Asia developed different material and social adaptations in their migration patterns, property regimes, and social structures.

Herders and Farmers

All through East Asian history, there has been an uneasy and unstable relationship between the inhabitants of the agrarian civilizations, including China Proper, and those of the pastoral zones of Central Asia. Almost nowhere was the pastoral economy self-sufficient, since it is very difficult for people to live off animal products alone. Pastoralists in places where they could not practice supplementary agriculture needed to trade for goods such as grain to supplement diets, tea as a stimulant drink and a way of consuming dairy products, silk for summer and ceremonial clothing (Jagchid and Hyer 1979, 37), and wood to build the frames of the yurts. By contrast, farmers who traded with pastoralists were much less dependent on that trade (Lattimore 1940, 69).

Central Asia is not ecologically uniform, and agrarian-pastoral relations varied over space and time. In Mongolia, only very limited agriculture is possible anywhere to the northwest of the 350-millimeter precipitation line, and pastoralists there have historically differed from the agricultural Chinese in religion, philosophy, language, and political structure. However, they still depended on long-distance trade with the agricultural peoples and regimes of China Proper (Lattimore 1940, 60–61; K. Chang 1968, 356–62). And because nomads depended on trade more than farmers did, climate variations complicated the changing cooperative and hostile relations between them. When there was more rainfall, the grass grew luxuriantly, herds could expand, and ambitious political leaders could unite related clans into confederations and eventually large polities (Pederson et al. 2014). In these warmer, wetter periods, empires based in China Proper also pushed agriculture and agriculturalists northward, often as military colonists, and some nomadic populations settled down and became farmers. During the Medieval Climate Optimum, for example, the Khitan people built the Liao empire in what are now parts of Mongolia and Manchuria; they controlled both agricultural and pastoral areas and mixed their indigenous institutions with adopted Chinese forms (H. Lin 2011). During colder, drier periods, however, the agricultural/pastoral frontier shifted southward, and nomadic polities sometimes collapsed for lack of fodder, leaving agriculturalists in the frontier zone literally high and dry (Pietz 2015, 60). If the nomads could not be provisioned from China Proper, they often resorted to wars of conquest, sometimes conquering northern China or, in the Mongols' case, all of China Proper and parts of Zomia. Even when originally no-madic peoples ruled northern China, however, economics and society within the pastoral areas changed little; agriculture and sedentary life were impossible, and society evolved within the limits set by ecology.

Xinjiang, by contrast, has a more complex, vertical ecology (Toops 2004), and the separation between agrarian and pastoral societies was more purely ecological, less historical and linguistic. In the Tarim Basin of southern Xinjiang, runoff from surrounding mountains feeds a se-ries of large oases rimming the basin, home to farmers growing grain, fiber, and other crops and to trading cities along the southern branch of the Silk Road. In the middle of the basin is the virtually uninhabited Taklimakan Desert. There is little pasture, so there are few pastoralists. In the Junggar Basin of northern Xinjiang, by contrast, the slopes of the Tianshan range to the south and the Altay range to the north sport abundant good pasture, so herding peoples have long made their home on these grasslands, migrating along an elevation gradient of pastures (Svanberg 1988, 111; Cerny 2008). As in the south, rainfall is not suffi-cient for agriculture, but both surface and underground runoff from the mountains feeds the farmers' fields in the oases (Toops 2004, 266), and pastoralist-farmer exchanges were more local. Nomads were mostly Turkic Muslims, with some Mongolian speakers, while city dwellers and farmers were almost exclusively Turkic Muslims.

Tibet contains both agricultural areas—particularly in the valleys of the Yarlung Tsangpo River watershed in the south—and pastoral areas on the high plateau to the north, so Tibetan pastoralists exchanged with Tibetan farmers. Pastoralists and farmers were united by a common Buddhist religion, a common written language, and closely related spo-ken languages, and since 1642 many of them were linked by allegiance to the lamaist clerical state (M. Goldstein 1989, 1–2), but there were still clear linguistic and cultural differences, not to mention mutual cultural stereotyping.

Ecological variations thus gave rise to different possibilities for envi-ronmental adaptation, including varying patterns of migration, property rights, and environmental values.

Migration Patterns

Kazakhs in Xinjiang until the present century moved with their animals to take advantage of seasonal pasture conditions (Lattimore 1929, 688–89; Svanberg 1988, 127–32). Depending on ecological gradients, they might migrate as many as two hundred or as few as thirty-five kilometers each way, but they moved from low elevations (sometimes on the edge of the desert) in winter to midaltitudes in spring and the highest pastures in summer before reversing course, reusing the spring pastures (regrown in the interim) in the fall (Cerny 2008, 14–28).

On the Mongolian steppes, some nomads migrated long distances, moving seasonally between different kinds of pastures; others migrated only locally (Lattimore 1940, 73–74; Jagchid and Hyer 1979, 21–26; Williams 2002, 68–70). Herders did not necessarily follow the same migration routes from year to year, though they were obligated to stay within the territories controlled by their own clans. Besides seeking good grass, migrants looked for places where summer breezes would blow away the insects that might bother the animals and where in winter there would be maximum sunlight and shelter from severe winds. In spring and autumn, they moved frequently; particularly in autumn they needed to find relatively lightly grazed pastures where the animals could fatten for the winter (M. Goldstein and Beall 1994, 40).

At Pala on the western part of the Tibetan Plateau in the early 1990s, nomads lived at a valley-floor elevation of about forty-nine hundred meters, very near the limit for grazing. Almost all grass growth happened between May and September, so animals had to eat dead, dry grass for about eight months out of the year. Nomads moved away from their home base only to an autumn pasture where animals were excluded for the rest of the growing season; there the herds could fatten up for winter and spring on a whole summer's growth of grass. This also allowed the home pastures to put on a last, late-summer growth that would be available, in dry form, when the animals returned in the winter (M. Goldstein and Beall 1990, 58–65).

Cycles of migration were thus shaped more by the limits of local ecology than by any cultural differences among ethnic or linguistic groups, reinforcing the fact that pastoralists must adapt to the environment, rather than try to modify the environment for their own convenience.

Property Regimes

In any pastoral ecosystem, the most important household property is livestock, not land. Everywhere in Central Asia, livestock belonged to individual households, so a rich household was one with large herds and flocks.

In Mongolia and the pastoral parts of Xinjiang, there was no private property in land. As Owen Lattimore (1940, 66) puts it, "No single pasture could have any value unless the people using it were free to move to some other pasture, because no single pasture could be grazed continuously. The right to move prevailed over the right to camp." Land was not property but territory associated with a political unit based on patrilineal kinship, often a smaller unit nested within a larger, kin-based political federation (Lattimore 1940, 67; Svanberg 1988, 127–32).

In a few places, pastures were not held in common. The herders of Pala in western Tibet owned their animals, but they were serfs of the Panchen Lama, who owned the pastures. The lama's agents allocated specific home and autumn pastures to families and reallocated them every three years to account for growth or shrinkage of the human and animal populations (M. Goldstein and Beall 1990, 69).

The Ecology of the Central Asian Pastoral Household

In nomadic households, energy and resources flowed between people, animals, pastures, and rights to use territory (figure 1.13). Until the Qing, these household ecologies were not spatially fixed but moved across space with the changes from season to season and year to year.

The larger systems in which the household was embedded were not stationary as in China Proper but moved over time. At the apex of this system was the interface between the pastoral and agricultural zones, which was characterized by relations of trade, tribute, and warfare at different times.

Buffers in Pastoral Ecology

In Central Asia as in Zomia, infrastructure could not replace ecological buffers, and thus intensification was impossible. Resilience in Central

Figure 1.13. Ecological flows in a pastoral nomadic household. Drawing by Kaitlin Banfill.

Asian ecosystems depended on ecological, institutional, and cultural buffers.

ECOLOGICAL BUFFERS

Diversity and redundancy characterized the most important ecological buffers in Central Asia:

- Livestock diversity. Different animals are vulnerable or resilient to different disturbances. Horses cannot graze grass very short, but they can eat through snow better than cattle or sheep (Jagchid and Hyer 1979, 21–22). Yaks can also eat through snow and can survive at altitudes where horses are poorly adapted (Ekvall 1968, 15). Cattle graze closer than horses, and sheep and goats closer than cattle, so sheep and goats can still benefit from a pasture that has been grazed over recently by larger animals (Ekvall 1968, 15). Goats and camels can browse on brush, and goats and yaks tolerate cold better than other animals (M. Goldstein and Beall 1990, 104).

- Spatial and temporal diversity. Household members and animals can be in different places at the same time or in the same place at different times. Nomads could choose pastures and migration routes within their territory and seek out the currently best pastures (Ekvall 1968, 34–36; M. Goldstein and Beall 1990, 72–79). Hay and dairy products of summer could be stored, especially since sheep and goats give milk only for a few months in the summertime, and even yaks give very little in winter (M. Goldstein and Beall 1990, 87; G. Tan 2016, 17, 40).
- Herd size. Herds can get too big and overgraze (Cerny 2008, 72, 182–83), but in historic times, weather events could dramatically reduce the size of a herd (M. Goldstein and Beall 1994, 39), so it was best to allow herds to breed fast during the good years to buffer against winter mortality events.

INSTITUTIONAL BUFFERS

The most important institutional buffers for pastoral resilience involved redistribution of resources.

- Pasture commons. Certain pastures at good seasons or in good years could accommodate multiple households' animals, avoiding pressure on those areas that were not growing as well at the time. Access would ideally even out over the seasons and from year to year.
- Human population control. In central Tibet, households allowed two or sometimes three brothers to marry a single wife and ordinarily allowed only one daughter to marry (M. Goldstein 1971). Others became nuns or maiden aunts (G. Tan 2016), while some sons became monks. This depressed population-level fertility and population pressure on resources.
- Markets and trade. Markets can work to either increase or decrease resilience, but across agricultural-pastoral boundaries they primarily increased resilience. In lean times, when pastoralists had to slaughter animals that would not survive the winter, they could trade more hides, horns, and other products for grain, tea, and other necessities (Ekvall 1968, 35). When markets disappeared, however, herders were liable to resort to raiding and wars of conquest.

- Frugality. Like farmers in China Proper, nomads were frugal; they used every part and product of their animals—meat, hides, milk, dung, horns, transport, and trade value (see M. Goldstein and Beall 1990, 80–107; G. Tan 2016; and Cerny 2008).
- Harmony and respect for the land. While farmers sought to improve the land, nomads stressed that they should respect the land and the landscape and not try to modify it. As Lattimore (1940, 238) states, "There is a genuine, sensitive and much deeper feeling that man should accommodate his needs and the use he makes of the land for himself and his herds to what one might call the needs and rights of the land itself," and as a male nomad told Melvyn Goldstein and Cynthia Beall (1990, 48–49), "We build no canals to irrigate pastures here, nor do we fence and sow our pastures with grass seeds to enhance yields. . . . It is not possible to try to control and alter the Changtang [plateau]. We do not try—instead we use our knowledge to adjust to it."
- Landscape diversity. For herders in eastern Inner Mongolia, "height and density of grass is seldom their only consideration in evaluating rangeland preferences. . . . Trees, hill slopes, and even patches of sand are explicitly considered desirable. Grass is good to eat, residents acknowledge, but animals also need browse matter, moisture, shade, protection from wind, and exposure to many kinds of forage. . . . Landscape diversity is a critical dynamic of seasonal pasturage, but it also determines the quality of grazing within each season" (Williams 1996, 678).
- Aesthetic appreciation. Nomads appreciated a wide-open and unconfining landscape. Mongols felt uncomfortable in cities, as did Chinese on the steppes, and aesthetics related closely to land ethics. Chinese referred to the open steppes as *huang*, or wild, and celebrated *kai huang*, or opening the wilderness to agriculture. Mongols, by contrast, referred to farming as *gajir hagalahu*, or shattering the land (Khan 1996, 128n7). In a land that cannot be improved, one must learn to live in harmony with it.
- Suspicion of "hard work." Unlike the rural Chinese, the nomads were surprisingly relaxed about work, as Sechen Jagchid and Paul

Hyer (1979, 137) report for nomadic Mongols. A Tibetan nomad of Pala summed it up: "We have a very easy lifestyle. The grass grows by itself, the animals reproduce by themselves, they give milk and meat without our doing anything, so how can you say our way of life is hard? We don't have to dig up the earth to sow seeds nor do any of the other difficult and unpleasant tasks that the farmers do" (M. Goldstein and Beall 1990, 48). The "lack of a work ethic" went hand-in-hand with recognizing the ecological and energetic limits of productivity on the grasslands. It was futile to try to do too much. Chinese rulers, at least through the Ming, recognized this, though they considered it regrettable. The Qing rulers were not so patient, and they introduced territorial administration to many Central Asian areas (Di Cosmo 1998, 300–303), thereby restricting mobility and beginning to erode buffers, as well as decreasing the resilience of the grassland environments. When the Communists, with the aid of fossil fuels, began their "scientific" program to modernize, rationalize, and develop the grasslands and their pastoral peoples, it did not work either, and the price today is being paid in the currency of land degradation, overgrazing, and desertification.

Despite their differences, since the middle of the twentieth century these widely varied social-ecological systems have all been transformed by the developmental policies and programs of the Chinese Communist Party. In the service of class struggle, nation-building, and above all development, the party has imposed its totalizing and homogenizing vision on the formerly disparate livelihoods of the peoples within the PRC borders. Specific policies and practices have changed from era to era and vary slightly from region to region, but the party has continued to implement its vision of world-historical progress everywhere, guided by the three basic principles of development, revolution, and science.

2. Development, Revolution, and Science

> Natural science is one of humanity's weapons in its fight for freedom.
> In order to achieve freedom in society, [one] must use social science
> to understand society, to reshape society, and to carry out social
> revolution. In order to achieve freedom in the world of nature, [one]
> must use natural science to understand nature, overcome nature,
> and reshape nature.
>
> MAO ZEDONG, speech at the Natural Science Research Society
> of the Border Region, 5 February 1940

Upon its victory in 1947–50, the Chinese Communist Party set about reconstructing the war-torn country both socially and physically, fixing and transforming both the disorder-prone social system and the disaster-prone ecosystem. To undertake this monumental task, the Communists adopted a strategy based on three important concepts: development, revolution, and science. In adopting these concepts, the party set in motion new turns of the Malthus-and-Boserup ratchet and the adaptive cycle.

DEVELOPMENT

The Chinese Communists, like the Nationalists before them, were and are committed to economic growth and social improvement. The Common Program, the principal general policy document issued in 1949 by the Chinese People's Political Consultative Conference, explicitly defines "development" or "development of production" as a primary goal to be achieved through the establishment of socialist relations of production: "The basic principle for the economic construction of the People's Republic of China is . . . to achieve the goal of developing production and a flourishing economy . . . to promote the development of the entire society and economy" (CPPCC 1949, Article 26).

China would achieve this goal through social revolution and a planned economy, taking sides in the Cold War, following the lead of the Soviet Union, and explicitly rejecting the market-based institutions and ideologies of capitalist countries. Both sides in the Cold War maintained that they were competing to demonstrate the superiority of their socioeconomic systems. In ecosystem terms, however, this great struggle was a conflict between *two varieties of the same ideology* of modernist developmentalism (Scott 1998; Wallerstein 2010).

The post–World War II idea of development, like rubrics such as "modernization" or even "progress," posits that the main thing that has happened on earth since the start of the Industrial Revolution has been material growth and prosperity, achieved through harnessing the energy embodied in fossil fuels and the ingenuity embodied in technological advance (L. White 1943). According to the developmentalist historical narrative, this material growth occurred first in Europe and North America but was destined to spread to the whole world:

> For many years the industrialized nations of North America and Europe were supposed to be the indubitable models for the societies of Asia, Africa, and Latin America, the so-called Third World, and that these societies must catch up with the industrialized countries, perhaps even become like them. (Escobar [1995] 2012, xlv)

> Even those who opposed the prevailing capitalist strategies were obliged to couch their critique in terms of the need for development, through concepts such as "another development," "anticipatory development," "socialist development," and the like. In short, one could criticize a given approach and propose modifications or improvements accordingly, but the fact of development itself, and the need for it, could not be doubted. (Escobar [1995] 2012, 5; see also Wallerstein 2010, 168)

The keys to development were many, according to its advocates, but for almost all these theorists, modern science and technology were among those keys. W. W. Rostow (1960, 30, emphasis added), an early advocate of economic development, stressed that "traditional societies . . . have not usually been static. Often they developed all the preconditions for growth 'take-off' except one. They improved their agriculture with irrigation. They built roads and other forms of social overhead capital. They engaged in trade and even did a certain amount of manufacturing. *But what ultimately triggered their decline was the invariable absence of a scientific attitude toward the physical world.*" The Chinese Communists, like Rostow and other advocates of capitalist development, also emphasized the necessity of a scientific attitude and rapid development of mechanical technology. As Sigrid Schmalzer (2014, 78–79) succinctly summarizes, "Both [Leninism and US modernization theory] were committed to modernization through technological development, and both depended on deterministic expectations that development would proceed through specific 'stages.'"

Chinese Communist conceptions of development as a series of stages connect closely to their ideas about ecology. Well before the concept of communism arose, Chinese historians and philosophers had connected civilization and morality with the agrarian ecology of China Proper. Other social-ecological formations—including the extensive, mostly stateless (Scott 2009) agro-silvio-pastoral systems of Zomia and both the pastoral system to the north and the pastoral-agricultural codependency of Central Asia to the northwest—were inferior. They had neither the settled, intensive resource base nor the literate, moralistic ruling class of China Proper. Limited "civilizing" was possible (Harrell 1995b), but ecology remained an important barrier to completing the civilizing project.

The Communists continued half this argument but rejected the other half. The Marxist-Leninist teleological narrative of progressive stages of history—the CCP version of the developmentalist vision—confirmed the traditional idea that peripheral areas were "backward" but at the same time compelled the regime to modernize or develop them, ecological barriers be damned. And modern technology meant that ecology was no longer destiny. Places never farmed before could be farmed; pastures and forests that had coexisted with low-intensity farms could be transformed into more farmland; parts of the steppes could be plowed up; cities could be built where none existed before.

There were, however, significant differences between the capitalist and Communist developmental ideologies. For the capitalists, the "developed world," as the rich countries were now called, would lead the way for the "developing world" (a euphemistic way of referring to poor countries), reflecting the idea that their poverty was temporary and remediable. This transformation sometimes happened, as when Taiwan's development benefited from a combination of US military aid and state promotion of small industry and high value-added products for export.[1]

By contrast, for the Communists and for certain noncommunist leftist thinkers associated with the "critique of development" (Escobar [1995] 2012), with "world-systems models" (Wallerstein 2010), or with "dependency theory," the continued poverty of some countries (which the Chinese Communists referred to as Asia, Africa, and Latin America) was a *consequence* of the wealth of others.[2] They were not "developing" under the benevolent hand of the rich countries; they were "underdeveloped" because of the oppressive colonial and then neocolonial exploitation by rich countries.

Capitalist and Communist developmentalist ideas also differed over *how* to develop—by capitalist investment or by revolution (figure 2.1). The material goals were the same: a comfortable life in which mechanical energy produced most material goods, most people were occupied in nonphysical labor, and material prosperity enabled people to enjoy life free from having to struggle to continue existing. In a classic passage, Karl Marx (1845) wrote that "in communist society, where nobody has one exclusive sphere of activity but each can become accomplished in any branch he wishes, society regulates the general production and thus makes it possible for me to do one thing today and another tomorrow, to hunt in the morning, fish in the afternoon, rear cattle in the evening, criticise after dinner, just as I have a mind, without ever becoming hunter, fisherman, herdsman or critic."

For the Chinese Communist Party, the poverty of underdeveloped countries was the consequence of structural inequalities that kept the poor countries poor on the international scale and kept the laboring people poor within the country. Revolution was the answer to poverty and misery; that is, *revolution was a means to development*. As Michel Oksenberg (1973, 13–14) pointed out, "One major purpose of the Chinese Revolution is to build a prosperous and strong country. Economic development is a major aim of any program undertaken in China."[3]

REVOLUTION

On the surface, the Communists were revolutionaries, pure and simple. In their telling, China was poor and backward because for thousands of years it had been ruled by a feudal class of rentiers whose self- and class interests lay in exploiting peasants' and other working people's labor to

Two Models of Development

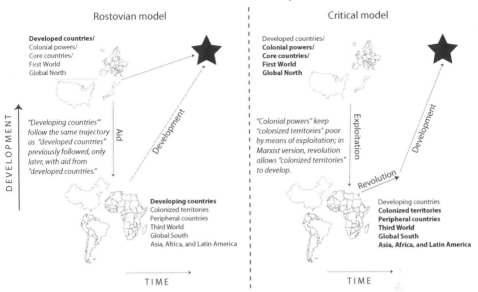

Figure 2.1. Two models of development compared schematically.

extract the maximum amount of surplus value. They kept themselves comfortable and in control while keeping the peasantry in a state of misery. A Marxist-Leninist version (figure 2.2) of the ecological flowchart of the Han village community (see figure 1.3) shows more goods and services traveling from the peasant household to exploiting landlords, rapacious tax collectors, and price-setting traders.

In addition, the official narrative dubs the period from 1840 to 1949 "semifeudal, semicolonial," as Western imperialism compounded the effects of feudal exploitation, resulting in further immiseration and continued social and ecological chaos, keeping the social-ecological system in what amounted to an extended omega phase of the adaptive cycle. The solution to this misery was social revolution, a fundamental transformation of the class basis of the social and political order. As the exploited overthrew their former exploiters under the party's leadership, they would not only become masters of their own fate but would also liberate their underused or unproductively diverted labor power and lift themselves out of poverty. Placing the forces of production in the hands of the producers, revolution would lead to strength, prosperity, and stability.

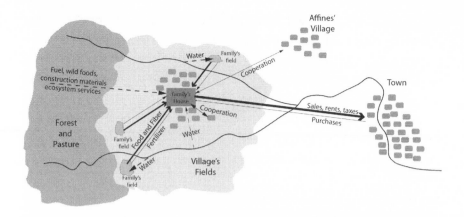

Figure 2.2. A Marxist-Leninist version of the flows of goods and services in a pre-Founding Han village. Solid arrows are private goods; dashed arrows are common-pool goods. According to Marxist-Leninist theories, the exploitation represented by the thick arrow keeps peasant households impoverished and prevents development.

The Marxism-Leninism that guided the party's revolution is a profoundly human-centered philosophy, rarely paying much attention to environment or ecology. Growing out of the eighteenth- and early nineteenth-century European tradition of political economy, Marxism has always focused on human relations, seeing material things and the natural environment primarily as forces of production, or things that humans use as material goods to produce or consume.[4] In Marxist-Leninist teleology, human history is a dialectical progression of modes of production, and in such a model material things are factors of production that can be altered by human organization and human will. Although material relations rather than ideas determine the course of history, limited natural resources in themselves are not obstacles to material progress, and ecosystemic relations are even less capable of being obstacles. The real obstacles lie in forms of human relations that divert human labor power from its potential for liberation toward the aggrandizement of a parasitic exploiting class. *The constraints on development are social and political, not ecological.* When the oppressive, exploitative class structure is transformed, the productive forces will be liberated, and people will then be able to construct a new material world (Trotsky 1924). Liberating the productive forces is partly about liberating the heretofore dormant energy locked up in natural resources, including water, land, forests, and minerals.

This is a deeply humanist view, putting people at the center of the physical and moral universe, granting them not only the ability but the moral right to use elements of the material world for their own benefit. As such, it bears similarities to Confucian humanism (Taylor 1998) as well as to various forms of humanism stemming from the Enlightenment. But Chinese Marxism-Leninism sometimes went further. Perhaps the most anthropocentric of all worldviews, in its extreme forms it rejected not only gods but also laws of nature as nothing but elements of false consciousness employed in the service of reactionary ruling classes. Demography and ecology—fields that view humans in relationship to other biotic and abiotic elements of the earth system—become, like religion, opiates of the people, things that the exploiting elites used to hoodwink the masses or perhaps even hoodwink themselves. Mao's campaign against the famous demographer Ma Yinchu, who advocated population control in the 1950s, is an example. For Mao, population increase was a positive development of the nation's productive forces (namely labor), as embodied in the slogan "More people, greater strength" (Ren duo, liliang da) (Shapiro 2001, 31–35). Even more extreme was the view expressed in the Great Leap Forward slogan "Ren you duoda dan, di you duoda chan," or "The product of the earth will be as great as the courage of the people" (Liu Xirui 1958). In other words, the only limits on the productivity of the ecosystem are those imposed by exploitative class relations and human timidity.

As with demography, so with environmental degradation. Communist thinking during the high socialist period was so hyper-anthropocentric, paying so little attention to the ecosystem, that Chinese writers would seriously maintain that pollution was a capitalist problem, despite the view out the window. They began with the premise that many eco-Marxists hold, that the logic of capitalism, driven by the profit motive, led to environmental pollution—a reasonable argument given the history of capitalism and the environment. However, they then extended that logic to its inverse, namely that since economic actors in a (state) socialist system were not driven by the profit motive but rather by the good of the people, *socialist* economic growth, unlike its capitalist counterpart, would not lead to environmental degradation (Whitney 1973, 107). As a 1972 article on toxic waste put it,

> Every day large quantities of the "three wastes"—waste gas, liquid, and residue—stream forth from industrial production. In capitalist countries, because the capitalists seek high profits and production is in a state of anarchy, these "wastes" which pollute the air and poison the rivers, pose an increasingly serious menace to the people's health.
>
> . . . In our country, the "three wastes" have done little harm to the people. This is because in a socialist country like ours, which is proceeding in all cases from the interests of the people, we can rely on the superiority of the socialist system to take various measures to prevent pollution harming the people. (Chi Wei 1972)

Chinese Marxism-Leninism's faith that the socialist system could liberate productive forces and avoid waste was a variety of the nineteenth- and twentieth-century anthropocentrism that James Scott (1998) has called "high modernism"—the idea that humans, freed of the constraints of religion and superstition and empowered by the powerful tools of science, could fundamentally remake the world to better serve human material needs. Scientific knowledge and engineering skills that developed the steam engine, motorized travel, flight, skyscrapers, and the wonders of electric everything could also redesign the social and natural worlds to human advantage. Scott (1998, 88) describes this as "the aspiration to the administrative ordering of nature and society." High modernism, in a sense, is a kind of antimaterial materialism, a mind-over-matter utopianism that subordinates the material and natural worlds to the powers of hyperrationality and design (Glover, Hayes, and Harrell 2021). Marxism of all flavors declares itself materialist, considering ideas to be manifestations or epiphenomena of material forces, but the materialism of Marxist-Leninist regimes did not always recognize the very material limits that the quest to increase material prosperity might face.

Humility in the face of the natural world is not a modernist character trait; even recognizing that we must live within the constraints of the natural world is foreign to high modernism, with its "nearly limitless ambition to transform nature to suit man's purposes" (Scott 1998, 94). The high modernist faith in hyperrationality and the powers of engineering design, combined with the Marxist faith in the ability of revolution to liberate the productive forces, supplied the Chinese Communists with many spurious solutions to the very real problems of China's ecosystem in the mid-twentieth century.

Revolution thus became more than simply a fundamental shift in the way the productive forces were owned and organized. It became a kind of utopian vision in which not just the laws of society but also the laws of nature could be rewritten. It led not only to the "utopian urgency" that Judith Shapiro outlines in *Mao's War against Nature* (2001) but beyond that to the utopian fantasy that the laws of nature were as much a creation of ruling classes as were the laws of property and, like the laws of property, could be rewritten.

SCIENCE

Marxism and science have a complex historical relationship. Karl Marx advocated belief in science and admired Charles Darwin's work on evolution.[5] In his influential 1880 polemic, *Socialism: Utopian and Scientific*, Marx's sidekick Friedrich Engels contrasted the "utopian socialism" of the eighteenth- and early nineteenth-century French philosophers to the "scientific socialism" advocated in Marx's writings and his own. He stressed that natural science must abandon the "metaphysical approach," which dominated the thought of Newton and Linnaeus, for a "dialectical approach" pioneered by Hegel and put on a firm materialist footing by Marx (Engels [1880] 1970). The Chinese Communists' predecessors in the Soviet Union proclaimed dialectical materialism to be the basis of scientific thought, and under their patronage Soviet scientists not only achieved great results in physics, chemistry, rocketry, and other physical sciences but also gave birth to Lysenko's pseudoscience, which severely set back both evolutionary biology and agricultural science. In Stalin's final years, they also criticized Einstein's relativity theories as idealist and contrary to the laws of dialectical materialism, a critique CCP ideology picked up at the same time (Danian Hu 2007, 549–52).

Many of the contradictions in the Chinese Communists' attitude toward and use of science stem from this uneasy marriage of natural science and Marxism. Science is a quest for truth, but it takes place in a social and political context (in Marxist terms, as part of a mode of production) and is thus never a *purely objective* quest for truth. Science is influenced by the wishes, prejudices, equipment, funding sources, and social networks of the scientists (Latour 1998, 2002). The Chinese Communists approved of scientific knowledge and practice only when they contributed to the project of revolutionary development. Pure science, the quest for knowledge for its own sake, was almost uniformly condemned as bourgeois and useless (Schmalzer 2014, 75). Mao himself viewed science through the lens of dialectical materialism:

> One's knowledge depends mainly on one's activity in material production, through which one gradually understands natural phenomena, natural properties, natural laws, and the relations between humans and nature. (Mao 1937)

> Where does people's correct thinking come from? Does it drop from the sky? No. Is it innate in the brain? No. It comes from social practice. It can only come from these three kinds of social practice: the struggle for production, the class struggle, and scientific experiment. (Mao 1963)

Mao also expressed the necessity for science to be applied to the practical problems of revolutionary development:

> Then comes the second stage in the process of cognition, the stage from consciousness [back] to matter, the stage from ideas [back] to existence; this is taking the knowledge gained in the first stage and putting it into social practice; to see whether or not the theories, policies, plans, or methods can obtain the anticipated success. Generally speaking, those that succeed are correct and those that fail are incorrect; especially humanity's struggle with nature is like this. (Mao 1940)

> Class struggle, production struggle, and scientific experimentation must be linked up. Doing only production struggle and scientific experimentation but not grasping class struggle cannot arouse people's spirit, and production struggle and scientific experiment won't be done well. (Mao 1964a)

In other ways, however, the relationship between science and revolution has been paradoxical and contradictory, because the term *kexue*, like its rough English equivalent, "science," means different things in different contexts. Roberto González's (2001, 20) typology of the different uses of "science" in contemporary English can help us understand the uses of the term *kexue* in Chinese:

- Science as a collection of certain truths that have been verified
- Science as high technology
- Science as a social enterprise
- Science as "what scientists do"
- Science as a form of inquiry that makes use of a particular method
- Science as the achievement of Superior Western Man

A contradiction between "science as a form of inquiry that makes use of a particular method" and "science as a collection of truths that have been verified" was particularly evident at times of the "high tide of political mobilization," such as the Great Leap Forward (Skinner and Winckler 1969). Propaganda at those times extolled the creativity and scientific abilities of ordinary people and condemned the bourgeois strivings of professional scientists. In some scientific fields, such as seismology and earthquake prediction, practice sometimes realized the ideal of "using the folk and the foreign together" (*tu yang bing ju*). Folk knowledge and peasant inventions were combined with technical, laboratory science in a national effort to prevent serious damage from earthquakes (F. Fan 2017). But in many cases, the *practice* of science fell short of this revolutionary ideal. Scientific practice was less the application of the scientific method to local conditions than the imposition of "a collection of truths that have been verified." Local scientific experimentation in agriculture often followed the example of local experiments in policy implementation: rather than encouraging local people to carry out experiments, agricultural planners took results from a particular local experiment and imposed them (as "a collection of truths that have been verified") uniformly and thus often inappropriately throughout the country (Heilmann 2008, 14–18).

This paradox perhaps resulted less from leaders' hypocrisy than from a contradiction between two elements of Marxist-Leninist ideology. Even though it celebrated the transformative potential of revolutionary (worker and peasant) classes, those classes were observably "behind" in the Marxist-Leninist historical teleology. Even under the semifeudal, semicolonial mode of production in the late Qing and the Republic, educated people, urbanites, scientists, and Communist leaders were all "ahead" of peasants, who were mired in "feudal superstition," illiteracy, and ignorance. Ethnic minorities in Zomia and Central Asia were, *a fortiori*, even more ignorant and in need of the scientific knowledge that could liberate them from centuries of misery. Reliance on peasant wisdom was thus inherently limited. As Sigrid Schmalzer (2016, 38) points out, "Through the celebration of *tu* science [i.e., experimentation done locally by producers rather than in the laboratory by trained scientists], the Chinese socialist state created for the Chinese nation a subaltern voice. Although that voice was deeply inspiring to many people, it did not necessarily do justice to the people it claimed to represent. Nor did it ever touch the developmentalist modernization paradigm that presumed human mastery over nature; indeed, in this respect Maoism was thoroughly consistent with the fundamental assumptions of modern technoscience everywhere."

There was also a contradiction between the expertise of science and the "redness" or revolutionary enthusiasm of various nonprofessional groups, from party revolutionaries to loyal peasants (Baum 1964). If science is, according to González's fourth definition, "what scientists do," then science is often what *professional* scientists do, and Mao and many of his followers were uneasy about bourgeois elements in society, particularly highly educated people such as scientists. On the one hand, their knowledge ("a collection of truths") and training ("a particular method") were necessary to the development of agriculture and industry. On the other, since Marxist epistemology states that knowledge comes from social practice and since professional scientists' social practice was formed in the university, the classroom, and the laboratory, their knowledge was suspect. Equally disturbing was the fact that many of the most prominent scientists, in agriculture, demography, and engineering—sciences that related directly to questions of environment and human ecology—had been trained at least partially in Europe or the Americas. Even though many of them were fervent Chinese nationalists, their science still carried the stigma of bearers of "the achievement of Superior Western Man," something made even more suspect by the

United States and its imperialist military allies' encirclement of China (Zuoyue Wang 2014). Thus, the ideal of scientists who were "both red and expert" (*you hong you zhuan*) was not always achievable (Levenson 1962, 17–18; Baum 1964). Scientists, like other "experts," were both necessary and dangerous, and so was their science.

INSTITUTIONAL CONTEXT

In addition to conflicts within Marxist-Leninist ideology, the PRC regime has faced contradictions between ideological goals and bureaucratic structures. No ideology is implemented in a structural vacuum, and all Communist ideologies, more utopian or more gradualist, have succeeded or failed in a social context where dreams of the future coexist uneasily with offices of the present. This "conflict between defenders of the faith and defenders of the institutions" came to a head during the Cultural Revolution, but it was present from the time the People's Republic was established (Lieberthal and Oksenberg 1988, 7).

It was ironic but inevitable that the Chinese Communist Party, having taken over the country with a peasant army recruited on the basis of simple grievances about an immoral economy and xenophobic resistance to an invader (Johnson 1962), had to turn to bureaucracy to implement its teleological visions of history.[6] It had, however, built up a considerable organizational apparatus during its rise to power (Esherick 1995, 71), and it solidified and formalized these in the first years of its rule. Drawing on the precedent established by Stalin in the Soviet Union in the 1920s, "Mao [Zedong] rationalized the need for superstructural hegemony during the initial period of socialist construction in China by invoking the necessity for a non-revolutionary dialectical leap to bring the backward Chinese economy into harmony with its advanced socialist superstructure" (Baum 1964, 1049–50).

By 1952 or earlier, a full structure of party, state, and military organizations was in place at central and local levels, with organization charts, pay grades, promotion standards, and even detailed layers of perquisites like the ability to buy soft-sleeper train tickets or claim apartments with indoor plumbing. This meant, as Kenneth Lieberthal and Michel Oksenberg (1988) have set out, that any policy, environmentally related or otherwise, would be formulated, implemented, revised, and perhaps eventually discarded on the basis of the interaction of leaders and offices with sometimes congruent, sometimes competing interests.

Bureaucratic rule also built an inherent conservatism into the system; customs and policies once in place were difficult to dislodge; innovation often encountered resistance. A rigidly organized bureaucracy can lead the adaptive cycle of any social-ecological system into a K phase where more and more energy (in both the physical and the motivational sense) goes into maintaining the system, and the structures themselves begin to inhibit those revolutionary changes that were the reason they were established in the first place. Bureaucracy thus becomes an institutional buffer against further change but requires more and more energy to maintain itself. When bureaucracies oriented to orderly development are in charge, this can severely inhibit efforts to mitigate the environmental degradation brought on by that development, leading to "pathologies of natural resource management" (Holling and Meffe 1996). It can also lead to rigidity traps (Holling, Gunderson, and Peterson 2002, 96–98) in which a system is so highly interconnected and stable in the short run that it becomes impervious to change, even in the face of small-scale disturbances. In this situation, the system may retain characteristics that even the regime's own leaders consider undesirable, such as the "capitalist roaders" within the party whom CCP radicals attacked during the Cultural Revolution, or the "black [heavy industrial] development" that advocates of "green ecological civilization" criticized in the 2010s. Such rigidity traps can temporarily render a system more resilient to small-scale disturbances, and thus the bureaucracy does not make the small-scale adjustments that would reinforce it against larger disturbances.

Finally, bureaucracies are prone to issue generalizing orders, promoting *yi dao qie*, or "cutting with one knife" solutions, better known in English as "one size fits all." Central orders often impose panacea solutions (Ostrom and Cox 2010), appropriate for one place or one time but misfits for another, or create cross-scale mismatches (Cumming, Cumming, and Redman 2006). Well-intentioned policies may have adverse results; things get done, but often they are the wrong things.

CCP leaders were well aware of the ways bureaucracy impeded innovation and revolutionary progress, and in order to "overcome bureaucratic lethargy and resistance," Mao and his followers employed voluntarist strategies of "political campaigns, ideological broadsides, suppression of dissent, purge of recalcitrant individuals, deification of the top leader, and so forth" (Lieberthal and Oksenberg 1988, 23). In addition, to counter the bureaucratic tendency toward top-down governance, managers often employed the technique of "from point to field" (*you dian dao mian*), trying a new policy in one or a few places, evaluating the results, and then deciding whether and how to scale up to general implementation (Schmalzer 2016, 33–34). They also invoked—though not always successfully—the Maoist idea of the "mass line," in which the party's role was not to dictate from first principles but to take the ideas of ordinary people, refine and consolidate them into policy, and then direct and implement them in local communities (Blecher 1979; Lin Chun 2019).

At the same time, there are beneficial aspects to bureaucracy. Alternatives to bureaucratic governance, including Maoist voluntarism, also have their faults and present the danger of "becoming panaceas for what ails failing bureaucracies" (Pritchard and Sanderson 2002, 160). Campaign governance, the Maoist alternative to bureaucratic rule, had its own pathologies stemming from both political intimidation of local populations and carelessness born of "utopian urgency" (Shapiro 2001). The sometimes nonsensical application of popular participation models in the Great Leap Forward, for example, contributed to system collapse, mass famine, and permanent resource degradation (Dikötter 2011; Yang Jisheng 2008, 46–53).

Over the history of the People's Republic, the balance between bureaucratic rationality and antibureaucratic voluntarism—between routine and campaign (Winckler 2018)—has varied in a pattern coinciding roughly with the compliance cycle of state incentives and popular reactions (Skinner and Winckler 1969; Winckler 2018), as shown in table 2.1.

At times when bureaucratic rationality ("routine") has dominated, so have gradualist plans for development, muted emphasis on revolutionary struggle, and a conventional, laboratory- and university-oriented view of the process of science. When antibureaucratic voluntarism ("campaign") has dominated, so have utopian ideas of development, heightened emphasis on revolutionary struggle, and a view of science oriented toward popular wisdom and distrust of laboratories and professionals. The rationalist model has dominated since the 1980s, as campaigns have become less frequent and less intense (Winckler 2018, 3). Innovation has thus come more from within the bureaucracy itself than from popular

Table 2.1. Two philosophies of development, revolution, and science

RATIONALIST MODEL	VOLUNTARIST MODEL
Capital	Labor
Workers	Workers and peasants
Material forces	Ideological fervor
Expert	Red
Cosmopolitan (yang)	Chinese (tu)
Bureaucracy	Masses
Routine	Campaign
Stalinist	Maoist

Note: These binaries are Weberian ideal types, not completely mutually exclusive and heuristic rather than realistic.

resistance, although there has been more latitude for popular protest in the environmental realm than in certain others (Lora-Wainwright 2017).

Across these cyclical variations, there have always been contests between the interests of different branches of the bureaucracy—between central, regional, and local, as well as between different functional bureaucracies (Lieberthal and Oksenberg 1988, 22–30). In recent years this has played itself out in infighting between branches more oriented to development and branches more oriented to environmental protection or restoration (Mertha 2008; Tilt 2010, 2015).

PERIODIZING PRC ECO-HISTORY

Conventional historiography, based on these alternations between voluntaristic and bureaucratic governance, has led to the conventional periodization of PRC history, illustrated on the left side of figure 2.3. This periodization divides the first seventy-plus years of the People's Republic into two eras: Maoist (here called "high socialist" or sometimes "collective," when referring to the economy), often characterized by an emphasis on revolution over development, and post-Maoist (here referred to as "Reform"), often characterized as abandoning revolution and socialism in favor of development. The first era is conventionally divided into a relatively sane period of Soviet-style socialist construction from the Founding until 1957; a disastrous experiment with utopia in the Great Leap Forward, leading to the famine of 1959–61; then a brief interlude of moderation until 1965; and finally the ten "lost years" of the Cultural Revolution, from 1966 to 1976. Reaction to the Cultural Revolution led to the reformist program under Deng Xiaoping and his successors, bringing forth rapid, unprecedented economic growth and an opening to the world, while avoiding the "Western-style" democracy that most modernization theorists would have assumed was necessary for true modernization. Some versions of this conventional periodization mark the beginning of another subperiod with the ascendancy of Xi Jinping to the position of paramount leader in 2012, an era marked by the reimposition of stricter authoritarian controls on politics and society, though not by a return of campaign-style governance (Winckler 2018).

This cycle or alternation between governance modes has not, however, always turned in phase with the relationships between policy and environment, so ecological historiography leads us to a different periodiza-

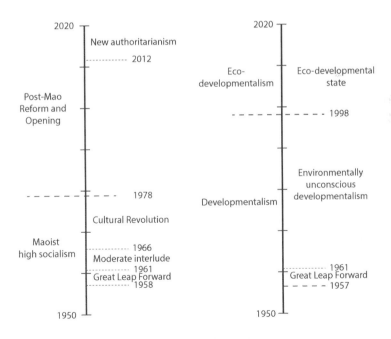

Figure 2.3. Conventional (*left*) and ecosystemic (*right*) periodization of PRC history.

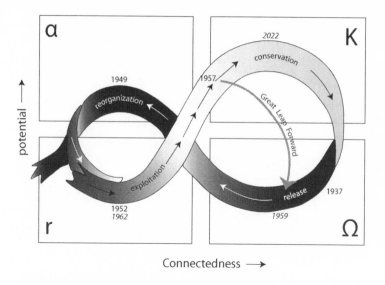

Figure 2.4. PRC history conceived as the adaptive cycle.

tion, illustrated on the right side of figure 2.3 and transformed into the adaptive cycle in figure 2.4.

Rationalism and Voluntarism in the Early Years

From the establishment of the People's Republic until the winter of 1957–58, the CCP regime set out to bring the country out of its extended omega phase and into an r phase of steady growth. The Common Program of 1949 explicitly advocated development or development of production as primary goals to be achieved through the establishment of socialist relations of production by laying out specific, rationalist programs for both agriculture and industry.

With regard to agriculture, the Common Program states that "the People's Government should organize peasants and all labor power that can carry out agricultural work, to . . . [develop] agricultural production and supplementary production. . . . Every step of land reform should be integrated with the revival and development of agricultural production" (CPPCC 1949, Article 34).

The Common Program goes on to specify raising agricultural yields through developing new varieties and increasing fertilizer use and through expanded repair and construction of waterworks, along with land reclamation and agricultural migration (CPPCC 1949, Article 34). In industry, the program emphasizes the development of "heavy industry, including, for example, mining, iron and steel, power, machinery construction, electrical, and important chemical industry, in order to build a foundation for industrialization of the nation" (CPPCC 1949, Article 35).

By 1952, with this Land Reform program mostly completed and most of China's agricultural and industrial infrastructure repaired or restored, the regime was ready to move on with plans for development. The first five-year plan, covering the period from 1953 to 1957, gave the goal of development a much more explicitly revolutionary flavor. It attributed the backward condition of the Chinese economy to contradictions between the development of the productive forces and the prevailing capitalist relations of production. The regime's preferred way to increase economic production was through revolutionary changes in both industrial and agricultural relations of production: "The first five-year plan for the development of our country's economy must thus include plans for step-by-step socialist transformation of agriculture, industry, and capitalist commerce" (People's Government 1955). The plan goes on to state that economic development must emulate the Soviet Union and make building heavy industry the primary project for facilitating the general growth of the economy.

During this initial period, the party did manage to increase both agricultural and industrial production beyond their prewar levels and begin building up a heavy-industrial base. The results were mixed. On the positive side, they built infrastructural and institutional buffers to supplement and replace some of the dwindling ecological buffers left over from the late Qing and the Republic, increasing the resilience of many agricultural systems against natural disturbances. At the same time, however, they duplicated at a very accelerated pace some of the problems that had plagued the Qing as it moved from its r phase to its K phase. Peace and very effective public health measures led to dramatic decreases in mortality, which combined with continued high fertility rates to produce rapid population growth. Thus, in spite of increased agricultural yields, food production barely kept pace with population growth, and continued increases in food production would require building more agricultural infrastructure, which required more and more maintenance. At the same time, the vulnerability of social-ecological systems was exacerbated by the severe pollution and environmental degradation that came with building heavy industry (Shapiro 2001).

The Triumph of Voluntarism in the Great Leap Forward

Recognizing the dilemma of supplying a rapidly growing population without the means to increase capital investment in agriculture, the party intensified its mobilization of human labor, which it continued to see as an underutilized factor of production. This led to the Great Leap Forward. The regime mobilized masses of people to increase labor inputs in both agriculture and industry, first through accelerated waterworks construction on the North China Plain in the winter of 1957–58, then through the formation of people's communes, which allowed rapid increases in cultivated land and radical changes in agricultural techniques, as well as crude programs of rural industrialization aimed at the goal of overtaking the United Kingdom in steel production. The ideology behind this program was long on hope and logic and short on connection to empirical reality: the way to solve problems of agricultural and industrial productivity was to think one's way out of them. In explaining the Great Leap slogan of "More, faster, better, frugal" (Duo kuai hao sheng), Mao ignored ecosystems' ability to accommodate or buffer against disturbances: "If we don't go as fast as a satellite, if we don't actually do more and faster, then it's impossible to be more and faster, then we can't do it" (Mao 1958a).

Xushui County, Hebei, was a pioneer site of labor reorganization for waterworks construction (Dikötter 2011, 46–49). Mao's remarks to farmers and cadres there in August 1958 exemplified the utopian thinking of the Great Leap: by reorganizing society, one could extract resources from the natural environment without limit and thus, in short order, achieve food security, gender equality, and Marx's dream of "hunt in the morning, fish in the afternoon . . . criticize after dinner":

> For sure it's better if you have surplus grain. If the state doesn't want it, members of the agricultural collective will eat more themselves. You could even eat five meals a day!
>
> Here you have thoroughly liberated women! It's true! Everybody eats in the canteen; every cooperative has a kindergarten.

If you can't eat all your grain, what are you going to do? If there's surplus grain, then later on you can grow a little less, do half a day's [farm] work, and the other half day you can do culture, study science, make a ruckus in culture and leisure, run universities, run secondary schools, what do you think? (Mao 1958b)

Not only waterworks but new agricultural techniques such as deep tilling and dense planting, the double-wheeled, double-bladed plow, and massive new irrigation works were touted as breakthrough technologies that would end thousands of years of agricultural shortages. The slogan "The product of the earth will be as great as the courage of the people" originated in an August 1958 letter from an investigation team sent to Shouzhang County, Shandong, and published in *People's Daily*.[7] It touted these new methods, anticipating huge increases in yields: "The whole county is conducting a campaign to achieve the high goal of 10,000 *jin* per *mu*. One *mu* will produce 50,000 to 100,000 *jin* of sweet potatoes, one *mu* will produce 10,000 to 20,000 *jin* of corn or millet; local cadres and masses regard this kind of quota as very ordinary" (Liu Xirui 1958).

None of the measures motivated by this cornucopian ideology of labor materialism could be sustained for even a few years. As a result, the Great Leap, lasting from 1958 through 1961, destroyed both the ecosystem buffers that had existed previously and many of the institutional buffers recently created through efforts to organize agricultural and industrial production. Some of the infrastructure built during the Great Leap did not replace ecosystem buffers but actually increased vulnerability of local ecosystems. The Great Leap thus became a massive, artificial, countrywide disturbance and led to rapid collapse and a precipitous omega phase in the form of declining production, widespread famine, and social chaos.[8]

Environmentally Unconscious Growth through "Normal Socialism" and Early Reform

The institutions thrown into chaos by the Great Leap Forward recovered with remarkable speed, but much long-term damage remained. The Great Leap had transformed the ecosystem in important ways, through deforestation, expansion of agriculture into unsuitable areas at both local and regional scales, construction of unstable and disaster-prone waterworks, and abject failure of its attempts at rapid industrialization. As a result, the ensuing r phase of slow growth in production, beginning with the recovery from the Great Leap and extending until the late 1970s, took place in an environment of rapidly expanding population, unsustainable agriculture, polluting industry, depletion of unrenewable and slow-renewable resources, and increasing bureaucratic rigidity. Despite the fact that the Cultural Revolution of 1966–76 probably cost over a million lives (Walder 2014), its ecological effects were relatively minor. Industrial growth continued to be concentrated in heavy industries such as mining, metallurgy, and military hardware (including the Third Front program's industrial projects in the interior) (Meyskens 2020), and agriculture continued to be based on increasing labor inputs, with the important exception of the installation of tube wells and gasoline-powered pumps on the North China Plain, a program begun before the Cultural Revolution and unrelated to its ideological objectives.

Beginning in late 1978, the party, newly under Deng Xiaoping's leadership, carried out the economic transformation known as Reform and Opening (Gaige Kaifang, henceforth simply Reform). This program gradually changed the Chinese economy from state socialism to bureaucratic capitalism, or a market economy with heavy state investment and state ownership, known in China as "socialism with Chinese characteristics." Though these changes were momentous for economic growth and for the personal life of every Chinese citizen, they did not bring about a fundamental shift in the ecological system or a new phase in the adaptive cycle. Instead, they continued, with many depauperate ecosystems, the slow transition from r phase to K phase that had been interrupted by the disastrous "epicycle" of the Great Leap Forward. The big change with Reform was acceleration of economic growth, which eventually led to acceleration of two important kinds of ecosystem degradation: increasing chemical and mechanical inputs to agriculture, as well as burgeoning air, water, and soil pollution due to overuse of agricultural chemicals, proliferation of toxic industrial pollutants, and massive increases in fossil fuel–based energy production.

Ecosystem vulnerabilities during this phase of rapid industrialization led to an environmental crisis (Smil 1993). Media and official documents reported all kinds of environmental disasters: deforestation, desertification, erosion, wetland damage, surface and groundwater depletion, eutrophication, water pollution, air pollution, soil contamination, habitat loss, contributions to global warming, and others. Although the regime recognized the existence of these problems even before Reform, all through the 1980s and most of the 1990s it refused to recognize their severity, until it received a rude wake-up call in 1998. Severe floods in the middle Chang River provinces of Hubei, Hunan, and Anhui caused dislocation of an estimated 29 million people and flooding of 146 million hectares of farmland in Hunan alone (Kuang 1999, 86). The state awoke to the realization that unbridled development, without concern for the environmental changes it caused, would lead at least to regional collapse (it already had in a few places) and perhaps to a systemic crisis.

The Turn to Eco-developmentalism

Thus began, in 1998, the Chinese Communist Party's ideological turn from pure developmentalism to eco-developmentalism (Haddad and Harrell 2020). Responding to public pressure as well as scientific evidence, the state began to formulate a new guiding ideology summed up in the labels "environmental culture" (*huanjing wenhua*) and particularly "ecological civilization" (*shengtai wenming*). Using these labels and others, leaders began to deploy the basic concepts of development and science in different ways, recognizing the unsustainability of the headlong rush to development that had characterized the previous epoch of Reform and Opening.

"Ecological civilization" is, at its base, an attempt to solve the aforementioned contradiction between Marxist-Leninist teleology and empirical evidence. Although revolution as violent transformation has disappeared from the state's agenda, the teleology of development and progress, in which revolution once played so big a part, continues to be a primary basis of the party-state's legitimacy. But development has badly messed up the ecosystem. The solution has not been to abandon the ideology but to amend it to make environmental restoration a stage in the teleological progression (Yifei Li and Shapiro 2020, 6). If feudal society is characterized by agricultural production, then capitalist and early socialist society is characterized by industrial production, which is what has caused environmental degradation, not only in China but around the world. In order to build an environmentally sustainable mode of production, the next stage must turn away from industrial civilization and shift toward something called "ecological civilization" (Pan 2003; Ren Z. 2013). Pan Yue, deputy director of the State Environmental Protection Administration and later vice-minister of environmental protection, expresses this with an implicit but obvious nod to Marxist dialectics:

> Three hundred years ago, the West turned from a traditional agri-cultural civilization to a traditional industrial civilization. . . . Tra-ditional industrial civilization brought flying-speed development of science and technology and of the economy and a huge rise in humanity's material standard of living. But the limits it brought with it become more and more obvious [numerous examples omit-ted here]. People have begun to reevaluate the traditional industrial economy and to consider its many kinds of maladies, in order to escape the many kinds of crises that it has created, to replace the black with the green, and to replace traditional industrial civiliza-tion with ecological industrial civilization. Because of this we can say that the ecological crisis created environmental culture and that the core of environmental culture is ecological civilization. (Pan 2003, 7)

Figure 2.5 sets out Pan's evolutionary progression in schematic form.

In this critique, Pan (2003, 8) explicitly criticizes the party's past ideas of development and science: "Environmental culture emphasizes that the earth's resources are limited and that science cannot allow people to mistakenly think that our ability to change nature to benefit humanity is limitless. If we take 'human plans overcome nature' [Ren ding sheng tian] to its logical conclusion, it will put humanity into an existential predicament."

Pan (2003, 10) also inverts the Marxist-Leninist idea that the relations of production are the primary determinants of progress, as well as the Maoist idea that "the product of the earth will be as great as the cour-age of the people": "Environmental culture emphasizes that relations between humans and nature control relations among people. Correctly adjusting relations between people and nature amounts to correct coor-dination of human social relations and amounts to the pursuit of human social harmony and progress."

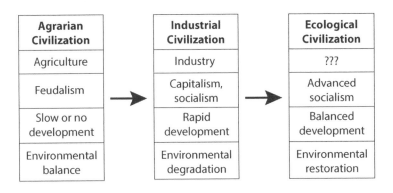

Agrarian Civilization		Industrial Civilization		Ecological Civilization
Agriculture		Industry		???
Feudalism	→	Capitalism, socialism	→	Advanced socialism
Slow or no development		Rapid development		Balanced development
Environmental balance		Environmental degradation		Environmental restoration

Figure 2.5. Revisionist Marxist-Leninist schematic of historical teleology.

Still, party theoreticians need to keep ecological civilization securely within the teleological tunnel that the party has drilled for its ideology and thus preserve the resilience of the ideology itself. They must affirm that in China's "socialist market economy," human-environment relations and social institutions are fundamentally different from those of capitalism, even while strongly critiquing the environmental damage of China's actually socialist past. They also have to blame "the West" and offer the world an alternative to failed "Western democratic" attempts to contain environmental degradation (Yifei Li and Shapiro 2020, 6). Here is Pan again:

> Even though traditional socialism strove to criticize and supersede capitalism, its mode of development, like that of capitalism, was built on the foundation of Western industrial civilization. . . .
>
> Ecological civilization refers to an integral whole that respects the objective laws of the harmonious development of people, nature, and society and produces material and spiritual results; it refers to a fundamental attitude of cultural ethics built on a foundation of harmonious coexistence of people and nature, people and people, people and society, virtuous cycles, all-around development, and sustainable flourishing. (Pan 2006, 13, 16)

One is tempted to dismiss these quotes as propagandistic gobbledygook or at best what Joseph Esherick (1995, 46) has called the "transparent fiction of revolutionary ideology." But it is probably wiser to take ideology seriously as a motivator of policy (Sorace 2017, 59–63). This propagandistic gobbledygook is, after all, directed at a real environmental crisis, and its author, as vice-minister of the environment, knew a lot about the hard facts of pollution, resource degradation, climate change, and other real environmental issues. Pan's slightly nonsensical equivocations result directly from tight strictures on political thought—it must be Marxist and must proclaim the superiority of socialism with Chinese characteristics while still addressing concrete environmental problems.

The actual material basis of an ecological civilization is never really spelled out, but it seems to be a combination of a service-heavy economy, efficiencies in production, and green technologies. It might be called a postindustrial economy if we disregard the fact that it will still depend on industry (and also agriculture) to supply the material lives of its people. Nevertheless, the regime has committed itself, as an authoritative *People's Daily* commentary from 2013 pointed out:

> The origin of ecological civilization lies in reconsideration of development, as well as in raising the level of development and in superseding industrial civilization. It is through reexamining China's development from new heights of civilizational development that our Party has included building an ecological civilization in the "five united into one" matrix of Socialism with Chinese Characteristics. The ecological civilization goal of "strive to build a beautiful China, and realize the sustainable development of the Chinese people," as included in the report of the Eighteenth Party Congress, through its clear form, and its rich content has expressed the direction of 1.3 billion Chinese people and attracted the attention of the world. (Ren Z. 2013)

In other words, ecological civilization, *like revolution before it*, is a means to promoting development, and therein lies the basic contradiction in the idea of ecological civilization (Schmitt 2016, 94–102). Ecological civilization as a goal is only politically permissible if it does not impede the continued *development* (i.e., economic growth) that legitimates the party's rule. Although theoreticians freely admit that at present there are sharp contradictions between development and environment, they attribute these to a variety of causes, from the still-early stage of China's development (not really at the point where we can move from an industrial to an ecological civilization) to well-known perverse incentives in the cadre evaluation system (Whiting 2001), that discourage local officials from any actions that might slow economic growth (Ren Z. 2013). Nevertheless, Chinese policy and practice really have turned toward taking the environmental effects of development seriously, and the government has entered a new era by beginning to put a gradual brake on the process of environmental degradation. To celebrate this new era, beginning in 2017 Chinese universities and regional governments established at least eighteen centers for the study of Xi Jinping's thoughts on ecological civilization (Josephine Ma and Xie 2021).

....

Ideas of development, revolution, and science have continued to dominate Chinese Communist environmental ideology and practice. Development (*fazhan*) has been the consistent anchor of the whole tripod— as Deng Xiaoping pronounced, "the only solid truth." From cautious imitation of Stalin through utopian fantasy and disaster, to retreat into planned economy, to adopting much of capitalism, to dreaming of a green future, development has always been the core. Revolution no longer plays an explicit role in plans for future development. But because no one dares not to be Marxist in public, revolution continues to hover in the background; even though officials rarely use the word, they must still couch development in a teleological narrative of human progress toward eventual communistic harmony and plenty. Science remains the primary means to achieve development. Now, however, science no longer incorporates peasant wisdom or local knowledge (except for a few peasant radicals and ethnobiologists). Instead, a network of increasingly world-class universities and research institutes not only addresses both "pure" and "applied" problems but also acts as a signifier of development—of the advanced state of China's urban civilization and its education system and of China's progress toward resuming its rightful status, as Edwin Winckler (2018, 3) so pithily puts it, as "the world's main country."

PART II

Land, Water, and Food

3. Feeding a Starving Nation, 1949–1957

> Every step of Land Reform should be integrated with the revival and development of agricultural production.
>
> CPPCC COMMON PROGRAM, 1949

When the Chinese Communist Party founded the People's Republic of China in 1949, it inherited a starving nation. With about a quarter of the world's population but less than a tenth of its farmland, China in late Qing and Republican times had become known as a "land of famine," where various regions had been devastated by starvation every few years since at least the early nineteenth century. This overall degradation of China's environment, especially that of China Proper, was an indirect result of overshooting the land's carrying capacity at the height of the Qing's power and prosperity and the consequent descent of the social-ecological system as a whole into a state of misery.

THE QING EXPANSION: OVERSHOOT LEADS TO MISERY

At the beginning of Qing rule three hundred years earlier, things were different. Although population, cultivated area, and irrigation works had all grown and forests had shrunk for millennia before the Qing's seizure of power in 1644, the Qing empire's estimated 150 to 200 million people in the mid-1600s could easily be fed with the technology available at the time.[1]

The Qing state during its century of greatest glory, from around 1680 to 1780, might well have been the world's most sophisticated and efficient governing machine. By the 1680s it had effectively taken control of China Proper (Wakeman 1985, 1124–27). It also deepened and extended its rule in Zomia.[2] Taking over parts of Central Asia took longer (Barfield 1989; Perdue 2005), but it was necessary for geopolitical reasons and increasingly for ecological reasons as well.

In China Proper at the beginning of the dynasty, there was still opportunity to raise agricultural productivity. American crops—corn, sweet potatoes, and potatoes—would grow where traditional East Asian crops

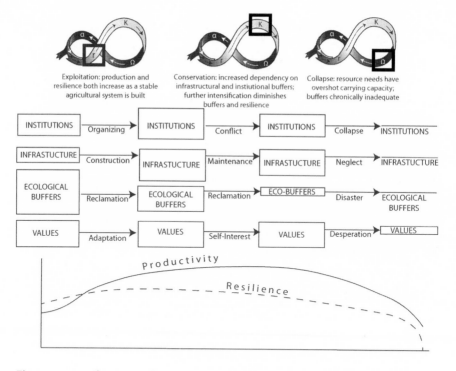

Figure 3.1. Intensification, buffer removal, and the adaptive cycle in lowland communities.

would not; new varieties of rice had higher yields or shorter seasons; new and improved waterworks could increase irrigated and double-cropped areas; people could work even harder. All these, combined with efficient governance and effective systems of disaster relief, simultaneously increased productivity and buffered against disturbance. Relative peace and prosperity allowed population to grow at an unprecedented rate, almost tripling to well over four hundred million by 1850.[3]

Eventually, however, further increases in productivity led to decreases in resilience. Slow variables undermined the abilities of both the regime and local communities to maintain the necessary buffers against weather and other disturbances. This process first threw local- and regional- and then empire-scale systems into a series of back loops that effectively stuck much of China Proper in the omega and alpha phases of the adaptive cycle (see figure I.6) for more than a century. Floods and droughts became ever more common on the North China Plain; erosion in the northwest took land out of production (Mostern 2019); continued enclo-

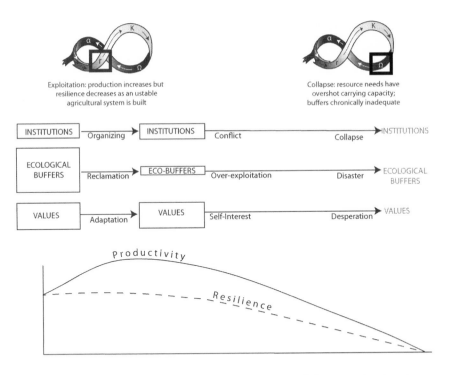

Figure 3.2. Intensification, buffer removal, and the adaptive cycle in mountain migrant communities.

sure of wetlands and lake margins in rice-growing areas removed buffers against heavy rainstorms, causing increased flooding there. Buffers of all kinds diminished first at local and then at wider and wider scales. Famines and disasters became more and more common (figure 3.1).

By the late eighteenth century, the lowlands of China Proper were simply full, and migration to mountainous regions had increased. Mountain migrants cleared forests and planted American crops in the rich soils, initially achieving very high yields. But these soils were soon exhausted of nutrients, eroded away, or both, and migrants moved on to the next stretch of forest. Meanwhile, erosion led to land- and rockslides, and the sediments released often clogged both irrigation systems in the adjacent lowlands and formerly navigable rivers and streams, further diminishing both productivity and the ability to transport relief supplies. Mountain social-ecological systems, having no infrastructural buffers, collapsed even faster than those in the lowlands (figure 3.2)

With both the lowlands and the mountains of China Proper overshooting their carrying capacities and declining into states of misery, Qing colonization sped up in Zomia and Central Asia, not just for geopolitical reasons but as further outlets for surplus population. As Zomian population grew rapidly throughout the Qing, livelihoods in many areas changed from extensive swidden agriculture to intensive fixed-field farming. Mining, logging, and cultivation of cash crops, including opium in the nineteenth century, brought about social changes and political transformation of native kingdoms, which were gradually incorporated into bureaucratic domains. Disasters increased in Zomia in the late Qing and Republican periods, but the remoteness and vertical topography of much of the region prevented the complete transformation of local livelihoods or large-scale increases in disasters.

In Central Asia, areas along the Mongolian frontier that had alternated between agriculture and pastoralism throughout history became exclusively agricultural as Chinese farmers moved in from overpopulated areas in the north of China Proper and many Mongols shifted to agriculture. After the Qing conquered the Junggar confederation of northern Xinjiang in 1760, authorities moved in military colonists, civilians from northwestern China, and Turkic peasants from southern Xinjiang (Perdue 2005, 342–45). This shifting of the population required agricultural intensification, including expansion of oasis cultivation into previously pastoral areas. As in Zomia, however, northern Xinjiang's pastoral livelihoods and southern Xinjiang's traditional oasis agriculture largely remained in place until the Communist revolution.

The Qing period and the short Republican interlude that followed it thus represent a classic turn of the Malthus-and-Boserup ratchet. Lillian Li (2007, 73), writing of overshoot and the descent into a back loop of misery on the North China Plain, well characterizes the transformation of China's entire social-ecological system: "To a great extent, then, the ecological crisis of the nineteenth century was a product of the very successes of imperial engineering of the eighteenth, not of its failures." The Qing's early political and ecological success, followed by largely ineffective "modernization" schemes during the final years of the Qing and the Republic, thus presented the Chinese Communists with a starving nation in 1949, and only through fossil fuel–based industrialization could they turn the ratchet once more, to a previously unthinkable level of population and resource use.

THE PROBLEM: PEACE, POPULATION, AND FOOD

At the Founding of the People's Republic, in addition to centuries-old problems of land and water, years of war had left many infrastructural and institutional buffers in China's agro-ecosystems in disrepair. At the same time, food supplies had to keep up with rapid population increase. Peace and public health programs both contributed to a dramatic decrease in mortality in the early and mid-1950s, from an estimated crude death rate of over thirty per thousand in 1950 to below twenty per thousand in 1957, largely attributable to cutting infant mortality in half (Banister 1987, 79–82) through intensive programs of improved sanitation (Banister 1987, 50–59) and mass vaccination (Brazelton 2019). To compound the population pressure, the party adopted explicitly pronatalist policies from the beginning, over the objections of some demographers who were derided as Malthusians and thus, by implication, anti-Marxists (Shapiro 2001, 36–51). In common with many other regimes, Marxist and non-Marxist, the Chinese Communists believed that more population, meaning more potential labor power, would be a source of national strength, and they promoted this belief through slogans such as "More people, greater strength" (Ren duo, liliang da) (Shapiro 2001, 31).

Fertility remained high as well (Banister 1987, 228), so population did grow rapidly. In 1953 China's first complete census in about two hundred years counted 582 million people, already an estimated increase of over 30 million since the Founding, and the population increased further, reaching an estimated 633 million on the eve of the Great Leap Forward in 1957 (Banister 1987, 42). This was a cause for not only joy but urgent concern. Most Chinese communities wavered uneasily around the line between misery and plenty. China had to crank up the Malthus-and-Boserup ratchet—it had to increase food production fast, and it did. From 113 million tons of carbohydrate staples produced in 1950, output had nearly doubled by the middle of the decade.[4] Record harvests ranged from 180 to slightly over 200 million tons from 1955 through 1958 (figure 3.3) (Liu D. et al. 2011, 3; Field and Kilpatrick 1978).

China achieved these gains in staple production despite numerous obstacles, some natural and some created by policy. There were limited options for increasing the output of food energy. Since the caloric supply was severely limited even in good years, and since carbohydrate-rich staples produce the most calories per unit of land, most farmers already ate almost nothing but carbohydrate-rich foods, and none too much of even them (Bramall 2000, 35–36; Smil 1993, 81–82; S. F. Du et al. 2014). Consequently, for the first thirty-plus years of its existence, the CCP regime emphasized staple production.

Even to increase staple production without improving diet breadth or quality was difficult. Chinese farmers already applied as much organic fertilizer, mostly human and animal manure, as they could readily save or collect, and Haber-Bosch nitrogen and other chemical fertilizers were too expensive to produce for a country with a very small industrial base (Perkins 1973, 59; Naughton 2007, 259). High-yielding Green Revolution crop varieties still lay a couple of decades in the future (Naughton 2007, 258–62; Schmalzer 2016, 73–99), and at any rate they would require farmers to use those same expensive chemical fertilizers that China could not yet produce or afford to import.

CCP propaganda, following the Soviet example, initially touted mechanization as a solution: a memorable New Year picture from 1949 (figure 3.4) contains a cute jingle about how the working class makes farm machinery to help the farmers achieve bumper crops:

Chengli gongren zheng nuli
Wei za' nongmin zao jiqi
Zhiyao zamen yongshang ta
Fengyi zushi wannian xi

City workers persevere
To build machines for us peasants here
If we just use them, then we'll have—
Plenty to wear, enough to eat,
ten thousand joyous years.

In reality, however, widespread mechanized agriculture was not an option in 1950s China: mechanization required capital investment, and the party's Stalinist model of industrialization called for minimizing investment in agriculture in order to put more capital into heavy industries. Even including water conservancy, agriculture accounted for

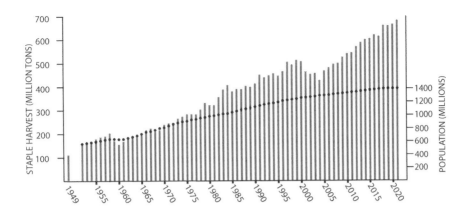

Figure 3.3. Staple production (bars) and population (dots), 1949–2021. Height of bars above the dotted line indicates grain supply per capita greater than the ratio in 1952. Data from Field and Kilpatrick 1978; SSB 2018a, 2018b, 2021, n.d.; Banister 1987; Kuaiyi Licai Wang n.d.

Figure 3.4. New Year's poster from 1949, with the song "Tractor" celebrating prospective agricultural mechanization. Courtesy of the Burke Museum of Natural History and Culture, University of Washington, catalog #2002-120/2.

only around 4 percent of the government's investment budget in the first five-year plan period of 1953–57 (Riskin 1987, 56–57; Naughton 2007, 231; Eyferth 2022, 6). Industries produced far more military weapons, railroad cars, and factory machinery than they did tractors. As a result, few of those miraculous machines rolled off China's urban assembly lines (E. Friedman, Pickowicz, and Selden 1991, 168–70); by 1957 China had only five thousand large and medium tractors, almost all of them used on state farms in previously nonagricultural regions, such as central and northern Manchuria. As late as 1962 the whole country had a measly total of one thousand of the small tractors useful in most of China's intensive agriculture, amounting to about one tractor per eight hundred villages (Nongye Bu 2009, 44). If China were to increase and stabilize its food supply, and if it had rejected the option of large-scale capital investment, only two factors of production were left: land and labor.

Figure 3.5 illustrates the thinking behind Chinese leaders' agricultural policy in the 1950s. The ultimate goal, shown at the right side of the diagram, was to increase agricultural output, which could be accomplished in two ways: increasing the yield per cultivated area or increasing the cultivated area. Double cropping increases the yield per cultivated area by increasing the sown area; all the other factors in the middle column—waterworks, traditional fertilizers, chemical fertilizers, new varieties, pesticides, and machinery—potentially increase yields per sown area. However, chemical fertilizer, new varieties, pesticides, and machinery all require the capital inputs that the planners were unwilling or unable to produce or manufacture. This left labor mobilization as the main strategy for increasing the food supply.

INCREASING LAND AND WATER INPUTS

Restoring fallowed land and repairing waterworks neglected or abandoned during the war years may have added as much as ten million hectares (about 7 or 8 percent more) to the cultivated area between 1949 and 1952 (Nongye Bu 2009, 6), but after that, increasing cultivated area meant reclaiming previously uncultivated land. Even though cultivation had expanded greatly during the Qing (often leading to ecosystem degradation), there was still some land that might be reclaimed. This included, at a larger scale, some parts of traditionally pastoral areas in northern

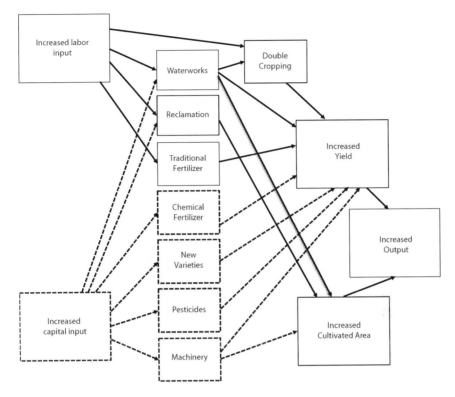

Figure 3.5. Actual (solid boxes and arrows) and potential (dashed boxes and arrows) inputs into agriculture under the Stalinist industrialization programs of the 1950s and the voluntarist campaigns of the Great Leap Forward.

Xinjiang, brought under cultivation in the 1950s; the Great Northern Wilderness in northern Manchuria, starting before the Founding and throughout the period of collective agriculture; and the wetter edges of Inner Mongolia, beginning with the Great Leap Forward.

Xinjiang's agriculture had long been dependent on irrigation canals that diverted into oases the water of snow-fed rivers flowing from the region's mountains. These irrigation systems had undergone cycles of repair and neglect for centuries (Kinzley 2018, 23–26); in 1949 they were in almost complete disrepair in many places (Wiemer 2004, 164). In the 1950s, the CCP regime enlisted the demobilized soldiers of the Xinjiang Production and Construction Corps (Bingtuan) to repair irrigation works and open up steppe lands to grow grain and cotton in northern Xinjiang, as well as to establish some state farms around the oases of southern Xinjiang. This increased Xinjiang's cultivated area from about 1.5 million hectares in 1952 to around 3.1 million in 1960. It also exhausted the potential for agricultural expansion in Xinjiang: cultivated area remained between 3 and 3.5 million hectares into the twenty-first century (Millward and Tursun 2004, 89–90; Wiemer 2004, 169; Kinzley 2018, 160). Any further gains in production would have to come from intensification.

The 1950s also saw the beginning of mass reclamation efforts in northern Heilongjiang (northern Manchuria) near the Inner Mongolian and Soviet borders, the area that came to be called the Great Northern Wilderness (Bei Da Huang). The central part of Heilongjiang, along the middle and lower reaches of the Sungari (or Songhua) River, had been developed for agriculture by late Qing, Russian colonial, warlord, and Japanese "puppet empire" Manchukuo regimes (Lary 2017, 45–46; Chiasson 2017). Nevertheless, much land was still not farmed in the late 1940s. The party began experimental reclamation programs after seizing the area in 1947 (Ding L. 2003, 14) and expanded them after they won the Civil War and established the People's Republic. They turned the energies of both demobilized soldiers and active military units to reclaiming lands for agriculture in three major areas of former wetlands and forested lands: in the valleys of the Nen River and its tributaries along the border with Inner Mongolia, in the plains along the Amur (or Heilong) River, and in the Three Rivers Plain at the extreme northeastern tip of the country (Baidu 2019) (map 3.1). By 1955 they had reclaimed about 224,000 hectares of farmland on 63 state farms with a total population of 81,000. Reclamation efforts sped up in 1956, and by 1966 the official reclamation district (*nongken diqu*) included 2.2 million hectares on 124 farms scattered about the three areas in the northern parts of the province (S. Liu, Zhang, and Lo 2014, 104–5). Most of them were farmed with large machinery, some of it donated by the Soviet Union (Ding L. 2003, 15). The reclaimers' heroic efforts are now memorialized in the Great Northern Wilderness Spirit of "bitter struggle, courageous reclamation, attention to the whole situation, selfless contribution" (*jianku fendou, yongyu kaituo, gu quan daju, wusi fengxian*) (Baidu 2019; Ding L. 2003, 15). Their ecological legacy of erosion, wetland loss, deforestation, and species endangerment persists to this day (Tao 1983; Muldavin 1997).

These large-scale efforts, however, probably only added a few million

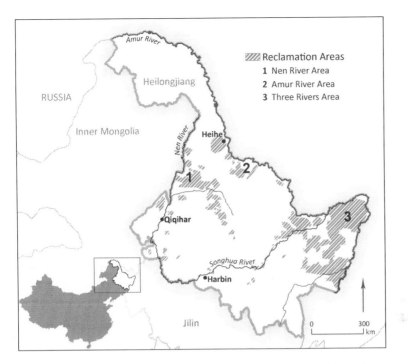

Map 3.1. Heilongjiang Reclamation District. Map by Lily Demet Crandall-Oral.

hectares to China's cultivated land, and in most of China Proper postwar rehabilitation had restored most of the good land that was restorable. This meant that farmers needed to intensify their work—to increase yields on lands already cultivated.[5] New varieties and techniques might bring marginal gains, but the largest increases would come from converting nonirrigated lands to irrigated fields. Farmers who had previously grown only a summer crop might add an early crop in spring; in much of Sichuan, for example, where they had grown a less thirsty crop such as rapeseed in spring and rice only in the summer, they could now grow two crops of rice. Irrigating more farmland thus became a prime means of expanding food production.

To increase agricultural productivity and ensure population safety, however, adding water when and where it is lacking is only half the project. The other half is getting rid of water where it is unwanted. As much as lack of water and droughts have depressed farm production throughout Chinese history, floods have wiped out crops and sometimes large populations of farmers. And just as irrigation has been a political tool to gain popular support for regimes throughout history, so has flood prevention. Conversely, deliberately causing floods, or diverting floods from one area to another, has been a tool of warfare since ancient and medieval times (Ling Zhang 2016), most recently when the Nationalists breached the Yellow River dikes at Huayuankou in 1938 to halt the advance of the Japanese imperial armies (Pietz 2015, 104–8).

In ecosystem terms, irrigation, if it is to work well and persist, means replacing ecological buffers with infrastructural and institutional ones. Irrigation can boost productivity, to be sure, and some infrastructural and institutional buffers, particularly water conservancy measures such as reservoirs (Huang S. 1989, 66), dikes, and floodgates (Siu 1989, 178) can also contribute to resilience. But these buffers create rigidity traps and require added effort to maintain. Sometimes, in spite of these added efforts (whose burden falls primarily upon farmers), resilience decreases, as dam collapses on the Ru River in the 1970s amply demonstrate. Thus the trade-off between productivity and resilience (see figure I.5) has been a constant dilemma for the planners, who try to keep increases in the food supply ahead of population growth, and especially for the farmers, who do the work and take on the risks of intensification.

RATIONALIZING LABOR INPUT

The three major means of increasing the food supply—adding to the cultivated area, increasing yields through intensification in the already cultivated area, and preventing floods—depended almost entirely on increasing the available supply of labor. In the early and middle 1950s, land redistribution and labor reorganization were the primary means of increasing labor inputs to agriculture, with the latter playing by far the more important role. In the Great Leap Forward, more radical labor reorganization failed to deliver promised benefits, and in the early 1960s the regime, seeing no alternative, finally turned to capital inputs.

Redistributing Land, Political Power, and Labor

The Chinese Communists began experimenting with land reforms in the 1920s, and during the Civil War, from 1945 to 1949, they carried out extensive land reforms in the "old liberated areas," mostly in the macroregional peripheries in the north of China (Hinton 1966; E. Friedman, Pickowicz, and Selden 1991). In 1950–52, they extended the Land Reform program to all Han agricultural areas and some ethnic minority areas in Zomia that practiced intensive agriculture and had systems of landownership similar to those of the Han. In other minority areas in Zomia and Central Asia, the party assumed political and military control but left traditional systems of local governance and landholding in place until the so-called Democratic Reforms of 1956 and in some places until even later.

Communist Land Reform began by classifying the households of every village according to their relationship to agricultural means of production. In a simplified schema,

- landlords owned land but did not work it, renting to tenants or hiring agricultural laborers or both
- rich peasants rented out some land or hired some laborers but also worked on the land themselves
- middle peasants owned land and worked it themselves, only renting small amounts of land to or from others
- poor peasants either owned no land or did not own enough to support themselves and thus had to rent land from landlords or rich peasants
- agricultural laborers worked for wages, usually meager.

Land Reform was designed to take land away from exploiters (landlords and rich peasants) and give it to poor peasants and agricultural laborers, who worked the land but did not have enough of their own. At various times and in various places, land redistribution was more or less egalitarian, with land, draft animals, and other means of production being taken from rich peasants as well as landlords; at other times and places it left the "rich peasant economy" in place.[6]

Land Reform was partly motivated by political control and Communist notions of social justice: not only did the poor peasants deserve a better life and the landlords a worse one, but the party also wanted to eliminate any real or potential local political rivals, break the power of previous local leaders, and put its own functionaries, including local loyalists, in place. However, Land Reform also had a developmental motivation. According to CCP logic, Land Reform would "liberate the

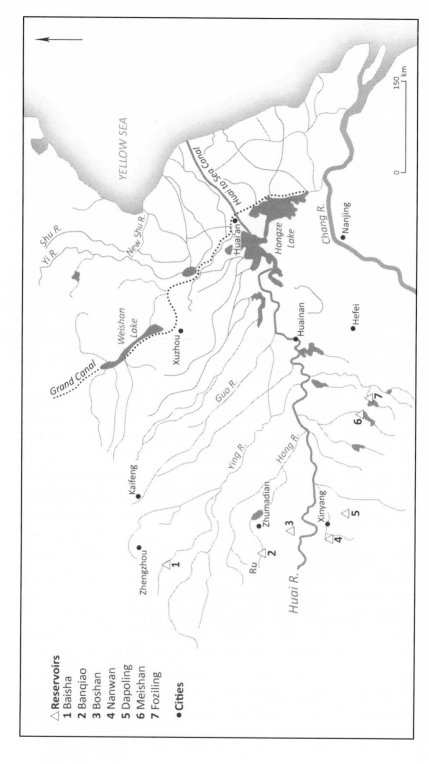

Map 3.2. Reclamation in the Huai River basin, 1950s. Map by Lily Demet Crandall-Oral.

productive forces," channeling peasant labor into greater agricultural production by increasing yields and expanding cultivated area, and it would also allow the surpluses created to go into national construction rather than into wasteful consumption by exploiting classes. Farmers would work harder and their labor would be used more efficiently, so agricultural output would grow.

One way efficiently reorganized and mobilized labor could help boost agricultural outputs was through water conservancy. Even before Land Reform was finished in many areas, the regime was already mobilizing labor for large-scale water conservancy projects, particularly in the Yellow and Huai River basins in North China. Plans for the Huai River Conservancy Project had begun in fall 1949, after a flood of the Yi and Shu Rivers in northern Jiangsu wiped out the entire fall harvest from 8 million *mu* (530,000 hectares) of land (map 3.2). Taking advantage of party organizations in place from recent victorious battles in the area, the regime mobilized local people to build a drainage canal, later called the New Shu River, directly to the sea (Yang L., Qiang, and Li 2017, 123).

In summer 1950, however, there were floods and waterlogging all over the Huai River basin. Over thirty-one million *mu* (two million hectares) of farmland flooded, and nine million people were affected in the Anhui portion of the Huai basin alone. Clearly a more comprehensive project was in order (Zhai 2020). Beginning that fall, millions of rural people were mobilized during the agricultural off-seasons. Spurred on by Mao's exhortation that "[we] have to fix the Huai," work on the Huai conservancy continued through 1952 in eastern Henan, northern Jiangsu, and northern Anhui, always proceeding on the principle of "using labor in place of funds" (*yi gong dai zhang*), pursuing water conservation goals with minimal capital expenditures (Pietz 2015, 133–39). Over the three-year course of the Huai Conservancy Project, farmers working during the off-season built at least eight major and many minor reservoirs (including the shoddily built Banqiao Dam, whose collapse caused the great disaster of 1975), plus drainage canals, sluice gates, irrigation canals, and other waterworks (Zhai 2020). During flood emergencies, they sometimes performed remarkable work in short periods of time. When emergency containment dams on the Zao River in Jiangsu required twenty-seven thousand cubic meters of stone, the party secretary of Suqian County reportedly mobilized local people at 3:00 p.m. to tear up paving stones and household thresholds and deliver them by water to the construction site, and by 11:00 the next morning the dam builders had more than enough stone (Yang L., Qiang, and Li 2017, 121).

Success, however, was still partial. From June through September 1952, there were four major rainstorms in the Huai basin. Although the dikes on the Huai mainstream and its major tributaries all held, 25 million *mu* (1.7 million hectares) of farmland, most of it in Anhui, suffered waterlogging. Some critics charged that overemphasis on big projects and flood prevention caused those in charge to neglect smaller projects that would deal with waterlogging (Yang L., Qiang, and Li 2017, 124). At the same time, major rains that affected both the Chang and Huai basins in 1954 brought less death and destruction than those of 1931, even though 760,000 hectares of farmland flooded (Zhai 2020).

The Huai continued to be a problem for many years, with severe floods in 2020 and 2021 (Sina News 2021). The attempt to control it was laudable, and there was some progress as the technology progressed from people digging up their thresholds to modern mechanized construction. Floods were still likely to continue indefinitely, however, perhaps exacerbated by influences of climate change that became increasingly visible in the late 2010s. Though technological lock-in and reliance on infrastructural and institutional buffers undoubtedly played a part, we must remember that the place is flat, crossed by numerous rivers and streams. Rainfall is extremely variable and tends to come in great gushes. Neither labor mobilization nor modern machinery was ever likely to change that.

How much both Land Reform and the resultant ability to mobilize labor actually contributed to increasing agricultural outputs is hard to say. According to the official line, a lot:

> From 1950 to 1952, this was a large-scale social revolution in Chinese history, ending the feudal land system that had persisted for over 2000 years, allowing peasants to receive over 700 million *mu* [46 million hectares] of land, avoiding 70 billion *jin* [35 million tons] of rent paid to landlords, and realizing "land to the tiller." . . . Comparing 1949 with 1952, the total value of agricultural production increased from 27.1 billion *yuan* to 47.1 billion *yuan*, staple output increased from 113 million tons to 163 million tons, cotton from 440,000 tons to 1.3 million tons, oil products from 2.56 million tons

to 4.19 million tons, livestock in the stable from 60.01 million head to 76.46 million head, pigs in the pen from 57.75 million head to 89.77 million head. (Guo S. 2009, 44)

Official statistics leave little doubt that staples harvests actually did grow (Nongye Bu 2009, 17). However, they probably would have grown anyway without Land Reform, since the country was at peace and many waterworks could be and were repaired. It is clear, though, that *how* Land Reform was carried out made a difference. Very egalitarian versions of Land Reform, in which not only landlords' assets but also those belonging to rich peasants were confiscated and redistributed, did less to spur growth in agricultural output than those versions that allowed the "rich peasant economy" to persist. This difference is evident both in a wide sample of county-level units (Bramall 2000, 35–37) and in a single township that tried several types of Land Reform between 1946 and 1950 (E. Friedman, Pickowicz and Selden 1991, 97). Overall, we can reasonably assume that land redistribution did have some positive effect on agricultural productivity (Perkins 1969, 108). Still, the effects were not dramatic, and it was clear to party leaders that Land Reform had not rationalized agricultural labor inputs enough. Thus, in 1953 they turned to various forms of cooperative and collective agriculture, in hopes that still more efficient use of labor would allow farmers to improve their land and their cultivation methods, particularly by mobilizing to bring irrigation to areas previously dependent on rainfall.

Three Stages of Collectivization

The first stage of collectivization organized farm households into mutual-aid teams, in which they exchanged labor and shared draft animals and farm implements. After the completion of Land Reform, individual families still retained ownership of the land they farmed, paid taxes individually, and sold grain at state-set prices. In 1954 and 1955, many areas proceeded to "beginning-level agricultural producers' cooperatives" (*chuji nongye hezuoshe*), in which households pooled their land as well as their capital resources and labor but received a share of the harvest based partly on how much land they contributed to the pool and partly on how much labor they contributed. This was a very complex system

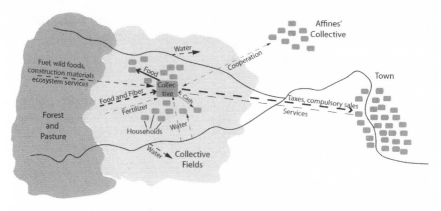

Figure 3.6. Flows of goods and services in a collectivized household in China Proper.

that was often difficult to implement. When problems arose, rather than backtracking, as some people considered sensible, the central leadership decided in late 1955, spurred on by a speech from Mao Zedong, to proceed to the "socialist high tide" of full-scale collectivization everywhere (K. Walker 1966), including many minority areas in the Zomia and Central Asia that had not undergone the Land Reform.

By June 1956, over 60 percent of rural households had been further collectivized in "higher-level agricultural producers' cooperatives" (*gaoji nongye hezuoshe*) (K. Walker 1966, 36), which were very similar to Soviet collective farms (*kolkhozy*). Several tens of households contributed all their land and draft animals, so recently distributed in the Land Reform, to the collective, and collective leaders assigned labor to all adult members, who were paid in "work points" according to how long they worked at what jobs. At the end of each year, collective accountants deducted taxes and fixed-price sales to the state (of such things as staples, food oils, cotton, and other controlled crops in areas where they were grown), while the collective retained enough seed for the following year and put away a little bit for investment and welfare funds. The collective's officers then distributed the remainder of the harvest, plus some money from sales to the state, to each household. Part of each household's income was paid at a uniform rate for each member's age and sex, and part was paid according to the work points earned by all its members.[7] Collectives lent small amounts of land back to individual households as so-called private plots (in Chinese, *ziliu di*, roughly "land retained for themselves"), where they were free to grow vegetables or other crops that they could consume directly or sell at local markets. This system allowed the state to expropriate much of the increased agricultural product in exchange for services such as education and health care, with the remainder of the surplus going to the primitive accumulation that would support heavy industry (figure 3.6). Except for the disastrous experiment with large-scale communization during the Great Leap Forward years of 1958–60, this system largely remained in place until 1979–83, depending on the area.

Collectivization and Water Projects

Collectivization at least partly facilitated the most obvious and most important change in Chinese agriculture during the early and mid-1950s: the dramatic increase in irrigated land, from approximately seventeen million hectares in 1949–52 to between twenty-seven and thirty-five million hectares in 1957.[8] This increase resulted almost entirely from increased capture and diversion of surface water (Perkins 1973, 61–62; Smil 1993, 45; Nongye Bu 2009, 7).

Some of this expansion of irrigated area, particularly in the very early years, came through repair of waterworks damaged or neglected in the war years from 1937 to 1949. But after 1955, newly built reservoirs and diversion projects brought irrigation water to many previously unirrigated areas in the north (Smil 1993, 44; Nickum 1998, 883). As David Pietz (2015) has pointed out, efforts at increasing irrigated area in North China were of two kinds. The first sort were grand projects. These were reminiscent of age-old ideas that imperial control was closely connected to waterworks, particularly those involving the Yellow River and the North China Plain, but they also bore the imprint of Stalinist rational high modernism as practiced in the Soviet Union. As the Huai River Project was proceeding, the regime also began to plan for major Yellow River projects. In 1950–52, mobilized labor built the People's Victory Canal (map 3.3), which diverted about sixty cubic meters of water per second from the Yellow River near Xinxiang in Henan to the Wei River.[9] Water from the People's Victory Canal irrigated about forty thousand hectares of land (Matalas and Nordin 1980, 900). But much bigger plans were in the offing; in 1954 the state announced the General Plan to

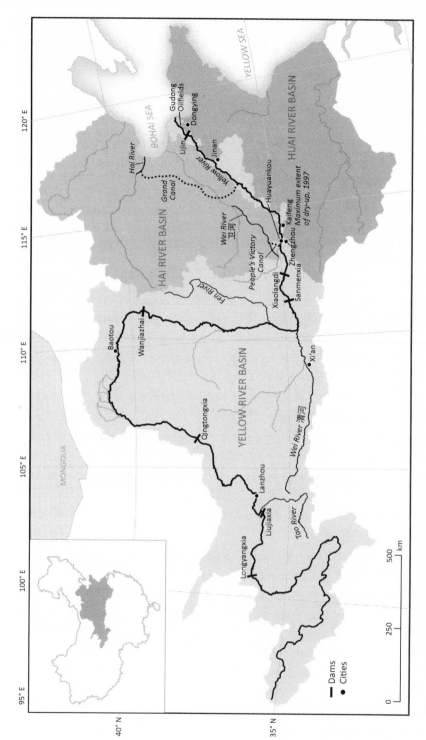

Map 3.3. Yellow River basin. Map by Lily Demet Crandall-Oral.

Fundamentally Control Yellow River Flood Disasters and Develop Yellow River Waterworks (Genzhi Huanghe Shuizai he Kaifa Huanghe Shuili de Zonghe Jihua).

Controlling the Yellow River has always been a huge challenge, for at least four reasons. First, although it only has about a tenth the flow of the Chang River to the south, it is still a big river, more than twenty-five hundred kilometers long, with a yearly average discharge volume of fifty-eight cubic kilometers. Second, its flow is very seasonal and flashy; over 60 percent of the rain that feeds it falls from June through August (Smil 1993, 39). Third, it carries a huge amount of sediment, most of it picked up in its middle reaches as it flows through the loess plateau in Northwest China. Its annual sediment load in the 1950s varied from six hundred million tons in 1957 to over two billion tons in 1958, and as much as one-quarter of that was deposited in the riverbed (Huang Ronghan 1988, 80; Walling 2008, 6; Ren M. 2015, 3). The Yellow River's estimated sediment concentration of twenty-five to thirty-seven grams per liter in the 1930s was fifty to seventy times the concentration in the Chang or the Mississippi and twenty-five times the concentration in the Ganges (Ren M. 2015, 2; Holeman 1968, 744). Fourth, as northern Chinese agriculture intensified across the millennia and silt deposits turned it into a "hanging river" (figure 3.7), floods became an ever-greater concern, so successive regimes built dikes to contain the floodwaters (Pietz 2015, 64–65), and the river changed course periodically (Ling Zhang 2016). Because its bed between the dikes is so high, the Yellow River downstream

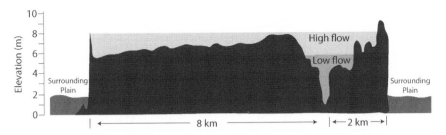

Figure 3.7. Profile of the Yellow River on the North China Plain. Adapted from David Pietz and Mark Giordano, "Managing the Yellow River: Continuity and Change," in *River Basin Trajectories: Societies, Environments, and Development,* edited by François Molle and Philippus Wester, 99–122. © 2009 CAB International, Oxford. Reproduced with permission of the CAB International through PLSclear.

from Huayuankou has no major tributaries (water would have to flow uphill to enter its bed) and a very small catchment area of about twenty thousand square kilometers (Huang Ronghan 1988, 80). The other rivers of the North China Plain flow roughly parallel to it (Pietz 2015, 19–20; Borthwick 2005).

The 1954 General Plan for the Yellow River was designed to address two issues at once: first, preventing the Yellow River from flooding, and second, converting the rainfall-fed fields of the North China Plain to irrigated agriculture. Both efforts, if successful, would allow much more food to be grown. With Soviet engineers' input, the plan proposed two huge dams, at Sanmenxia and Liujiaxia, on the Yellow River itself; three other dams that would irrigate a million *mu* along the whole course of the river; 638,000 check dams to prevent local erosion; and afforestation of 21 million *mu*, or 1.4 million square kilometers (Pietz 2015, 161). In fact, however, much of the plan was not carried out, and for the parts that would ultimately be implemented, construction did not begin in earnest until 1957, coinciding with the beginning of the Great Leap Forward.

If the General Plan for the Yellow River represented the party's high-modernist developmental strategy (as well as some spiritual inspiration from the historical connection between imperial success and controlling the river), the second sort of water-control projects in the north illustrated the voluntarist or local (*tu*) side of the CCP developmental ideology (Schmalzer 2016, 38), which involved promoting small-scale projects that benefited local communities and required only very limited state investment. As the countryside was being collectivized between 1954 and 1956, small irrigation and flood control projects were the perfect way to take advantage of the opportunities to mobilize labor that collectivization was intended to provide. According to official reports, North China collectives built 27,000 small reservoirs on local streams and hand-dug 4.5 million wells to reach the high water table that prevailed at the time on the North China Plain. These projects probably made the biggest contribution to the overall 25 percent increase in irrigated land in China (Pietz 2015, 181; Smil 1993, 44). However, many of the dams and dikes they constructed were shoddy and unstable, and it is not clear how many of them survived, or for how long.

Expansion of the irrigated area was not confined to North China. In

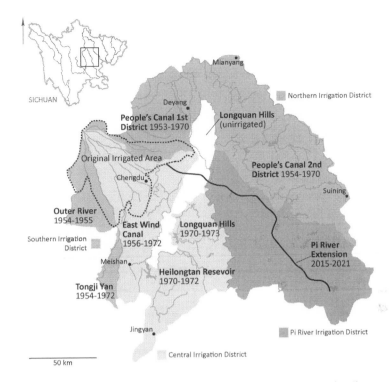

Map 3.4. Expansion of the Dujiangyan Irrigation Area, 1953–2021. Map by Lily Demet Crandall-Oral.

the Chengdu Plain, the ancient Dujiangyan irrigation system (map 3.4) was still functioning at the time of the Founding, and PRC engineers wasted no time mobilizing local people to enlarge it (Shuyou Cao, Liu, and Er 2010, 5–6). They built the People's Canal (Renmin Qu) in early 1953, extending the irrigated area to the northeast, and enlarged it in the winters of 1953–54 and 1954–55, almost doubling the area irrigated with the diverted waters of the Min River. In 1956 they mobilized over thirty thousand people to build several branches of the East Wind Canal (Dongfeng Qu), extending irrigation to the south and to some hilly areas east of Chengdu that had previously contained mostly dry farms. In addition, they repaired existing irrigation works, drawing water from the Outer River below the headworks (Shen G. 1979; Chen Jinlin and Bo 1993, 222–34).

Other than expanding the irrigated area, the change that probably made the biggest contribution to increased yields was one that was almost certainly facilitated by collectivization. This was the switch from single- to double-cropped rice in the south. In some places, such as northern parts of the Sichuan Basin, this meant investing labor in irrigation projects (Chen G.n 2014), but everywhere, even where water was originally sufficient in the early spring season, double cropping makes huge labor demands on the grower. In all but the most southerly parts of China, in order to fit in two crops during the growing season, farmers must plant a seedbed about four weeks before the first crop is to be harvested, then harvest and thresh the first crop, till and harrow the fields, and transplant the seedlings from the seedbed into the field for the second crop, all in searing heat. Before 1949, most rice was double-cropped in Fujian, Guangdong, and eastern and southern Guangxi, with smaller amounts in Jiangxi, Hunan, Hubei, and Zhejiang. Between the Founding and 1957, double cropping began in Sichuan, Anhui, and southern Jiangsu and was greatly increased in Jiangxi, Hubei, and Hunan (Perkins 1969, 44). By 1957, over 19 million *mu* (1.2 million hectares) in Sichuan were being double-cropped (Perkins 1969, 44).

....

Between 1950 and 1957, on the eve of the disastrous experiment that was the Great Leap forward, China had made modest progress in food production. Although cultivated area expanded in a few places, such as Xinjiang and Heilongjiang, increasing food production relied almost entirely on increased labor input per unit of land. Land Reform and collectivization redistributed labor and may have made it more efficient, and certainly this labor reallocation permitted the two major changes in the application of agricultural technology—constructing waterworks in the north and increasing the double-cropped area in the south. I think it is fair to say that with some exceptions (such as the reclamation in Heilongjiang), labor redistribution and increased labor inputs *increased* the resilience of most local agro-ecosystems. Per capita staple crop production increased by somewhere between 7 and 16 percent from 1952 to 1957.[10] Flood control and drainage projects, though hardly foolproof and sometimes shoddy, probably reduced both the frequency and the severity of local drought and flood disasters, though they certainly could do little to contain major floods like those in the middle Chang region in 1954 (Courtney 2018a). Reservoirs, where properly constructed, allowed farmers to allocate water more efficiently—storing it when there was too much to use and when floods threatened and using stored water when they needed extra. But these modest gains fell far short of solving China's developmental problems. Output was barely staying ahead of population growth, and population was not going to slow down in the absence of a major disturbance (the Great Leap Forward was such a catastrophic disturbance, but it did not last very long, and population recovered quite quickly). Although the state extracted almost all the surplus product of agriculture, there was very little of it, and this inhibited the state's plans to carry out "socialist primitive accumulation" (Eyferth 2022) to pay for the industrialization that would also, in turn, allow agriculture to really modernize. And more important, any potential gains from collectivizing household agriculture, constructing small irrigation projects, and increasing the double-cropped area seemed to have been exhausted by 1957. Still reluctant to invest capital in agriculture, the regime doubled down on labor reallocation. Thus the Great Leap Forward.

4. Three Years of ~~Natural~~ Disasters, 1958–1961

> For Raoyang [Hebei] villagers entering 1958, . . . breakfast was millet gruel. Coarse grain cooked for lunch was eaten again at dinner. There was little meat, fowl, fish, or fresh vegetables. Homes lacked chairs, sofas, beds, radios, and clocks. Women did washing using a rock and icy water drawn from a distant well. Cooks continuously fed straw to a brazier in the middle of the house. The fumes and dust dirtied the dwelling and filled the lungs. . . . There was no running water and no electricity.
>
> EDWARD FRIEDMAN, Paul Pickowicz, and Mark Selden,
> *Chinese Village, Socialist State*, 1991

> History has passed on, but the profound lessons and experiences it has left us deserve to be etched into our memory.
>
> LI CHUNFENG, "Prelude to the 'Great Leap Forward,'" 2010

The Great Leap Forward, which formally started in the spring of 1958, did not just happen suddenly. Rather, the Great Leap continued and expanded on many policies and programs that had already been tried at times in the early and mid-1950s. The Great Leap, however, took those policies and efforts to new extremes. Preludes to the leap itself took place in 1956 and 1957, including imposition of the collectivist model on Zomia and Central Asia and mobilization for ever-larger water control projects in China Proper

PRELUDE 1: REFORMS IN ZOMIA AND CENTRAL ASIA

While the Chinese Communist Party carried out Land Reform and the early stages of collectivization in China Proper between 1948 and mid-1955, it mostly left the local social systems of Chinese Central Asia and most of Zomia alone. Beginning in late 1955, however, while the socialist high tide of collectivization washed over all of China Proper and some lowland parts of Zomia, the party also moved to make fundamental changes in the governance and property systems in the mountains of

Zomia and in Central Asia. In the Democratic Reforms (Minzhu Gaige), a purportedly peaceful process, the party replaced traditional local elites with party and government cadres (most of them recruited from the minority groups themselves). Rebellions followed, in the Khams region of eastern Tibet (Shakya 1999, 136–43; M. Goldstein 2014, 75–141) and in the Liangshan or Cool Mountain region of the Nuosu Yi (Liangshan 1985, 153–57; Luo 1998), often led by some of the very same local elites who had previously been induced to serve the new regime. Both rebellions were contained, though not totally crushed, within two years, but the mere fact that they occurred sheds light on the perhaps not-so-democratic nature of the Democratic Reforms. As did Land Reform in China Proper, the Democratic Reforms replaced previous forms of property ownership and labor organization, but they skipped the independent smallholder stage that had followed Land Reform in China Proper, going straight to agricultural producers' cooperatives in the winter of 1957–58.

In Chinese Central Asia policies were more complicated. Since much of Inner Mongolia belonged to the "old liberated areas" and the Inner Mongolian Autonomous Region had been established under CCP control as early as 1947, the party initially carried out Land Reform there in 1947–48, redistributing the land and herds of both agricultural landlords and pastoral "herdlords" (*muzhu*).[1] However, the attempted reforms in most pastoral areas failed, and herds and pastures remained under traditional ownership regimes until the 1960s. The party did carry out Land Reform in Inner Mongolia's agricultural areas and in Uyghur farming communities in Xinjiang's oases, as they did in China Proper (Bulag 2002, 114–21). But Kazakh and Mongol herders in northern Xinjiang were not collectivized until the Great Leap Forward (Millward and Tursun 2004, 87–88). In central Tibet, which remained under the uneasy joint administration of the People's Liberation Army and the Dalai Lama's theocratic regime until 1959, land reform, followed by collectivization, did not take place in agricultural areas until 1960 (Shakya 1999, 253) and in pastoral areas until 1966 (M. Goldstein and Beall 1990, 140).

PRELUDE 2: WATER CONSERVANCY CAMPAIGNS IN NORTH CHINA

Meanwhile, by late 1957 development in China Proper had stagnated, and Mao was impatient. A year earlier, he had tried an experiment called "A hundred flowers all bloom; voices of a hundred schools compete" (Bai hua qi fang; bai jia zheng ming), which called for open discussion of the problems facing the nation. It ended badly for him in spring 1957, when some people went so far as to voice fundamental criticisms. Many of these contentious voices were scientists representing the rationalist side of developmental thinking (Schmalzer 2016, 47). As a result, the party began the Antirightist Campaign in June 1957, and it specifically targeted intellectuals, scientists, and party cadres who had hewn too closely to the ideas of rational development. In addition, harvests in 1957 were poor. Despite the expansion of irrigated area since 1949 and efforts put into collectivization during the socialist high tide of 1954–55, the area sown to staples actually diminished by over three million hectares from 1956 to 1957; over thirty-one million hectares had been affected by floods, and as a result it was estimated that about twelve million tons of harvested staples (perhaps 6 percent of the anticipated harvest) were lost to floods and droughts in 1957 (Wu Zhijun 2006, 12). In addition to endangering the already precarious food supply, disappointing harvests hampered the efforts to squeeze agriculture in order to capitalize heavy industry. Given the similarly disappointing pace of industrial growth, the regime continued to champion the strategy of "using labor to replace investment" as a way to increase harvests without investing much money in agriculture. Labor mobilizations previously undertaken to control the Yellow, Huai, and other rivers on the North China Plain (see maps 3.2 and 3.3) offered promising models, and in September 1957 the party center announced the Campaign to Build Agricultural Water Conservancy (Xingxiu Nongtian Shuili Jianshe Yundong). From November 1957 through January and February 1958, massive armies of rural people headed out to tame the waters (figure 4.1).

One slogan of the campaign identified "three primarilys": "Primarily storage, primarily small scale, primarily managed by the cooperatives" (Li C. 2010, 114), but what the projects lacked in size they made up for in number and total volume. Villagers and sent-down cadres spent many

Figure 4.1. Peasant teams constructing the Ming Tombs Reservoir near Beijing, 1958. Photo by Joseph Needham, reprinted courtesy of the Needham Research Institute.

weeks in the fields, carrying buckets or baskets of earth on the ends of carrying poles. By January 1958, 100 million villagers had taken part in the effort, putting in an estimated 13 billion days of work (Li C. 2010, 114); the 7 million people who took part in Hubei, for example, were about half the adult labor force of the whole province, and they reportedly built 700,235 water projects, 563,223 of them for irrigation, 49,902 for drainage, and 92 for reclamation by diking (Wang R. 2008, 124).

Reports written in the spring of 1958 were ecstatic about the results. Manual labor had moved 25 billion cubic meters of earth and stone, which, if it had been used to build a road a meter thick and 66 meters wide, would stretch from the earth to the moon (Li C. 2010, 114).[2] As a result, Chinese farmers expanded total sown area by five million hectares (i.e., 5 Mha), brought irrigation to 24 Mha (about a fifth of China's total farmland), improved a further 9.6 Mha, and solved waterlogging problems on 13.3 Mha of low-lying land. They improved a further 6.7 Mha of poor land and planted trees on 16 Mha of previously unforested land, contributing to erosion control (Wang R. 2008, 123–24). Fu Zuoyi, the vice-premier in charge of agriculture, exulted that it had taken

4,000 years of Chinese civilization to bring 15 Mha under irrigation; in a mere four months, the campaign had doubled that amount (Wang R. 2008, 123–24). China would now be able to solve its food problem once and for all and industrialize rapidly to boot.

There were, however, problems, as Li Chunfeng has pointed out:

At the same time, there existed a psychology of blindly pursuing speed and anxiously looking for results. [People] did water conservancy everywhere without paying attention to local conditions. . . . This led to simplistic and crude water conservancy equipment, impossible quality control, and water control projects that fundamentally could not achieve the utility they should have had. There were some places that just stressed "storage" and did not stress "drainage," destroying the previously existing surface water network and compounding waterlogging and salinization. In addition, [the projects] used a large amount of arable land, consumed immense amounts of labor, materials, and money, influenced agricultural outputs, harmed the activist spirit of the masses, and brought about unnecessary waste and damage. (Li C. 2010, 114)

These were not the only problems: larger-scale projects inundated villages and required villagers to relocate; lack of attention to worker safety caused unnecessary deaths and injuries; nonagricultural occupations, particularly the women's labor that sustained household economies (Eyferth 2022), were interrupted when everyone was mobilized to go to the construction sites; many projects did not include plans for future maintenance; some projects appropriate to places where they were first undertaken were copied in places where they were not suited, causing ecosystem damage and ruining drainage systems (Wang R. 2008, 124). Nevertheless, the enthusiasm of high officials and fear of being labeled rightist meant that these difficulties went unacknowledged (Wang R. 2008, 124).

At the same time, some projects did work despite all these problems, and the experience of mass mobilization led project enthusiasts to begin announcing that labor mobilization could lead to a "leap forward" (*yuejin*) in agricultural production. The water conservancy campaign thus became the first act of the tragicomic opera that was the Great Leap Forward (Wang R. 2008, 124; Li C. 2010, 113). More water conservancy efforts were to follow in the next two agricultural off-seasons in the north, complemented by similar efforts in other parts of the country. The drama of the world's greatest famine had begun, one in which short-term considerations of production not only outweighed but obliterated any longer-term considerations of resilience.

THE GREAT LEAP ITSELF

The party leadership did learn lessons from the water conservancy campaign, but not the ones that speak to us today. Instead, they took it as evidence that ideological fervor and labor mobilization could lead to monumental changes in the country's ability to exploit natural resources (MacFarquhar 1983, 19–30). They quickly extended this lesson beyond water to land and to the crops grown on it. If temporary massive collective labor reorganization could mobilize enough people to construct hundreds of thousands of water projects and double China's irrigated area, permanent large collectives would allow labor to be more regimented (literally) than ever before. This would create a pool of surplus rural laborers who could be mobilized to undertake ever larger and more ambitious programs to increase productivity, in a few years transforming a millennia-old landscape into a modernist paradise. Furthermore, because the success of the water conservancy campaign had relied on voluntarist strategies of social reorganization and ideological mobilization, the subsequent all-out effort of the Great Leap itself would use even more extreme social and ideological means to rewrite the landscape.

Socially, collectives were made bigger and more collective. Several former high-level cooperatives merged into one huge collective, with a population in the tens of thousands, usually encompassing either a standard market town and its surrounding villages or even more than one standard marketing area (Skinner 1964–65, 382–99). Over the course of the summer of 1958, these megacollectives came to be called people's communes. And the communes were much more communal than the cooperatives they had absorbed. Previously existing "private" plots were abolished; all the land, domestic animals, and agricultural implements of the entire commune came under collective ownership and management, and income was distributed throughout the large area. Even more radically, the household was abolished as an economic unit. People ate in the collective canteens, and, ideally at least, collective nurseries took care of children while their parents were working. The most radical communes in a few areas even razed private housing and built collective dormitories.[3] In this system, the party could mobilize not only the male labor that had built the dams, dikes, and ditches in the previous winter but also the female labor that had been tied up in housework or handicrafts and producing "nothing of value," that is to say, nothing worth any work points (though work in collective settings such as canteens or child-care centers did earn a few points [M. Brown 2017, 45–46]).

All this radical reorganization of landscapes and people was directed toward establishing communism (with a small *c*): in the words of Marx's *Critique of the Gotha Program* (1875), each would contribute according to ability and be compensated according to need. Effectively abolishing the household as a unit of consumption as well as production would finally realize the Marxist promise of women's liberation, both from the mindless drudgery of domestic labor and from the tyranny of fathers and husbands (Manning 2005). And most important, communism would be possible because there would be plenty to go around. Agricultural surpluses would feed industry, peasants would become workers, and China would overtake the United Kingdom in industrial production in three years and catch up with the United States in fifteen or less. Little capital would be necessary initially—labor would create capital at the same time it transformed the agricultural landscape.

There was an undeniable logic to this plan. What it failed to consider were the biology of plants, the psychology of people, and the limits on both the productivity and the resilience of ecosystems. Specifically, the "leapers" made four grave systemic mistakes, which I have elsewhere called "The Four Horsemen of the Ecopocalypse" (Harrell 2021). First, they pushed almost everything too far, assuming that more of something was always better. Measures that were helpful in moderation were disastrous when taken to excess. Deep tilling and dense planting are the most obvious examples. Second, they managed ecosystems at inappropriate scales, creating what Graeme Cumming, David Cumming, and Charles Redman (2006) have called "cross-scale mismatches." Collectivization might have brought advantages at the scale of tens or even hundreds of households but not at the scale of tens of thousands.

Mobilization and hurriedly built projects led to short-term gains in harvests in 1958, but these could not be sustained. Third, they imposed uniform, "panacea" solutions from the top down (Ostrom and Cox 2010); things that might have worked in one place rarely worked everywhere; the double-wheeled, double-bladed plow illustrates this. Fourth, as a result of the other three measures, the Great Leap planners disrupted the feedbacks in ecosystems. By emphasizing one element of a system to the exclusion of others, they stopped the system from functioning (B. Walker and Salt 2006, 9). Important examples were the continued water conservancy campaign and the effort to eliminate the "four pests" (rats, sparrows, flies, and mosquitoes).

Carrying Things Too Far: Deep Tilling and dense Planting

Henan Province called out great armies of laborers numbering in the millions to eat and sleep in the fields, carrying out deep tilling of the earth day and night in a huge way. Xiayi County took deep tilling particularly seriously, mobilizing 200,000 laborers . . . organized into brigades, battalions, and companies, divided among 60 great battlefields . . . attacking to turn over the earth day and night. The masses in Henan adopted the slogan "Harvest a plot, turn over a plot, till deeply plot by plot." (Zhong L. 2011, 16)

Like mass water conservancy projects, deep tilling and dense planting (*shen fan mi zhi*) did not start with the Great Leap Forward.[4] As early as 1954, collective leader Ma Tongyi of Changge County, Henan, reported that his collective was able to increase wheat yields dramatically by tilling to a depth of a foot and a half. They first spread 60 percent of available manure on the surface, then turned over the fertile topsoil layer and piled it on one side, then dumped the remaining 40 percent of the manure on the underlying subsoil, breaking up clods and mixing the fertilizer in. They then put this layer on top and worked down, layer by layer (Zhong L. 2011, 16).

This method came to the attention of Mao and other leaders, who promoted it enthusiastically at the start of the Great Leap. The *People's Daily* published supportive editorials in May 1958 and again in September, when deep tilling became official policy not just on the North China Plain but all over the country. Several factors promoted this. First, the labor was available. It was the slack season in the north, but even where agriculture continued in the winter, massive-scale collectivization allowed military-style mobilization. Collectivization of household chores such as cooking and child care made women available to join the brigades for deep tilling and other work (Pietz 2015, 207) or to stay home and tend the fields while the men were out doing construction or deep-tilling work (Ruf 1998, 104). Second, the spirit of the Great Leap Forward, embodied in such slogans as "Don't be afraid of what you can't do, only be afraid of what you can't imagine" (Bu pa zuobudao, zhi pa xiangbudao), along with the recent attacks on people with cautious or empirically based thoughts, made it dangerous to point out empirical limits on just about anything, including soil fertility or the availability of labor. Third, because thinking and policy were not tied to local conditions, leaders at all levels were compelled to try the same panacea methods everywhere—what's good for Changge County in Henan must be good for Meishan County in Sichuan (Ruf 1998, 201) or Dongguan County in the Pearl River delta (A. Chan, Madsen, and Unger 1984, 25). Fourth, political pressure compelled even scientists to explain just how deep tilling could increase soil fertility. A conference held in Changge reported that deep tilling

> can increase the depth of the organic layer, adjust the water and air content of the soil, increase the ability to hold water and nutrients, prevent desiccation and waterlogging, and improve the soil structure. Dead soil becomes live soil; bad soil becomes good soil. It is advantageous to the growth of the root structure of the crops, allowing them to sink deep roots, thus preventing lodging, crowding out weeds, and eliminating insect pests. . . . [We need to] deep-till all of the 1.6 billion *mu* [106 million hectares, or about 80 percent of the country's agricultural land] across the country that can be deep-tilled. The depth should ordinarily be up to a foot and a half, but on bumper-yielding fields it should be two feet or more, and fertilizer should be applied down to the soil horizon. (Zhong L. 2011, 15)

In North China in particular, however, the deep-tilling campaign went too far. An inspection team from the Northwest Agricultural College visited the campaign's point of origin in Changge in July. They wrote a piece featured in *People's Daily* on 27 July stating that it was possible, by tilling to a depth of *one and a half meters*, to increase sweet potato harvests to twenty thousand *jin* per *mu*, or double the maximum yield achieved in optimal conditions of chemicalized agriculture in the 2010s (Huobao 2014). By the fall, millions of farmers were deep-tilling all over the country, even in impossible conditions: "By November, frozen ground in the North impeded the progress of deep tilling. In order to assure that deep tilling proceeded smoothly, and to fulfill their responsibilities, provinces in the North employed all sorts of methods. For example, in Jilin a few areas used big picks, earth rammers, and other tools to turn over frozen earth, or used explosives to blow it up. Many places in Inner Mongolia used fires to warm the earth so they could continue to deep-till it" (Zhong L. 2011, 18).

Local propaganda in Xushui, Hebei, stated that the campaign had reflected the "limitless wisdom and historically unprecedented diligence of the broad masses," as over seventy-five thousand people braved temperatures of −8 to −10 degrees C to "go on the attack." Their county slogan was "To change winter to spring, if our thought isn't frozen, the land won't be frozen" (Bian dongtian wei chuntian, zhiyao sixiang budong, di jiu budong), and the masses said, "Weather now always follows policy" (Tianqi xianzai dou suizhe zhengce zou) (Zhang T. and Guo 2008, 145). By 19 November, 680 million *mu*, or about 30 percent of China's agricultural land, had been deep-tilled (Zhong L. 2011, 17).

Dense planting followed the same "more is better" logic. The increased fertility supposedly achieved by deep tilling would enable farmers to plant many more wheat or cotton seeds, or rice seedlings, in a field than they had planted previously. In southern Fujian, "scientific" farming meant decreasing the space between rice seedlings from the traditional fifteen centimeters to seven centimeters, thus quadrupling the number of plants in a plot (Huang S. 1989, 60). In similar ecological conditions in the Pearl River Delta, "they pushed a system of planting called 'Sky Full of Stars' where a field would be so overplanted the seedlings starved each other out. . . . The peasants knew it was useless, but there was simply no way to oppose anything, because the orders came from so high above" (A. Chan, Madsen, and Unger 1984, 25). In Meishan, Sichuan, farmers were instructed to plant up to eight hundred thousand rice seedlings per *mu*, which would mean spacing the clusters about three centimeters apart (Ruf 1998, 201).

Top-Down Panacea Solutions: The Strange Case of the Double-Wheeled, Double-Bladed Plow

The notorious double-wheeled, double-bladed plow illustrates what Judith Shapiro (2001) has called "dogmatic uniformity" or what Elinor Ostrom and Michael Cox (2010) warned against as "panaceas," or taking something that had worked well enough in some particular place at some particular time and compelling it to be used uniformly everywhere, always. The plow in question was a Soviet model, originally designed for large-scale nonirrigated farming on the broad Russian plains and the Central Asian steppes, where it was very efficient and effective. It was made entirely of steel, unlike traditional Chinese plows that were made out of wood with a single iron or steel share and no wheels. It had been tried earlier—Mao had admired it as early as 1950. As with water projects and deep plowing (for which it was said to be suited), the Great Leap provided an ideal opportunity to use the plow universally. The party, through the *People's Daily*, promoted it aggressively, even in rice-growing regions such as Zhejiang and Hunan (Zhu X. and Hu 2009, 62). Predictably, it had its problems. In Wenzhou, southern Zhejiang, farmers found that it was too clumsy for many rice fields and too heavy for a single water buffalo to pull, and it often got stuck in the mud even when more than one animal pulled it. It was impossible to use in shallow or rocky soils. Sometimes, farmers complained, it plowed too deep, turning the fertile topsoil layer under and causing the roots to take up too much nitrogen in later phases of their growth, leading to too much vegetative growth and less grain (Jin Z. 2011). Some farmers tried to turn it into a harvesting machine, but many just discarded it (Zhu X. and Hu 2009, 62).

From Mao himself on down to local authorities, people predicted that all these measures would contribute to a leap forward in agricultural outputs, and those who doubted the possibility were condemned as rightists (Shapiro 2001, 76–80). Grain yields would multiply by a factor of ten or more, to as high as 35,000 *jin* per *mu* (262,000 kilograms per hectare) for corn in Shandong or 10,000 *jin* per *mu* (75,000 kilos per hectare) for rice (Shapiro 2001, 76–80). Results were positive at first but then crashed. In 1958 China's farmers produced a bumper harvest of 200 million tons of staples, but in following years output went down, to around 170 million tons in 1959 and a disastrous 140 million tons in 1960. Output of other agricultural products, including cotton and pigs, increased in the initial phases of the Great Leap but then declined drastically from 1958 or 1959 on (Dikötter 2011, 139, 141). Combined with increased requisitions of grain by the state, some of them used to pay off loans from the Soviet Union, people starved all over the country; estimates range from twenty-five million to more than forty million excess deaths in 1959–61.[5]

Not all this death and misery were due to ecosystem disruption or overshooting long-term carrying capacity. Areas where provincial and local leaders were particularly "red," and thus eager to speed the advance toward communism and/or please their superiors, suffered much worse than areas where leaders tied their actions more closely to empirical reality (Bramall 2011). Areas where the state could more easily requisition grain also suffered more severely (Garnaut 2014). People in the cities did not starve, because the state fed them with grain taken from already hungry farmers. In many areas, farmers who believed that they had solved the grain problem once and for all planted less area to grain in 1959 than they had in 1958, thus contributing to the decline in output. Social and political systems broke down. Still, ecological overreaching, spurred by political enthusiasm, was the most important factor. In some places deep tilling diminished rather than enhanced soil fertility, and certainly crops planted at very high densities did not have enough nutrients or water—competing plants starved each other out. Instead of "The product of the earth will be as great as the courage of the people," the hard ecological lesson learned from Great Leap overreaching was less catchy but more profound, something like "No matter how great the courage of the people, the earth still has its limits" (Ren you duo da dan, di rengran you xian), or in less parallel terms, "The fertility of the land has limits, and the people's courageous labor eventually produces diminishing returns."

Disruption of Ecosystem Feedbacks:
Dams, Dikes, Ditches, and Sparrows

Emphasizing single variables, managing at inappropriate scales, and imposing top-down panaceas all led to disruption of crucial ecosystem feedbacks. Large-scale water conservancy campaigns from the slack seasons of 1957–58 continued for the next two winters and also spread to the south of China. The *Hubei Daily* proclaimed that, after massive waterworks projects that used more than two hundred million cubic meters of earth, stone, and concrete, "the natural landscape of Hubei is beginning to undergo a fundamental transformation" (quoted in Zhang T. and Guo 2008, 143).

Sadly, though, not all this landscape transformation worked out well, either locally or regionally. In a case from the Taihang Mountains, a village with sixty-two adult laborers decided to build a reservoir with a storage capacity of ninety thousand cubic meters. They worked day and night, sleeping and eating in the fields, and finished it in a few months. However, it was not built to last, and the dam was in a state of almost total collapse after two years. In Funing County, Hebei, a big rainstorm in July 1959 wiped out twenty recently built small reservoirs (Zhang T. and Guo 2008, 143–44).

Viewed at a larger scale, "the replumbing of the North China Plain created a host of soil problems, including waterlogging and soil alkalization/salinization. The construction of reservoirs and channels obstructed drainage. Pernicious salts and other minerals rose to the surface in saturated soils with devastating effects on soil fertility. By the early 1960s, one estimate calculated that two-thirds of irrigated land in North China was threatened by salinization due to poor drainage" (Pietz 2015, 224).

Large-scale, high-tech, mechanized projects guided by the rationalist side of CCP ideology also disrupted ecosystem feedbacks. The Sanmenxia Dam on the Yellow River was not strictly a Great Leap project; it had been planned with Soviet help since 1954, and construction began in 1957 and continued through the Great Leap years until the first phase of construction was finished in early 1961. By the end of 1961, it had become evident that underestimating the amount of sediment the dam would trap had caused serious problems. The reservoir expanded faster than expected because of all the sediment deposited on the bottom, causing unexpected increases in farmland flooding, widening the channel upstream near the confluence of the Wei (渭河) and Yellow Rivers, backing up the Wei, and causing rising water tables and widespread salinization (Pietz 2015, 225; Shapiro 2001, 62–63). By early 1962, sediment deposits had already reduced the capacity of the reservoir behind Sanmenxia

Dam by half. The saga continued for decades; even into the 2020s the dam still didn't work as well as intended.

We must be careful, however, not to condemn everything associated with the Great Leap or Great Leap thinking. In other places, waterworks construction was not so disruptive nor were the results as vulnerable or temporary. Work continued, for example, on the Dujiangyan irrigation system in western Sichuan (see map 3.4) during the Great Leap, apparently seamlessly continuing projects carried out in earlier years. By the end of 1958, the total area irrigated below Dujiangyan had expanded to 5.9 million *mu* (390,000 hectares). Many projects begun in late 1959 at the height of the famine were temporarily discontinued, but they were completed after 1965, after the food supply had recovered (Shen G. 1979; Chen Jinlin and Bao 1993, 222–34).

In the rice-growing regions, many projects were soundly constructed and contributed to agriculture over the long term. Construction brigades in the Pearl River delta built bridges, roads, and dikes (Siu 1989, 178). Lin Village in Fujian built a large reservoir that inundated a lot of its arable land but was nonetheless welcomed because of increased yields on the remainder (Huang S. 1989, 66). Songjiang County in eastern Jiangsu dug a major irrigation canal and two major drainage ditches (P. Huang 1990, 222–23).

It would thus be facile to dismiss all Great Leap water projects as wasted time, labor, and suffering (Wang R. 2008, 129). Like the enlargements of the Dujiangyan system, thousands of small- and large-scale water projects built during the Great Leap were still functioning in the twenty-first century. Instead, success or failure probably depended more on the degree to which the projects maximized one variable in an ecosystem to the detriment of others or managed at the wrong scale, imposed panaceas, or disrupted system feedbacks. When water projects went all out for storage and forgot about drainage, storage contributed to waterlogging and salinization. When people built small dams and weirs by cutting down forests and/or flooding farmland, they ended up with less land to farm, as well as erosion in the deforested areas. When Sanmenxia engineers blithely figured that the silt problem would take care of itself, they lost farmland to both inundation and salinization. By contrast, projects that succeeded, from village ponds to the massive enlargement of the Dujiangyan irrigation area, did not overemphasize one variable.

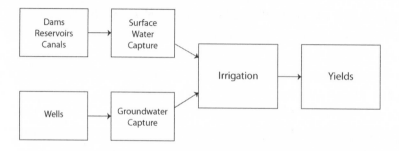

Figure 4.2. Unidirectional causality is apparent in Great Leap Forward thinking; it is simplistic and excludes any feedback or externalities, as shown in this diagram.

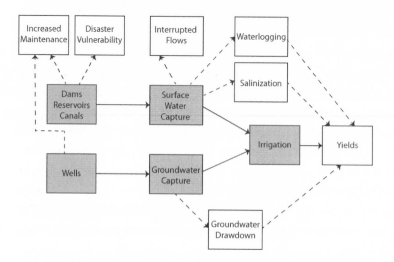

Figure 4.3. A more realistic model of Great Leap Forward policy shows that increasing one input in a system (here, irrigation water) and disregarding externalities interrupts ecosystem flows. Shaded boxes replicate the unrealistic model shown in figure 4.2. Solid arrows indicate beneficial changes; broken arrows indicate detrimental changes.

Figure 4.2 shows the simple, unidirectional thinking of the Great Leap Forward.

If increases in inputs (dams, reservoirs, canals, and wells) are modest, sometimes the planners can ignore negative externalities (as figure 4.2 does) and increased yields will be stable. However, if inputs increase past a certain limit without regard to negative consequences, those externalities become more and more salient, resulting in the situation depicted in figure 4.3, where the inputs have the opposite of their intended effect.

The "Eliminate the Four Pests" (Chu si hai) campaign illustrates even more clearly the dangers of maximizing a single variable. Pests are a real problem for agriculture and public health, and since chemical pesticides were scarce, as early as 1955 leadership formulated a goal of mobilizing labor to eliminate rats, mosquitoes, bedbugs, and sparrows. Of these, sparrows were thought to be the most harmful to agriculture, because they constantly pecked around piles of drying or stored grain, taking good food away from people. Like so many other Great Leap programs, the campaign to eliminate the four pests began as a local experiment in Xushui, Hebei, but it spread nationwide with the high tide of communization, promoted by an editorial in the *Guangming Daily* in September 1958. All over the country, sparrow-killing teams, often composed of elementary and middle school students, mobilized for the slaughter. In Hebei, these children were reported to have killed precisely 234,856 sparrows, often by exploding firecrackers and banging on drums, gongs, and pans to scare the sparrows into flying until they fell exhausted out of the sky and the children could move in for the kill. Only later did scientific research show that while sparrows did consume some grain, the mainstay of their diet was insects, including plant pests, and with the sparrows gone, the bugs flourished. Again, disrupting the flows within an ecosystem, in this case the relationships between predators (sparrows) and prey (insect pests), led to disastrous results that were the diametric opposite of what was intended (Zhang T. and Guo 2008, 144–45).

The consequences of excess inputs, managing at inappropriate spatial and temporal scales, top-down panacea solutions, and cyclic disruptions were horrific. Food shortages began in spring 1959, when "the green did not meet the yellow" (*qing huang bu jie*)—the new harvest didn't come soon enough after the old grain was exhausted, undoubtedly due partly to waste, increased requisitions, and perhaps overeating when free food in canteens replaced meals cooked out of families' own stores. With ecosystem disruption, reduced planting area, and eventual demoralization or exhaustion, harvests for the 1959 agricultural season fell; national staple yields, which had achieved a record of 200 million tons in 1958, were down to 170 million tons in 1959 and to 140 million tons in 1960, leading to the world's worst famine ever, anywhere. The story of the human suffering has been told exhaustively elsewhere (see Becker 1996;

Yang J. 2008; and Dikötter 2011), and there is no need to repeat it here. The triumph of voluntarist enthusiasm and hyperrational logic (Glover, Hayes, and Harrell 2021) over empirical science—including ecological science—and often over common sense, disrupted the feedbacks of agro-ecosystems to the point where tens of millions starved.

Land and Water in Central Asia

Uyghur farmers in the oases of Xinjiang had undergone Land Reform and collectivization on the same schedule as the farmers in China Proper. During the Great Leap, communization, intensification, and unrealistic goals and policies also subjected them to hunger and even starvation. Hunger was not as rampant in Xinjiang, however, and a million Han fleeing famine in China Proper took refuge there, incidentally also helping dilute the Turkic majority of Xinjiang's population (Millward and Tursun 2004, 93).

The nomads of the northern Xinjiang steppes (mostly Kazakhs and some Mongols) saw a fundamental change in their ecology and livelihood with the Great Leap Forward. Like everyone else, they were formed into communes beginning in 1958, and their herds, heretofore managed according to traditional tenure arrangements, were collectivized. They were also often merged with nearby Uyghur farmers into giant communes (Millward and Tursun 2004, 93–94). Many herders, though they continued to take care of animals, did so from newly sedentarized villages. The scheme apparently did not work out: livestock were lost, people starved, and by 1962 many nomads had fled to the no-longer-friendly Soviet Union or had been moved out of their traditional herding grounds near the border to sedentary villages farther away from it (Salimjan 2019, 6–9).

Mongol nomads appear to have fared better than their Kazakh counterparts, a fate often attributed to the political savvy of Mongol Communist leader Ulanhu. Although the famine did reach agricultural areas of Inner Mongolia, it was not as serious there as in many parts of China Proper. In the pastoral areas, Ulanhu managed to delay actual communization of herds and pasturelands—township administrative units (*sumu*) became communes in name, but forms of herd management did not change, and in fact numbers of animals continued to increase until

1965, on the eve of the Cultural Revolution, after which herds were finally communized (Bulag 2002, 231).

Collectives Come to Zomia

In Zomia, communities that had been practicing swidden agriculture or mixed agro-forestry continued to do so for the most part throughout the early 1950s, as their political and economic systems also continued relatively undisturbed. But the political changes beginning in 1956 brought different kinds of disturbances to the ecosystem. In the Baiwu Valley in Yanyuan County, Liangshan Prefecture, the first disturbance came with the reaction to the ethnic rebellion that broke out in late 1956. In order to better control the population and prepare for collectivization, in 1957 the local authorities implemented "people's residential points" (*jumin dian*), where people previously living in isolated houses or small housing clusters scattered about the landscape were concentrated into villages of several tens of households. In order to build new houses, they built furnaces to roast local limestone for use in making cement and whitewash, in the process cutting down forests to fuel the furnaces.

At the same time, state cadres promoted intensification of cultivation. In the Baiwu Valley, Nuosu Yi farmers were required not only to open up more fields for cultivation but to grow crops every year on all the mountainside fields formerly farmed in a swidden rotation. Instead of leaving residues on the fields after burning, which helps soil fertility, they had to clear all the debris away in order to be able to plant crops more densely. In 1958, along with the rest of the country, they formed a people's commune, and for a while one local brigade centered in the village of Yangjuan had two public canteens where people could eat as much as they wished. But the original plenty did not last long; soon the canteen was rationing grain according to the labor power of the members. Everyone was classified as "full laborer," "half laborer," "nonlaborer" (students and children too small to contribute work), or "burden" (*fudan*), which included babies. The canteen had four different-sized wooden bowls that would hold the single-meal ration for each category of laborer or nonlaborer. Once some schoolboys got hungry and decided to smash the smallest bowl, which was about the size of the rice bowls conventionally used in restaurants and by urban families. The cadre in charge of the canteen then replaced that bowl with an even smaller one. In 1960 food was so short that six people died of what were officially reported as illnesses but that people widely regarded as starvation.[6]

Nuosu people in Sha'er Village, Mianning County, had originally lived in a high mountain environment to the east, but in the 1956 Democratic Reforms they were moved to an unoccupied spot in the lowlands of the Anning River valley. Communization and the collective canteens soon brought starvation, and some people began to suffer from edema; a few died. Relying on their local knowledge, they decided that "turnips are not basic grain" (grain had to be turned over to the collective), so they moved back to their old homes in the high mountains, where there were plenty of turnips and other wild vegetables; they thus avoided further starvation and demonstrated how environmental diversity can enhance local resilience, even in the face of large-scale disturbance. Some of them did not return to the lowlands until the 1990s.[7]

Akha (Hani) villagers of Mengsong Township in Sipsongpanna were also collectivized and organized into work teams that were assigned to specialized tasks: some to dry-rice cultivation, some to vegetable gardening, some to herding, and others to cooking in the canteen and other chores. As in China Proper, they were paid work points. This reorganization disturbed many people, and about a third of the village residents fled to Burma, but they soon returned. Nevertheless, grain ran short in the gap between yellow and green many times from 1958 until the end of collective agriculture in 1982 (Sturgeon 2005, 148–49).

The disturbances to previously low-intensity agro-silvio-pastoral eco-systems in Zomia had effects far beyond tragic, if temporary, disruptions in the food supply. In agro-forestry communities, converting swiddens to permanent cultivation and expanding cultivation farther into forests harmed both plant and animal biodiversity. Elders in the Baiwu Valley recalled several species of formerly common trees that either disappeared or were rare by the first decade of the 2000s (Harrell et al. 2022) and also lamented the disappearance of bears, wolves, and large cats.

On a larger scale, Great Leap policies—expanding cultivated land and fueling steel furnaces—contributed to deforestation of about 3 percent of China's total land area (Fanneng He et al. 2008, 68). Zomia, being one of the two most heavily forested regions of the country, suffered proportionately. But the Great Leap was only the most extreme example of shortsighted forestry policy that sacrificed long-term goals for short-term expediency. Forestry in the early years of the People's Republic was focused on timber demand rather than healthy forest growth; even though forest cover expanded greatly after 1980, by the early twenty-first century forest quality had really not recovered. Forests were too young, average trees were too small, and standing biomass per hectare was on average very low (Zhang Yuxing 2008).

....

There was a logic to the Great Leap Forward. In fact, Great Leap thinking was in one way too logical, the ultimate triumph of rationalist logic over empirical reality (Glover, Hayes, and Harrell 2021). There were ways to raise agricultural production by expanding irrigation, and there were perhaps ways to enhance soil fertility without adding chemical fertilizers. Put in ecosystem terms, the problem was about degrees and limits. Because tilling a foot deep might improve the penetrability of soil and give plants access to more nutrients, it followed only in the most hyperlogical, antiempirical way that tilling three feet or five feet deep would enhance these effects. Because there was water to capture for irrigation, it followed only in the most hyperlogical, antiempirical way that capturing *all* the water in the ecosystem would allow irrigation on even more land. Because labor could be mobilized during the winter agricultural off-season in North China to build irrigation works and other infrastructure, it followed in only the most hyperlogical, antiempirical way that people all over the country could work 24/7. Because collectivization could rationalize labor allocation, it followed only in the most hyperlogical, antiempirical way that building enormous collectives would make labor allocation even more rational and efficient.

For many years after the Great Leap, Chinese official history referred to the time of the Great Famine as the "three years of natural disasters" (*san nian ziran zaihai*), claiming that unprecedented acts of nature had cut already marginal agricultural production by almost a third. Later research by meteorological historians has shown that in 1959, the year of the first drastic drop in production, weather was better than average, though bad weather in 1960 probably exacerbated the already desperate situation. Despite even worse weather in 1961, grain production picked up again because of the abandonment of the extreme policies of the Great Leap (Swamy and Burki 1970, 51), leading to the end of the famine (Kueh 1984). The irony is that the overreaching, top-down panaceas, as well as disruption of system feedbacks during the Great Leap, made China's agro-ecosystems more vulnerable to the bad weather conditions that did occur, showing us that there are natural phenomena but only human disasters. Great Leap policies had obliterated the short-term resilience of China's agro-ecosystems, and the farmers, who had contributed so much labor to the effort, paid the horrible price.

5. Normal Socialist Agriculture, 1962–1978

> Emphasize staple production in overall development.
>
> CCP AGRICULTURE SLOGAN, 1970S

The disastrous harvests, mass starvation, and general disorganization from 1959 through 1961 brought the central Chinese Communist Party leadership to the realization that no actual leap forward was going to happen, that the utopian dreams of a few years earlier had been just that. Even though their propaganda continued to blame the famine primarily on "three years of natural disasters" (Smil 1999a, 1620; Bramall 2011, 1007) and secondarily on the abrupt withdrawal of all Soviet aid in 1960, their actions show that they knew the Great Leap had been irrational or hyperrational overreaching. In response, they did three kinds of things. They de-communized agriculture, tried to expand the area of staple cultivation, and invested large amounts of capital in agricultural intensification.

DE-COMMUNIZING THE PEOPLE'S COMMUNE

The Great Leap experiment in communism ended quickly, though the people's communes still kept their name. In late 1960 the "unit of account" within which labor was allocated and income distributed devolved from the commune to the brigade (in many cases corresponding to the former higher-level agricultural producers' cooperative) and by the end of 1961 to the production team, a group usually consisting of twenty to forty households (D. L. Yang 1996, 71–81). Over most of the country, the production team remained the unit of account until agriculture was decollectivized altogether between 1979 and 1982. Communal canteens and child-care centers had been disbanded earlier; all vestiges of small-c communism were removed for the time being and dismissed not as foolish but as premature. China would retreat to socialism for a while.

EXPANDING THE AREA PLANTED TO STAPLES

Since China was still striving to produce enough calories to prevent further famines and because staple foods yielded the most calories per unit of land, planners adopted the strategy of maximizing the production of staple crops, encapsulated in the slogan "Yi liang wei gang, quanmian fazhan," usually translated as "Take grain as the key link, develop overall" but perhaps less quaintly rendered as "Emphasize staple production in overall development." Consequently, people subsisted overwhelmingly on grain-based diets (Popkin 1994, 290; Smil 1993, 81). This policy had both positive and negative effects. Positively, because it was Mao Zedong's idea, it not only led local leaders to follow it but also afforded a degree of protection to agricultural scientists otherwise liable to being attacked as rightist or bourgeois during the Cultural Revolution. As a result, staple production increased. Negatively, as with so many other policies, local officials often either simplified the policy to "emphasize staple production" without the coda "in overall development," a position satirized at the time as "Emphasize staple production; sweep away everything else" (Yi liang wei gang, qiyu sao guang) (Zou H. 2010, 51). In effect, this distortion of the original policy intention was another instance of a strategy from earlier during the Great Leap—maximizing a single variable in the ecosystem (in this case, staple cropland area)—that could boost short-term productivity at the cost of medium-term resilience, leading to visible ecosystem damage in relatively short order.

The expanded reclamation of Heilongjiang's Great Northern Wilderness for agriculture illustrated the pitfalls of unilaterally maximizing staple production. Although much land had been brought under cultivation before the Great Leap, Heilongjiang added another 4 million hectares of staple cropland during the Great Leap and the subsequent collective period (Jay Gao and Liu 2011, 478). Land reclaimed during the 1950s had mainly been planted with wheat (Tao 1983); only about 350,000 hectares had been planted to rice. Unirrigated fields continued to expand after 1960, from 9.8 to 12.7 million hectares, but there was also a big push to convert to single-crop irrigated rice. Between 1958 and 1980, almost 1.4 million hectares were converted. About 46 percent of this acreage was previously dry-cropped fields, but 16 percent was reclaimed primary and secondary forest, and more than 30 percent was reclaimed wetlands (Jay Gao and Liu 2011, 479). This program did manage to increase agricultural production in Heilongjiang dramatically, making the Great Northern Wilderness (Bei Da Huang) into the Great Northern Granary (Bei Da Cang). In Bei'an County, for example, by 1980 about 126,000 hectares of grasslands and 119,000 hectares of forests, along with some wetlands, had been converted to agriculture (Tao 1983).

Ecosystem problems started very soon. By 1980, areas where flooding had been the major weather concern were already experiencing spring-time drought. Black soils in previously forested areas in the watershed of the Wuyur River, a tributary of the Nen, had already lost a full meter of depth due to erosion, and the remaining soils had low organic content (Tao 1983, 25). Researchers estimated that 65 percent of Bin County, just to the east of Harbin, had suffered erosion damage since being opened up in the 1950s—rich black topsoil layers 80 centimeters or more thick had thinned to a few centimeters or eroded away completely (Cheng S. 2012). A 2010 map of the county showed over 12,000 erosion gullies; in a matter of months a 20-centimeter-wide erosion gully that began as a furrow could widen to where "you could drive a train through it." To the west at the edge of the Song-Nen Plain, density of gullies more than doubled from 1965 to 2005 (Zhang S. et al. 2015); in much of that region, erosion gullies averaged 200 to 350 meters per square kilometer (Zhang S. et al. 2015). Areas with erosion also showed large losses in total soil carbon and soil organic matter (Yang Zi et al. 2017). All in all, it was estimated that 27 percent of the 1 million square kilometers of the black soil belt of Manchuria was affected by erosion; the sediment runoff had raised the bed of the Sungari River as much as 50 centimeters in places (Zhao F. 2010). It had become clear that in the rush to grow staple grain everywhere, areas suited to forest or to grazing were being logged, plowed, and planted, costing the land not only soil but ecosystem services generally, along with grazing animals' manure, which could have restored organic richness to the soil (Tao 1983, 41).

The same single-minded concentration on expanding grain area led to a bizarre project at Dian Lake in Yunnan, where a Great Leap–style campaign to "enclose the lake and create rice fields" (*wei hai zao tian*) mobilized more than one hundred thousand farmers for eight months in 1970 to enclose shallow areas of the lake and covert them to rice fields, in order to "emphasize staple production." Although farmers managed to grow some rice in the newly reclaimed areas, much of the land was too soft, yields were low, and the ecology of the lake was severely disrupted (Shapiro 2001, 117–31). In Fanchang County, Anhui, efforts to convert forestland to rice fields were similarly unsuccessful, and of course the forest's ecosystem services were lost in the process (Ross Doll, pers. comm.).

Similar but smaller-scale expansion of cultivated area happened in parts of Zomia. In the Baiwu Valley in Liangshan, forests cut to fuel lime kilns were replaced by farmed fields, but the fertility did not last long, and soil erosion probably increased (Harrell et al. 2022). Yields on these new fields were meager; a local Chinese-language ditty summed up the process as "Plant a whole hillside in the spring; harvest one basketful in the fall" (Chun zhong yi pian po, qiu shou yi luokuang). Also, presumably because rice was considered a more "advanced" or desirable crop than the potatoes and corn the local farmers had been growing on their alluvial fields, they were directed to plant rice for two seasons in the early 1960s. Yields were miserable, and they never tried to grow rice again.

Villagers in other parts of the Zomia were also forced to expand cultivated area. For Akha in Mengsong Township in Sipsongpanna, collective agriculture both diminished yields (villagers rightly claimed that they knew more about how to grow grain than did the cadres who directed the new system) and forced collectives to sell grain to the state that they previously would have kept for their own consumption. As a result, they had to construct additional wet-rice fields, on land originally devoted to other crops, simply to continue feeding themselves, thereby reducing patch diversity (Sturgeon 2005, 147–51).

CAPITAL INPUTS TO AGRICULTURE

Having exhausted the potential of reorganizing and intensifying labor to increase agricultural yields, and realizing that even heroic mass campaigns could reclaim only limited land areas, the leadership finally turned to capital inputs in a serious way. They invested in fossil fuel–powered irrigation, artificial fertilizers, and other chemicals, facilitated limited agricultural mechanization, and promoted high-yielding "Green

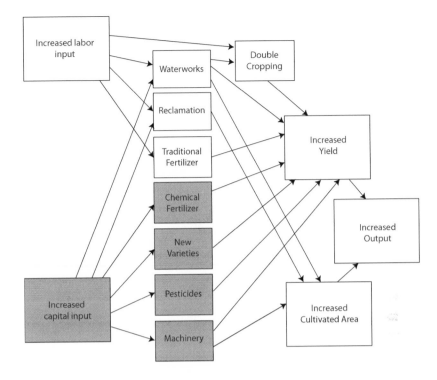

Figure 5.1. Capital inputs (gray boxes) added to the strategies for increasing grain production beginning in 1962 (compare figure 3.5).

Revolution" varieties of wheat, rice, and corn. From the early 1960s until the early 1980s (including the Cultural Revolution years of 1966–76), despite expansion of the area planted in staples, growth in food production would come almost entirely from these capital inputs (Bramall 1995; B. Stone 1988) (figure 5.1).

Agricultural intensification during this collective era was much more successful than during the socialist construction of the 1950s or the wild utopian experiments of the Great Leap Forward. No longer did the state mobilize people (at least not very much) to do crazy things that either disrupted ecosystem flow cycles or immediately pushed elements of ecosystems far beyond their physical limits. And in the medium term, this capital-focused strategy worked. Intensification increased production enough to feed the burgeoning population, which grew from 644 million in the depths of the famine in 1961 to 1 billion at the beginning of Reform in 1981 (Banister 1987, 42–43). But there were always trade-offs, of two kinds. Temporal trade-offs, such as increasing groundwater irrigation, *postponed* negative effects. They traded away longer-term sustainability in the interest of short-term production increases, depleting nonrenewable or slow-renewable resources. Spatial trade-offs, such as introducing chemical fertilizers and other local capital inputs, *externalized* the damage to water quality, food safety, and human health generally. In balance, it is difficult to condemn policy makers and scientists for making these trade-offs: the whole country had just experienced massive famine, and even after the postfamine reforms and abandonment of most of the nuttiest schemes, local food supplies were still precarious. However, twenty-first-century China began to pay the price in terms of groundwater drawdown, surface and groundwater pollution, soil loss and contamination, depletion of fisheries, environmental disease, and pervasive public anxiety over food safety.

Pumping Groundwater to Expand Irrigation

Expanding irrigation would not only increase the reliability of harvests in widely variable climates of North and Northwest China; it could also enable farmers to begin cultivating high-yielding Green Revolution varieties of wheat and corn. But there was little surface water left to exploit. By 1962, according to official statistics, the gargantuan efforts of the Great Leap Forward era had only increased the permanently irrigated area by about 11 percent, from 27.3 to 30.5 million hectares (Nongye Bu 2009, 7). In addition, some previous projects, such as the People's Victory Canal, completed in 1952, had to be partially shut down: they delivered large amounts of surface water in times of drought but did not provide adequate drainage in times of flood, making waterlogging a serious problem. More ominously, the water table in the area served by the canal, previously about 3 to 4 meters below ground, had risen to an average of 1.3 meters (Pietz 2015, 239), leading to salinization of at least 19,000 hectares, or about half of the canal's irrigated area (L. Cai 1988, 84; Nickum 1988, 89). In addition, irrigation gates became clogged with sediment that was supposed to have been sequestered upstream at the stalled Sanmenxia Dam project (Pietz 2015, 239). At the same time, withdrawals of surface water from the Yellow River had increased so much that in spring 1972 the diminished river no longer flowed all the way to the sea (Pietz 2015, 241), a harbinger of yearly flow stoppages during the 1990s. Clearly, there was not much more potential for surface irrigation in North China.

The answer to these problems was to tap the North China Aquifer, one of the world's largest sources of groundwater, stored in a complex, multilayered system. The top layer is unconfined water, fresh in the western part of the plain but brackish or saline closer to the sea (S. Foster et al. 2004, 82; Chen Su et al. 2018, 1401). The second layer is confined saline water, and the third and fourth are both confined fresh water. Some shallow wells had already been tapping the shallow aquifer since the 1950s, but most of them were inoperable by the 1960s because of shoddy construction. The decision to invest capital in agriculture, the recent increase in the supply of petroleum, and the capacity to build diesel-powered pumps combined to enable the state to embark on a massive campaign to drill tubewells that would tap both the shallow unconfined and the deeper confined freshwater aquifers.

Beginning in the early 1960s and accelerating after a major drought in 1968, tubewells were drilled in villages all across the North China Plain, financed about 25 percent by government subsidies and the remaining 75 percent by local commune funds (Huang Ronghan 1988, 82). By 1975 an estimated 1.05 million tubewells had been drilled in Hebei, Henan, Shandong, and Beijing (Vermeer 1977, 193, cited in Pietz 2015, 245); by 1982 the number had increased to somewhere between 1.4 and 2.0 million, pumping 25 to 35 cubic kilometers of water and irrigating about 11 million hectares in North China, 8 million of that on the plain, in some places supplemented by surface water irrigation (see Huang Ronghan 1988; B. Stone 1988, 772; Nickum 1988, 89). Because Cultural Revolution rhetoric dominated politics during most of this period, this project was often referred to as the Groundwater Campaign (Dixiashui Yundong) (Pietz 2015, 245), but in fact it depended more on technological upgrades and capital investment than on the mass labor mobilization of campaign-style development.

Drilling wells dramatically increased irrigated area and water delivery. This, along with increased inputs of fertilizer and the introduction of new high-yielding varieties, led to major increases in grain output: between 1972 and 1980, wheat yields increased by 127 percent in Henan, 54 percent in Shandong, and 27 percent in Hebei (Huang Ronghan 1988, 82), and yields continued to increase thereafter. There are no known instances of famines in North China after the recovery from the disasters of the Great Leap. In addition, some water became available for industrial and residential use; for example, water was diverted from the Yellow River to the city of Tianjin during dry years of the 1970s (Pietz 2015, 241). The dangerous silting and salinization of farmland that had plagued earlier attempts to deliver surface water were partially resolved as well: the groundwater had no silt and was all used locally anyway, and withdrawals from the shallow aquifer initially lowered the dangerously high level of the water table to its previous three to four meters, which solved the problem of salinization in many areas (L. Cai 1988, 82).

These medium-term gains, which were a huge part of China's efforts to solve its food problems, were not, however, without their long-term costs. The water table (the level of the unconfined aquifer) fell below its previous levels and kept going down. In 1975 it ranged from nine to eighteen meters below the surface; by 1985 it had fallen to thirty-two meters below ground level, and it continued to fall thereafter. At Cangzhou, near the coast, where the unconfined shallow aquifer is saline and tubewells primarily tapped the deep aquifers, the potentiometric surface of the deep aquifer had fallen by fifty-five meters by 1975 and by seventy-six meters by 1980 (Chen Su et al. 2018, 1405).[1] This rapid drawdown had already resulted in the formation of a large cone of depression, an area where depletion of groundwater leads to subsidence of the overlying land. Such cones proliferated and deepened in the ensuing decades (Chen Su et al. 2018, 1402). Everywhere on the North China Plain, recharge rates were slower than drawdown rates, with the result that the water would eventually run out (S. Foster et al. 2004; Yanxin Wang, Zheng, and Ma 2018). Still, exploiting this decades-long source of irrigation water brought decades of food security to North China, making it difficult to criticize the original decision to drill tubewells, especially after the Great Famine and unsuccessful efforts to increase food supply by surface water irrigation, all at a time when population was still rapidly expanding.

Still More Surface Water Projects

At the same time, there was still room for some projects that delivered surface water to previously unirrigated fields. Probably the most famous of these projects was the Red Flag Canal, built to channel water from the Zhang River through a precipitous stretch of the Taihang Mountains to formerly dry fields in Lin County, Henan (Yuan et al. 2007). In a certain sense, the Red Flag Canal was the last of the Great Leap Forward projects built by popular mobilization, even though it was constructed later, between 1961 and 1969. It was designed and built entirely by farmers, with little if any input from engineers, college students, or other experts. A documentary film about the Red Flag Canal (Zhongyang 1970) shows almost self-parodying scenes of Communist labor heroes hanging off cliffs, blasting rock while not wearing any protective gear, and moving mountains with picks and shovels, all to the sound of rousing martial music with lyrics about bringing water to the parched fields (Jack Chen 1973).

Stabilizing water availability on the North China Plain was not the only short-term success story during the collective era. Both surface and groundwater irrigation helped increase yields in other parts of China as well. The Dujiangyan irrigation system on the Chengdu Plain (see map 3.4) continued to expand. A tunnel was even dug under the Longquan Hills in 1970–72 to irrigate areas outside the plain itself, and by the late 1970s it was sending water to more than 10.6 million *mu* (700,000 hectares) (see Shen G. 1979, 106; Sichuan Sheng 1994, 87; and Chen Jinlin and Bo 1993, 228). The resulting replacement of dry field crops with irrigated rice raised the average yields in the now-irrigated area from about 500 *jin* per *mu* to over 1,200 (Wu Minliang [1983] 2010, 63; Whiting et al. 2019, 270).

At the same time, however, continuing the earlier emphasis on surface water sometimes brought disastrous results. The Huai River floodplain was once again the scene of tragedy. The original Huai River Conservancy Project in the 1950s had built about one hundred reservoirs; another one hundred were added during the 1960s, taking up a lot of space that had originally been available for flood diversion (Fish 2013). One of the original dams, at Banqiao on the boundary between Biyang and Yicheng Counties on the Ru River, a tributary of the Huai (see map

3.2), had been completed in 1953, built with local labor and minimal state investment. It was enlarged to a reservoir capacity of 492 million cubic meters in 1956 (Zhang Jing 2013, 56). When an unusual typhoon moved inland in August 1975, over 1,600 millimeters of rain fell in three days, about half that amount in a six-hour period on 7 August. Communications with the outside were cut off (Jiang H. 2012). Dam breaches had already happened in nearby areas, and soon water in the reservoir was exceeding its designed capacity by almost half. According to survivors' accounts, just after midnight the rain stopped and the sky cleared, and shouts went up everywhere, "The water is retreating!" But then the ground shook beneath their feet, and a wall of water 5 to 9 meters high gushed through a gaping breach in the top of the earthen dam, advancing 6 meters per second downstream toward Zhumadian City and the Beijing-Guangzhou railway line. It swept away everything in its path, including people, livestock, buildings, and equipment; a 50-ton locomotive was found 5 kilometers from the station where it had been parked. People not in the direct line of the water wall were driven to rooftops and treetops, but many of them were nevertheless swept away when the flood levels topped the roofs and trees. The official human death toll was more than eighty-five thousand; this was later revised downward to twenty-six thousand, but no one really knows (Jiang H. 2012).

Heroic rescue efforts prevented death from starvation (though people were reduced to eating bark and leaves), as well as major epidemics. But more than 17 million *mu* (1.2 million hectares) of agricultural land was flooded in twenty-nine counties and cities, and eleven million people were displaced (Zhang Jing 2013). Not until state authorities dynamited the Bantai sluice gates on the Henan-Anhui border downstream did the floodwaters finally recede. Once again, the strategy of maximizing water storage at the expense of flood diversion capacity and maximizing productivity at the expense of resilience had taken a tremendous toll.

Chemicals Galore: Industrial Fertilizers, Pesticides, and Plastic Mulch

The second leg of capital investment in agriculture was chemical inputs—fertilizers, pesticides, and plastic mulch—all of which contributed to boosting agricultural yields. Used correctly (which they were not always), agrochemicals would increase yields when applied to whatever varieties were selected, but what was more important was that they allowed the introduction of high-yielding Green Revolution varieties, which attain their full potential only with the application of one or more of these inputs.

Traditional grain agriculture, particularly paddy rice cultivation, depended heavily on recycling organic nutrients, above all animal and human manures but also plant-based nutrients such as the "bean cake" fertilizer made from the residue of tofu production. Studies of soil chemistry conducted during both the 1930s and the late 1950s, however, revealed that while nutrient supply was fairly constant, it was not sufficient. Of the basic triad of nutrients, only potassium was deemed adequate. Nitrogen was deficient wherever rice, corn, wheat, rapeseed, millet, corn, and cotton were grown, and phosphorus levels were well below the optimum in rice-growing areas (J. Jin 2012, 1006–7).

This situation changed little through the years of socialist construction and the Great Leap Forward; chemical fertilizer was another capital input that the state chose to expand only slowly. But after the Great Leap and the subsequent famine, the state drastically changed course on its use of chemical inputs. China had used only 72,000 tons of chemical fertilizers in 1952, a number that increased more than eightfold, to 630,000 tons, in 1962, but the real expansion came afterward. By 1980, usage had increased another twentyfold, to almost 12.7 million tons, of which about 75 percent was nitrogen fertilizer, another 20 percent was phosphate fertilizer, and the remainder was potassium and combined fertilizers (Nongye Bu 2009, 8). The volume of nitrogen applied per hectare increased from about 5 or 6 kilograms to around 30 kilos setting the stage for even faster expansion in the Reform era (figure 5.2).

Having failed to eliminate pests by getting farmers and schoolchildren to go out and bash them, the state turned to chemical solutions there, too. In 1950 China produced around five hundred tons of pesticides and imported another thousand tons, a trivial amount consisting mostly of inorganic insecticides and fungicides such as sodium fluoride, lead arsenate, and various mercury compounds.

By 1960 production had been ramped up to about 260,000 tons, consisting mostly of organochloride insecticidal compounds such as benzene hexachloride (BHC) and parathion (the two active ingredients in a popular insecticide known in China as 66 or 666), along with DDT.

Figure 5.2. A sign for US agrochemicals used on dragon fruit plantations in Guangxi, 2018. Photo by the author.

Copper sulfate was the most commonly used fungicide, and herbicides included 2.4-D and 2.4.5T, the major ingredients of the infamous Agent Orange used by the US military in Vietnam.

Chinese farmers and authorities soon discovered the dangers of chemical pesticides. In 1970 rice and other foods in Zhejiang were found to be contaminated by organic mercury compounds, so the provincial authorities forbade their use there, and a national ban followed in 1972. DDT and BHC were both found to have had deleterious effects in other countries, and they were banned in Japan and the United States in 1971. China continued to use them until 1983, when their manufacture, use, and sale were prohibited and they were replaced by less obviously harmful pesticides (Koitabashi 2009), although their residues continued to contaminate China's farmlands even into the twenty-first century.

Another important contribution of modern chemicals to Chinese agriculture was a white or clear plastic film covering or mulch known as *dimo*, or "ground membrane." *Dimo* comes in long rolls, and farmers spread it over newly planted seeds (poking holes for the plants to grow through) to preserve moisture, warmth, or both, thus enabling earlier planting in China's typically cold, dry spring weather (figure 5.3). *Dimo* was first widely used in the US South in the 1950s and in 1958 was introduced to China, where it was used to lengthen the growing seasons for early rice cultivation in the central and southern parts of the country (Ingman 2012, 4). By 1965, along with other capital inputs, *dimo* was used in every province to make the growing season begin earlier and reduce the need for irrigation water (Ingman 2012, 19). It also allowed the introduction of Green Revolution plant varieties. Like other capital inputs, however, *dimo* also has an environmental cost. As Mark Ingman (2012, 5) describes it, "China commonly uses an extremely thin type of polyethylene plastic that is eight hundredths of a millimeter (0.08mm) thick. Installing this plastic with hand tools or machinery requires as much as 30% of the plastic mulch to be buried in the soil to seal the edges of field rows. The buried plastic usually gets shredded through tillage and thus contributes to . . . *baise wuran*, the 'white pollution' that litters their landscape, endangers livestock, and pollutes rivers and lakes."

Figure 5.3. Nuosu Yi farmers spreading *dimo* on Green Revolution corn seedbeds, 2012. Photo by the author.

The Dream of a Red Tractor

Tractors had been the cornerstone of Soviet collective agriculture, and the Chinese Communist Party initially touted them as a solution to the intense physical labor that Chinese farmers had historically endured, as well as a means of ensuring high crop yields (see figure 3.4). However, because of the priority of heavy-industrial investments in steel, railroads, and cement, only large tractors (in small numbers) were produced before the 1970s, mostly for use on state farms in the Heilongjiang Reclamation District and nearby. The capitalization of agriculture in the 1960s brought small, hand-operated tractors to rice regions of China and sped the plowing of rice fields to the point where the labor squeeze between rice crops in a double-cropping season became more manageable. For example, in Songjiang County in the Chang River delta, the state had pushed double cropping in the mid-1950s, but it had simply been too laborious for farmers. When small tillers became available, however, beginning in 1969, tilling times decreased dramatically, and "double-cropped rice was pushed across the board in the county, dramatized in the slogan, 'Wipe out single-cropped rice!' (*xiaomie danjidao*). The acreage under single-cropped rice declined dramatically, from 513,989 *mu* in 1963 to a low of 19,146 in 1977" (P. Huang 1990, 225).

From the mid-1960s on, mechanized tillage in the countryside began to increase exponentially, surpassing draft-animal power around the end of the collective era in the late 1970s (Naughton 2007, 263). In 1962, at the end of the Great Leap, there had been only 51,000 large tractors and an astonishingly few 1,000 small tractors in use in China's fields; by 1970 that figure had risen to 125,000 large and 78,000 small tractors; in 1980 there were close to 2.5 million tractors, 745,000 large and 1,874,000 small, still amounting to fewer than 4 large and small tractors per brigade, or a single tractor for each 70 households (Nongye Bu 2009, 44). Clearly, almost all of the huge investments in mechanized power for agriculture during the collective period went into irrigation pumps (B. Stone 1988, 772); human and ungulate muscle power still worked the fields. The contribution of mechanized fieldwork to output increases was minimal compared to those of irrigation, fertilizer, and Green Revolution varieties, and the dream of a red tractor was put off until later.

Science and Green Revolution Varieties

Worldwide, the Green Revolution of the 1960s through the 1990s was responsible for huge increases in food and other agricultural production. At the same time, it has also been blamed for increasing dependency on chemical and mechanical inputs and, in many areas, for increases in social inequality, impoverishment of peasantry, massive urbanization, and slum growth (Niazi 2004), as well as certain kinds of agro-environmental degradation (Vitousek et al. 2009). Common narratives in European languages—both pro and con—center on the work of American agronomist Norman Borlaug and the research done at the International Rice Research Institute in the Philippines, established in 1960, and the International Maize and Wheat Improvement Center in Mexico, established in 1966. These "green revolutionaries" used conventional plant-breeding techniques to develop new varieties of rice, corn and wheat (the world's three primary grain staples).[2] The new varieties were intended to be more responsive to fertilizer, pesticide, and water inputs and yield dramatically more than conventional varieties (Sumberg, Keeney, and Dempsey 2012).

This narrative pays little attention to China, but thanks to the path-breaking work of Sigrid Schmalzer (2014, 2016), we now know clearly that very similar research and experimentation on hybrid, high-yielding, or high-response varieties were going on in China at the same time. Schmalzer describes the paradoxical geopolitics of China's own Green Revolution as follows:

> In 1969, *People's Daily* bemoaned the pursuit of "green revolution" in India, defining it as "the so-called 'agricultural revolution' that the reactionary Indian government is using to hoodwink the people." The article made clear just why the green revolution represented a "reactionary" choice: the Indian Minister of Food and Agriculture had reportedly "cried out in alarm that if the 'green revolution' . . . does not succeed, a red revolution will follow."
>
> Does this mean that socialist China opposed the new technologies of the green revolution or agricultural modernization more generally? No. Contrary to common perception, even the most radical leaders in socialist China embraced the causes of science and

modernization, and so in some important ways, the green revolution in red China looked strikingly similar to the green revolution as [USAID director William] Gaud imagined it. The goal there as elsewhere was to transform the material conditions of agriculture through mechanization, the introduction of new seeds and the application of modern chemicals in order to increase production and raise standards of living. (Schmalzer 2016, 3)

One reason that the inputs described above—including stable and mechanized irrigation and drainage, fertilizer, pesticides, and plastic mulch—were so effective in increasing China's agricultural production in the 1960s and 1970s was that the newly developed hybrids of wheat, corn, and rice were much more responsive to (and dependent on) inputs of water and fertilizer than are conventional varieties (Naughton 2007, 259). Conversely, the reason Chinese farmers could grow the new varieties successfully during that time was that the inputs of mechanical energy and agricultural chemicals were available.

The agronomist Yuan Longping's laboratory in Hunan made a series of crucial technological breakthroughs in the early 1970s, enabling traits of hybrid vigor to be passed down through many generations of rice plants using male-sterile genetic lines (Schmalzer 2016, 75–77; B. Stone 1988, 793). Within five years, hybrid varieties began to spread widely through many rice-growing regions, though the speed of adoption varied. By the time the decollectivization of agriculture was essentially complete in 1984, 24.7 percent of the total rice area in the country was planted in hybrid varieties, with over 56 percent of Sichuan's rice area in hybrids. This was the beginning of a trend that continued steadily through the subsequent Reform period, seemingly unaffected by the decollectivization itself (J. Huang and Rozelle 1996, 341; B. Stone 1988, 800–802). Yields from hybrid varieties averaged about 14 percent more than those of conventional varieties, though this ratio varied a lot by province—again in Sichuan, hybrid yields were about 53 percent higher. Combined with the availability of stable water supplies and agricultural chemicals, adoption of these new varieties led to yield increases of around 1.2 tons per hectare. When Yuan Longping died in 2021 at the age of ninety, he was mourned as a national hero (Chik 2021), an honor not given to Norman Borlaug in any nation.

China first imported Green Revolution varieties of wheat and corn from Mexico in 1968, but these soon developed problems, leading to their replacement by domestically developed varieties or Chinese-Mexican crosses (B. Stone 1988, 793–95). By as early as 1978, Green Revolution hybrids accounted for 60 percent of corn and 40 percent of sorghum in China. Improved varieties of potatoes and sweet potatoes also replaced traditional varieties in much of the country in the 1960s and 1970s (B. Stone 1988, 796–97), although they are not technically Green Revolution varieties (because their productive potential is not dependent on stable inputs of water and fertilizer).

Clearly, something worked, or more accurately, many things worked together, spurred by the commitment to agricultural investments and agricultural research. According to official statistics, staple production expanded from about 190 million tons in both 1957 and 1965 (before and after the Great Leap disasters) to 240 million tons in 1970, 285 million in 1975, and 320 million in both 1980 and 1981 (Nongye Bu 2009, 17). However, even this impressive growth in staple output was barely enough to keep up with population, which grew from 633 million in 1957 to 820 million in 1970, 918 million in 1975, and 983 million in 1980 (Banister 1987, 42–43). This meant that the impressive absolute growth in staple harvests translated from per capita availability of about 300 kilograms per year, or 820 grams per day, in 1957, to about 325 kilos per year, or 890 grams per day, in 1980—a little more, to be sure, but not enough to ensure food security yet.

The trend toward ever-higher-yielding varieties and increased area planted with those varieties continued across the divide of decollectivization throughout the 1980s (B. Stone 1988, 799; J. Huang and Rozelle 1996, 363–64), eventually leading to staple harvests of 510 million tons by the end of the twentieth century, finally far outstripping population growth. However, the increases during the high socialist period still call into question the historical narrative, promoted not only by official Chinese sources but also by foreign capitalist triumphalism, that science and technology stagnated during the years of Maoist revolution, to recover only with the partial opening of markets and partial freedom of expression that came with the period of Reform and Opening.[3]

....

Certainly, many other areas of the economy took off after the 1980s reforms, and Chinese citizens, even under the increasingly repressive regime of the early 2020s, still have a lot more opportunity to learn, travel (at least until the COVID-19 pandemic) and consume. But the high socialist period was not one of total scientific and technical darkness. Agricultural science and technology progressed greatly during the years of collective production—they had to, given the growing population and the need to feed it. Agricultural modernization after the Great Leap is analogous to China's development of nuclear weapons, culminating in the first successful test in 1964. If the regime believed that defending itself was imperative (whether nuclear weapons actually defend a country is immaterial), it also believed that it had to feed its people. Scientists working on both projects did their work even in the times when "politics was in command."

6. Solving the Warm and Full Problem, 1978–1998

> Deng Xiaoping's post-1978 rural reforms were not undertaken because of a long-suppressed admiration for private farming, but in order to lift the nation from chronic food shortages and massive malnutrition.
>
> VACLAV SMIL, "Who Will Feed China?," 1995

By the late 1970s, China's "normal socialist agriculture" had corrected most of the short-term ecological damage from the Great Leap Forward and had even slightly increased its per capita staple crop production. But it was still far from being able to "solve the warm and full problem" (*jiejue wenbao wenti*). Food supplies were still precarious, China was unable or unwilling to import more than 2–3 percent of the grain it needed (X. Dong, Veeman, and Veeman 1995, 325), and despite the initial successes of the nationwide "late, spaced, and few" (*wan, xi, shao*) program to encourage people to have fewer children, the population was still growing. Agricultural growth had leveled off: staple crop production increased only 7 percent from 1978 through 1981, or about 2 percent per year (Nongye Bu 2009, 17). There were also rumblings of discontent in the countryside. A combination of work point systems that rewarded political enthusiasm rather than labor input (E. Friedman, Pickowicz, and Selden 2005, 176; Mobo Gao 1999, 177) and a lack of economic decision-making power that made peasants into "bare labor power" (Sun Zexue 2006, 80) was beginning to provide them with less and less incentive to participate in productive work, while mass mobilization also seemed to have run its course. Something still needed to be done. Opportunity to do something happened when Mao Zedong died in 1976.

After a brief interregnum under Mao's chosen successor, Hua Guofeng, Deng Xiaoping seized control of the party and government in late 1978, and the so-called Reform and Opening (Gaige Kaifang) began. Ecologically, in the early Reform and Opening period, policy maintained the direction established with the retreat from the Great Leap, namely increasing capital inputs into agriculture and promoting agricultural science (see figure 5.1). But economically, the Reform and Opening program was a sharp break: it decollectivized agriculture, opened markets, and diversified agricultural products. The results were spectacularly successful. While population growth continued to slow down, agricultural yields increased explosively for a variety of reasons, both economic and agronomic. Between 1980 and 1998, staple production increased by 60 percent, or 47 percent per capita, despite a slight decrease in the area sown to staples, while the total sown area for all crops (including nonfood crops such as cotton and tobacco) increased from 146 to 153 million hectares (Nongye Bu 2009, 17). As a result, food production diversified greatly—large increases in output of practically every kind of food, including oils, sugar, vegetables, meat, eggs, and fish, meant a huge increase in the quality and variety of urban and rural diets. Table 6.1 shows outputs of a representative sample of crops, livestock, and fish in 1980, 1990, and 1998.

POLICIES THAT LED TO CHINA'S FOOD SELF-SUFFICIENCY

At least four new policies undertaken as part of the Deng Xiaoping–era reforms contributed directly to the project of feeding the people:

- The Planned Birth Policy (aka the One-Child Policy) accelerated the fertility decline and thus alleviated worries about population-food balance.
- Decollectivizing agriculture and reopening agricultural markets boosted farmers' production incentives and income opportunities.
- Overall commercialization of the economy gave farmers new opportunities to grow diverse crops and gave consumers added breadth in their diets.
- The retreat from "emphasize staple production" led to new opportunities for farmers to plant and sell crops more suited to local conditions.

By 1998, then, when China turned from Deng Xiaoping's "development [as] the only solid truth" to an eco-developmental compromise, the "warm and full problem," if not completely eliminated, was confined to a few areas. However, like any set of programs that intensifies production in a system that is already highly intensive, each of the policies of the

Table 6.1. Increases in production of representative foods
in China, 1980–1998

FOOD TYPE	1980 OUTPUT (MILLION TONS)	1990 OUTPUT (MILLION TONS)	1998 OUTPUT (MILLION TONS)	INCREASE 1980–1998 (%)	PER CAPITA INCREASE 1980–1998 (%)
STAPLES	320	446	512	60	47.2
EDIBLE OILS	7.7	16.1	23.1	200	157
SUGARCANE	22.8	57.6	83.4	265	208
APPLES	2.36	4.32	19.48	725	570
LEAFY VEGETABLES	No data	195	384	97 (1990–98)	89 (1990–98)
BANANAS	.061	1.46	3.52	5670	4535
CITRUS FRUITS	.712	4.85	8.59	1106	870
EGGS	2.8 (1982)	7.94	20.2	621	488 (1982–98)
PORK	11.3	22.8	38.8	243	191
BEEF	.269	1.07	4.79	1680	1322
COW'S MILK	1.14	4.15	6.62	480	377
FISH (capture)	3.15	6.28	17.3	449	353
FISH (aquaculture)	1.34	5.18	21.8	1526	1201

Note: Figures from Nongye Bu 2009, various tables.

Reform era placed strains on the agro-ecosystems, leading to several kinds of long-term resource degradation and loss of resilience. These included neglect of rural infrastructure, hardening rigidity traps, and chemical pollution.

Controlling Demand: The Planned Birth Program

Mao didn't have to die for China's leaders to back away from his earlier dictum of "more people, greater strength." In fact, by the mid-1960s urban fertility was already declining, primarily in response to developmental factors such as increases in education levels and state welfare benefits. Rural fertility, however, was still high, and in the early 1970s the regime had instituted the "late, spaced, and few" policy, requiring couples to start childbearing late and leave longer intervals between births, as well as have fewer children. This was effective in some rural areas, buoyed by such ditties as "One is not too few; two are just right; three are too many" (Yige bu shao, liangge zhenghao, sange duoliao), recited in Xiaoshan County, Zhejiang (Harrell et al. 2011, 27). In other areas, because of work point incentives and perhaps traditional values, rural fertility remained high throughout the 1970s (Harrell et al. 2011, 33).

At a nationwide scale, however, population growth had definitely slowed, and the "late, spaced, and few" program was primarily responsible (Lavely and Freedman 1990; Hesketh, Lu, and Xing 2005, 1172). The crude birth rate dropped from 31.4 per thousand in 1970–75 to 22.4 per thousand in 1976–80, with total fertility dropping from 4.77 to 3.00 (UNPD 2022a). The population grew by 2.4 percent per year from 1970 to 1973 but by only 1.4 percent per year from 1978 to 1981. Thus, as Susan Greenhalgh (2003) has pointed out, it was strange and rather irrational when a team led by a rocket scientist declared that China had a population crisis and proposed the Planned Birth Policy (Jihua Shengyu Zhengce) to reduce total fertility to 1.4 by 1990 through limiting most couples to one birth (T. White 1994).[1]

The Planned Birth Policy was not motivated primarily by worries about population and food supply, even though food supply was still precarious and had recently grown only slightly faster than population—by 2.3 percent per year from 1976 to 1980 (Nongye Bu 2009, 17)—meaning that China was not building up a significant food surplus. Instead, the primary motivation for the Planned Birth Policy was worry about slow economic growth: resources that might otherwise have gone to development were instead going to feed a still-growing population. A second motivation was environmental: increasing population would lead to deforestation, pollution, and resource overuse (Greenhalgh 2003, 174–75). Population policy and agricultural policy, clearly closely connected from a standpoint of popular welfare, were running on separate tracks.

The Planned Birth Policy, whether necessary or not, did coincide with further declines in fertility. Stories of resistance and accommodation to the policy arose (see Wasserstrom 1984; Greenhalgh 1990; Huang Shu-min 1989, 175–85; and Hong Zhang 2005, 260–61; 2007, 255–6). Based on that information, we can surmise that without the policy rural people in particular would have borne more children. From 1980 to 1998, population growth rate declined from 1.6 to 0.6 percent per year, with most of the decline coming after 1990. During the same interval, the crude birth rate dropped from 21.7 in 1981 to 13.2 in 1998 (UNPD 2022a). A further drop in total fertility to 1.51 by 2000 more than offset the effects of the large cohort of women of childbearing age at that time (UNPD 2022b). With the increases in food production summarized above, such a strict fertility control program was probably unnecessary in retrospect (Feng Wang, Cai, and Gu 2012). But it did contribute marginally to the per capita increases in food supply. Far more important were incentives for farmers to produce more food and a greater variety of foods.

Decollectivization as a Microeconomic Incentive

Decollectivization was not originally Deng Xiaoping's idea. The party had experimented with versions of it at various times in the past, after they had gone too far with their original collectivization drive in the mid-1950s and again when they had gone *way* too far with the radical communism of the Great Leap Forward. There had always been opposition from leftist fundamentalists, and Mao had usually ended up taking their side. The late 1970s were different—farmers began to take the initiative in working out arrangements with their own local cadres. Households would take responsibility for a particular piece of land, delivering to the team a little more than their share of the team output would have been (and thus offering the cadres the opportunity to skim a bit of the extra), in return for keeping the surplus for the household to consume or sell. In the words of Kate Xiao Zhou (1996, 58), "Since there was finally an incentive to work, efficiency improved dramatically and a work ethic came back into farming. Formerly cadre-bound farmers put it clearly: 'Now is different from the past. We work for ourselves.' In a 1982 interview, farmers in Tongxin [Hubei] told me that they would go to the fields without waiting to hear the whistle from the team leaders: 'You cannot be lazy when you work for your family and yourself.'"

These arrangements were technically illegal, but they spread quickly (E. Friedman, Pickowicz, and Selden 2005, 250–53; Ruf 1998, 126–29) and were officially if informally permitted starting in 1978 in Anhui and Sichuan, two provinces where agricultural production had lagged and where there were food shortages in some areas (K. Zhou 1996, 60). Despite debate and resistance among the central party leadership, contracting production to households (*baochan dao hu*), also known as the household responsibility system (*jiating zeren zhi*) had become official policy nationwide by 1983. Being able to make their own decisions, as well as reaping the rewards of their own labor directly, the farmers worked harder and produced more, sometimes aided by agricultural price reforms that allowed them to grow and market a greater variety of crops (Mobo Gao 1999, 172; Siu 1989, 278–84). Farmers could follow the "entrepreneurial ethic" that has been a part of Chinese family economic culture for centuries (Harrell 1985), or, in more orthodox Marxist terms, decollectivization "liberated the productive forces."

Marketization as a Macroeconomic Reform

The microeconomic incentive of decollectivization, however, would not have been enough by itself. Most households had enough grain once they could control their own farming, and surpluses would have been no use had farmers not been able to sell them. Fortunately for the farmers, however, the central authorities were cautiously transforming the economy from plan to market. One of the first places these changes were effective was in the agricultural sector.

For the previous two decades, farmers had been restricted by the "unified rural economy" (Sun Zexue 1992) in which, although they often had enough to eat and manufactured most of what they used (see Mobo Gao 1999, 59–60; Leonard and Flower n.d.; Pamela Leonard, pers. comm.), they had little purchasing power. This was partly because "there was no cash around in rural China" (Mobo Gao 1999, 177)—money was simply not a part of everyday life (Xin Liu 2000, 85). Even if they had had money, there was little they could have bought with it. Nor did they have anything to sell: any surplus they produced went to the collective, and if they had had anything to sell, there was no place they could have sold it, since rural markets had long been abolished. Several reforms at the beginning of the 1980s changed this situation in a hurry.

Individual contracts and eventually the household responsibility system meant that if farmers produced a surplus over what they needed for themselves and what they were required to sell to the state, they would have something to sell in the newly reestablished markets. They could not only sell over-quota staples at higher prices than the state paid for the compulsory sales; they could also produce and sell something besides staples, perhaps something that would bring them more income. Three related policy changes made this possible. First, the state increased the prices paid in compulsory staple sales, as well as those for other agricultural products. Second, it reduced the number of agricultural commodities subject to state monopsony and monopoly. Third, it reestablished rural markets for produce and other commodities (see Mobo Gao 1999, 177–79; Ruf 1998, 129; Xiao 2018, 28). Finally, China reentered the world grain market at a larger scale, purchasing around 5 percent of its consumption needs in the early 1980s, further removing pressure on Chinese farmers (X. Dong, Veeman, and Veeman 1995, 325; Xiao 2018, 28).

Retreating from the Emphasis on Staples

Because it could now produce enough food, China in the 1980s could begin to produce a much greater variety of foods, including those previously listed in table 6.1. Reductions in purchase quotas for staples, along with rising prices paid for "secondary agricultural products" (everything else), provided incentives to specialize (Xiao 2018, 28), while agricultural extension services provided needed knowledge. In some cases, this allowed farmers to revive specializations they had engaged in before the all-out emphasis on staple production.

Examples abound. In Meishan County, Sichuan, a production brigade began to revive citrus cultivation in the late 1970s, taking back land newly contracted to households to plant three thousand tangerine trees, compensating the households with jobs in the newly established orchard: "For three years, the orchard staff tended the saplings with care, spraying them with pesticides purchased by the brigade and fertilizing them with manure from more than a dozen hogs raised expressly for that purpose. In 1980, the young trees fruited and were carefully transplanted again, over a wider area. Sales of tangerines brought new cash revenues to the brigade, but a decision was made to graft roughly half the trees into larger and more profitable oranges" (Ruf 1998, 132). By 1982, the brigade had conducted its first substantial harvest, reinvesting the profits in more orange trees, and by the middle of the decade they were shipping fruit all over the country, eventually enlarging the orchard to eighty-five hundred trees.

In Wugong, southern Hebei, a brigade also decided to plant a fruit orchard and contracted it out to the highest bidder. As a bonus, the successful bidder could also grow medicinal herbs, sunflowers, beans, peanuts, yams, melons, and a variety of vegetables; "Beijing residents would no longer be limited to turnips and cabbages in the winter" (E. Friedman, Pickowicz, and Selden 2005, 252).

In Yanyuan County in the mountains of Zomia, newly privatized farmers in many townships were encouraged to plant apples, after a study determined that the climate and soils in the region were ideal. Extension agents provided rudimentary instructions on how to plant, and in the late 1980s farmers of all local ethnicities started to go into the apple business. By 1993, in the geographically remote township of Baiwu, a survey showed that 57 of 119 households (divided into two production teams) had planted apples, and others were considering it. By the mid-1990s, many former subsistence farmers were making money from apples (Harrell 2001, 132–33).[2]

The apple boom, however, continued only in certain parts of the county. In the township of Weicheng, close to the county seat and to the main highway to larger markets, apple growing continued to prosper into the 2020s. In poorer and more distant Baiwu, by contrast, the fruits were small, the farmers had little market information, and transport costs on primitive roads were prohibitive in terms of both time and money (J. Ho 2004). By 1999, truckers were no longer coming to buy apples, and by 2001 farmers had cut down all but a few trees and had begun to produce hybrid corn to sell to commercial pig feed producers (Warren 2005).

The boom and bust in apple growing in Yanyuan illustrates the perils of the market transition during the Reform era. In 1993 I asked local cadres and entrepreneurs in Yanyuan about the potential danger of a glutted market. They laughed; China had more than a billion people, and they had been deprived of diet variety for decades. By the late 1990s, however, market forces really were taking over, and areas without economic comparative advantage were soon priced out of the market for various crops.

Not just trees and vegetables but also animal agriculture and aquaculture boomed under the reforms, as shown in table 6.2. Pork, the most commonly consumed meat in most Chinese diets, was already widespread before the reforms, partly because pig manure was the most valuable and widely used fertilizer. During the collective period, many brigades and teams had piggeries, but by the middle 1980s almost all pigs were in household pens, as they had been for millennia before collectivization (only in the 1990s did pig-raising CAFOs [concentrated animal feeding operations] begin to proliferate [Schneider 2011, 6]). But only about a third of the meat from the 319 million pigs that Chinese farmers raised in 1979—10 million tons of pork—was sold on the market. The rest either remained alive in the pen or were slaughtered as part of the traditional New Year festivities and either consumed on the spot or preserved for later consumption (Schneider 2011, 6). By 1998, however,

Table 6.2. Animal agriculture and animal agricultural products, 1979–2008

YEAR	1979	1988	1998	2008
LIVE PIGS (million)	319	342	422	462
PIGS SOLD (million)	187	275	502	610
PORK (million tons)	10	22	46	56
LIVE CATTLE (million)[a]	71	98	124	105
CATTLE SOLD (million)[a]	3.0	8.6	36	44
BEEF (million tons)[a]	.23	.96	4.8	6.1
LIVE SHEEP/GOATS (million)[b]	183	201	269	280
SHEEP/GOATS SOLD (million)[b]	35	68	173	261
MUTTON (million tons)[b]	.38	.80	2.3	3.8
POULTRY (million tons)		2.7	10.6	15.3
EGGS (million tons)		7	20	27
DAIRY PRODUCTS (million tons)	1.3	4.2	7.5	38

[a] Cattle (niu) and beef (niurou) include water buffalo and yaks.

[b] Sheep and goats are subsumed under one category (yang) in many Chinese statistics.

Source: Nongye Bu 2009, various tables.

even though the number of pigs had increased by only a quarter, the amount of pork sold on the market had more than quadrupled.

Despite the increase in pork consumption, pork lost ground as a percentage of meat consumed in China. While beef and mutton together accounted for only 5 percent of the total red meat on the market in 1979, by 1998 they accounted for almost 8 percent, even as per capita pork consumption (minus pigs slaughtered for household consumption) had increased from about ten kilograms to around thirty-four kilos. Similarly, poultry, eggs, and dairy products accounted for larger and larger proportions of dietary animal protein.

As on land, so in the water, both salt and fresh. From 1978 to 1998, output of freshwater aquaculture increased from 800,000 to more than 13 million tons of fish, and marine aquaculture from less than half a million tons to more than 8 million tons (Nongye Bu 2009, 39–40). In 1978 China produced 450 tons of shrimp; by 1991, that figure had exploded to 200,000 tons. Most of that was for export, and almost all of it was produced by "specialized households" (*zhuanye hu*) like those living in the municipality of Zhanjiang on the Leizhou Peninsula in southwestern Guangdong. These shrimp farmers took out contracts to raise shrimp in coastal ponds converted from rice fields or from mangrove swamps (Y. Huang 2012, 48). In addition, China's fivefold increase in capture fisheries (see table 6.1) accounted for about 60 percent of the worldwide increase during that time (FAO 2010, 6).

In short, by 1998 there was no question that China could feed itself, with minor supplements of imported foods. This was a tremendous achievement in a land where food shortages and famines had been part of life for hundreds of years and where the greatest famine the world had ever seen was less than four decades in the past. There were costs, however, in pollution and in ecosystem resilience and sustainability.

THE ECOSYSTEM COSTS OF SOLVING THE WARM AND FULL PROBLEM

China by 1998 was finally able feed itself, at least in the short term, because it intensified production in agricultural fields, animal pastures and enclosures, aquaculture ponds, and capture fisheries. Pushing food production to sufficient levels to solve the "warm and full" problem also took production to levels where it varied inversely with resilience (see figure 1.5). Neglecting ecosystem consequences, this intensification led to a series of ecological disasters.

Neglect of Infrastructure and Elimination of Buffers

One positive aspect of "normal socialist agriculture" was that teams and brigades took responsibility for maintaining infrastructure, whether irrigation ditches, roads, or even village lanes. As individual households' income became detached from collective output, there remained little incentive in many places to take care of the infrastructure. Mobo Gao (1999, 173–74) describes the sorry condition of his home village when he returned in 1992, a decade after decollectivization: "There used to be two paths running vertically and horizontally across the middle of Gao Village, like a cross. In 1992 when I visited the village the two paths had gone and there had been a complete transformation of Gao Village's physical appearance. There was no proper path any more in the village. Dirty ditches and smelly pigsties were everywhere, blocking every way possible. There is no village centre, nor is there a playground for children. In short, there is no longer any focal point for public life in the village."

More generally, there were problems maintaining many of the irrigation works built during the collective era, and they gradually fell into decay when, "with the implementation of the responsibility system, everyone farmed their own land, everyone took care of their own paddies" (Liu C. and Jing 2016), and many people neglected repair and maintenance. Over the decades this resulted in more and more leaky dams, rotting pipes, and eroding irrigation works (Liu C. and Jing 2016; Z. Xu 2017, 292).

Capturing More Surface Water in Rigidity Traps

To feed itself, China captured and harnessed more waters, both surface and subsurface. In addition to stabilizing and increasing agricultural yields, water capture also served expanding demands of industries and urban populations, and it generated hydroelectric power. These demands on water had two kinds of ecosystem costs. First, with demand exceeding supply, particularly in the north, supplies began to run out—rivers ran dry and water tables sank. Second, almost all the workings of the newly prosperous economy were dependent on using more water than was sustainably available. This created rigidity traps—even though everyone knew levels of water use were unsustainable, there was no easy way to avoid those levels of use without giving back the gains that water capture had partly facilitated.

In 1997 a film director wanted to shoot a historical scene taking place in the Central Asian desert. He was short of funds, however, so he found a place closer to home: the desiccated bed of the Yellow River near the Shandong provincial capital of Jinan (Shang 2009). The year 1997 was when the Yellow River did not reach the sea for 226 consecutive days, the culmination of a trend of flow stoppages that began with 100 days without functional flow in 1965. The river dried up completely for the first time in 1972, but the trend accelerated alarmingly in the 1990s, when even upstream flows at Sanmenxia were insufficient to supply ecosystem services for at least 70 days every year and the river did not reach its mouth at all for an average of about 180 days per year (Borthwick 2005). Not only did the length of dry-ups increase steadily from the 1970s to the 1990s, but the dry-up point moved upstream. It was 310 kilometers from the sea in 1972, but in the driest year of 1997 it was 704 kilometers from the river mouth, near the city of Kaifeng (Pietz 2015, 276). Every year in the 1990s, the famed springs of Jinan dried up as well, since they are naturally fed by the river (Shang 2009).

Although industrial and residential water use more than doubled from the late 1980s to the early 2000 aughts, agriculture accounted for over 92 percent of the withdrawals between 1988 and 1992 and over 84 percent from 2002 through 2004, taking about thirty cubic kilometers of water per year, or about 60 percent of the river's "natural" runoff in its lower reaches (Pietz and Giordano 2009, 111). Between 1949 and 1997, the amount of irrigated land in the Yellow River basin (including the watersheds of the Huai and Hai systems) increased from about 480,000 to 3 million hectares, fed by a system of 3,147 dams and reservoirs and 4,500 diversion works, together holding about 29.5 cubic kilometers of water, and 29,200 river pumps, which could extract a further 4.8 cubic kilometers (Pietz 2015, 275). Because the Yellow River has no tributaries in its lower reaches, the mainstream water supply is not replenished when water is extracted for irrigation, and evaporation soon takes care of the remainder.

Although the Yellow River is the most iconic instance, dry-up plagued the entire river system of North China in the 1980s and 1990s. For example, the Zhang River, where the famous Red Flag Canal begins, frequently ran dry downstream of the canal diversion, and farmers, deprived of water, attempted to blow up the canal several times. In the drought year of 1992, two villages fought a war over the water, using both mortars and bombs. Villagers also attempted to blow up the Yuefeng Canal at least thirty-seven times (Shandong Shuili 1993, cited in Pietz 2015, 258–59). The mainstream and several tributaries of the Huai River to the south and four major tributaries of the Hai River to the north also dried up several times in the 1980s and 1990s (Pietz 2015, 270, 274).

Consequences of excessive surface water capture reached far beyond peasant artillery duels. Importantly, less water in the river meant less force to transport the heavy sediment load that makes the river appear "yellow" (actually brown). Of the estimated one billion tons of sediment entering the lower reaches of the river in the 1990s, about four hundred million tons were neither flushed to the sea nor impounded by reservoirs upstream, settling instead in the lower riverbed, raising its level and paradoxically causing floods in the wet seasons even though the river was not flowing at all in the winter and spring (Pietz and Giordano 2009, 114). In addition, the lack of pressure from the nonexistent river flow led to seawater incursion in coastal areas (Pietz and Giordano 2009, 113), not only rendering agriculture impossible there but also depriving of water the city of Dongying, near the river mouth, and the nearby Gudong oilfield when the river was not running. The oilfield had to halt operations, and armed police had to guard the city's reservoir to prevent water theft (Shang 2009, 4–5).

Because industrial discharge increased with the economic growth of the 1980s and 1990s at the same time flow in North China's rivers was being reduced by withdrawals, water pollution got worse. Between 1980 and 1995, the proportion of river water in the Huai River basin designated as class IV ("mainly applicable to . . . industrial use and entertainment which is not directly touched by human bodies" [MEE n.d. a]) or worse increased from 45 to 88 percent. The Huai, after a series of increasingly severe pollution disasters beginning in the 1970s, famously turned black in 1994. Despite intensive and temporarily successful efforts to clean it up, it experienced another pollution crisis in 2001 (Economy [2004] 2010, 1–6). No better example could exist of a social-ecological system trapped in its own rigidity: North China's agriculture and industry both depended on diverting, drying, and polluting the very rivers that gave it life.

Continuing Drawdown of the North China Plain Aquifer

Even with all its negative consequences, surface water capture was not enough to irrigate North China's still-expanding agriculture in the late twentieth century. Farmers continued to rely on water from tubewells (another rigidity trap), withdrawing about twenty-seven cubic kilometers per year from the Hai River basin (the northern part of the North China Plain) alone. This exceeded the natural recharge rate by at least half (S. Foster et al. 2004, 82), so the water table of the shallow aquifers fell about one meter per year (S. Foster et al. 2004, 86), and the potentiometric level of the deep, confined aquifers continued to fall as well (Chen Su et al. 2018). In particular, the cones of subsidence around the cities of Shijiazhuang on the southwestern part of the plain and Cangzhou to the east both deepened between 1980 and 2000. Shijiazhuang's water table was about fifty-five meters below the surface in 2005, and Cangzhou's was over one hundred meters down (Chen Su et al. 2018, 1404). In total there were about 240 cones of depression on the North China Plain in the mid-2010s, occupying an area of seventy thousand square kilometers, about the size of Ireland (Yanxin Wang, Zheng, and Ma 2018, 1301). The water table did not just fall; it partially collapsed the land along with it. In the central part of the plain, as well as in the coastal megacity of Tianjin, the land subsided more than two meters between 1970 and 2000, contributing to the instability of buildings (S. Foster et al. 2004, 87; Yanxin Wang, Zheng, and Ma 2018, 1314). In addition, particularly in the coastal regions, the aquifer, like the surface waters, became more saline, and more inland regions of the aquifer were polluted by untreated municipal and industrial discharge (Yanxin Wang, Zheng, and Ma 2018, 1310–12.).

Arid regions in the northwestern parts of China Proper, around Yinchuan on the Yellow River before it enters the loess plateau, also suffered water shortages and pollution, as did both the Tarim and Junggar Basins in Xinjiang, where oasis agriculture had expanded under the Bingtuan (Xinjiang Production and Construction Corps). Salinization became a problem as upstream impoundment of the surface waters prevented recharge of the aquifers and depressed the water table (Yanxin Wang, Zheng, and Ma 2018, 1314). Similarly, land subsidence due to excessive groundwater withdrawals affected over fifty cities in China, some of them in areas as distant from the North China Plain as Shanghai, Wuxi, and Suzhou in the Chang River delta (Yanxin Wang, Zheng, and Ma 2018, 1314; Dinghuan Hu et al. 2004).

It is difficult to imagine how China would have solved its "warm and full" problem without expanding both surface and groundwater withdrawals for irrigation or without constructing flood-control infrastructure. Getting water to the fields in the unreliable climates of northwestern and North China in particular, as well as lessening the occurrence of floods in those same areas, meant that harvests not only stabilized but also expanded. Irrigation also allowed farmers to plant Green Revolution varieties that have higher-yield potentials, provided they get enough water and fertilizer. But China has fed itself at a cost—ecological, social, epidemiological, and even geological. And because those costs often take the form of rigidity traps, repairing the damage has been and will continue to be difficult.

Intensification and Vulnerability to Disease: Shrimp

Shrimp, perhaps more than any other food commodity China produces, shows how intensification leads to vulnerability. Although shrimp had been farmed in China for hundreds of years as part of low-intensity polycultures, three economic factors converged to make shrimp farming extraordinarily attractive to rice farmers in coastal areas beginning in the mid-1980s. First, food markets opened up in the early 1980s, and shrimp was one of the foods that contributed to the blossoming of diet diversity after the austerities of the collective era. Second, it became difficult to make any money from rice farming, since the regime was then paying low, set prices for the crop. Third, China was opening up to international trade, and there was a global shortage of shrimp, especially after a virus wiped out most of Taiwan's shrimp farms in 1987.

As a result, the regime instituted several economic incentives, such as tax breaks on imported fertilizer and equipment, to promote shrimp aquaculture (Feigon 2000).

Large numbers of rice farmers in coastal areas of Guangdong, Fujian, and other provinces converted their fields to shrimp ponds, increasing harvests dramatically and making big profits (Y. Huang 2012, 52). But they made two mistakes. First, they paid little attention to preventing diseases, particularly viruses, from infecting the shrimp. They often built inlet and outlet channels next to each other or used a single channel to both take in fresh water and flush out waste. They also declined to use certified virus-free larval stock or disease-free feed.

Second, and more important, in pursuit of higher yields and greater profits, they overproduced, rendering their enterprises even more vulnerable to disease outbreaks. They switched from local varieties to the high-productivity black tiger species and from local, seasonal feeds such as clams to scientifically formulated feed pellets. These changes enabled them to intensify production from around 100 kilograms per *mu* to as much as 750 by the early 2000 aughts (Y. Huang 2012, 66). These crowded conditions meant that the shrimp were more susceptible to diseases (Y. Huang 2012, 329).

The result was the so-called Great Shrimp Disaster of 1993, when the white spot virus wiped out almost the entire black tiger–based shrimp aquaculture industry. Cultured shrimp production fell from 210,000 tons in 1991 to 50,000 tons in 1993 and remained at that level for the next several years (Y. Huang 2012, 330). Thus, a combination of reduced diversity (shrimp as monoculture replacing shrimp as part of polyculture), overproduction (boosting yields in response to market demands), and top-down implementation of received science reduced the resilience of the shrimp ponds to almost zero, and the crop was wiped out.

The crop recovered in the early 2000s, with a second species switch, from black tiger to Pacific white shrimp, but scientists and producers did not learn the lesson of overproduction. As Yu Huang (2012, 71) puts it, "Many scientists and extension officials embraced a developmentalist mindset that sought to maintain high yields unencumbered by environmental constraints. In the aftermath of the disease outbreak, instead of calling for a change in aquaculture practice that scaled back its intensity and cut off the international trading of live species, the scientists and officials ironically encouraged both practices as they considered that only the virus, and not the culture environment, to have caused [white spot] disease."

As a consequence, even though scientists and farmers continued to attempt to exclude pathogens from their ponds, their operations had little resilience if pathogens did get in; disease outbreaks once again became prevalent starting in 2003 (Y. Huang 2012, 83). As long as productivity is the primary concern for aquaculturists and leads to overstocking, the industry will be on a continuous hamster wheel of pathogen and chemical; there is no permanent solution other than less intensive farming.

Agricultural Chemicals, Municipal and Animal Wastes, and Weird Lakes

When I rode a tour boat across Lake Tai from Wuxi in Jiangsu to Huzhou in Zhejiang in 1985, brown-sailed junks, only a few of them sporting motors, plied the blue waters in search of abundant fish. The water was clear, large plants dominated the aquatic flora, and there were many species of fishes and large invertebrates (Guan, An, and Gu, 2011, 50). Activist Wu Lihong remembered swimming in the lake as a boy in the 1960s (Savadove 2011). Within a few years, however, the lake was green in the summer, choked with blue-green algae, and many of the fish were gone. Eutrophication had set in.

Eutrophication in Lake Tai and elsewhere is mainly caused by over-nutrification, specifically with nitrogen and phosphorus, leading to a chain reaction that can happen quickly in a shallow lake like Lake Tai. Blue-green algae, which need phosphorus for growth, begin to bloom in surface waters, turning the formerly blue water into what is often described as "green paint." The algae take oxygen from the water, depriving fish and other animals—including the important zooplankton that both control algae and provide food for small fish—of the ability to breathe. They also release toxins harmful to human and animal health. They render the water turbid, making it difficult for fish and other animals to see and to capture food, which has become scarcer anyway. This disrupts the food web, and when the blue-green algae die, they release additional toxins and contribute to the further breakdown of the food web (Carpenter and Cottingham 2002).

China's chemical fertilizer usage, which had already increased greatly between 1960 and 1980, exploded again in the Reform period, increasing from 12.6 million tons in 1980 to 41 million in 1998 (Nongye Bu 2009, 8) and relying partly on imports (Yuxuan Li et al. 2013, 973). Much of this fertilizer was probably unnecessary (Z. Zhu and Chen 2002, 123); even nutrient-greedy Green Revolution varieties of rice grown in the Lake Tai region can only take up a limited amount of nutrients. Estimates placed the uptake of nitrogen and phosphorus from fertilizers in various regions of China Proper at between 30 and 60 percent, depending on the area and the crops grown (Z. Zhu and Chen 2002, 118; Xuejun Liu 2013, 459–63). Some of the remainder stayed in the soil, but much of it leached into the ground and eventually into the waters, contributing to high nitrate content in groundwater, eutrophication of nearby lakes, and transport to river deltas, where it contributed to red tides (Z. Zhu and Chen 2002, 120–23). In addition, phosphorus was wasted, depleting the phosphate deposits from which it is mined (Xuewei Liu et al. 2020).

Overuse of agricultural chemicals was not the only culprit in water pollution and lake eutrophication. Other, nonagricultural sources of nitrogen and phosphorus included municipal waste—the Lake Tai region is heavily urbanized, and hundreds of thousands of tons of sewage from the cities of Suzhou, Wuxi, Huzhou, and others rimming the lake, none of it treated, drained into the lake in the 1990s (Guan, An, and Gu 2011, 48). The estimated 120,000 tons of phosphate laundry detergent consumed yearly in the Lake Tai basin contributed about 16 percent of the phosphates released into the lake, at least until phosphate detergents were banned in 1999 (Wu Xinhua 1998, 110).

Eutrophication began in Lake Tai in the mid- to late 1980s; by the late 1990s more than half the area of the lake was eutrophic (Wu Xinhua 1998, 110), and in the 1990s and again in 2007 algal blooms blocked the intake of the municipal water supply of the large city of Wuxi for ten days. Beginning in the 1990s, local and national governments took aggressive measures to counteract nutrient flow into the lake but to little avail (Guan, An, and Gu 2011, 50), since eutrophication exhibits hysteresis—it is much more difficult to reverse after it occurs (Ludwig, Walker, and Holling 2002, 32–33; Carpenter and Cottingham 2002).

Lake Tai was of course not the only freshwater lake in China affected by overnutrification and consequent eutrophication. Chao Lake in Anhui, a similarly sized, shallow lake, began to turn green at around the same time and experienced disruptions to its fishing industry and disturbances in the municipal water supply of Hefei, Anhui's capital (Dong R., Zhao, and Yun 2002; Lai, Hu, and Fang 2016; Yanping Wang et al. 2020). Scholars have attributed Chao Lake's eutrophication almost entirely to agricultural and municipal pollution, since industry in its region is not as developed as in the area around Lake Tai (Wei Y. et al. 2010, 102). Beginning around the same time, other lakes, including Dianchi in Yunnan, Hongze Lake in Jiangsu, Wuhan's municipal East Lake, Poyang Lake in Jiangxi, and Baiyang Dian in Hebei, also developed the complex of algal blooms, plant and animal die-off, and turbid, toxic waters, largely as a result of overuse of fertilizers and other agricultural chemicals (Wu Xinhua 1998, 110). Once again, overuse of a good thing was the source of the problem. The Green Revolution had enabled China to feed its people securely between 1980 and 1998. Unfortunately, it also fed the blue-green algae.

....

The early Reform period from 1979 through about 1998 was a paradoxical time for China's food and water systems. Mass starvation no longer visited the "land of famine," and the carbohydrate-rich, protein-and-variety-poor "peasant diet" was rapidly being replaced by more nutritious and varied fare. China accomplished this admirable change with short-sighted measures, however, leading to ever-tightening rigidity traps and ever more apparent pollution of its rivers and lakes, along with chronic shortages of water for multiple and increasing needs. Environmental Cassandras had been sounding the alarm for years, but at the end of the century their cries finally entered the mainstream, and China embarked on its current course toward eco-developmentalism, marking another turn in the Chinese people's relationship to their water and their food.

7. Every Last Drop, 1998–2022

There is a lot of water in the South and not much water in the North. If it's possible it would be OK to borrow a bit of water.

MAO ZEDONG, 1952

Heavy rains in summer are the norm in the middle Chang River valley. In 1998 the "plum rains" of late June started on time, lasting as usual about three weeks, but were quite heavy in places, causing some local floods near Poyang Lake in Jiangxi and around the Hunanese capital of Changsha. The weather there then dried up for a while, but unusually heavy rains began in the *upper* Chang basin in early July, lasting through the summer. Exacerbated by snowmelt from glaciers on the Tibetan Plateau, eight flood waves swept downriver into the middle Chang lowlands between the beginning of July and the end of September. Meanwhile, short but torrential rainstorms returned to the middle Chang region in late July, adding to the problems brought by the flood waves. With each wave, water levels in most of the middle basin rose, even though there was very little rain in the middle basin itself throughout August (Zong and Chen 2000, 173–74).

The resulting floods devastated much of Hunan, Hubei, and Jiangxi. A United Nations relief mission visiting in mid-September reported that "conditions of the flood victims remain precarious. . . . Along the Yangtze river basin, 2.9 million people, who sought refuge on dykes up to two months ago, survive in overcrowded, insufficient shelters under deteriorating sanitary and health conditions. These people are now [September 1998] homeless and have lost all their sources of income. The water surrounding them on all sides is not expected to recede for another two–three months. As autumn and winter approach, the flood victims on the dykes are totally exposed to the wind, rain and the cold weather" (UNDAC 1998, para. 14).

According to official statistics, the floods directly or indirectly affected 223 million people and destroyed 4.9 million homes, although remarkably only 1,320 lives were lost, in contrast to the death tolls from big floods in 1954, when the government deliberately flooded rural areas to save the cities and about 150,000 people died (Courtney 2018b, 3). Economic losses were greater: 21 million hectares of farmland were flooded (about the size of Kansas), and about 15 million farm families lost their crops (UNDAC 1998, para. 2). Direct economic losses were estimated at ¥166 billion, or about US$20 billion at the 1998 exchange rate (Zong and Chen 2000, 166).

There were at least ten major breaches of dikes designed to protect local communities from flooding. Beginning on 26 June, 125,000 residents of De'an County in Jiangxi were isolated by floodwater, and the next day there was a breach fifty meters wide in the dikes of Changsha, finally plugged after a two-day struggle. Breaches continued in all three provinces until at least 8 August, causing flooding and evacuations.

From the relief efforts of the People's Liberation Army and other armed forces emerged the tales of heroism that, like cleanup, rebuilding, and blame, inevitably follow fights against natural disasters:

> Warrior Luo Weifeng had gone into the waters seven times and rescued seven people. When he jumped into the floodwaters for the eighth time, an old man grabbed on to him for dear life and wouldn't let go. In the midst of danger, [Luo] grabbed onto a big tree, using all his strength to hold onto the trunk with both legs, and got the old man to sit on his shoulders. For a whole night, if the water rose an inch, he just held the old man an inch higher, using his own shoulders to support the old man, persisting until dawn. (Xunyang 2017)

When General Dong Wanrui, who had been commanding the relief efforts around Jiujiang, stood on the train platform to say farewell to the heroic, mud-slathered soldiers who had put their bodies on the line to repair dikes and rescue people washed away by the flood, the iron-faced old soldier shed tears (Xunyang 2017).

The weather did not spare other parts of China—there were floods in Guangdong in June, and the plains of the Sungari (Songhua) and Nen Rivers in Heilongjiang and Jilin experienced the worst floods in a century, with eight thousand square kilometers of farmland flooded (UNDAC 1998, para. 70–72). More than two thousand oil wells were flooded around the oil industry city of Daqing, where water levels rose more than two meters above flood stage (Bao G. 2002).

There was extensive damage to agriculture in both the middle Chang River region and the northeast—farmers lost standing crops to both floods and subsequent waterlogging, and livestock died in large numbers (UNDAC 1998, para. 44–46, 72). But despite huge local agricultural losses, relief grain from local storage and from less affected areas prevented immediate starvation. More important, in the longer term China's food system had become robust and resilient enough that there was no reported starvation or famine in the flooded areas or anywhere else as a result. China could feed itself now, even with disastrous floods. Thirty or even twenty years earlier, there might well have been a famine.

Even though water was not causing starvation, it was still causing suffering, and the enormous amounts of suffering set in motion a process of profound self-examination. Why were the floods so severe? Yes, it rained a lot in the middle Chang River basin in June, and more fell in July. Yes, it rained continuously in the upper Chang basin all summer, bringing flood surges down the river. But the meteorological and hydrological records showed some puzzling numbers. At Yichang, where the river emerges from the Three Gorges (not yet dammed in 1998), carrying all the runoff from the upper basin, the water flow was between 51,000 and 63,000 cubic meters per second in each of the eight flood surges, lower than the 66,800 in the last great floods, in 1954, and much lower than the historical maximum of 71,000 in 1895. Downstream at Wuhan, however, flow was more than 68,000 cubic meters per second, much higher than the great flood year of 1931 (Z. Ye 2015; Courtney 2018b) and second only to 1954. In addition, the river levels at several downstream stations were all the second highest on record (Zong and Chen 2000, 174). In other words, the rainfall in the area was not unusual for the 1990s, and the amount of water arriving in the middle Chang basin from upstream was large but hardly unprecedented. But more water flowed through the basin than almost ever before, bringing higher water levels, frequent dike breaches, severe flooding, and mass suffering. What had happened?

INTENSIFICATION, LOSS OF RESILIENCE, AND THE 1998 FLOODS

After the floods, the Chinese media blamed deforestation in the upper basin of the Chang River (in Sichuan and Yunnan in particular), which caused increased sediment transport into its middle reaches, reducing the storage capacity there (Zong and Chen 2000, 173–74). At the Founding, Sichuan and Yunnan were two of the most heavily forested provinces in the country, with between 20 and 34 percent forest cover (Ling 1983; Fanneng He et al. 2008, 65). However, the early years of the People's Republic brought considerable deforestation to the area, particularly in the steep mountain areas, and forest coverage in the two provinces probably reached its minimum around the early 1960s (Fanneng He et al. 2008, 68; Fanneng He, Li, and Zhang 2015, 78). Subsequently, however, conservation and reforestation programs significantly *increased* forest cover in both provinces between 1960 and 1980, and especially between 1980 and 2000 (Fanneng He, Li, and Zhang 2015, 78). In addition, a detailed study detected no overall increase in the sediment yield of several large rivers in southwestern China between 1953 and 1987, despite extensive early deforestation. Although land clearing for agriculture brought about considerable erosion, most of the sediment released did not reach the major rivers (Schmidt et al. 2011).[1] Despite equivocal evidence, critics then and later blamed upriver deforestation for the downriver floods (see, e.g., Tan Shukui 1999, 495; Wu Fengsheng 1999; F. Yu et al. 2009, 210), and a few months after the floods the regime banned all commercial logging in Sichuan, Guizhou, Yunnan, and Tibet (Brandt et al. 2012, 359).

Whether upstream deforestation contributed to the 1998 floods or not, land-use intensification in the middle Chang basin itself—the very intensification that had contributed to the solution of the "warm and full" problem—certainly played a part in reducing the resilience of the middle Chang ecosystem to natural events such as 1998's heavy local rains and upstream flood waves that sent water pouring into the region.

The basins of attraction, both metaphorical and literal, all got shallower when land use intensified around middle Chang River region lakes, increasing sediment deposition in the lake basins and reducing the lakes' previous capacity to accommodate water overflows (Fanneng He, Li, and Zheng 2015, 78; Zong and Chen 2000, 176; F. Yu et al. 2009, 215). Perhaps more important, between the Founding and 1980, as the combined population of Hubei, Hunan, and Jiangxi more than doubled, from 77 million to 169 million, diking, farming, and urbanization had reduced the combined surface area of the six largest lakes in the middle Chang region by 40 percent, from 13,332 to 7,980 square kilometers (table 7.1). Hubei officials blocked the inlets of Xiliang and Diaocha Lakes, reducing their surface areas by 80 and 94 percent, respectively. Of an estimated total of 1,066 lakes on the Jianghan Plain around the city of Wuhan, only 180 remained (Zong and Chen 2000, 177; He N. 1998, 57–58). Dongting Lake, once the largest freshwater lake in all of China, had by 1998 shrunk to only slightly more than a third of its 1650 size (map 7.1), much of that shrinkage occurring after the Founding, as the population of Hunan grew from 32 million in 1953 to 54 million in 1982 and 65 million in 2010 (see table 7.1; map 7.1) (see F. Yu et al. 2009, 215; Banister 1987, 304; SSB n.d., table 1.1).

Storage capacity of the rivers, like that of the lakes, had fallen victim to developmental pressures before the 1998 floods. Marshlands in the floodplains of Hubei and Hunan had almost all been converted to fields, eliminating the ecosystem services they had previously provided, and islands within the river channel, some of them populated by as many as one hundred thousand people, were diked off for cultivation, further diminishing the channel's water capacity (Zong and Chen 2000, 177; F. Yu et al. 2009, 215). Compounding all this, increases in sedimentation in the lakes and in the river channel raised the water level an estimated two

Table 7.1. Changes in surface area of major lakes in the middle Chang River basin, 1949–1980

LAKE	1949 AREA (KM²)	1980 AREA (KM²)	AREA LOST 1949–1980 (%)
Poyang	5,100	3,900	34
Dongting	4,300	2,700	38
Liangzi	1,500	680	55
Hong	720	400	45
Xiliang	1,100	210	81
Diaocha	720	75	90
Total	**13,440**	**7965**	**41**

Source: Data from Zong and Chen 2000, 177.

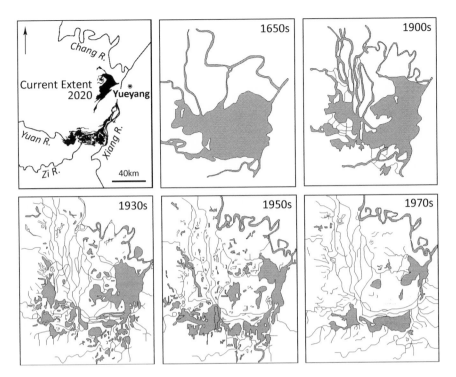

Map 7.1. The shrinking of Dongting Lake, 1650s to 1970s. Redrawn from "Analysis of Historical Floods on the Yangtze River, China: Characteristics and Explanations" by Fengling Yu, Zhongyuan Chen, Xianyou Ren, and Guifang Yang, *Geomorphology* 113, nos. 3–4: 7. © 2009, with permission from Elsevier.

meters—in the Jingjiang reach, between Shashi and Wuhan, the Chang River became, like the Yellow River, a "hanging river," with its water level inside the dikes higher than the surrounding plain. This created demands to raise the dikes and narrow the channel even more (He N. 1998, 57). Finally, additional reclamation of lands to accommodate the growth of cities exacerbated a trend, beginning as early as the late Qing, of less storage capacity and thus greater flood vulnerability within the cities (Z. Ye 2015).

In short, it did not take either locally unprecedented rainfall or record discharges of water from upstream to make the 1998 middle Chang River floods catastrophic. Development had diminished resilience to both heavy rains and heavy upstream discharges, rendering the floods of 1998 as severe as those of 1954 and 1931, even though the precipitating events were not as large. Similar intensification of land use, particularly expansion of cultivation, also combined with heavy rains to cause catastrophic floods in the northeast in the same summer.

THE 1998 FLOODS AND THE TURN
TO ECO-DEVELOPMENTALISM

The floods of 1998 were not the first large-scale disaster to hit the People's Republic, and they were far from the most destructive in terms of human life. Efforts to deal with previous floods, such as the 1954 inundations in the middle Chang River basin and the 1963 floods in Tianjin, had spotty success at best, but the regime always approached them in the spirit of conquering nature through infrastructure, mass mobilization, and calling on some people to sacrifice for the greater good (Paltemaa 2011). However, the 1998 floods struck a different chord in a China that was both hell-bent on rapid economic development and less in the grip of "man-against-nature" Marxism-Leninism. In October 1999, a group of Chinese literary scholars and social critics issued a report addressing a few "received truths" that they considered dangerous and misleading. They cast doubt on the idea that environmental damage was an inevitable by-product of increases in the standard of living brought about by development. They also questioned the idea that China's current environmental problems were simply the result of misguided developmental policies of the former high-socialist regime and its Stalinist and Maoist models of development: "The relationship between the environment and development is often constructed as an either/or dilemma: If we desire development, then the environment will have to be sacrificed. If we want to protect the environment, then development will have to be sacrificed and poverty will result. This developmentalist logic and discourse have been internalized as 'common sense' or 'common knowledge' by many, preventing any serious and meaningful discussion of the relationship between development and the environment" (Han S. et al. 2004, 239).

Floods and Forests

In response to the 1998 floods and other obvious results of environmental degradation, the state accelerated the hitherto slow change away from Deng Xiaoping's "development is the only solid truth" toward taking environmental degradation more seriously. Broad, overall changes took a while, but there were immediate effects of the (perhaps mistaken or at least oversimplified) perception that upstream deforestation was the primary cause of the 1998 Chang River floods. Not only was logging banned throughout the southwest, but the regime also began to address damage caused by the expansion of agriculture between the 1950s and the 1980s, implementing various forest and grassland restoration programs that aimed to restore the "natural" condition and ecosystem services of forests and grasslands.

The two biggest reforestation programs were officially announced in late 1998, piloted in Sichuan, Shaanxi, and Gansu the following year, and implemented widely in 2000. One of these, the Natural Forest Protection Program (Tianran Lin Baohu Gongcheng), called for drastic reductions in timber harvests, conservation in more than ninety million hectares of forestlands, reforestation of thirty-one million hectares, logging bans in most of Sichuan, Yunnan, Guizhou, and Tibet, and alternative employment opportunities, including reforestation, for former logging workers (State Forestry Administration n.d. a, b; Shixiong Cao et al. 2010; G. Miao and West 2004).

The other major reforestation program, called Returning Farmland to Forest (Tui Geng Huan Lin), had a fourfold logic.[2] First, campaigns to increase grain production by expanding cultivated area during the era of collective agriculture had not been very productive and were no longer needed, since the "warm and full" problem was basically solved. Second, much of the land newly brought under cultivation in those times was marginal for agriculture, since it was often steeply sloped or otherwise unproductive but had previously provided ecosystem services such as erosion control and flood control. Those ecosystem services needed to be restored, as the floods had demonstrated. Third, national planners wanted to increase forest cover at the macroscale. And fourth, despite their low productivity, these lands had provided some income for farmers and herders, who would have to be compensated, at least temporarily, for the loss of income. Thus, the state would pay farmers to plant trees on their marginal farmland and also pay them a yearly subsidy to compensate for lost income, in effect a form of "payment for ecosystem services" (State Council 2002).

The program had varied success. In some places it resulted in both

reforestation and stable or increasing rural incomes; in others, plantations failed because of improper species or site selections or because
local people did not receive the promised income and thus ignored the
requirements for successful reforestation (Zinda et al. 2017). For example,
three different sites in mountainous western Sichuan showed three different kinds of results. In Baiwu Township in Yanyuan County, attempts
to introduce exotic species failed, while replanting with native pines
succeeded, but local officials ignored the fact that some trees did not
grow and instead reported uniformly high rates of successful planting
(Trac et al. 2007). In Sanjiang Township, Wenchuan County, planting
orchards (rather than what we would normally consider forests) on
previously farmed land brought income to local farmers, but reporting
them as forests (allowed under China's classification system) meant
that the ecosystem benefits were exaggerated in the reports (Trac et al.
2013). In Jiuzhaigou County, including the famous Jiuzhaigou National
Park, some areas were appropriately reforested, but official zeal for the
Returning Farmland program meant that some areas that had previously
been meadows before they were converted to farms were now monocultural forests, with deleterious effects on the renowned biodiversity
of the region (Trac et al. 2013; Urgenson et al. 2014; Harrell et al. 2016).

Overall, the Returning Farmland to Forest program has probably contributed to the increase in China's forest cover, which expanded from
20 percent in 2000 (Robbins and Harrell 2014, 4) to nearly 23 percent in
2020 (State Forestry Administration 2020, 2); rough estimates show that
the program reforested about twenty-four million hectares, a little more
than 2 percent of China's total area (Cheng Chen et al. 2015; Yuchen Gao
et al. 2020). Quality of forests is another question—many commercial
orchards were counted as forests as long as they met China's criterion
of 20 percent or more canopy cover, and most of the replacement forests
were monocultures. That the Returning Farmland to Forest program
fell short of its potential points to problems in China's environmental
governance. Officials are eager to pad their résumés with quick results
(such as counts of trees planted or hectares covered), and starting a program counts as an "administrative accomplishment" (*zhengji*) almost as
much as bringing it to a successful conclusion would. The mentality of
"cutting [everything] with one knife" (*yi dao qie*) often results in the implementation of a panacea solution that is inappropriate for a particular

site, whether it is planting the wrong species or planting trees in what should be a meadow. Still, in evaluating environmental management in China, we should not let the perfect be the enemy of the good.

Pastures and Grasslands

State concerns about overuse of productive lands were not limited to previously forested areas; they were also worried about overgrazing on grasslands, and so they followed the Returning Farmland to Forest program model in implementing the Returning Pasture to Grassland program (Tui Mu Huan Cao) in many pastoral areas, particularly in regions inhabited by Mongol, Tibetan, and Kazakh herders. The ecological results of this program were much more equivocal. Quantitative scientists and planners almost universally blamed local pastoralists for "unscientific" management of their lands, disregarding the importance of flexible management practices that local herders had employed before collectivization. Bureaucrats also adopted the neoliberal assumption that private land management was more likely to be effective than commons management. This and earlier grassland rehabilitation programs removed herders from their pastures, taking away traditional means of livelihood without any viable alternative (Ptáčková 2020, 71–74, 92–108; Salimjan 2021). Fragmented pastures reverted to shrubland or changed into to a patchwork of lush, fenced-off paddocks with no animals in them, surrounded by pastures even more heavily grazed and degraded than before (Williams 1996; Cerny 2008).

THE PERPETUAL PROBLEM OF WATER IN THE WRONG PLACE AT THE WRONG TIME

While the floods of 1998 demonstrated what happens when there is too much water in the wrong place, China also faced the problem of not enough water in the right place and chose to address that issue with developmentalist solutions. By 1998, the Yellow River had run dry for ten years in a row, including its longest-ever dry-up in 1997. In response, the state doubled down, using fixes to fix the previous fixes. Having intensified water use with dams, dikes, and diversions and having removed ecological buffers and only partly replaced them with infrastructural and institutional buffers, China reaped rewards in the form of more and better food, but it also reaped the whirlwind of Yellow River dry-ups, floods on the Chang River and throughout the northeast, and general water shortages. The government's response was not to remove infrastructure that had exacerbated both floods and water shortages but rather to build more infrastructure, hoping to both replace at least some of the lost resilience and satisfy the expanding economy's ever-growing demands not only for water but for energy to fuel the growing economy. The results were many more dams and some of the world's biggest water diversions.

The Next Steps with the Yellow River: Correcting Sanmenxia with Xiaolangdi

The Sanmenxia Dam (see map 3.3) had been a mistake. Even Zhang Guangdou, one of the prominent engineers involved in both Sanmenxia and the giant Sanxia project on the Chang River, admitted in a 2004 television interview that "Sanmenxia was a mistake, and I am not the only one to say that" (quoted in Pietz 2015, 302). Qi Pu, senior engineer of the Yellow River Water Resources Research Institute and a big advocate for water control projects, told a group of touring ecology students that "the idea to construct Sanmenxia was fundamentally flawed" (Wang Lina and Wang 2011). Rapid silt buildup behind the dam rendered the project almost useless for power generation and also caused salinization and flooding of about 3.3 million hectares in the Guanzhong Plain around Xi'an, more than 100 kilometers upstream on the Wei River. In that area, in Hua County, siltation had raised the riverbed high enough to bury an important bridge, so a new one had to be built. By 2003, the Wei, like the Yellow and certain stretches of the Chang, had become a "hanging river," with its bed higher than the surrounding plain (Tan Ye 2003). In 2005 autumn floods on the Wei displaced farmers for several months (Wang Lina and Wang 2011).

In addition, the Sanmenxia Dam, together with increased withdrawals for irrigation, altered the downstream regime of the Yellow River, particularly in its lower course as a hanging river flowing across the North China Plain. Even though a lot of sediment was trapped behind the dam, the river flow slowed so much that its ability to carry the sediment that

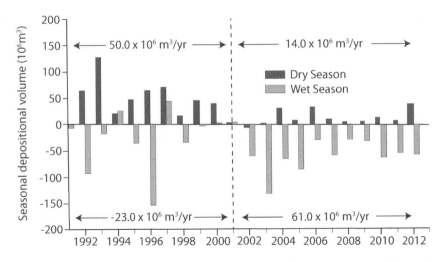

Figure 7.1. Changes in sediment regime after water sediment regulation began in 2002. Previously, more sediment was deposited downstream of Xiaolangdi than was eroded by streamflow. After 2002, erosion began to dominate. Redrawn from Naishuang Bi, Zhongqiang Sun, Houjie Wang, Xiao Wu, Yongyong Fan, Congliang Xu, and Zuosheng Yang, "Response of Channel Scouring and Deposition to the Regulation of Large Reservoirs: A Case Study of the Lower Reaches of the Yellow River (Huanghe)," *Journal of Hydrology* 568:976. © 2019, with permission from Elsevier.

did pass the dam was reduced, and it deposited more sediment downstream, especially when the river was drying up (Bi et al. 2019, 973). Typically, however, instead of removing the dam, authorities decided on a fix to fix the fix and built another major dam at Xiaolangdi, about 130 kilometers downstream from Sanmenxia (see map 3.3), and it became operational in 2000. Two years later, they began a process called water sediment regulation (WSR). For about two weeks every year in the flood season, usually sometime in June or July, sluice gates in the three dams at Wanjiazhai (700 kilometers upstream, on the loess plateau), Sanmenxia, and Xiaolangdi were opened, creating a cascade of artificial floods, first sluicing relatively clear water and later on releasing water with concentrated sediments (X. Li et al. 2017, 162). If there was heavy rainfall in the Wei River valley, the gates might be opened again to alleviate upstream floods (Wang Yongchen 2011). Despite their force, these artificial floods have not prevented buildup of sediment in the Xiaolangdi Reservoir itself, and so the water in the artificial floods carries less sediment than previously. This whole series of effects has changed the downstream behavior of the river once more, switching the predominant regime from deposition to erosion (figure 7.1) (Bi et al. 2019, 977).

Using water sediment regulation at Xiaolangdi to correct the mistakes at Sanmenxia and further regulate the river seems to have worked, as the downstream water and sediment regime has been brought to an equilibrium (Bi et al. 2019, 973). The state has also managed to impose regulations on water use, despite competing claims from various upriver and downriver provinces (Pietz 2015, 296), and the Yellow River did not dry up after 1999, though floods in the valley of the Wei River remained a problem. But there have been other consequences.

As with so many dams, Xiaolangdi inundated valuable farmland and displaced farmers from their homes. About twenty-eight thousand hectares of land were appropriated, and two hundred thousandvillagers formerly farming there were relocated, often to places where it was impossible to farm or where they were exposed to mudslides. Twelve hundred people from Liaowu Village were moved about 40 kilometers away, but they received only small amounts of land in compensation. Many of them became "peasant migrants" (nongmin gong) but achieved only spotty success finding factory jobs (Wang Yongchen 2011 Zhang Di 2021).

There were also more changes in the earth system. Nutrient balances downstream changed; there was a particularly notable increase in dissolved nitrogen (X. Li et al. 2017), and discharge to the ocean below Dongying on the Shandong coast increased dramatically during the WSR period each year, so fresh water flowed much farther offshore than before or after the yearly WSR (B. Xu et al. 2016, 105). The altered timing and volumes of both water and sediment discharge also caused morphological changes in the delta. Between 1976 and 1981, the delta had grown rapidly with the increased sediment load in the river, but after 1982 the growth slowed, partly due to decreased water flow and the concomitant decrease in sediment delivered to the sea (Ren M. 2015; Ren H. et al. 2015) and partly due to seawalls built to protect the Gudong oilfields near the shore. In 1996, when engineers altered the course of the last ten kilometers of the river to protect the oilfield, the newly formed delta lands began to recede again. When water sediment regulation started in 2002, parts of the delta began to grow once more because the floods of water and sediment released by WSR scoured the riverbed downstream of Xiaolangdi and delivered increased sediment to the river mouth (Houjie Wang et al. 2010; Xiao Wu et al. 2017), but the overall trend from 1985 through 2018 was loss of land (Zhengjia Liu, Xu, and Wang 2020, 6047; Ren H. et al. 2015).

Will this most recent generation of Yellow River fixing be more successful than its numerous predecessors? With repeated fixes to fix previous fixes, the system has become ever more dependent on infrastructural buffers, thus further hardening the rigidity traps built of concrete, steel, management schedules, and bureaucratic rivalries. The greatest threat to this delicate balance may be the fact that the artificial ecosystem services provided by dams, reservoirs, procedures like WSR, and even the plethora of multiauthor hydrology papers cited here do not have an exclusive claim on the still-scarce waters of the Mother River. In addition to addressing sediment buildup with WSR, any plan for the Yellow River still needs to address four more objectives (W. Jin et al. 2019). These include water supply (particularly for irrigation), hydropower generation (which grows with China's increasing material living standards and its desire to replace fossil fuels to combat global warming), flood control (the age-old problem), and ice control (the upper and middle reaches of the river in Central Asia often freeze up in the winter, causing ice jams that, when they break up, can cause flooding downstream). It is often impossible to satisfy all these demands at once: for example, WSR sends downstream large amounts of water that would otherwise be available for power generation and irrigation, particularly in dry years. One can create a lovely mathematical model (W. Jin et al. 2019) to determine how to optimize these potentially conflicting objectives, but which goals get favored depends less on hydrologists' models and more on the relative power of particular bureaucratic agencies and local governments. In a system where artificial buffers (infrastructural and institutional) have replaced ecological ones, ecological problems become human problems. So far, the trade-offs seem to be worth it, but the Ministry of Water Resources was concerned enough about overuse of both surface and groundwater in late 2020 to declare a moratorium on new water permits in many areas in the upper and middle Yellow River watershed (MWR 2021b). Meanwhile, severe floods in both the Yellow and Huai River drainages in summer 2021 showed that despite all the past decades' waterworks, sometimes there was too much water in the north (Guowuyuan 2022a).

Water Transfers: The Ambition of South-to-North and Repeated Fixes to Fix Fixes

Repeated fix-fixing fixes, however, did not address the fundamental reality that, as Chairman Mao pontificated in 1952, "There is a lot of water in the South and not much water in the North" (Nanfang shui duo, beifang shui shao) (quoted in Crow-Miller 2015, 180). Engineers had been working for millennia to solve the water problems of the north—both scarcity and flooding. By the end of the twentieth century, however, new stresses had appeared. Rapid urbanization in North China resulted in the use of more and more surface and groundwater for urban households and municipal activity (Crow-Miller 2015, 175), and continued irrigation was drawing down the North China Plain aquifer ever lower, as well as causing salinization in coastal areas, cones of subsidence, and other problems. Lakes and wetlands in much of North China were drying up because increasing amounts of water were being diverted to irrigation and urban use (Hongyan Liu et al. 2013; H. Xia et al. 2019). Even partially restoring dry-season flows to the Yellow River after the annual dry-ups in the 1990s meant that less water was available to meet these other increasing demands. At the same time, water was not priced to promote conservation. Both poor farmers and urban residents would complain about higher water prices, and local water bureaus, even though they were aware that users were wasting a lot of water, were reluctant to anger their constituents by raising water prices (Webber et al. 2017, 376).

In this situation, planners turned to the second half of Mao's comment about the north-south imbalance in water availability: "If it's possible it would be OK to borrow a bit of water" (Ru you keneng, jie dian shui lai ye shi keyide) (quoted in Crow-Miller 2015, 180). Water from the Chang River basin could be moved northward to ever-thirsty North China. Thus began plans for the world's largest water-diversion project, the so-called South-to-North Water Transfer Project (Nanshui Bei Diao).

Transferring large amounts of water was of course not a new idea, and the canals built for irrigation and urban supply on the North China Plain in the 1950s were early examples (Pietz 2015, 299). But moving water from the Chang River basin to the North China Plain is a bigger deal. When engineers began drawing up plans for such a transfer in the 1990s, there was considerable debate over whether the project was necessary.

Critics pointed out that with water being virtually free for farmers, as well as inexpensive for both urban residential and industrial users, there was a lot of water to be gained by conservation and more efficient use. Transferring water from the south, also at low cost to users (even though the project itself cost tens of billions of dollars), would create a moral hazard, reducing the incentive to conserve (Berkoff 2003). In addition, it was estimated that more than 8 percent of the water stored and diverted for one of the proposed routes would be lost to evaporation (Y. Ma et al. 2016), and there were concerns that roads, bridges, industry, and agriculture in the areas traversed might potentially pollute the water supply (C. Tang et al. 2014; M. Tang et al. 2018; Z. Miao et al. 2018). On the other side, proponents pointed out the extreme social and environmental opportunity costs of *not* building the project (Berkoff 2003; Chen Lin et al. 2012). Eventually, the "megaproject mentality" and the persistent bureaucratic incentives to prioritize building over managing prevailed (Webber, Crow-Miller, and Rogers 2017; G. Lin 2017), and the State Council approved the project in 2002 (Pietz 2015, 300), aiming to transfer a total of more than forty cubic kilometers per year along three "routes" (*xian*) (map 7.2).

The western route would be a kind of sci-fi project, constructing dams and reservoirs, drilling tunnels, and building aqueducts in rugged mountainous terrain. It would transfer water from several tributaries of the Tongtian River (the uppermost reaches of the Chang), the Yalong, and the Dadu to the mainstream of the Yalong, and from there through a 170-kilometer tunnel to the upper reaches of the Yellow River, before it enters its Great Bend through Central Asia. This route, if built, could carry up to 20 cubic kilometers per year (Berkoff 2003; D. H. Yan et al. 2012), but even with its megaproject mentality and its engineering expertise, the PRC regime remained hesitant, though advocates of this and even more ambitious schemes did not completely give up (Gao Baiyu 2020a), and in November 2019 Premier Li Keqiang advocated continuing feasibility studies and preliminary plans for the western route (Zhang Ruoting 2019).

As of mid-2021, any plans seemed far from approval or even detailed design (Wang Hanjuan 2021), but there were even wilder proposals, including, for example, one promoted by Wang Fusheng at the Gansu Academy of Social Sciences that would transport water from the headwa-

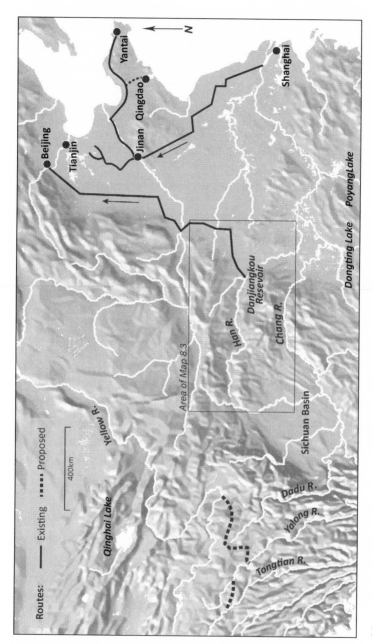

Map 7.2. South-to-North Water Transfer Project. Map by Lily Demet Crandall-Oral.

ters of the Nu River through a bewildering array of tunnels, aqueducts, and reservoirs to Xinjiang and the western parts of Inner Mongolia. Wang also proposed taking water from the Palung Tsangpo or Yarlung Tsangpo River in Tibet and diverting it through similarly convoluted routes to the arid regions of northwestern China and Xinjiang (Wang Fusheng 2020). Some actually existing projects also ran into difficulties. In the rush to bring more water to dry parts of northern Xinjiang, engineers trying to tunnel from the Irtysh River on the north slope of the Altay Mountains drilled unexpectedly into groundwater sources and ended up with water in their tunnel as they were still building it (Stephen Chen 2021).

In the middle and eastern routes of the South-to-North Water Transfer Project, by contrast, water was flowing by the mid-2010s. The middle route begins in the Danjiangkou Reservoir at the confluence of Dan and Han Rivers, straddling the border of Hubei and Henan (see map 7.2). The reservoir, originally built for water storage, was completed in 1974, but to facilitate the South-North Project its dam was raised by more than 14 meters, making the reservoir Asia's largest artificial freshwater lake (Baidu n.d. a). This allowed water to flow northward following the pull of gravity, all the way down to and across the North China Plain, under the Huai and Yellow Rivers, as well as several in the Hai River drainage area, to Beijing and Tianjin (Berkoff 2003, 5–6). Highways and bridges had to be rerouted or rebuilt to make way for the complex system of over 1,750 channels, tunnels, culverts, and aqueducts (C. Tang et al. 2014). Water began to flow in the middle route in late 2014, supplying between 8.5 and 9.5 cubic kilometers per year through 2021 (Guowuyuan 2022b), and there were plans to expand to about 13 cubic kilometers per year (Webber, Crow-Miller, and Rogers 2017, 371; Zhang Juan and Guan 2020).

The eastern route takes water from the mainstream of the Chang River below Nanjing and, following much of the route of the Grand Canal, delivers water to North China. Unlike the middle route, however, the eastern route cannot operate by gravity alone. Instead, water must be pumped up about sixty-five vertical meters to where the canal crosses the Yellow River, whence it can flow downhill. After it reaches the North China Plain, its northern or western branch supplies water to Tianjin and surrounding areas, while its eastern branch reaches the Shandong

Peninsula (Berkoff 2003, 5–6). Water started flowing along the eastern route in late 2013, and within a few years a total volume of fourteen cubic kilometers flowed along the eastern route each year, bringing the South-to-North Project's total annual volume to about twenty-four cubic kilometers (Webber, Crow-Miller, and Rogers 2017, 371).

Moving all this water from the Chang River watershed to North China not only eased pressure on municipal supplies in many northern cities but also partially slowed and in some places stopped the drawdown of the various North China aquifers (Gao Yuanyuan et al. 2018). But there were still major ecological trade-offs that probably decreased resilience and at minimum required fix-fixing fixes. Problems of the middle route included population relocation, pollution, and water shortages in the sending area, while the eastern route created problems of pollution, energy expenditure, and salinization. In addition, along both routes, failure to implement a rational pricing structure for water caused conservation to lag behind supply, so that despite all the effort and expense, there were reports that destination areas were not using all the water they received (Webber, Crow-Miller, and Rogers 2017, 377).

There were also social costs. Raising the dam and thus enlarging the reservoir at Danjiangkou required resettling an estimated one-third of a million people, more than for any water project besides the Sanxia Dam. Not only did the reservoir itself inundate valuable farmland, but in order to preserve water quality in the Han and Dan Rivers, additional farmers were moved out of designated water-source conservation areas, while farmers who were allowed to stay were encouraged to reduce pesticide and fertilizer use. Propaganda appealed to people in all the affected areas to sacrifice for the greater good (Webber, Crow-Miller, and Rogers 2017, 376).

Impounding water in the raised Danjiangkou Reservoir and sending it north reduced the volume in the Han River (into which the reservoir empties from the other end) by 25 to 30 percent, leading to shortages in the sending area and causing problems of supply, navigation, and pollution (Webber, Crow-Miller, and Rogers 2017, 377; Lian, Sun, and Ma 2016, 963). The source region of the middle route receives average annual rainfall of about eight hundred millimeters (China Maps n.d.), about half from June through August and only 20 percent in the crucial growing months of March through May; there is also considerable variation from year to year (Sui et al. 2012). The volume of water that Danjiangkou can store thus varies greatly, and seven times between 2002 and 2014 there was not enough to meet projected demands in either sending or receiving areas, particularly in May, when irrigation demands are high in both places and summer rainfall has not yet begun to replenish the reservoir (Hai Liu, Yin, and Feng 2018).

More Plumbing Fixes

Anticipating these problems, between 2010 and 2014 Hubei authorities built a fix-fixing fix, the Taking the [Chang] Jiang to Fill the Han (Yin Jiang ji Han) water diversion, which took about 3.7 cubic kilometers of water from the Chang River at Jingzhou between the Sanxia Dam and Wuhan, diverting it through a 67-kilometer open canal to the Han River at Gaoshibei Township, a little more than 100 kilometers upstream from Wuhan (map 7.3). This helped alleviate shortages of irrigation, municipal, and industrial water in the lowest section of the Han River (Baidu n.d. f). However, it did nothing to provide extra water for the reach between Danjiangkou and Gaoshibei.

These perceived shortages in the middle section would only get worse in the early 2020s, when the water began to flow in the Taking the Han to Fill the Wei (Yin Han ji Wei) project. This new effort was designed to divert about 0.9 cubic kilometers of water per year from the upper Han River at Huangjinxia and tunnel it for 16 kilometers to the gorge of the Ziwu River at Sanhekou. There, another 0.9 cubic kilometers per year would be added to the flow and channeled into a 75-kilometer tunnel as much as 2,100 meters deep under the main Qinling range to emerge in the Wei River valley, then be diverted into a series of channels serving agricultural, industrial, and residential needs in the cities of Baoji, Xianyang, Xi'an, and Weinan and surrounding farm areas. It was also designed to help alleviate low-water conditions in the Yellow River, of which the Wei is the largest tributary (see Baidu n.d. b; Liu Yanqin 2019; Hu Yue 2020). Since the diversion point for this project is upstream of Danjiangkou, however, shortages in that region may get more serious. Further local fixes, such as better monitoring of water levels or injecting summer surplus water into aquifers for later use can redress some of the imbalance brought about by this particular fix (Liu Yanqin 2019), but

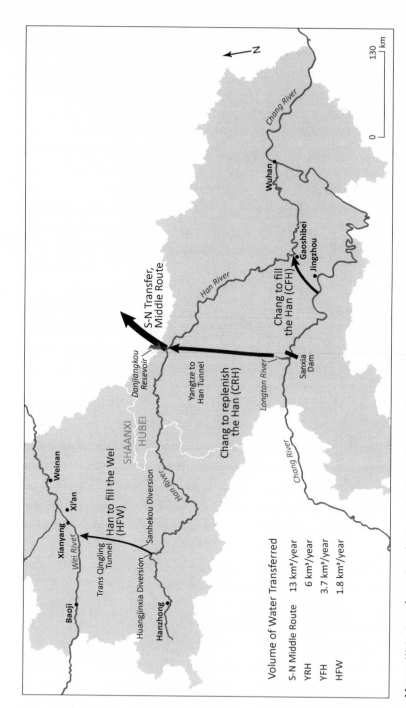

Map 73. Water transfer projects between the Chang, Han, and Wei River basins. Map by Lily Demet Crandall-Oral.

discussions in the early 2020s predictably centered around yet another diversion, a fix to fix the fix to fix the fix.

There is still lots of water in the Chang River mainstream, so a project for "taking the [Chang] Jiang to replenish the Han" (*yin Jiang bu Han*) began in July 2022 (*Hubei Daily* 2021; Baidu n.d. g). It would take about 3.8 cubic kilometers per year from the Longtan River, a left-bank tributary that flows into the Sanxia Reservoir 7 kilometers above the dam, through a 194-kilometer tunnel to the Han River immediately below the Danjiangkou Reservoir, a site chosen to avoid lowering the quality of water carried to North China users by the middle route (Danjiangkou 2021). This would alleviate some of the water shortages on the Han downstream of Danjiangkou.

Pollution was another persistent concern. Since both the middle and eastern routes (particularly the latter) of the South-to-North Transfer Project passed through agricultural and increasingly industrialized areas, including mining regions, planners early on began antipollution measures. These included restricting the use of agricultural and industrial chemicals in the watershed of the Danjiangkou Reservoir, mandating buffers of fifteen meters along open-air parts of the channels, and formulating contingency plans to prepare for serious accidents such as chemical spills (see Zhuang et al. 2019; C. Tang et al. 2014; M. Tang et al. 2018; L. Sun et al. 2016). Not only the aqueducts, however, but also the Han River downstream of Danjiangkou was threatened with higher rates of pollution, since its flow volume was reduced so substantially—at least until the Taking the Chang to Replenish the Han project was built (Webber, Crow-Miller, and Rogers 2017, 377). Also, because the Han contributed more than any other tributary to the flow of the lower Chang, the middle route deprived the lower Chang of some water, and diverting water to the eastern route subtracted even more. As a result, the South-North Project may have contributed to the decline in freshwater flow near the Chang River's mouth and thus to saltwater intrusions into the delta areas (Webber et al. 2015).

Finally, governance and the economics of water diversion and use affected ecological outcomes. Authorities from the State Council on down thus adopted a complex series of ad hoc governance structures in order to address two problems: pollution and water use inefficiency. A top-down regulatory scheme adopted in early stages of the project failed to prevent polluted water from entering and transiting the systems, and some areas along the route were using transferred water to dilute sewage (Z. Miao et al. 2018). In response, authorities replaced this regulatory structure with a system of payments for watershed services—paying local governments to finance antipollution measures that regulation had failed to implement. It is not clear how well these measures worked (Sheng and Webber 2019, 1250). At the same time, authorities had to deal with an equally vexing problem of pricing: originally the corporation set up to regulate and sell water was charging prices higher than local governments charged for local water, so some of the South-North Project water was going unused; the solution was either to raise the price of local water or mix local and South-North Project water and sell at a uniform price (Sheng and Webber 2019, 1653; Webber, Crow-Miller, and Rogers 2017, 377).

Although it stands out for its size, scope, and interregional nature, the South-to-North Water Transfer Project and its offspring are not alone. Since 2000, many other transfers have been completed or proposed. In 2014, 172 large projects were put on the agenda, and 142 of those were under construction by 2019 (Zhang W., Hong, and Rogers 2022). The State Council announced a further 150 large water projects in fall 2020, and in order to unify plans, the Ministry of Water Resources issued a directive in January 2022 laying out basic plans for waterworks during the fourteenth five-year plan, extending through 2025 (Xinhua 2022). To construct what was now dubbed the "national water network" (*guojia shuiwang*), they announced a budget of about ¥3 trillion, of which 36 percent would be spent on interbasin transfers, another 36 percent on flood control capacity in reservoirs and lakes, 13 percent on local water supply, and 15 percent on new irrigation projects (Zhang W., Hong, and Rogers 2022).

Some examples of large-scale planned projects include the Qiandao Hu Project, bringing water from the Qiandao Reservoir in western Zhejiang to the Hangzhou metropolitan area, transferring an anticipated 1.3 cubic kilometers per year (Baidu n.d. c) beginning in September 2019 (Xinhua 2019b). Work started in 2019 to move about the same amount of water from the West River in western Guangdong to the Pearl River Delta megacity, which ironically gets more than 1,500 millimeters of precipitation per year and is itself built on a reclaimed estuary but was still short of water for household and industrial use (Guangdong 2019). Yunnan began drilling tunnels in 2020 for a 630-kilometer canal to bring about 3.2 cubic kilometers of water per year from the Jinsha River in the west to Lijiang, Dali, Kunming, Yuxi, and Honghe (Yunnan 2018; Xinhua 2020a). Perhaps most incredibly, a plan that mercifully had been laid aside in 2010 was revived in November 2019, when participants at a conference in Ürümchi discussed a project to bring desalinized seawater two thousand kilometers inland to Xinjiang to "solve its water shortage problems" (CCTV 2019).

....

Along with other existing and proposed water transfer projects, as well as continued massive dam building, the South–North Water Transfer Project clearly illustrates three more general discontents of ecosystem intensification: fix-fixing fixes, rigidity traps, and moral hazards. Any project that meets perceived needs to intensify production by reducing or eliminating ecological (and sometimes institutional or cultural) buffers will have side effects, as we have amply seen with water transfers beginning as early as the period of socialist construction. Beginning around the turn of the twenty-first century, the PRC regime, in its slow move toward a more ecologically conscious developmentalism, became ever more reluctant to just let these side effects stand, whether they were pollution, loss of ecosystem resilience, or human suffering because of population relocations.

At the same time, dependence on infrastructure, both as an intensifier and as a replacement for earlier ecological buffers, creates rigidity traps. Whatever its dangers—dam collapse, chronic water shortages, increased pollution—if a water project is essential to maintaining production, there is little or no chance that authorities will decide to alleviate the dangers by removing their cause, which is the project itself. Although small-scale water projects, such as many of those constructed during the Great Leap Forward, were abandoned as impractical or just shoddy, I know of no large-scale project involving massive state investment that has been abandoned, even though its negative ecosystem effects might outweigh its production advantages. The continued repurposing of the Sanmenxia Dam is a prime example of this.

Finally, creating local abundance of any scarce resource becomes a disincentive to conserve that resource, leading to a moral hazard. The loosening of water conservation programs in North China after the water began to flow in the South-North Project illustrates this point.

None of the massive water projects of the early twenty-first century led to large-scale disasters comparable to, for example, the Banqiao Dam collapse in the 1970s. The engineering, the materials, the labor, the monitoring, the governance, the compensatory fixes are all better. For better or worse, China is now locked in to the rigidity trap of its replumbed landscape. The following chapter tells the story of dams, another important part of this plumbing project.

8. Dammed If You Do, 1993–2021

On the occasion of the safe and timely startup of the first group
of turbines at the Baihetan hydroelectric station on the Jinsha River,
I offer my enthusiastic congratulations. . . . Your future work . . .
will make an even bigger contribution to realizing our goals of peak
carbon and carbon neutrality, and promote the complete greening
of our social and economic development.

XI JINPING, June 2021

Water is not just a life-giving (and sometimes life-threatening) fluid. Falling with the force of gravity, it is also a source of energy through hydroelectric power, usually harnessed by dams. Hydroelectric development seems to be a stage of industrialization everywhere. In North America and Europe, it peaked in the mid-twentieth century (Tilt 2015, 41): the largest projects in the United States and Canada were completed between the 1930s and the 1970s. China did build dams in the early years of the People's Republic, and although they generated some power, their primary purposes were water storage and flood control. Only beginning in the late 1980s did China build bigger dams, primarily to produce electric power.

After the problematic Sanmenxia Dam, begun in 1954, China did not build any really big megadams (with installed hydropower capacity of two gigawatts or more) until the Gezhou Dam on the middle Chang River, finished in 1988, and it was another decade before the next two were finished: Lijiaxia, in the upper reaches of the Yellow River in 1998, and Ertan, on the Yalong River in southwestern Sichuan in 1999 (Wikipedia 2019).

Since then, China has made up for lost time and then some. Of the world's 71 existing megadams in 2021, 19 came on line in China after 2007, while only 5 were built in the rest of the world—3 in Brazil and 1 each in Russia and Malaysia. By 2021, China sported 4 of the world's 6 true maxi-megadams, with 10 gigawatts or more of installed generation capacity (map 8.1). These included the biggest of all, the Sanxia (Three Gorges) Dam, with 22.5 gigawatts capacity, and the second largest, Baihetan, on the upper Chang or Jinsha River, the subject of Secretary Xi's

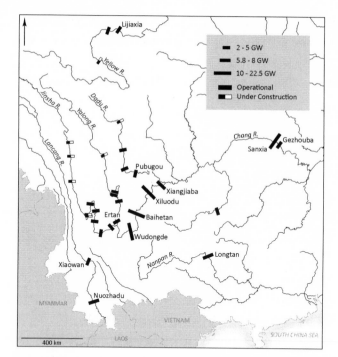

Map 8.1. High-capacity dams. Map by Lily Demet Crandall-Oral.

Figure 8.1. Trackers or haulers in the Chang River gorges, 1913. Photo by D. C. Graham. Wikimedia Commons.

congratulatory message above, scheduled to realize its full generating potential of 16 gigawatts in 2023. At a smaller but still significant scale, at the end of 2020 China had more than 150 dams with generating capacities of more than 300 megawatts, bringing its total installed capacity to 369 gigawatts (MWR 2021a, 27).

THE MOTHER OF ALL DAMS

Everything about Sanxia, or Three Gorges, is *big*. It is one of two dams on the mainstream of the middle Chang River, along with the much smaller (but still mega-sized) Gezhou Dam, about 40 kilometers downstream. Sanxia sits at the foot of the eponymous Three Gorges, famous in Chinese history for their precipitous and spectacular scenery and for the labor of the half-naked haulers who trudged the riverside trails pulling barges against the current from the plains of Hubei to the Sichuan Basin in the west (figure 8.1). The dam stretches 2.3 kilometers across the river and measures 115 meters thick at the base and 40 meters thick at the top; large ships going either up- or downstream journey through a system of six massive locks, a transit taking about three hours (Bryan Tilt, pers. comm.), while smaller craft can ride a faster "ship elevator" (Li Pengxiang et al. 2016). The dam contains 134 million cubic meters of earth fill, along with 17 million cubic meters of concrete and 590,000 tons of steel. It holds back a reservoir stretching 600 kilometers up the gorges to the outskirts of Chongqing, also ranging tens of kilometers up several tributaries. Its total surface area is only 1,084 square kilometers, however, because it is built in narrow gorges, and its estimated volume is 39.3 cubic kilometers. The difference between the reservoir water level of 175 meters above sea level and that of the river below the dam is about 113 meters. It is, quite simply, a giant dam (SCSHKX 2006; China Three Gorges 2002).

Whether to Build Sanxia: An Almost Open Debate

The dream of Sanxia goes back to Sun Yat-sen, who in 1919 proposed damming the Three Gorges for transportation and hydropower, which would open up commercial links between Sichuan and the middle and lower Chang River basins (Meyskens 2021). In the 1940s, the Republic of

China (ROC) government, making postvictory plans for the anti-Japanese war, invited US civil engineer John "Jack Dam" Savage to consult on the design of a dam (Meyskens 2021; Embassy n.d.; Wolman and Lyles 1978, 232), but no work was done because of the worsening situation of the Civil War. Mao Zedong revived the dam dream in the 1950s, inviting Soviet experts to consult on design and emphasizing the necessity of protecting Wuhan and the Jianghan Plain after the disastrous floods of 1954 (Mei Wu 1993, 58). Zhou Enlai presented a dam proposal at two of the conferences that planned the Great Leap Forward in 1958, but bigger troubles again intervened (Embassy n.d.).

With its push toward the "Four Modernizations" in the early Reform period, the Chinese Communist Party once again revived plans for a Sanxia dam, this time emphasizing the need to develop massive hydropower resources and a plan to start construction in 1986 (Mei Wu 1993, 60). However, the Chinese People's Political Consultative Conference, usually a decorative body with no real institutional power, became a hotbed of opposition and began organizing field trips to the proposed dam region, as well as convening forums of experts (Mei Wu 1993, 60). Their report recommended delay, so the Ministry of Waterworks and Electric Power appointed an "argumentation panel" (lunzheng zu) of 412 "experts," divided into fourteen areas of expertise, to compile a comprehensive report on the pros and cons of building the dam (Ai Sixiang 2011). At the same time, a Canadian hydropower consortium conducted an independent study, recommending that construction proceed.

Meanwhile, panel members and others raised doubts on a variety of grounds. For example, economist Guo Laixi, originally a proponent, changed his mind because of the plans to resettle more than a million people displaced by the reservoir, pointing out not only the hardships for the migrants but also the burden placed on local governments in destination areas (Ai Sixiang 2011). Botanist Hou Xueyu stated in committee deliberations that "after the dam is built, dense populations on both sides of the river will need to expand to the mountainsides . . . whereupon they will ruin forests by reclaiming land on steep hillsides to plant . . . after which erosion and declining soil fertility will become more serious, the area of eroded lands will increase, with landslides and avalanches, and both droughts and floods will become more serious" (quoted in Ai Sixiang 2011).

Economist He Gegao complained that the massive project would use up too much of the state's budget for expanding power generation, calculating that building three smaller (but still gigantic) projects at Xiluodu, Xiangjiaba, and Goupitan would generate a little more power than Sanxia, with 990,000 fewer refugees and 330,000 *mu* (22,000 hectares) less farmland flooded (Ai Sixiang 2011). Most tellingly, the conclusion to the report pointed out that the three objectives of flood control, power generation, and transport were sometimes in concert but sometimes in contradiction (Ai Sixiang 2011). Predictably, all three of those other still gigantic dams had been completed by 2013.

In the end, only 9 of the 412 experts refused to sign the panel's report (which of course recommended proceeding with construction), but there was other opposition, both in the media and within the corridors of power. The most prominent critic was Dai Qing, a journalist and party insider, adopted daughter of former state president Marshal Ye Jianying. Distressed by the possibility of the dam, Dai compiled interviews with both opponents and proponents in a book entitled *Changjiang! Changjiang!* (*Yangtze! Yangtze!* in English translation) (Dai 1994). Even in the atmosphere of relatively free expression immediately before the massacres of 4 June 1989, she could not publish in Beijing but ultimately found a publisher in Guizhou, far from the centers of power, and funded her book by crowdsourcing among environmental activists (Mei Wu 1993, 68–69). Dai's interviewees on both sides of the dispute took pains to couch their arguments in the language of science, as well as to decry the intrusion of politics into what ought to have been a debate about science and modernization (Boland 1998, 29). Weighing in on the dispute, 272 representatives in the National People's Congress signed a petition asking for the project to be delayed, and Vice-Premier Yao Yilin issued a statement that it would be postponed "into the next century" (Mei Wu 1993, 63).

After 4 June, however, there was little room for explicit or implicit critiques of any government policy. Dai was arrested and served ten months in prison, and her book was withdrawn from circulation in the fall (Mei Wu 1993, 70). Even then, however, the issue was not completely settled, though severe floods in 1991 gave the project a boost (Mei Wu 1993, 64). Plans for the dam were finally brought before the National People's Congress at its fall 1992 session, and very uncharacteristically for the "rubber stamp" legislature, there was actual debate and a far-from-unanimous vote: 1,767 deputies in favor, 177 against, and 664 abstaining. So they built the dam, beginning in 1993 (Embassy n.d.). It took sixteen years, but the reservoir was rising and Sanxia was generating power in 2009, and it ramped up to full power in 2011. During construction and afterward, many of the warnings voiced in the debate of the 1980s came true, but not all of them.

Mitigating Some of the Dam's Socioeconomic Effects

Some problems needed attention before the Sanxia Dam was finished or the reservoir started to fill. Since the water behind the dam would inundate an area with a rich cultural history, the State Council appropriated ¥505 million for a massive salvage archaeology effort. Over the course of several years, archaeologists from Hubei, Sichuan, Chongqing, and elsewhere excavated 723 sites and conducted surface archaeology and rescue missions at a further 346 sites. Over 60 sites were paleolithic and paleontological, over 80 neolithic, and over 900 historic. In total, they recovered over 200,000 artifacts, 13,000 of them considered to have particular historical or cultural value; the finds ranged from 6,000-year-old stone figurines to Song dynasty porcelains (Liang F. 2009, 85–86). The archaeological project was the occasion for replacing the Chongqing City Museum, previously housed in a crumbling Nationalist-era office building, with the gleaming new Chongqing China Sanxia Museum, featuring artifacts from Sanxia excavations in many of its spacious, modern exhibits (icity 2015).

Some rescued objects were too big to put in museums, so several hundred historic temples, traditional houses, bridges, and other structures from the area to be flooded were moved to higher ground nearby, in a series of "reconstruction districts" (*fu jian qu*), outdoor museum parks where visitors could view structures rescued from inundation, including "temples, pavilions, houses, bridges, portals, city gates and walls" (Zhou Jun 2015, 321). One such district, the Phoenix Mountain Historic Structures Complex (Fenghuangshan Gujianju Qun) in Zigui County, included twenty-four buildings, four bridges, two city gates, one portal, and one old well, all from the Qing dynasty (Wikipedia 2013).

Just as massive but much more important socially and ecologically was the program to relocate more than a million residents of the area that would be flooded. The People's Republic already had some experience with "waterworks refugees," having resettled several million during the waterworks-building frenzies of the 1950s through the 1970s, including about a million from Sanmenxia on the Yellow River, Danjiangkou on the Han River, and Xin'anjiang on the Qiantang River. In those days, governments simply trucked people out to villages on higher ground and gave them a bag with some money; there were no resettlement or relocation programs. Conflicts arose between refugees and natives of the resettlement areas, and because the good land was already taken, resettlers resorted to farming steep slopes, setting off the familiar cascade of deforestation, erosion, and siltation. Many people returned to higher ground nearer to their original homes, and in the Sanmenxia case, a positive feedback loop ensued—not only did resettlers put increased pressure on the higher elevation areas they were moved to, but the reservoir itself also caused landslides that made the new areas even more hazardous, eventually forcing authorities to move the refugees again (Li Heming, Waley, and Rees 2001, 197–201).

Determined to avoid more such fiascoes, the regime planned ahead for the Sanxia project. The situation was daunting. Both urban and rural people had to move: eight county seats would be inundated along with vast areas of farmland. Local governments would need help building new cities above the water level of the projected reservoir, and farmers would need to be moved, compensated this time with new land and help with house building. But there was not enough land available locally to accommodate all the displaced farmers; some of them would have to move to cities and towns and take up nonfarming occupations, while others would move as far away as coastal cities and provinces (Li Heming, Waley, and Rees 2001, 204; Du Jinping, Li, and Li 1999; X. Xu, Tan, and Yang 2013). The shortage of farmland was compounded by environmental concerns: in 2001 the Natural Forest Protection Program went into effect, preventing settlers from farming steep slopes or otherwise causing deforestation and its consequent effects. Polluting local industries, which might otherwise have employed resettlers, were shut down (Wilmsen 2015, 41).

Despite the planners' foresight and good intentions in responding to an environmental and social fait accompli, resettled populations initially had it rough. Brooke Wilmsen (2015, 52), in a longitudinal study of resettlers' fates in Baodong and Zigui Counties, found that "the TGP . . . required people to move and so they went. They went to new locations and they began rebuilding. But life was hard. In 2003, only a few short years after displacement, impoverishment was widespread and the resettlers were suffering multiple deprivations. In response, they drew on their assets and cobbled together a living from available opportunities—mostly working as manual laborers to contribute to the reconstruction effort."

However, the ensuing decade brought a pleasant surprise. Authorities recognized that all was not well and made several adjustments to policies and practices. As early as 2003, policy turned from bare resettlement to economic development of the affected areas, as companies from Jiangsu and Beijing were encouraged to invest in local industries (Wilmsen 2015, 44; X. Xu, Tan, and Yang 2013, 119). Bryan Tilt (pers. comm.) described the relocated city of Zigui as having "a skyline about like Portland's." By 2011–12, Wilmsen's (2015, 46–49) further study of families interviewed in 2003 showed that their inflation-adjusted incomes exceeded prerelocation levels, and people expressed much more satisfaction with their postrelocation lives.

Still, problems remained, and the regime, as reflected in state media, was very willing to acknowledge the difficulties and address them constructively (map 8.2). In May 2011, the State Council "announced that 'problems that demand prompt solutions exist,'" and it formulated new policies aimed at resolving remaining problems, including a second move for farmers who were forced by the initial relocation to farm unsuitable lands (Hu Yinan 2011). Most of the new policies were aided by the fortuitously simultaneous push toward urbanization of rural populations in general.

Changes in the River's Sediment Regime

The real questions with Sanxia didn't end when people and artifacts were moved out of the way. Such a massive disturbance to the land- and waterscape also brought about a whole series of ecological changes. Many of the biggest ecosystem changes (other than the flooding of more than twelve hundred square kilometers of cities, towns, and farms) happened

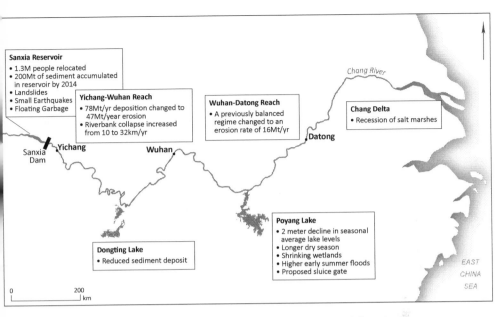

Map 8.2. Hydrological effects of Sanxia Dam. Map by Lily Demet Crandall-Oral.

because the dam and the reservoir altered the hydrological regime of erosion and sediment deposition. Between 1956 and 2002, before Sanxia was built, there was essentially no net sedimentation or erosion in the stretch of river that Sanxia would later inundate (S. Yang et al. 2014, 475). Planners of the dam, perhaps reacting to the silty debacle of Sanmenxia three decades earlier, predicted that Sanxia, by slowing the river's flow and allowing sediment to settle, would cause the reservoir to accumulate as much as four hundred million tons of sediment per year, but the actual value in the first few years was only about half that (S. Yang et al. 2014, 476). The earlier predictions had failed to account for the effects of soil conservation and reforestation programs on the Chang River and its tributaries, of other dams being built upstream, or of slight but significant decreases in annual precipitation, all of which reduced the amount of sediment flowing into the reservoir area. By the mid-2010s, the water storage capacity of the reservoir was predicted not to diminish as fast as the planners had originally thought, extending the projected water conservancy and hydroelectricity generating capacities well into the twenty-second century (S. Yang et al. 2014, 475; X. Xu, Tan, and Yang 2013, 120). With the completion of more large dams on the

Jinsha (the upper Chang) and its tributaries in the next few years, the sediment inflow to the reservoir was projected to decrease even more.

Downstream effects were a less pleasant surprise. Logically, less sediment trapped by the reservoir should mean more sediment and thus less erosion downstream. However, because less sediment was flowing into the reservoir than predicted and since the reservoir was accumulating a higher percentage of what sediment it did receive, there was much less sediment flowing out of the reservoir to the lower reaches of the Chang River. This changed the regime of the river downstream. Between Yichang, immediately below the dam, and the major city of Wuhan, a *deposition* rate of seventy-eight million tons per year before 2003 changed to an *erosion* rate of forty-seven million tons per year afterward, and from Wuhan to Datong below the confluence with Poyang Lake, a previously balanced regime changed to an erosion rate of sixteen million tons per year (S. Yang et al. 2014, 479). This, in turn, caused several kinds of major effects:

- The riverbed downstream eroded faster than expected, a situation worsened by sand mining for concrete manufacturing, meaning that the water level was also lowered (S. Yang et al. 2014, 479), and the flood danger might have decreased. In addition, water level in the reservoir could be regulated, with drawdown in the winter when river flow was lowest, leaving room to contain upstream flood surges in the summer. This provided some hope that catastrophic floods were much less likely than before, but major floods in summer 2020 shed doubt on the flood control efficacy of the dam. Authorities ordered some community dikes to be breached in order to use farmland as flood diversion basins and avoid large-scale floods in Wuhan. Clearly, the Sanxia Dam by itself could not completely control flooding.
- Erosion downstream of the dam caused an increase in riverbank collapses, particularly between Yichang and Wuhan, from an average of nineteen per year, affecting 10 kilometers of riverbank before the dam closure, to an average of twenty-five per year, affecting more than 31.7 kilometers, between 2003 and 2007 (X. Xu, Tan, and Yang 2013, 121).
- Since very little sediment was discharged to the sea at the riv-

er's mouth after the dam was built, the ocean current that flows along the coast began to wash most of the sediment away; the salt marshes of the delta receded instead of growing, raising possible problems for coastal and estuarine ecology and management (S. Yang et al. 2014, 484).

Building Sanxia did not just affect the river itself but also contributed to the continuing saga of Dongting and Poyang Lakes, both downstream of the dam. Since the river surface below the dam was lowered and the annual flow rate decreased because of withdrawals from the reservoir, the hydraulic connections between the river and the two lakes were disrupted, leading to a huge reduction in the amount of sediment deposited in Dongting (S. Yang et al. 2014, 481), as well as to some panic around Poyang.

The Undead Problem of Poyang Lake

Before 2003, the main concern in the Poyang Lake region (map 8.3) was flooding. In recent decades the centuries-long process of diking and enclosing land had accelerated. In earlier years those efforts had mostly served agriculture, but urbanization increasingly became the driver. In addition, sediment deposition in the lake increased, exacerbated by upstream deforestation in the five major rivers that feed the lake, as well as by increases in the flow of the river itself, also brought about at least partially by deforestation (Shankman and Liang 2003). Investigators predicted that building Sanxia would increase flows in the river in the early summer, when the reservoir level was being lowered, at the same time as heavy rainfall in the Poyang watershed would raise the lake level; in combination, these factors would increase the danger of summer lake floods, even as disastrous river floods like those of 1998 would become less likely.

However, the average lake level declined more than predicted. Lake levels fluctuate greatly with the seasons and from year to year depending on rainfall. In early summer months of an average year Poyang Lake's surface area is more than thirty-seven hundred square kilometers, while it declines to around nine hundred square kilometers at its lowest level in the winter (R. Wan, Die, and Shankman 2018, 1). But there appeared

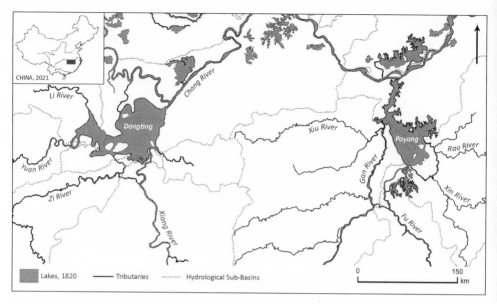

Map 8.3. Hydrologic basins of Dongting and Poyang Lakes. Map by Lily Demet Crandall-Oral.

to be a long-term decline in the lake level after the Sanxia Dam was built. Before Sanxia impoundment, midlake water levels at Duchang typically varied, from twelve to twenty-two meters above sea level in September and from twelve to eighteen meters in October, but between 2003 and 2012 the corresponding levels were ten to twenty meters in September and ten to sixteen meters in October, or an average drop of about two meters from pre-dam levels. The mean annual level also dropped, although only by about half as much (Mei et al. 2015, 2), and the dry season began to start earlier in the fall and extend later into the winter than previously (David Shankman, pers. comm.). This led to landscape changes, including shrinking wetlands and expanding grasslands, potentially affecting the habitats of a lot of native animals, particularly the rare Siberian crane, which winters in the Poyang area (Mei et al. 2015, 1; R. Wan, Die, and Shankman 2018, 1; Burnham et al. 2017). A Ming dynasty causeway, dubbed the "Thousand Eyes Bridge," emerged from underwater and has been restored as a tourist attraction (Mei et al. 2015, 1; Baidu 2014). On 6 August 2022, despite the water level having been very high earlier in the year, the lake surface dropped below twelve meters, sixteen days earlier than the previous record early start of the annual "dry period" (Kang J. 2022).

Changes due to Sanxia were not, however, the only factors contributing to falling lake levels, and there was considerable controversy over which were the most important. Climate variability played a role of some sort—the first decade of the twenty-first century was unusually dry, so less water flowed into the lake from its tributary rivers.[1] The changing hydrological regime of the Chang River below Sanxia also contributed, since lowering the riverbed increases the gradient between Poyang and the river (Mei et al. 2015; Jian Hua Gao et al. 2014, 550). Perhaps the most important factor was sand mining around the mouth of Poyang Lake, which intensified after 2001, when it was prohibited on the mainstream of the river. This mining deepened and widened the outflow channel from Poyang to the river, reducing friction and thus allowing water to flow out more rapidly into the now-lowered riverbed (Mei et al. 2015; X. Lai et al. 2014; Peng Wang, Zhang, and Qi 2019; Shankman, pers. comm.).

Alarmed by the dropping water level and its implications for conservation, among other things, authorities in 2009 decided that the logical solution to a problem caused by a technological fix was a fix to fix the fix. They proposed spending $1.9 billion to build a dam three kilometers long across the mouth of the lake to retain water when the level was low, with several sluice gates to release water when it was high. This would supposedly protect fishing, shipping, and wetlands (Ives 2016).

Then, in the summer floods of 2020, despite the long-term trend of declining average water levels and shrinking surface areas, Poyang Lake rose to its historically highest level, 22.52 meters, expanding the lake area to its largest extent since Sanxia was built (BBC Chinese 2020). Contrary to the experts' predictions, both river floods *and* lake floods could happen at the same time when ecological buffers were removed. In early 2021, as crises of both low and high water levels continued, the Jiangxi government announced detailed plans for the location, basic design, and functioning of the dam and its sluice gates, which would "solve dry-up but not control floods," and it appeared that the project would go forward (Xinhua 2019a; Liu Yiman 2021; Diao 2021). As of this writing, the future of Poyang Lake, of floods and droughts, engineering and politics, tourism and bird conservation, remains completely uncertain.

Earthquakes, Landslides, Pollution, and Garbage

Hydrological changes are not the only biophysical consequences of building the Sanxia Dam. When full, the thirty-nine cubic kilometers, or thirty-nine billion tons, of water in the reservoir is enough to cause earthquakes; minor quakes from 2003 to 2018 were about seven to eight times as frequent as from 1952 to 2002. Almost all the quakes were very small—only thirty-six registered a magnitude of 3.0 or more (strong enough to be felt locally). These small earthquakes were probably caused by collapses of karst caves and mines undergoing increased stress as the reservoir filled and not related to faults in the region (Y. Yao et al. 2017, 1098). However, the largest quake, a magnitude 5.1 tremor in 2013, did occur along a local fault and was probably related to the presence of the reservoir water, which caused increased pressure on the fault and increased water infiltration into the fault region. During the same period, surrounding regions not directly affected by the reservoir experienced no increase in earthquake frequency after the reservoir was filled (Rong Huang et al. 2018).

The Sanxia area has always been prone to landslides, especially when its steep slopes have been cleared for cultivation. But the frequency of slides doubled from the 1990s to the first decade of the reservoir, probably due to the waterlogging of newly inundated soils (X. Xu, Tan, and Yang 2013, 122). Not only did new slides cause considerable loss of lives, property, and living space, but many old slides were reactivated by a combination of rainfall and the periodic draining and filling of the reservoir for flood control (M. Xia, Ren, and Ma 2013; Junwei Ma et al. 2016, 319).

The biggest threat was some kind of seismic or other event that would trigger catastrophic dam collapse. Engineers designing the dam were of course aware of the problem and certainly must have known about previous dam collapses in China, so Sanxia was not plagued by the shoddy construction of the 1950s and 1960s that contributed to the Banqiao disaster in 1975. But there were still signs that a breach of the dam was not entirely out of the question. In summer 2019 Chinese netizens published two satellite pictures from Google Maps (which is blocked in China), apparently showing that the concrete sections that make up the dam had shifted position, changing the shape of the dam as a whole and leading to a vigorous, worldwide discussion on Twitter, Weibo, and WeChat (Weixin) about whether the dam was in danger of collapse, as well as to renewed advocacy for its removal. Officials dismissed such concerns, first stating that the shape had not changed—suggesting instead that there was something wrong with the pictures (Jingchuhao 2019)—and then later conceding that the sections had indeed shifted but that this was to be expected, given the resilient nature of the dam design, an explanation that some experts then ridiculed (Wang Weiluo 2019; Liang J. 2019). Like building the Poyang sluice gates, removing Sanxia would be a big project that may never get done, and there is no reliable way to evaluate the chances of a disaster in a dam of that size, because there is no other dam of that size.

Meanwhile, there were more immediate problems with water quality and garbage. Pollutants found in industrial and residential waste flowed downriver from above Sanxia and along the many tributaries that empty into the reservoir, and smaller amounts were dumped from the more than seven thousand ships registered to operate there. When the reservoir was full, flow in the mainstream and in the lower reaches of many tributaries slowed considerably, so pollutants that emptied into the river (now the reservoir) remained there longer. This exacerbated their effects, causing eutrophication and algal blooms in several tributaries and their estuaries, beginning with the Xiangxi River in summer 2003 (Yinghui Li, Huang, and Qu 2017; X. Xu, Tan, and Yang 2013, 119). By 2010, eutrophication had been found to affect 34 percent of river sections throughout the watershed (though not the mainstream of the Chang River), and there were twenty-six algal blooms in the watershed in 2010, compared to only three in 2003, when the river was first blocked (X. Xu, Tan, and Yang 2013, 119). The government spent tens of billions of *yuan* addressing these problems, and industrial wastewater discharges had decreased considerably by 2015, but residential discharges grew, driven by population growth, rising material affluence, and inadequate sewage-treatment systems (Yinghui Li, Huang, and Qu 2017, 1318). Parts of the Sanxia Reservoir thus acquired a patina of green, joining ranks with such natural freshwater bodies as Tai, Chao, and Dian Lakes.

Algae were not the only undesirable thing in the water—there was also a lot of floating and submerged garbage. Plastic packaging, plastic goods, and other throwaway items became ubiquitous in China beginning in the 1990s, and the citizenry, recently enabled to buy and use vast quantities of consumer goods, was only beginning to learn the lessons of the three Rs—reduce, reuse, recycle—which was ironic, given the extreme frugality of traditional peasant livelihoods. After the reservoir level reached its maximum height of 175 meters in late 2010, authorities mobilized 68,000 workers in 21,000 boats to collect an estimated 78,000 tons of floating debris (figure 8.2), consisting mainly of plant residues, woody debris, and, most significantly, plastic (Ning and Zhang 2010, cited in Kai Zhang et al. 2015, 118). A 2015 study found between 3.4 and 13.6 million pieces of floating microplastic per square kilometer, or between 3 and 13 pieces per square meter. Concentrations in the estuaries of the tributaries were slightly lower, ranging from only 2 pieces per ten square meters up to 11 pieces per square meter (Kai Zhang et al. 2015, 118), from one to three orders of magnitude more than found in the Mediterranean, the Great Lakes, or the estuarine areas of Chesapeake Bay around the same time (Kai Zhang et al. 2015, 122). Trash cannot pass the dam, so it accumulates in the areas closest to the dam end of the reservoir. In addition, although the numbers are not well studied, a lot of large pieces of plastic also floated on the surface at Sanxia. Perhaps there is a shiny, silver-colored plastic lining, however—in 2018 researchers at Sanxia University applied for a patent for a "water surface garbage cleaning device" operated by remote control, replacing manual garbage skimming (Global IP News 2018).

Sanxia, immense as it is, provided less than a quarter of the hydroelectric generating capacity in the Chang watershed by 2020. Four other maxi-megadams harnessed the energy of the upper Chang (called the Jinsha River upstream from Yibin). Including the world's second-largest hydroelectric station, at Baihetan, these dams had a combined installed generation capacity of over forty gigawatts, and their combined reservoir capacity was a bit larger than Sanxia. There were also six other mere megadams on the Jinsha, four existing and several proposed megadams on two of its major tributaries—the Yalong and Dadu—and many smaller (but still massive) dams, ranging in capacity from one hundred megawatts to more than one gigawatt, on the Jinsha, Dadu, and Min Rivers (the last being a tributary that joins the Chang River at Yibin). In addition to the familiar local effects of population relocation, habitat loss, pollution, landslides, and increased seismic activity (Dongfeng Li

Figure 8.2. (Top) Cleaning up garbage from the Sanxia Reservoir, 2013. Photo courtesy of Bryan Tilt. (Bottom) Collecting garbage at Three Gorges Dam, 2014. Photo by Yoshi Canopus, https://creativecommons.org/licenses/by-sa/4.0/deed.en.

et al. 2018; Xiao-rong Huang et al. 2018; Wikipedia 2019; Y. Dong 2014), they also had system-wide consequences for the entire Chang watershed. All these dams (beginning with Ertan on the lower Yalong, completed in 1999) evened out seasonal discharge, something that can benefit development but is often detrimental to riverine wildlife (Xiao-rong Huang et al. 2018, 2757–58), and also had major effects on sediment load. Before the dams were built, sediment load in the Jinsha had increased, probably due to deforestation (which slowed considerably around 1999–2000 for reasons unrelated to dam building). After the dams were built, however, sediment load downstream decreased considerably, contributing to the decreased sediment deposition in the Sanxia Reservoir and thus indirectly to the scouring of the riverbed in the lower Chang River and the shrinking of its delta (Xiao-rong Huang et al. 2018, 2762; Dongfeng Li et al. 2018, 44, 47).

Fish and Other Important Creatures

Sanxia and other large and small dams were big factors in drastic declines in many species of fishes and other aquatic animals (Liu Fei et al. 2019; Kang N. 2020).[2] Dams have disrupted the migration routes of many anadromous fishes, whose annual reproductive cycles often coincide with variations in lake levels and river flows. The Chinese sturgeon (*Acipenser sinensis*) formerly spent the bulk of the year in the saltwater Bohai and East China Seas, migrating upstream to spawn in the lower reaches of the Jinsha River and the upper-middle Chang River mainstream. Cut off from its spawning grounds by construction of the Gezhou Dam beginning in 1981, the sturgeon established a new spawning ground below the impassable dam, but its reproductive rates declined. In 1983–84 surveys revealed 2,174 spawners; by the late 1990s this figure had declined to 373, and in 2013 researchers found fewer than 100. Frequency of spawning declined as well, and the date of spawning onset was a month later than before dam construction. As a result, the International Union for the Conservation of Nature declared it "critically endangered" in 2022 (IUCN 2022a), and at the same time it declared the smaller, nonmigratory Yangtze sturgeon (*Acipenser dabryanus*) "extinct in the wild" (IUCN 2022b) and the Chinese paddlefish (*Psephurus gladius*) extinct altogether (IUCN 2022c). Other, smaller fishes suffered severe declines as dams blocked their migration routes. Fisheries of the Chinese tapertail anchovy (*Coelia nasus*), the Reeves shad (*Tenualosa reevesii*), and the pufferfish (*Takifugu* spp.), collectively known as the "three savories of the Chang" (*Chang Jiang san xian*), all declined to near zero once the dams were built on the river's mainstream (Liu Fei et al. 2019).

Inland migratory fish were similarly affected. The "four great domestic fish" (*si da jia yu*) of legend used to migrate as juveniles from spawning grounds in the river to nearby lakes, where they fattened through the high-water months of summer. During the dam and sluice construction era from the 1980s to the 2010s, however, their egg production in the Chongqing-Yichang section of the river, now mostly submerged in the Sanxia Reservoir, dropped from between twenty and thirty billion to less than a hundred million. The numbers picked up somewhat due to efforts to restore flows after 2011 but remained at about one-twentieth of 1960s levels (Liu Fei et al. 2019). A possibly worse fate befell a ray-finned fish, the *Coreius guchenoti*, which may have become extinct by 2020 (Kang N. 2020).

A 2009 survey of endangered fishes listed one species presumed extinct, two extinct in the wild, five extremely endangered, and forty-one endangered. Fishing moved down the trophic chain as some species fared worse and worse; fishers caught smaller species and smaller specimens of those large species that did survive. In the 1960s and 1970s, the average weight of any of the "four great domestic fish" brought in was around 10 kilograms; by the 2010s it had declined to 500 grams. The average size of tapertail anchovy caught in the lower Chang River declined from 114 grams and 31 centimeters to 70 grams and 25 centimeters, and the average age at catch dropped from four years to two (Liu Fei et al. 2019).

Dam building was not the only factor in these extinctions and declines in fish and fisheries along the Chang and its tributaries. Pollution also played a part, as did reclamation for agriculture and urban construction, overfishing, and invasion by exotic species (Liu Fei et al. 2019). Some mitigation measures brought back a portion of certain stocks, but Chang River fisheries in the early 2020s were at the point of an only very partial return. Unlike overfishing, which can be addressed with fishing bans such as the one instituted in January 2020 (Kang N. 2020), or pollution, which can be mitigated by controls on urban, industrial, and agricultural discharges, dependence on dams is the most rigid of rigidity traps. Despite calls from various sources, neither Sanxia nor any of the other megadams on the Chang River or its tributaries are coming down any time in the near future. Aquaculture and deep-sea fishing will continue to supply Chinese tables, but inland capture fisheries and the fish that fed Chinese diners for millennia are essentially kaput, sacrificed on the altar of cheap, "clean" electric power.

INTERNATIONAL RIVERS: THE DILEMMAS OF THE LANCANG, NU, AND YARLUNG TSANGPO

From the site of the proposed Guxue Dam on a tributary of the Jinsha River, one could drive along a rough road westward over a 4,100-meter mountain pass, descend to the town of Deqen, and continue down a tributary to the Lancang River. After about 70 kilometers one would arrive at the site of the proposed Quzika Dam, on the mainstream of the Lancang at an elevation of 2,100 meters. Like the Jinsha, the Lancang was the site of massive hydroelectric dam building beginning in the 1990s. Although upstream dams on both rivers were still in the planning stages in the early 2020s, there were already eleven big dams on the Lancang farther downstream. The 990-megawatt Wunonglong Dam, only 60 kilometers below Deqen, and the Lidi Dam, immediately below Wunonglong, both began producing electricity in late 2018 and 2019 (Wang G. 2019; Zou J. and Ma 2018). Farther downstream in Yunnan, a cascade of seven dams includes the Nuozhadu megadam, with an installed capacity of 5.8 gigawatts, and extends to the Jinghong dam in Sipsongpanna Dai Autonomous Prefecture on Yunnan's borders with Myanmar and Laos. At least fifteen more dams were planned along the course of the river from near its source on the Tibetan Plateau all the way to the border (Geren Tushuguan 2018).

Some effects of dam building on the Lancang were similar to those on the Jinsha and on the middle and lower Chang. Beginning with the Manwan Dam, finished in 1995, local farm communities were relocated, lost agricultural and forest land, and were forced out of agriculture. Landslides increased and hydrology changed upstream and downstream of the dams (Tilt 2015; Pu Wang et al. 2013). But building dams on the Lancang presented a further challenge—the Lancang, under the more familiar name of Mekong, flows out of Yunnan through Myanmar, Laos, Thailand, Cambodia, and Vietnam before emptying into the South China Sea (map 8.4). Dam building on the Lancang affected international relations, and to an extent international relations affected dam building in China.

In the early 2020s, individual Mekong countries had very different relationships to hydropower and dams. Although most parts of the Lancang River basin are poor by Chinese standards, after 2000 they became increasingly connected to the wider Chinese economy and society through the Great Western Development (Xibu Da Kaifa) program and its offshoots (Magee 2006a, b). China's "big five" state-owned hydropower companies planned, underwrote, and built most of the dams on the Lancang and its tributaries (Tilt 2015, 50; Urban, Siciliano, and Nodensvärd 2018), turning Yunnan into a net exporter of power to other parts of China and to Southeast Asia (Yu X., Chen, and Middleton 2019, 62). Relatively poor Laos prioritized hydropower development on the Mekong and its tributaries (Geheb and Suhardiman 2019, 8), join-

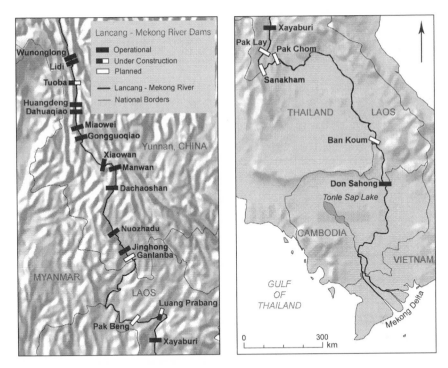

Map 8.4. Lancang-Mekong River dams. Map by Lily Demet Crandall-Oral.

ing Yunnan as a resource exporter and naming itself the "Battery of Asia" (Urban et al. 2012, 312), with a projected generating capacity of twenty-six gigawatts (IWP 2018). Like a petrostate, Laos often ignored environmental effects of hydroelectric installations, since they brought in needed foreign exchange (IWP 2018, 10; Räsänen et al. 2017, 39). Most Lao hydropower was exported to Thailand, which grew much wealthier in the late twentieth and early twenty-first centuries and became a net importer of hydroelectricity, using Lao hydropower to further develop its own poorer areas, particularly its northeast (Geheb and Suhardiman 2019, 8). Myanmar is a minor player in Mekong politics, with only 3 percent of the river's catchment within its territory. Both Cambodia and Vietnam, as downstream countries, generated some hydropower but primarily felt the ecological and social effects of hydropower development in Yunnan and Laos.

Lancang dams had complex effects on the hydrology and ecology of the downriver countries. The Mekong watershed sports some of the world's most productive inland capture fisheries, most notably in and around

Tonle Sap, a large lake in Cambodia. Most of these fisheries depend on the Mekong's great variations in flow between the low season from December through May and the flood season from June through November. Just as the Chang River backflows into Poyang Lake in the flood season, the Mekong backflows into the much larger Tonle Sap, increasing its surface area from about twenty-five hundred square kilometers in the dry season to more than fifteen thousand square kilometers during the floods (R. Stone 2011, 814; Mekong Dam Monitor n.d.). This variation drives the life cycle of many local fish species. Big dams upstream, however, reduced the yearly fluctuations (Mekong Dam Monitor n.d.), retarding floods and enabling irrigation in addition to evening out hydropower generation. The process, however, disrupted the fish life cycle, and local dams got directly in the way of fish migration (Räsänen et al. 2017, 28). These effects increased markedly in 2014 when Nuozhadu, the largest Lancang dam, came on line (Räsänen et al. 2017, 39).

Not only dams on the Lancang in China but dams in Laos itself affected the hydrology and ecology of the Mekong (see map 8.4). For this reason, conservation organizations (International Rivers 2017), along with Cambodian and Vietnamese governments (R. Stone 2011), sued to try to stop the Xayaburi Dam in particular. In July 2019, when the dam was undergoing tests before it began generating power commercially, the region was in the midst of a drought, with river flows at alarmingly low levels. The Thai government sent a letter to the Lao government requesting a suspension of the tests, which the Lao government granted. Nevertheless, in October of the same year the dam began permanent operation (Apinya 2019). Other dams on the Mekong and its tributaries in Laos also began producing power, while still others remained in the active planning phase in 2018 and 2019 (IWP 2018, 2019).

Dams in Laos, although they were far enough downriver that they did not affect China's ecosystems directly, were still entangled with China economically. As opportunities for domestic dam building began to shrink because there were fewer rivers left to dam, large Chinese dam builders and the banks that financed them felt the imperative to expand overseas (into Africa and even Latin America in addition to Southeast Asia). This trend accelerated with China's Belt and Road Initiative of infrastructure projects all over Asia (RFA 2016). As of 2017, Chinese firms were involved in building over 350 dams in 74 countries (Tan-Mullins,

Urban, and Mang 2017, 470). In Laos, Chinese firms were invested in about half of the 90 major dams on the Mekong and its tributaries either already built or planned to be built by 2030, including 4 on the Mekong mainstream (Tan-Mullins, Urban, and Mang 2017, 470). These firms thus had an indirect effect on Mekong ecology and local society, an impact that was of similar magnitude to the direct effect of building dams in Yunnan.

There is no road westward over the Kawa Kharpo range, from the 420-megawatt Lidi Dam on the Lancang, to the Nu River valley. However, one could probably walk, in two or three days, 40 kilometers up to a pass at an elevation of 3,900 meters and then down to the deep gorge of the Nu, the westernmost of the "three parallel rivers" of Yunnan, at 1,460 meters. Culturally and ecologically, the Nu River is not much different from the Lancang or the upper Jinsha: they all share steep topography, biodiversity, and ethnic diversity, and both the Lancang and Nu flow out of Yunnan into Southeast Asia, where the Nu is known in English as the Salween (Lamb 2019). As of mid-2022, however, there were no dams on the mainstream of the Nu in Yunnan or Southeast Asia. Chinese and international environmental activists and international journalists called it "China's last wild river" or even "China's Grand Canyon" (Lamb 2019, 20). According to one journalist,

Three great rivers rush through parallel canyons in the mountains of southwest China on their way to the coastal plains of Asia. At least 10 dams have been built on two of them, the Mekong and the Yangtze. The third remains wild: the remote, raging Nu, known as the Salween in Myanmar, where it empties into the Andaman Sea.

No dam stands in the path of its turquoise waters. It is the last free-flowing river in China. (Ives 2016)

In some ways, the sobriquet "wild river" is a misnomer, misleading especially for Westerners thinking of roadless, uninhabited "wilderness." There are two small hydropower dams near the headwaters of the river in Tibet and numerous small hydropower projects on its tributaries (Yu X., Chen, and Middleton 2019, 43). Even along the free-flowing majority of the Nu mainstream, communities of Lisu, Nu, and other Tibeto-Burman-speaking ethnic groups cultivate both the alluvial fans laid down by small tributaries and the slopes above, and a riverside highway follows

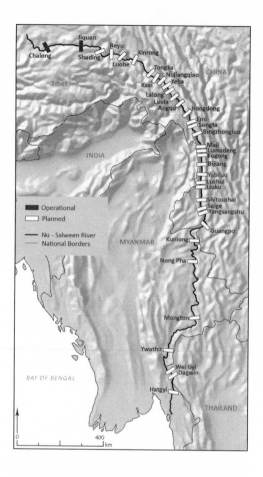

Map 8.5. Proposed dams on the Nu-Salween River. Map by Lily Demet Crandall-Oral.

the Nu River for several hundred kilometers (Harwood 2013). Beginning in 2005, local authorities, subsidized by the central government as part of its Great Western Development program, made major efforts to develop these relatively isolated communities (Tilt 2015, 84, 94; Harwood 2013). However, as of 2022 there were still no dams on the mainstream of the Nu/Salween anywhere below Tibet. In the sense that it still flowed freely and largely unimpeded, the Nu River was "wild."

That the Nu River did still flow freely seems remarkable (map 8.5). Plans for a cascade of thirteen or more dams emerged as early as 1991, and in 1999 the powerful National Development and Reform Commission approved the Nu River plan and took bids for planning units. In 2003 Huadian, one of the "five giants" of Chinese hydropower, was given the contract to develop the dams. This began what Bryan Tilt (2015, 105) has called "policy debates over the past decade [that] have had the back-and-forth quality of a tennis match" (see Magee 2006a, 158–83; Mertha 2008, 117; and Yu X., Chen, and Middleton 2019).

This back-and-forthing began within weeks of the Huadian contract. Opposition came from domestic and international NGOs, Chinese media, and some ecological scientists. In July 2003, UNESCO designated the Three Parallel Rivers area (encompassing parts of the watersheds of the Jinsha, Lancang, and Nu) a world heritage site but included only areas above 2,000 meters elevation, thus excluding the river itself and the riverside communities and leaving the river open to damming (Magee 2006a, 177). The Nujiang prefectural government quickly mounted a media campaign advocating hydropower development as the only way out of poverty. When Cha Cao'ou, head of the poverty alleviation program for the prefecture, pointed out that there were ways to alleviate poverty other than by building dams, he was fired (Yu X., Chen, and Middleton 2019, 54).

Meanwhile, Chinese environmental activists, including Yu Xiaogang, founder of Green Watershed, and Wang Yongchen, founder of Green Earth Volunteers, began to oppose the project publicly, to the point where "Nu River" was listed as the ninth most covered issue in Chinese media (Yu X., Chen, and Middleton 2019, 59; Mertha 2008, 112). Wang organized sixty-two scientists and celebrities to sign a petition opposing the construction of Nu River dams (Yu X., Chen, and Middleton 2019, 59). Compounding the obstacles for Huadian, China's Environmental Impact Assessment (EIA) Law was scheduled to take effect on 1 December 2003, meaning that a thorough review would be required if the project had not been approved by that time. Dam opponents in the State Environmental Protection Administration managed to draw out the process of approval until the EIA Law came into effect (Mertha 2008, 121).

In February 2004, Premier Wen Jiabao settled the issue—temporarily. He stated publicly that "such a large hydropower station project that draws high social attention, and has environmental controversy, should be cautiously studied, and scientifically decided" (Mertha 2008, 122). Yu Xiaogang, meanwhile, continued his antidam activism, arranging a meeting in August 2004 between people who had been relocated for the Manwan Dam on the Lancang and those who would be relocated if the Nujiang dams were built. He also took some Nujiang villagers with him to an international hydropower conference in Beijing (Tilt 2015,

174; Mertha 2008, 114), activities that led the Yunnan government to withhold his passport and restrict his travel for a time.

Proponents continued to agitate, since Premier Wen's statement was somewhat equivocal; for example, the Yunnan Power Company established a "Yunnan Power Prize" of ¥800,000 to support research by the Yunnan Academy of Social Sciences, which published a book of essays touting hydropower development as the way out of poverty. Dam proponents also encouraged a discourse of "hydropower nationalism," blaming NGOs and Western opponents of China for the temporary cancellation of plans and even for the premier's statement temporarily halting them (Yu X., Chen, and Middleton 2019, 56).

In early 2006, officials decided to proceed with a scaled-down project, building only four of the thirteen dams originally planned, at least two of which were supposed to begin construction before 2010 (Magee 2006a, 170–76). But Premier Wen again put a halt to the plans in April 2009, urging further study and refinement of the plans. When Wen retired in 2012 along with his boss, General Secretary Hu Jintao, plans were on again, and the State Council announced in January 2013 that construction would begin on the Songta Dam and that four others were officially approved.

The case was still not quite closed. Although some villagers were reportedly relocated in 2015 (E. Wong 2016), several NGOs mounted another campaign to halt construction, emphasizing the potential harm to biodiversity and local landscapes. As a result, Nu River dams were excluded from hydropower development programs during the thirteenth five-year plan, which extended to 2020 (Yu X., Chen, and Middleton 2019, 61). The flagging enthusiasm for the projects may have resulted not only from China's turn toward an eco-developmental philosophy of governance (Haddad and Harrell 2020) but also from a temporary electric power surplus: by 2018 the government's slogan had shifted from "western power sent eastward" to "Yunnan power for Yunnan use," and Yunnan began to actively export power to its Southeast Asian neighbors, including Laos, in 2015 (Yu X., Chen, and Middleton 2019, 62).

Hydropower advocates were still not about to give up on the Nu River. They argued that not only would dams bring prosperity to one of China's poorest regions but that in China's attempt to decarbonize its energy sector as quickly as possible every gigawatt-hour of hydroelectric power was one less gigawatt-hour of coal-fired power. Perhaps the Nujiang prefectural government expressed it best on its website in late 2018: "Today, the 'revival' of Nujiang development has already become a certainty, but a series of questions that come with it—relocation, ecology, geology—are accordingly obstacles that are difficult to get around. What we can foresee is that the debate between development and conservation with regard to Nujiang hydropower is bound to continue" (Nujiang 2018).

Although there were no dams on the Nu mainstream downriver from Tibet, the river valley still had more than one hundred small hydro projects (E. Wong 2016), including both dams and spillway installations. These had their own ecosystem effects, including habitat loss, alteration of river channels, and electric power for local mines, with all the ecosystem problems that mines bring (Yu X., Chen, and Middleton 2019, 57; Kibler and Tullos 2013).

When it flows across the border into Myanmar, the Nu becomes the Salween. In 2020 there were no dams on the river's mainstream there either, but there were plans for at least five (two others were canceled in the late 2010s). Preparatory work was done for some of the dams between 2014 and 2018. Of the five with active plans, at least two—both in the Shan State near the Yunnan border—and perhaps another in the south—were scheduled to sell power to China, while the others would sell their electricity primarily in Thailand. Regardless of the market, however, the contractors for all the dams were Chinese (Middleton, Scott, and Lamb 2019, 33). And although there were as yet no dams on the mainstream, four were completed on tributaries, with several others under construction (Middleton, Scott, and Lamb 2019, 34). The Salween was already even less of a "wild river" than the Nu upstream.

Both construction and power export from the Salween dams were complicated by the unsettled security situation in all the areas of northern Myanmar where dams were proposed, as several ethnic militias had been engaged in on-again, off-again rebellions against the central government for decades (Middleton, Scott, and Lamb 2019, 39). This instability complicated power-sharing and construction agreements between Chinese firms and the Myanmar government. More worrisome perhaps was the fact that the agreements concerning dam building were all about power—there was no comprehensive water resources agreement (Middleton, Scott, and Lamb 2019, 43). Even in the third decade of the twenty-first century, environment sometimes continued to take a back seat to economic development, as represented by hydropower. Whether the military coup of February 2021 in Myanmar would lead to finally building the dams or to more delay was unclear as of mid-2022 (T. Roney et al. 2021; Pianporn 2022).

China has also begun to exploit the hydroelectric potential of another international river—the Yarlung Tsangpo of Tibet, which becomes the Brahmaputra when it flows into northeastern India and then Bangladesh. Hydro development in the Yarlung Tsangpo watershed began with the 510-megawatt Zangmu Dam, which came on line in 2014 in Gyaca County, where the river flows rapidly through a narrow gorge (Liu L. 2019). By 2019, construction was proceeding on three more large dams within a few kilometers of each other along that stretch of the river, which would host about 2 gigawatts of power generation capacity, and several dams were planned. But these were mere warm-ups for what could be the main event, announced in late 2020—a project along a 50-kilometer stretch close to the disputed border with India where the river drops more than 2,000 meters. According to Yan Zhiyong, chairman of the Power Construction Corporation of China, the proposed project could potentially have generation capacity of over 60 gigawatts, "equivalent to three Sanxias" (Jia K. 2020; Lo and Elmer 2020; K. Huang 2020; Lo 2021). These plans did not proceed without protest from downstream countries worried about ecosystem disruption (Palma 2019). China's plans were not alone, however; in retaliation for Yan's 2020 announcement, there were reports that India was planning a 10-gigawatt station, by far its largest, just downstream, partly to guard against total Chinese control of the water flowing into the Brahmaputra (Yang Yong 2014; Reuters 2020). These projects were part of an overall, nationwide plan to raise China's installed hydropower generation capacity to 510 gigawatts capacity by 2050 (Sun Zhiyu and Hu 2020).

....

As of the early 2020s, then, Chinese dam development showed no sign of slowing down, especially at a time when China was aggressively trying to reduce its dependence on fossil fuels. Indeed, there is arguably even more pressure to build hydropower as China massively increases its capacity in variable renewables such as wind and solar, thanks to hydro's fast ramp rate and unmatched capability of providing a backstop power source capable of "firming" those variable generation sources faster than other generators (though extreme drought in 2022 called the reliability of hydropower into question as well [M. Wang 2022]).

Still, China's dam development in the twenty-first century is very different from that of the mid-twentieth. Most of the earlier dams were part of water control and water diversion projects, built in lowland areas in China Proper inhabited by Han people. They were designed to capture water for irrigation and to prevent recurrence of the floods that had devastated those areas for so many centuries. They were built mostly of earth fill transported on the two ends of shoulder poles carried by millions of conscripted peasants. Relocated people had to fend for themselves and were expected to sacrifice for the nation. Some of the dams, such as Banqiao, which collapsed in 1975, were not safe; others, such as Sanmenxia, had design flaws that made them virtually useless or worse. Scientists who opposed them often suffered greatly.

Fast forward to the 2010s and 2020s, and most dams were built in the deep gorges of rivers in Zomia, inhabited primarily by ethnic minority people. Although they have some irrigation and flood control functions, they were designed primarily to provide hydroelectric power for export to industrial cities in the east. They were designed and engineered to the world's best standards, and the chances that they would fail or not work were remote. Giant machines did almost all the physical work—the 1.2-gigawatt Yangqu Dam on the upper Yellow River in Qinghai was touted for being built by robots using 3D printing (S. Chen 2022). Relatively few people were relocated, and although those people suffered from being moved, they received some help and compensation (Tilt and Chen 2021). Opponents sometimes suffered, but many of them retained positions of influence, and in several cases they succeeded in delaying or canceling the projects.

Sanxia was both the last of the old-style dams and one of the first of the new. It was designed for flood control as well as power generation and was built in China Proper, occasioning mass removal of population. But it was also built by big rigs with the most modern design and engineering. It is unlikely that there will be another Sanxia, but it is certain that there will be more huge dams in Zomia and Tibet.

9. A Toxic Cornucopia, 2000–2022

China's emergence as a massive grain importer ... may well force a redefinition of security, a recognition that food scarcity and the associated economic instability are far greater threats to security than military aggression is.

LESTER BROWN, *Who Will Feed China? A Wake-Up Call for a Small Planet*, 1995

Despite its alarming title, Lester Brown's book did not claim that Chinese people were facing renewed threats of starvation or famine. Rather, he predicted that rising incomes, urbanization, and the resulting demand for a more varied diet, combined with population growth to a projected total of 1.6 billion by 2030, would mean that China's increasing demand for grain would exceed the available domestic supply, putting pressure on world grain markets and causing food insecurity worldwide.

It didn't exactly happen. Chinese diets, already broadened by the 1990s, continued to diversify, requiring more intensive agricultural inputs, and the population urbanized even faster than anyone would have predicted when Brown was writing. Despite all this intensification, in 2022 China still basically fed China. Although China imported large amounts of soybeans, almost entirely for use as livestock feed, and urbanization increased demands for water and upped pressure on farmland, China still produced almost all its food and most of its nonsoybean feed.

Consequently, there was no steady upward trend in world grain prices until Vladimir Putin's invasion of Ukraine in February 2022 and the associated blockade of Black Sea ports pushed up the price of wheat everywhere, including China (H. Gu and Patton 2022). It remained unclear as of late 2022 whether this trend would be permanent, but most people still considered that climate change, rather than either military aggression or China's insatiable appetite, was the greatest threat to world food security and indeed to security in general (IPCC 2019, 13).

HOW CHINA (ALMOST) FED ITSELF

Although China did feed itself, it did so at great environmental cost, at home and abroad. A more varied diet depended partly on imports of more food and especially more feed, if not nearly to the extent that Brown had predicted. Urbanization reduced the agricultural population and the available farmland, leading to consolidation of small farms into large farms, which were not always more productive. Continued dependence on agricultural chemicals compromised food safety, polluted the soil, and reduced the resilience of agricultural ecosystems. It also led to concerns over food safety—some legitimate and some overblown—particularly among the urban population, which began to turn to organic foods. Responding to all these threats, China's leadership intensified its pursuit of these "big ag" policies that hardened rigidity traps, increased vulnerability to diseases, dispossessed small farmers, and actually did little to address problems of food safety and agricultural pollution.

Continued Increases in Staple Production

Even though China had fairly definitively solved the "warm and full" problem in the early Reform period between 1980 and 1998, food production and consumption continued to grow in the early twenty-first century, both absolutely and per capita. Although rates of increase slowed a bit after 2000, production of almost every agricultural product continued to grow. From 2000 to 2021, total staple production grew from 462 to 683 million tons, raising per capita supplies from 366 to 484 kilograms. Grains accounted for 620 million tons, distributed among rice at 213 million tons, a 13 percent increase; wheat, at 137 million tons, up 36 percent; and corn, at 273 million tons, almost two and a half times the amount produced in 2000 (Statista 2022a). China did import some grain, though only a small fraction of what it consumed. Corn imports ranged from 2.7 to about 7 million tons between 2010 and 2018, amounting to only 1 to 3 percent of China's total corn consumption, but this figure then increased dramatically to 29 million tons in market year 2020, 23 million tons in 2021, and 180 million tons in 2022 (Index Mundi 2022). Wheat imports fell after a peak in the 1980s and 1990s, amounting to around 2 to 7 percent of total consumption from 2016 through 2020, while rice imports accounted for 1.5 to 3 percent of the total consumed (Index Mundi 2022).

In addition to eating more meat and vegetables and fewer staple grains, most people "upgraded" the kinds of staples they ate. Until very recently most people could not afford to eat the preferred grains all the time—rice in the south and wheat in the north. In the north, they resorted to less-favored coarse grains such as millet, sorghum, and corn, and in the south, to corn, potatoes, and sweet potatoes. As rice and wheat became sufficient, production of sweet potatoes and potatoes dipped from 36 to only 27 million tons, and millet and sorghum almost disappeared from the mainstream Chinese diet. In 1952 "coarse grains" were estimated to constitute around 70 percent of the grain eaten in China, but by 1992 that had decreased to 16 percent (S. F. Du et al. 2014, 15). From 1952 to 2008, output of millet shrank from 9.8 million to 814,000 tons and sorghum from 1.1 million to 183,000 tons (Nongye Bu 2009, 21). Output of these grains increased again in the 2010s, so that by 2019 China was producing about 3 million tons of millet and 2.5 million tons of sorghum, but they continued to have a minor role in people's diets (EPS China Data n.d.)

Chinese people did not of course eat an average 484 kilograms of staples yearly. Since most grains contain about 3,500 calories of food energy per kilogram (FAO 1972), if people ate all the staples that China produced, they would have consumed well over 4,500 calories per person per day, a ridiculous amount. Even if an estimated 50 million tons of grain were lost each year in storage, processing, and postconsumer waste (G. Liu 2014, 16), more than 4,000 calories per day per person would be available, more than all but the most physically active young adults need to consume. Large numbers of people would be morbidly obese. Instead, increases in China's corn production and imports, along with its massive imports of soybeans, went to feed the animals that made meat, fish, poultry, eggs, and milk products greater and greater proportions of Chinese people's diets.

Diet Variety

If the reforms of the 1980s and 1990s meant that, for the first time since the early Qing, most Chinese had enough food, by the early twenty-first century they had enough *good* food. Ordinary people could now enjoy varied diets that had been available only to elites from at least the mid-Qing to the early Reform period. The primary dietary changes were the huge increases in consumption of animal foods, including meat, fish, and milk products. Three kinds of increases facilitated this dietary quality improvement: more corn grown for animal feed, more acreage devoted to nonstaple crops and to aquaculture (since there were enough staples), and more imports—soybeans and, more recently, corn for animal feed, along with a variety of luxury foods.

Traditionally meat was a luxury for most farmers in China Proper. They killed a pig at the New Year, ate part of it fresh, and salted or otherwise preserved the rest for occasional special meals later. Sometimes they also slaughtered a chicken or duck. By the 2010s almost all Chinese people ate meat frequently. A national survey of adults showed that 86 percent had eaten some red meat in the past three days, 29 percent had eaten seafood, and 20 percent had eaten poultry (Z. H. Wang et al. 2015, 230).

Table 9.1 compares estimated per capita availability of animal foods and animal protein in China and the United States in the 2010s. In the 98 percent of Chinese communities that are not Muslim, "meat" means mostly pork. According to available statistics, in the early 2010s the average Chinese ate 61 kilograms of meat per year (pork, beef, mutton, and poultry), or around 167 grams (about a third of a pound) every day, around two-thirds of that (39 kilos per year or 107 grams per day) consisting of pork. This is about twice the yearly consumption of pork in the United States. Chinese on average ate almost as much fish as they did pork, which is more than half again as much fish as Americans ate. They consumed ten times as much lamb or mutton as their US counterparts, as well as a few more eggs, but less than a quarter as much beef, chicken, or dairy products.

Between 2000 and 2017, production of red meat (beef, pork, and mutton) grew from 47 million to about 65 million tons—54 million tons of pork, 6.1 million of beef, and 4.7 million of mutton (SSB 2018a, table 12-14)—and hovered around those levels for the next few years, as beef production increased but pork suffered during an epizootic of African swine fever in 2019–20 (Index Mundi 2022). The 2.5-fold growth in corn production provided a big part of the feed for 700 million pigs slaugh-

Table 9.1. Annual amount of animal-derived food and protein availability per day per capita in China and the United States, 2010s

TYPE OF FOOD	CHINA (WEIGHT, KG)	US (WEIGHT, KG)	CHINA (PROTEIN, G)[A]	US (PROTEIN, G)[A]
PORK	39[e]	23[c]	11.86	8.9
BEEF	6[e]	26[c]	2.34	12.79
MUTTON/LAMB	4[e]	0.4[c]	1.34	0.20
POULTRY*	10[e]	48[c]	4.79	20.94
OTHER MEAT			0.23	0.40
FISH	38[b]	22[b]	9.04	5.09
EGGS	22[e]	16[a]	6.52	4.82
MILK PRODUCTS	23[e]	125[d]	2.77	22.19
TOTAL			38.89	75.33

Note: Figures are for varying years, but all are from between 2012 and 2018, a period when there was not much change in food habits.

* In China, poultry consumed is about 70 percent chicken and 30 percent duck and goose (Pi, Zhang, and Horowitz 2014, 12); in the United States, poultry consumption breaks down to about 86 percent chicken and 14 percent turkey (Shahbandeh 2022).

Sources: [a] Our World in Data n.d. a; [b] Our World in Data n.d. b; [c] Shahbandeh 2022; [d] Bentley 2014; [e] Index Mundi 2022.

tered in 2017, as well as their 440 million relatives still alive at the end of the year. But pigs require protein as well, and China grew between 15 and 20 million tons of soybeans (SSB 2018a, table 12-10; Index Mundi 2022), while importing several times as much—between 74 and 100 million tons from 2018 to 2021, primarily from the United States, Brazil, and Argentina (Statista 2019; Index Mundi 2022). Direct imports of pork played a small role, in 2019 amounting to about 2.6 million tons (around 5 percent of total consumption) but rising in 2020 and 2021 to compensate for the domestic epizootic. Beef imports, by contrast, rose to 3 million tons in 2021, accounting for nearly 40 percent of total consumption (Index Mundi 2022).

The country had a lot more chickens in 2019 than in 1999, as evidenced by the sharp growth in production of eggs—from 22 to 31 million tons—

and of chicken meat, which almost doubled, from 8.6 to 14.8 million tons (Index Mundi 2022). Official statistics show that poultry consumption per capita rose from about 1 kilogram in 1990 to more than 15 kilos in 2019 (EPS China Data n.d.). Urban people consumed much more chicken and more eggs than rural people and ate a significant minority of that chicken meat in fast-food restaurants (C. Pi, Zhang, and Horowitz 2014, 14–15). However, these numbers, based on sales, probably underestimate how much chicken, duck, and goose meat rural people continued to eat. Almost every household traditionally had a few chickens, and in rice-growing regions ducks were an important part of the ecology, gleaning the paddies after harvest, eating snails and other pests, and depositing manure that helped with fertilization. So folks had an occasional egg and would kill a chicken or two every once in a while to serve guests or for a holiday occasion. Even in the mid-2010s, rural households on the Chengdu Plain, having moved from their traditional dispersed compounds into consolidated settlements, were no longer allowed to raise pigs, but they still had chickens in their yards, and neither these birds nor their eggs probably made it into official statistics.

Fish consumption grew even faster. China's output of "aquatic products" (finfish, shellfish, and edible marine plants) grew from 12.4 to 70.5 million tons (FAO 2022, 30). This meant that Chinese people in the mid-2010s consumed almost as much seafood as they did pork—around 41 kilograms per year, nearly twice the world average (FAO 2022, 82). Almost all of the increase was due to the explosive growth of China's aquaculture, which cultivated 49 million tons of aquatic animals in 2020, or about 56 percent of the world's total production (FAO 2022, 30). During the same period, China's marine capture fisheries plateaued—at around 12 million tons, or 15 percent of the world total (FAO 2022, 14), and its inland catch declined from above 2 million tons in the early part of the century to 1.46 million tons in 2020, probably reflecting not just overfishing but the impact of dam building and other waterworks on major fish species. Still, China's inland catch was about 13 percent of the world total (FAO 2022, 22).

While the more common varieties of fish became everyday fare for most urban Chinese, many other kinds of seafood remained luxuries. Some of the rarest and most expensive appeared mostly at banquets hosted and attended by officials and businesspeople. Intended to impress guests, the special fare included shark fin, sea cucumber, and, most expensive of all, live reef fish (not eaten alive, but kept alive in restaurant tanks until customers ordered them). Hosts might spend as much as US$600 for a single endangered Napoleon wrasse weighing about a kilogram (Fabinyi and Liu 2014, 222). Even middle-class consumers in Beijing and Shanghai strove to impress banquet guests with more affordable types of seafood (Fabinyi et al. 2016, 5).

From 2000 to 2007 milk consumption exploded in China, growing from 8.2 to 31 million tons, about 5 percent of that imported, and consumption remained at those levels through 2021 (Index Mundi 2022). Traditionally, pastoral peoples of Central Asia relied heavily on milk products from cattle, sheep, goats, camels, and horses as sources of protein and energy; they drank raw milk and made cheese, yogurt, and other secondary products that could be preserved for a longer time. But in China Proper, ungulate milk was simply not part of the diet before the 1990s. Although much of the Han population is lactose intolerant, ecological factors probably explain this fact better; there was not enough fodder to feed ungulates that did not perform agricultural labor, and growing grain for feed would have been a waste of land and water.

In the early twenty-first century, while traditionally pastoral peoples still consumed large amounts of milk, increases in milk production mostly fed those Han who never used to touch the stuff. Urbanites were much more likely to consume milk products than before, in particular milk powder for baby formula, which rapidly replaced breastfeeding in Chinese cities, as well as increasing amounts of yogurt, ice cream, and other secondary products (Sharma and Zhang 2014). China's leadership promoted milk consumption as a sign of prosperity and modernity. Premier Wen Jiabao stated in 2005, "I have a dream and my dream is that each Chinese person, especially the children, can afford to buy one *jin* [five hundred grams] of milk to drink every day" (quoted in Sharma and Zhang 2014, 15). The Ministry of Health's dietary guidelines in 2009 recommended that every adult should drink three hundred milliliters of milk per day, and the powerful National Development and Reform Commission also opined that "dairy consumption is one of the most important criteria to measure a country's living standard" (Sharma and Zhang 2014, 15–16).

Vegetable consumption presents a paradox in that statistics show vegetable gardens occupying about 12 percent of China's agricultural land in 2017, twice their percentage in 1995 (SSB 2018a, table 12-9). There was also a great deal more variety: gone were the days when North China markets offered nothing but cabbage and a few root vegetables throughout the winter; all but the most perishable vegetables had become available all year long in outdoor markets and especially indoor supermarkets. According to some nutritional studies, however, people actually ate fewer vegetables than they had thirty years before, presumably because they now ate more meat (Popkin and Du 2003; K. Liu, Chang, and Chern 2011).

Data on fruit, however, are unequivocal. Fruit was rare, expensive, and available only locally in the 1980s and 1990s; one of the nicest gifts one could bring back for family or friends from a trip out of town was a bag or box of fruit from the destination. In the 2010s this was still true, but the primary value of the gift was in the packaging and the prestige of the giver and no longer in the value of the fruit itself, which was likely to be available locally. Acreage devoted to fruit orchards increased from 1.6 million to almost 11 million hectares between 1998 and 2017 (SSB 2018a, table 12-8), and nutritionists who worried that the Chinese didn't eat enough vegetables celebrated the increase in fresh fruit consumption (Popkin and Du 2003; K. Liu, Chang, and Chern 2011). Fruit juices, only sporadically available in the late twentieth century, were everywhere, and even banquets at fancy restaurants offered watermelon, mango, or orange juice as an alternative or supplement to liquor or beer.

What non-Asians have usually thought of as a typical Chinese meal—bowls of rice, or various buns, breads, or noodles made with wheat flour, as the main source of food energy, accompanied by dishes made with vegetables or mixed vegetables and meat (E. Anderson 1990, 114)—had never been the norm for the poor in China, but by the early twenty-first century it was the norm for rich and poor alike. The long series of agricultural development programs, from collectivization and irrigation (which led to famine when overdone), through introduction of chemicals, to market liberalization, led first to calorically sufficient diets and then, when production continued to grow, to nutritionally rich and culinarily varied diets.

Variation in Diets

There remained, however, huge variations between the diets of urban and rural Chinese, as well as between the diets typical of different regions. In the 2009 and 2011 national nutritional surveys, the most urbanized third of the population was three times more likely than the most rural third to have eaten meat recently (Z. H. Wang et al. 2015, 229), but regional variations in consumption of meat, milk, and fish were probably even greater than those between cities and the countryside.

A detailed study of rural Yunnan, for example, found that people there consumed more meat, more grain, and fewer vegetables than the national recommendations but no fish, fruit, beans, dairy, or eggs (Q. Jiang et al. 2017, 5). By contrast, in rural coastal communities in the Leizhou Peninsula, rice and shrimp farmers ate a gruel made of sweet potatoes and rice at every meal, along with small fish, bought from local fishers, mostly cooked in soup. They had plenty of fresh vegetables, along with fruit from their own trees and from local markets, including coconut and jackfruit, as well as some bananas, longans, and pineapple. They ate shrimp from their own ponds around each quarterly harvest but otherwise had no facilities for storing them. Pork and chicken were reserved for holidays; they kept a few chickens but raised no pigs after 2009 (Huang Yu, pers. comm.).

In Zomia, as late as 2002 the villagers of the Baiwu Valley ate as staples primarily buckwheat, corn, and potatoes, all of which they grew in their own fields, and they supplemented that diet with meat (pork, mutton, chicken and very occasionally beef) on special occasions when they killed an animal for a holiday or to honor guests. They ate fruit (plums, apples, Asian pears) from their own orchards when it was in season but very few vegetables aside from turnips, and they ate no fish. By 2015 their diet had changed radically. They often purchased meat from local markets. They continued to eat two of their traditional staples—potatoes and buckwheat—but had stopped eating corn (except fresh off the cob for a snack). Those foods had been replaced by rice, which they purchased with money earned from selling hybrid corn to pig feed companies and Sichuan peppercorns (*huajiao*, *Zanthoxylem* sp.) to spice merchants.[1]

Xinjiang diets, reflecting the region's Central Asian mixed agro-pastoral economy, were different still. As in other parts of China, consumption of meat and animal products increased dramatically during the early reforms and continued to grow in the twenty-first century. Here, though, mutton was the largest source of animal protein for both urban and rural households. Rural households, being mostly Muslim, consumed little pork, unlike urban households, which included increasing numbers of pork-eating Han. Urbanites, like those in most of China, decreased their consumption of grain in favor of high-protein, high-fat meat and milk products. Both rural and urban people ate lots of melons and other fresh fruit, as well as the dried fruits (raisins, cherries, apricots, and others) that abounded in the region's markets (Cerny 2008, 243–44).

Restaurants and Prepared Foods

In addition to eating a more varied diet, early twenty-first-century urban Chinese began eating out often and consuming prepared foods at home. In the early 1980s there were few restaurants, all state run, mostly serving banquets for officials and for ordinary people on very special occasions, such as weddings. Government guest houses served meals for visiting officials and conferencing scholars staying there; eating at one of the few restaurants on the street often required a long wait in line. People entertained guests at home, with wives often taking a day or two off from work to find and prepare "company food" and borrowing chairs from neighbors to seat all the invitees. In the late 1980s, when economic reforms allowed small private enterprises, or *geti hu*, little restaurants began to appear. Some sold simple breakfast foods such as noodles, steamed buns, and rice gruel, accompanied by strong-tasting garnishes for flavor. Others offered lunch and dinner—sometimes there was a menu posted on a chalkboard, or sometimes the proprietors would tell the guests what they could fix and then make it for them.

By the early twenty-first century, all this had changed. Every city and town had streets lined with numerous restaurants, ranging from expensive and lavishly decorated palaces featuring private rooms, costumed or uniformed hostesses, and the latest innovations on local classic cuisines, to little storefronts where one could get a fortifying bowl of noodles before heading out for the day or grab a quick lunch of dumplings and a vegetable dish or two. China's regional cuisines attracted the attention of international food writers, up to and including Anthony Bourdain and Eric Ripert, who experienced them, it is safe to say, almost entirely in restaurants (DeJesus 2016). Almost everyone ate out now and then; home dinners for guests became a rarity and diminished urban women's burden of entertaining guests or even preparing dinner every single night—they could buy packaged foods or order delivery on their phones (Yijing Zhou et al. 2015).

The restaurant boom was probably not good for nutrition. A study in Hunan showed that restaurant dishes were higher in sodium, total fat, and saturated fat, as well as lower in potassium and protein, than the same dishes prepared at home (X. Jia et al. 2017), and since people tended to order "fancier" (that is, meatier) dishes when they ate out, the study probably minimized the differences. The growing presence of "Western-style" fast-food restaurants, including such stalwarts as KFC (X. Shen and Xiao 2014) and McDonalds (X. Shen and Xiao 2014; Y. Yan [1997] 2006) probably intensified these dietary trends; people also drank more soda and other sugar-sweetened drinks (Y-H. Lee et al. 2020; Li Donghua, Yu, and Zhao, 2014). Xiaofang Jia et al.'s (2017, 1312) study showed that chicken burgers from fast-food restaurants had even more calories and less protein than dishes from more conventional restaurants and that even the fried dough treats known as "oil sticks" (youtiao) from fast-food outlets had a higher sodium/potassium ratio and more calories than the same foods from street stalls.

In the aggregate and on balance, early twenty-first-century changes in Chinese diets were still probably beneficial. Until the 1990s, most rural Chinese (that is, most Chinese) were in what Barry Popkin has characterized as the third stage of the historical progression of human diets: the "pattern of receding famine," characterized by "starchy, low variety, low fat, high fiber" foods, which left people susceptible to various forms of malnutrition (Popkin and Du 2003). Solving the "warm and full" problem meant basically eliminating energy, protein, and micronutrient deficiencies (see table 9.1). At the same time, some Chinese populations entered Popkin's fourth stage of dietary history, when increased intake of fat, sugar, and processed foods, along with decreased participation in physical labor and increased life expectancy, led to rising rates of obesity and susceptibility to "diet-related [or nutrition-related], non-communicable diseases" (Popkin and Du 2003, 3899S). Despite these trade-offs, amounts of protein and fat in the Chinese diet were still probably more

optimal than those in the United States, both ecologically and nutritionally; people in the United States consumed more animal protein than
was good for their bodies or the ecosystem (Smil 2002).

THE TECHNOLOGY OF CORNUCOPIA

In addition to nutritional trade-offs, China's food revolution brought
ecological trade-offs. As with any intensification of production, China's
increases in and diversification of food production and consumption
led to slow-variable changes and disturbances that brought increased
pressure on the local agro-ecosystems, raising questions about their
future resilience.

Feeding the Livestock

Even though China's staple production continued to rise faster than its
population in the early twenty-first century, China did not have a surplus
of staples. Switching to an increasingly meat-, fish-, and dairy-based diet
required intensifying agricultural production far beyond what had been
necessary to solve the "warm and full" problem. It takes about five to
eight times as many calories, in the form of feed, to produce an equivalent amount of calories in the form of pork for human consumption, and
beef requires even more feed. Although dairy is the most efficient way
to convert plant calories to animal protein (Smil 2002), milk cows still
require a lot of feed and water. With the continued growth of pig, beef,
and dairy farming, China had barely enough of most of the elements of
pig and cattle feed and not nearly enough soybeans.

Pigs traditionally did not consume a lot of grain or other staples.
Peasant households slopped their hogs, feeding them mostly food scraps,
sometimes supplemented by grasses and other, otherwise useless plants
gathered on hillsides and elsewhere. Some farmers in Zomia even grazed
their pigs on pasture along with cattle, sheep, and goats. Grain and
beans were simply too valuable as human nutrition to feed to pigs, and
the supply of food scraps was of course very limited—one reason meat
played so small a part in the traditional diet. People ate anything humans
could readily digest, and only the few things that were not edible could
be spared to feed the pigs.

The new and growing appetite for meat could only be satisfied if there was a surplus of staples available to give to the pigs. By the mid-2010s, almost all pigs were raised on commercial feed, which typically contains grain for energy, soybeans for protein, and additives such as micronutrients and antibiotics (Sharma 2014, 18). In 2012 about 52 percent by weight of pig feed used in China consisted of corn, another 22 percent of other grains, such as wheat and rice bran, and 26 percent of protein meal. Although China continued to import about 80 to 90 percent of the soybeans used in animal feed, mostly from the United States, Brazil, and Argentina (Amanda Lee 2021), there was nowhere near enough corn available on the world market to meet the food needs of China's pigs and chickens (Sharma 2014), so China had to rely on domestically grown corn to fatten its pigs. Thus, in 2011 about 70 percent of China's corn went for animal (mostly but not entirely pig and chicken) feed, 20 percent for industrial uses, and only 5 percent for human food (Sharma 2014, 9).

Cattle in this new food system also needed feed. China had approximately seventy-nine million beef cattle and thirteen million dairy cattle in 2018 (FAS 2019). Most cattle were raised in traditionally pastoral regions of Central Asia, but pasture area was inadequate to feed the increasing numbers of cattle, and cattle-producing regions needed to stall-feed their stock with imported feed from agricultural areas or from other countries. China's dairy cattle consumed vast amounts of alfalfa, and between 2012 and 2017 almost half of this was imported from the United States and other less densely populated countries (Sharma and Zhang 2014, 22; FAS 2019). This dependency on foreign feed both exacerbated water shortages in California agriculture (Sharma and Zhang 2014, 10) and led the Chinese government to promote domestic alfalfa production, which covered as much as 3.3 million hectares in 2017, producing about 3.8 million tons of the forage (FAS 2019). In addition, the government for the first time began to advocate producing silage as a component of cattle feed, and the initial yield came in at around 50 million tons (FAS 2019).

Urbanization and the Squeeze on Farmland

Despite the plenitude of food in the early twenty-first century, government officials, farmers, and consumers continued to worry about just how sustainable this apparent cornucopia might be. They were particularly concerned with continuing encroachment of urban construction on farmland.

There is no doubt that during the era of collective agriculture, particularly when the slogan "Emphasize staple production" was in vogue, too much of China's land was under crops. Farmers were forced to plant staples on marginal land that produced very little, thus contributing to deforestation and its concomitant ills of erosion, geomorphic changes, and biodiversity loss. However, until the mid-1990s, Chinese government statistics grossly—and probably intentionally—understated the amount of land under cultivation, perhaps to emphasize the precariousness of the country's newly achieved self-sufficiency. Authorities eventually admitted this when they later published incongruous statistics showing 96 million hectares under cultivation in 1995 but 130 million—close to the estimates of foreign researchers—in 1996 (see Smil 1999b; Lichtenberg and Ding 2008, 61; and Nongye Bu 2009, 6).

Still, even as early as the mid-1980s the cultivated area actually was shrinking. Before 1990 this was mostly due to converting marginal or less profitable farmland back to forests, pastures, and orchards (Smil 1999b, 425), a process accelerated after 1999 by the various "returning farmland" (*tui geng*) programs implemented after the Chang River floods of the previous year, which aimed to convert up to fourteen million hectares of previously cultivated land to forests (Trac et al. 2013; Lichtenberg and Ding 2008, 61). By the early twenty-first century, however, the biggest threat, both real and perceived, to farmland came from urban expansion, as urban population grew from less than 20 percent of China's total at the beginning of the Reform era to an officially reported 65 percent in 2021 (Statista 2022b).

In light of this rapid expansion of cities, in the 1990s Chinese officials joined Lester Brown in worrying about China's ability to feed itself. To stem the trend of farmland loss, the government put into effect the Basic Farmland Protection Regulations in 1994 and the New Land Administration Law in 1999 (Lichtenberg and Ding 2008, 59). These laws required governments at county levels and above to designate areas in every township and village where farmland was to be protected from residential or industrial development. If they did convert land (which they often did), they had to replace it with new farmland somewhere else (Lichtenberg and Ding 2008, 60). They also drew a national-level "red line" of 1.8 billion *mu*, or 120 million hectares, below which total farmland area could not dip. In 2021 the Ministry of Natural Resources reported that the cultivated area was 1.9 billion *mu* (1.28 million hectares), still holding very slightly above the (admittedly arbitrary) red line (*Renmin Wang* 2021).

The effects of this shrinkage were exacerbated by the pattern of what land was taken out of agriculture for development and what land was newly planted to compensate for it. At macroregional scales, the greatest losses of farmland tended to be in the most productive core areas, where urbanization was happening the fastest, such as the Chang River delta, the Pearl River Delta, and the Chengdu Plain (Lichtenberg and Ding 2008). Local case studies support this: at Pinghu, for example, in the highly productive region of the Chang River delta between Hangzhou and Shanghai, almost all the potential "new" farmland that could be used to compensate for development was already being farmed by 2001, leaving little room for further trading (C. Wu, Ye, and Fang 2004, cited in Lichtenberg and Ding 2008). By contrast Jingzhou, in a much less industrialized and less urban region in western Hubei, also lost farmland in the same period but primarily when farmers converted grain land to fishponds (Y. Shi 2004, cited in Lichtenberg and Ding 2008).

On the Chengdu Plain, local authorities worked out a way to fudge these exchange requirements and not impede the expansion of the megacity of Chengdu into rich agricultural areas like Pi County.[2] They consolidated traditional dispersed housing (*linpan*) into concentrated villages and used former homestead land for crops, counting this "new" cropland against farmland taken out of production by urban expansion. Despite these measures, however, the municipality of Chengdu, the historical "Heavenly Precinct" (Tian Fu), bounteous even when the rest of Sichuan was experiencing famines (Whiting et al. 2019, 265), imported rice in 2011, for the first time in a century and a half (Liang Hao 2015).

In the face of these worries about both quantity and quality of agricultural land, China continued to feed itself. Surprisingly, despite anxieties about urban sprawl and widespread fudging of land-exchange regulations, the overall land under food production of one sort or another (especially if we include orchards, tea plantations, and aquaculture facilities) did not decrease appreciably from 2010 to 2019, so China's food security seemed not to be impaired in the medium run. In contrast to Brown's dire predictions of massive effects on world food availability and prices, the only ways in which China's food needs seriously affected world food production and marketing were through soybean imports (J. Huang et al. 2015, 12), which contributed to deforestation in Brazil (Maciel 2016), and through distant-water fishing, which was a primary stressor of the world's ocean fisheries (Hongzhou Zhang and Wu 2017). In 2014 China imported about a sixth of its edible oils, sugar, and dairy products but produced just about all of its fruits, vegetables, seafood, eggs, meat, and poultry and in fact was a slight net exporter of vegetables, eggs, and seafood (J. Huang 2015, 12–13). Jikun Huang et al. (2015, 17–18) concluded, after examining the effects of urbanization on farmland, irrigation water availability, and labor reallocation, that "the concern of a serious threat of urbanization to food security—especially grain security—is unfounded."

Still, officials continued to fret about food security. In December 2020, the Ministry of Agriculture announced that, in order to secure grain supplies, it would designate more than 6.6 million hectares of prime agricultural land as "grain zones," farmed with "modern" (i.e., consolidated, chemicalized, commercialized, and mechanized) methods, and it prohibited converting grain land to other crops. In addition, the ministry announced that it would raise subsidies for rice, corn, and soybeans. Officials were not worried about starvation or undernutrition but rather about the danger of not having enough animal feed to sustain the protein-rich diets that consumers had become accustomed to in the previous two decades (W. Zheng 2020). At the annual "legislative" Two Sessions in April 2022, undoubtedly spurred by the twin supply chain disruptions of the COVID-19 pandemic and Russia's invasion of Ukraine, Xi Jinping doubled down on the growing emphasis on food self-sufficiency. "Grain security is a 'major national issue' [guo zhi dazhe]; among ten thousand worries, eating is the biggest," he said, adding, "China must rely on its own strengths, and earn its own living" (quoted in Zhang Xiaosong and Lin 2022). Still, imports of basic grains and soybeans did not decrease (Index Mundi 2022), even when staple harvests in 2022 reached a record high of 686 million tons (Amanda Lee 2022).

Consolidation and the American Model

From antiquity through the 1980s, feeding China required the labor of the majority of Chinese people. The rapid industrialization and urbanization of the country at the end of the twentieth century, however, meant that the majority were no longer farmers, and a minority would have to ensure that China could still feed itself. Realizing this, in the first years of the new century localities experimented with diverse ways to make agriculture more labor efficient and to implement mechanization, including peasant cooperatives and subsidies for small farmers (Y. Huang 2015, 393; Doll 2020). But around 2008 Chinese authorities decided to follow post–World War II North America and implement CCCM agriculture—consolidated, chemicalized, commercialized, and mechanized. Big, highly mechanized farms would practice "scientific" (i.e., chemical) agriculture and vertically integrate the supply chain, often involving corporate control of various stages of agricultural production, processing, and marketing (Day and Schneider 2018, 1228; Q. F. Zhang, Oya, and Ye 2015, 299). Advanced mechanization, bigger farms, and advanced agricultural technology all required more capital investment, something that could be provided by the state, by domestic firms, or to a limited extent by international agribusiness giants (J. Ye 2015, 330).

At the end of 2008 the Communist Party's Central Committee and the State Council issued a joint document entitled "Some Opinions of the CCP Central Committee and the State Council on Promoting Steady Development of Agriculture and Continuing to Raise the Incomes of Farmers," which signaled that the regime was going all in on the North American model and eliminating the peasant economy of past centuries. In addition to making many suggestions for reforming agricultural markets, "Some Opinions" contained a section on "strengthening the system of supporting and serving modern agriculture," which advocated accelerating the progress of agricultural science, accelerating construction and regularization of agricultural lands, strengthening waterworks infrastructure, and accelerating agricultural mechanization, as well as increasing the scale of both animal husbandry and aquaculture operations. At the same time, realizing the problems with this model, the document's authors also advocated such "ecological" measures as using organic fertilizer, growing green manures, and even regenerative agriculture and arid cultivation techniques, as well as shoring up "unhealthy" waterworks (CCPCC 2008). Evidently, the authors of "Some Opinions" were already anticipating adverse effects of the model they were prescribing and thus suggested possible solutions while suppressing any doubts that the model should go ahead.

Nowhere was this scaling up more evident than in the livestock sector. By 2010 or so, the traditional model of a sow or two and their piglets, raised mainly for family consumption and perhaps a few sales, was rapidly disappearing, even as pork output continued to grow. Between 2000 and 2015, the share of pigs raised in "backyard farms," which included operations with up to 50 pigs—declined from 75 to 27 percent of the total, replaced by large-scale "specialized household" farms raising between 50 and 1,000 head and by even larger commercial farms based explicitly on the US model (Schneider and Sharma 2014, 19). Several surveys revealed that less than half of households were raising pigs at all, and it was estimated that the number of households raising pigs dropped by half just in 2008 (Li Jian 2010, 61). In a village in Yunnan in 2007, many pigpens sat empty, and the yearly average number of pigs sold had dropped from 3.11 to 0.66 per household in the previous ten years. Villagers cited labor shortages (people were migrating to the cities for work), low prices, disuse of pig manure (replaced by chemical fertilizers), and lack of veterinary services as reasons they no longer raised pigs (Li Jian 2010, 63).

The share of pigs raised on "commercial farms," owned by giant agri-corporations, including both domestic and international firms (Tyson, Brasil Foods, Nippon Meats, Danish Crown, Hormel, and others) and raising over 1,000 head in true CAFO conditions, grew to 15 percent of the total (Li Jian 2010, 63) during this time, and some commercial pig farms were real mega-operations. One farm in Guangdong that Mindi Schneider visited in 2010 raised 120,000 head of exotic-breed pigs that year. It sold both breeding stock and meat in Guangdong, Hong Kong, and Macao (Schneider 2017, 93). As Schneider and Shefali Sharma (2014, 31) remark, "A CAFO in China looks like a CAFO in Iowa, though sometimes at a larger scale with more connected buildings."

As with pigs, so with chickens; the big increase in production and consumption of poultry beginning in the 1990s led to large-scale standardization and scaling up of the poultry industry. Farms continued to expand (C. Pi, Zhang, and Horowitz 2014, 11), and China's poultry production became "the most industrialized protein segment in the country" (C. Pi, Zhang, and Horowitz 2014, 20). "Feeds, breeds, medicine, production facilities, growth environment, slaughtering/processing facilities, and food safety inspection" were all done according to uniform industrial standards (C. Pi, Zhang, and Horowitz 2014, 11). As in the pork sector, the farmhouse courtyard with a few chickens running around pecking at seeds, spilled grain, and various bugs and snails was rapidly passing from the scene. Egg production became even more centralized than broiler chicken operations: as early as 2005, the largest 2 percent of egg farms produced more than 70 percent of the eggs sold in markets (C. Pi, Zhang, and Horowitz 2014, 11). Even ducks, which constituted about 30 percent of China's poultry production, lived their short lives in indoor factory farms (C. Pi, Zhang, and Horowitz 2014, 13).

Cattle farming followed a different course. Since China had no history of widespread small-scale cattle farming or ranching, rather than consolidating small-scale farms into large enterprises, China in the early twenty-first century saw the growth of intensive, stationary ranching in places where extensive herding had previously been the norm. In the 1980s and 1990s, some herders or ranchers in parts of Central Asia close to China Proper sold milk to itinerant buyers who converted it to milk powder (Sharma and Zhang 2014, 13), about the only form of milk widely available in Chinese cities at the time—it was thought to be a nourishing breakfast for children and sometimes a nice bedtime drink for frail elders. The dairy products industry really began to boom after the introduction of high-temperature pasteurization in the late 1990s. In addition to expanding herds, introducing new breeds, and sedentarizing previously nomadic herders (Ptáčková 2020), dairies and feedlots grew from scratch, ranging from small farms with a few hand-milked cows to large, vertically integrated operations controlled by giant corporations, most of them domestically owned and almost all of them in formerly pastoral areas in Inner Mongolia (Sharma and Zhang 2014, 13–17).

Even grain farms got much bigger, particularly after 2010. As early as 1996, the regime was worried about self-sufficiency in staples and announced in a white paper that China should continue to produce at least 95 percent of the staples it consumed directly or indirectly. The regime reiterated this goal in several more white papers and in the twelfth five-year plan, for the years 2011–15 (Gong and Zhang 2016, 4). A 2019 white paper reiterated a "new staple security vision" (*xin liang-shi anquan guan*) announcing that the nation should "guarantee basic self-sufficiency in grain and absolute staple security" (*quebao guwu jiben ziji, kouliang juedui anquan*). In other words, China should not rely so much on imports that it jeopardized access to the staples that fed both people and livestock (State Council Information 2019).[3] In order to meet these goals, the 2019 white paper emphasized that China should pursue a strategy to "store staples in soil, store staples in skill" (*cang liang yu di, cang liang yu ji*), meaning essentially that there had to be enough land to produce the needed staple foods and feeds and that progress in agricultural technology was needed in order to guarantee high yields (State Council Information 2019).

Land had in fact become quite fragmented during the Reform period (in contrast to the preceding period of collective agriculture), and this of course inhibited mechanization and some kinds of investment (State Council Information 2019, 320). The solution was to consolidate farms by urging (or coercing) farmers to transfer their land rights to other farmers or to individual or corporate investors, who thus became the effective owners of larger farms. The most radical form of consolidation encouraged a new class of farmers known as "big firms" (*da hu*)—mostly outside individual or corporate investors—to acquire the rights to land formerly farmed by large numbers of smallholders and operate large, mechanized farms relying on hired labor (Day and Schneider 2018, 1233). Ross Doll (2020, 33; 2021) describes the process of a *da hu* takeover of much of the land in a rice-growing township in Anhui that began as an experiment in 2007 and spread rapidly to much of the rest of the country. The state first gave the township about US$40 million between 2007 and 2013 to grade the land, eliminate the existing barriers between paddy fields, and consolidate previous irrigation ponds, transforming the village landscape "into a repeating pattern of featureless, seemingly interchangeable plots" (Doll 2022, 1750). They then persuaded or coerced small farm families to lease their land rights to the *da hu*, and by 2018 over 60 percent of the land in the township, previously farmed by about 6,800 households, was leased to 130 *da hu* (Doll 2022).

Partly in reaction to low productivity and other problems of *da hu* (many of whom were investors with little farming experience), the state began to advocate the model of "family farms," in which households transferred their land-use rights to other households, which could accumulate enough land (in grain-growing regions, often three to six hectares or more) to be able to mechanize and specialize, usually involving hiring labor, at least at busy seasons (see J. Ye 2015, 330; Day and Schneider 2018, 1233; and Doll 2021, 51). These farms turned out to produce higher yields than the larger, *da hu*–managed farms (Doll 2021, 51). When land was contracted to external *da hu* or to local manager families, most residents of a village would no longer be farmers, and many of them would migrate temporarily or permanently to urban areas.

Agricultural Science and Agribusiness

The national fixation with science was an important part of the agricultural modernization program. The "Some Opinions" white paper on agricultural development (CCPCC 2008) stressed scientific research on all aspects of agriculture, including genetically modified organisms (GMOs), as well as the role of "dragon head enterprises" (*longtou qiye*, domestically owned agribusinesses) and extension services in developing and commercializing "new disease-resistant, stress-resistant, high-yield, high quality, and highly effective genetically modified organisms." In the early twenty-first century, Chinese agricultural scientists, agricultural economists, and rural sociologists published literally thousands of articles in international journals, almost all of them in English, regarding topics as diverse as climate, soil, pests, and crop varieties. Most of these articles directly or indirectly promoted the science of modern, highly consolidated, chemicalized, commercialized, mechanized agriculture on enlarged farms.[4] Despite this emphasis on scientific agriculture, however, it was not until June 2022 that the Ministry of Agriculture issued the first standards that would enable commercial cultivation of genetically modified corn and soybeans, presumably starting with the 2023 planting season (MOA 2022).

Making farms larger, more efficient, and more scientific also depended on vertical integration, in which consolidated management by agribusiness corporations replaced both the age-old smallholder model and the collective model of the pre-Reform era. Some of these vertically integrated agribusiness corporations would acquire rights to land directly from households or collectives and hire farmers to work it; more commonly, operators of larger farms (*da hu*) would enter into contractual arrangements with the "dragon heads," often agreeing to grow a certain crop and grant the dragon head monopsony rights. Dragon heads might provide raw materials such as seed, fertilizer, or chicks, or they might pass the responsibility for inputs on to the farmers. As Schneider (2017, 7) explains, "Dragon heads are not like ordinary commercial enterprises: they are responsible for opening up new markets, innovating in science and technology, driving farm households, and advancing regional economic development."

In some sectors, particularly animal agriculture, foreign agribusiness firms played roles similar to Chinese dragon heads, but they had a very small role in grain farming, since wheat and rice were subject to the 95 percent domestic "red line" established earlier. Chinese firms were subject to no similar restrictions and had few scruples about investing in agriculture overseas. In 2015 the Ministry of Agriculture issued its "Strategic Plan for Agricultural Going Out" (Nongye Zouchuqu Zhanlüe Guihua), which provided state subsidies to firms that invested in places as distant and varied as Argentina, Brazil, Australia, the Philippines, and Russia (Sharma 2014, 23; Schneider 2017, 16).

Chinese aquaculture experienced some of the same changes that happened on dry land, but in other ways aquaculture was slower to change. Shrimp farms on the Leizhou Peninsula remained small into the mid-2010s; the largest of them followed the model of specialized monocultural "family farms" that replaced traditional polyculture in the 1980s. State-owned and private agribusinesses "control[led] the upstream sector of credits and inputs for shrimp juveniles, compound feed, aeration machines and shrimp pharmaceuticals, as well as the downstream sector of processing, marketing and sales," while the small farmers still operated the actual ponds and assumed the risks of failure from disease (a common occurrence due to the overstocking necessary to make a profit) or from price fluctuations (Y. Huang 2015).

Specialization is another element of the North American model. Specialization was not new in the twenty-first century—in the Qing period farmers in parts of the Chang River delta grew only cotton or only mulberry trees (for silkworm cultivation). Even the emphasis on grain during collective agriculture was a kind of widespread specialization.

Beginning in the early Reform period, specialization increased, but real specialization on the North American model came in the twenty-first century. The village that Liu Yinmei (2018) calls "Nacun," in southwestern Guangxi, illustrates this trend. Farmers there had grown mainly rice before the Founding and for decades afterward. In the early collective period they also grew some beans, sweet potatoes, peanuts, and vegetables; later on they established a collective chestnut orchard. In 1986 the county government began to promote agricultural specialization to transform villagers from semi-self-sufficient peasants into entrepreneurial farmers, encouraging them to grow and sell lychee, with cassava plants interplanted between the trees. Pursuing profits, they even cleared forested mountains to plant the lychee trees. In 1997 they followed up with sugarcane, which replaced the chestnuts; quite naturally, planting all these commercial crops led to a decline in rice cultivation. By 2003 most farmers were buying both rice and vegetables in township markets. At this time, however, these specialized commercial crops were still part of a household farm economy (Liu Yinmei 2018, 30–31).

There were problems with lychee by the early 2000s, however—the price went down and expenses went up, and cassava was no longer profitable either (Liu Yinmei 2018, 32–34). Changes in international markets also made the remaining chestnuts unprofitable, and sugarcane, which was difficult to mechanize, declined with labor migration. In 2006 "Nacun" began to turn to large-scale, commercial, highly specialized agriculture when the Jinsui Company appeared in the village. Jinsui was a dragon head (or perhaps medusa head) enterprise with two decades' experience in mechanization, consolidation, and scientific agriculture in a variety of crops and products, along with reforestation, tourism, and finance. The company initially leased 1,200 *mu* (80 hectares) of hillside land to grow bananas. Along with other, smaller dragon heads, some villagers and outsiders who became *da hu* (Liu Yinmei 2018, 42) decided to try bananas on their own, expanding cultivation to about 10,900 *mu* (720 hectares) by 2014 (Liu Yinmei 2018, 37–39). But by 2014, in an example of the lack of resilience in monocultures, most of the banana plantations in the village had succumbed to *Fusarium* wilt (also called Panama disease or yellow-leaf disease in English and *huangye bing* in Chinese). Investors and *da hu* then switched most of their land to dragon fruit cultivation, citrus orchards, and revived sugarcane plantations. In

2017 these operations occupied 83 percent of the village's farmland in large, monocrop plots, the smallest of which was 120 *mu*, or 8 hectares (Liu Yinmei 2018, 39–40).

THE END OF CHINESE PEASANT AGRICULTURE

All these changes in scale, specialization, ownership, and financing of agriculture added up to perhaps the most profound change in the East Asian countryside since the first foragers settled down to grow millet in the north and rice in the south ten millennia ago. Despite many changes across the dynasties and more rapid changes since the Founding—in ownership of land, in organization of labor, in crops, and in technology—agrarian China had always been a place of small farms worked mostly by human and animal labor, aided at times by small machines such as the waterwheels of the Song dynasty or the small *bengbeng* tractors of the late twentieth century. Villages were full of farmers, and most of them consumed most of what they produced and produced most of what they consumed. Outside agents—landlords, merchants, or the state—expropriated only a minority of the products of their labor; similarly, farmers acquired most of the means of livelihood—food, fuel, construction materials—from within the village landscape and met only a minority of their needs with outside goods (Leonard and Flower n.d.). This was true even in the period of collective agriculture, when households pooled both their labor and its products—most of the products and the labor stayed local, and when the regime tried to change this in the Great Leap Forward, the whole system broke down and led to famine. Cultivators in agrarian China were, in short, peasants.

Beginning in the first decade of the twenty-first century, however, the social-ecological systems of rural China entered a new basin of attraction. The flows of goods, money, and labor no longer either stayed primarily within the resource circle or were exchanged locally with the town (figure 9.1). Forest and pasture had long ago shrunk and become less relevant to local economies and ecologies, as people switched from biomass to fossil fuels, no longer ate wild foods, and saw ecological buffers largely replaced by infrastructure. Those still engaged in farming worked operations greatly enlarged by leases and other transfers, while most households leased their fields to others and did not farm any land

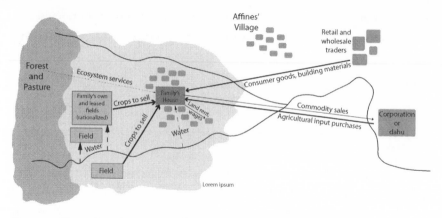

Figure 9.1. Flows of goods and services in the postpeasant Chinese household.

themselves. Food no longer came from a family's fields, even if family members still worked them; rather, they sold their crops or animals or leased their land to agribusiness corporations and/or *da hu* and purchased food at stores or markets. Fertilizer no longer came from pigpens and outhouses but rather from sacks purchased in stores or provided by investors. Kin relations with neighboring villages no longer involved cooperation in subsistence; affinity was no longer ecological. Scaling up was thus more than just bigger farms, more pigs or chickens, more machinery, more chemicals, and more investment. Beginning in the 2010s, rural Chinese, even as they decreased in number with migration to cities, became increasingly entwined in a national and even global system of ecological flows. The new rural ecology was less localized, less circular, more monetized, and far, far less self-sufficient.

Questions arise as to the resilience of this kind of a system. On the one hand, wider circles of exchange can buffer a local system against purely local disturbances—a local hailstorm or flash flood that wipes out one season's crops in a small area may bring hardship, but it will not bring starvation. On the other hand, both inability to produce for one's own consumption and reliance on outside inputs for everything from energy to building materials to basic foodstuffs makes people dependent on the functioning of complex, large-scale systems of production and distribution. In Pingpu, Anhui, for example, *da hu* investors and family farmers alike suffered from market fluctuations and diseases that ruined some of their crops. They not only lost money to the point that many did not renew their contracts, they often left perfectly good fields fallow or allowed planted crops to rot in the fields because there was no market where they could sell them at a profit (Doll 2021).

....

From 2014 through 2022, China produced record amounts of grain and was able to import record amounts of soybeans, corn, and alfalfa, so people were able to eat as much rice, wheat, and animal foods as they wanted, and they could find fruits and vegetables in abundance. At the same time, the externalities produced in this hyperintensified, highly interconnected, multiscale system threatened the viability and safety of the soil that nurtured the crops that in turn nurtured the people and their livestock, and it put severe limits on both the quantity and quality of the water used to nourish them. These contradictions lead us to question how long and at what cost China can maintain this abundance and variety of food.

10. Big Ag and Its Ecosystem Effects, 2002–2022

Between the rivers of manure that flow from industrial livestock operations
and contaminate rural waterways; the loss of soil nutrients
and food calories in the inefficient conversion of grains and oilseeds
into industrial meat; the erosion of agricultural knowledge and practice that
accompanies the dispossession of China's farmers; and the shifting values of
pigs, pork, and manure, this is a system that "wastes" the
rural in service of capital.

MINDI SCHNEIDER, "Wasting the Rural," 2015

As China's food system gave people more and better food, the country also came to depend more and more on consolidated, chemicalized, commercialized, and mechanized (CCCM) agriculture, leading to two kinds of effects. First, ecosystem degradation intensified. Chemical overuse, monocrop agriculture, concentrated animal feeding operations (CAFOS), and intensified aquaculture affected ecosystems well beyond the farms themselves and indeed beyond China's borders. Second, political and environmental activists, aware of these negative ecosystem effects and of the possible detriments to individual and public health, began to advocate for alternative food systems, including organic agriculture, community supported agriculture (CSA) programs, farmers' markets, and even vegetarianism and veganism.

CCCM AGRICULTURE—INCREASING PRODUCTIVITY AND DECREASING RESILIENCE

CCCM agriculture can degrade both its own ecosystems and the larger ones to which it is connected. Untrammeled mechanization, scaling up, and especially chemicalization consume industrial agro-ecosystems from within, degrading the very resources on which farmers depend, especially the soil where they grow the crops they sell or feed to their animals. Their wastes also pollute nearby air and water while consuming large quantities of fossil fuels that contribute to climate change. China's full-scale adoption of CCCM agriculture in the early twenty-first century, while continuing to provide the tasty variety of foods and boisterous food experiences described in chapter 9, brought about degradation of both terrestrial and aquatic food-producing ecosystems.

Chemical fertilizer, pesticides, antibiotics, crowding, and lack of diversity in agro-ecosystems all potentially trade long-term resilience for short-term productivity. In service of these short-term gains, state planners and scientists, spurred on by their general faith in linear science and Hu Jintao's ideology of the "scientific perspective on development" (*kexue fazhan guan*), actively promoted "conventional," linear agricultural science. The party Central Committee and State Council's "Some Opinions" document (CCPCC 2008) on promoting agriculture justified this linear thinking as the salvation of a system hampered by farmers who were "low cultural level, older, mostly women, [and] partial[ly] conservative [with] lack of enthusiasm and initiative for application of agricultural science and technology" (Dan Wang and Li 2010, 119; Hua 2015). The poisons brewed up by this faith in science continued to spill into the environment in China and elsewhere.

Chemical Fertilizer

Most Chinese farm soils in the mid-twentieth century were short on nitrogen and phosphorus. This dearth not only limited productivity but also led farmers to recycle almost all available nitrogen, whether in the form of human and animal manure or in various plant-based fertilizers such as "bean cakes" made from the residue of *doufu* (tofu) manufacture. Even after chemical fertilizers became widely available beginning in the 1970s, most Chinese farmers continued to use manure as their main fertilizer, supplemented with chemicals, for the next two decades.

Since the 1990s in particular, this cycle of nutrients has broken down. Farmers reduced or even eliminated manure as fertilizer (Li Jian 2010, 68), even though, with far more livestock, as much as four times more manure was available than in 1949 (Yuxuan Li et al. 2014, 977). By the early 2000s, it was estimated that farm manure accounted for only 25 to 40 percent of the fertilizer used in China (Li Jian 2010, 71; Yuxuan Li et al. 2014, 978). Where manure was available from CAFOs, crop farmers often could not use it all (Schneider 2015, 95), so the CAFOs had to dispose of it otherwise, often improperly (Li Jian 2010, 73). Excess nitrogen and phosphorus leached into groundwater and eventually flowed into streams, contributing to algal blooms that turned lakes green and deprived aquatic animals of oxygen. By 2010 manure from CAFOs accounted for 42 percent of chemical oxygen demand in surface waters (Schneider and Sharma 2014, 32), while agriculture as a whole, mostly through overuse and inefficiency, contributed about 60 percent of both nitrogen and phosphorus found in all water bodies in the country (Yuxuan Li et al. 2014, 973).

Farmers in China in the early twenty-first century not only replaced manure with chemicals but also often used too much of the chemicals, particularly nitrogen, and used them inefficiently. Many farmers believed that more fertilizer would result in higher yields, not realizing that there is a limit to the amount of nutrients that plants can take up. From the 1960s to the 2010s, nutrients applied in chemical fertilizer increased ten times while staple yields only tripled—clearly plants were not taking up all the fertilizer applied to them. In 2010 Chinese wheat farmers used about 11 percent more nitrogen per hectare than did farmers in the United Kingdom, but British farmers achieved yields about 1.6 times those of their Chinese counterparts (Yuxuan Li et al. 2014, 977). Similarly, China's corn farmers used 60 percent more nitrogen per hectare than US farmers but produced yields of 43 percent *less* per hectare (Yuxuan Li et al. 2014, 977). Much nitrogen was lost to the atmosphere through volatilization, and both nitrogen and phosphorus flowed into groundwater through leaching (Yuxuan Li et al. 2014, 977). In the Heilongjiang Reclamation District (see map 3.1), where large, mechanized farms predominated, it was estimated that between 10 and 14 percent of nitrogen applied to wheat, corn, and rice fields between 2003 and 2012 was leached, contributing about 80 percent of the organic pollutants in regional waters (Q. Yang, Liu, and Zhang 2017). In addition, after 2010 there were diminishing returns to marginal increases in fertilizer application: fertilizer inputs increased, but yields no longer increased with them (Q. Yang, Liu, and Zhang 2017).

Farmers the world over know that, unlike chemical fertilizers, manure and other organic fertilizers can increase soil organic carbon (aiding carbon sequestration, as well as improving soil quality), maintain or sometimes raise soil pH, promote beneficial soil microbial activity, help to maintain soil structure by increasing the percentage of water-stable aggregates, and improve water retention in semiarid regions (see Ozlu and Kumar 2018; X. Ding et al. 2012; Hongyuan Yu et al. 2012; M. Wei et al. 2017; and T. Fan et al. 2005). It is thus likely that the switch from organic to chemical fertilizers has harmed soil quality in several ways.

Pesticides

Chemical pesticides (insecticides, herbicides, and fungicides) are also key inputs in modern industrial agriculture, and China's large-scale industrial agriculture came to depend on insecticides and fungicides to prevent insect and disease outbreaks in expansive monocrop fields and on herbicides to eliminate weeds.

Monocrops typically lack the resistance to diseases and weeds that diversity provides in traditional polyculture. If Chinese agriculture was going to move toward large farms, monocrops, and mechanization, it was would need to adopt pesticides on a large scale, and it did, beginning in the 1990s. By 2014, China's 1.8 million tons of pesticides, up from about 260,000 tons in 1960 (Koitabashi 2009), accounted for about 36 percent of worldwide use (L. Fan et al. 2015, 361). Use per hectare overall was between 10.8 (Science 2013) and 14 kilograms (Huizhen Li, Zeng, and You 2014, 965), contrasted to about 2.2 kilos in the United States and 2.9 kilos in France, but not much higher than the 8.8 kilos in the Netherlands or the 13.1 kilos in Japan (*Science* 2013). These national averages, however, hid huge regional differences. Farmers in southern provinces applied much more pesticide per hectare: in 2009, 36.6 kilograms in Guangdong, 43.5 in Fujian, and a whopping 63.4 in Hainan, while Ningxia's rate of 2.3 kilos was barely higher than in the pesticide-conscious United States (Huizhen Li, Zeng, and You 2014, 965). In addition, the majority of pesticides used in China were insecticides, which generally have greater toxic effects on humans than herbicides or fungicides (Huizhen Li, Zeng, and You 2014, 965). Between 60 and 70 percent of insecticides used between 2006 and 2013 were organophosphates, replacing the organochlorides such as DDT and 666, which were ubiquitous (though in much smaller quantities) until they were banned in 1983. Some organophosphates were also banned in 2003, but others, along with carbamates, then took their place (Huizhen Li, Zeng, and You 2014, 966).

Even though pesticides probably helped increase crop yields in the short run, Chinese scientists and farmers in the twenty-first century became increasingly aware of their adverse environmental and human effects, and environmental scientists discovered that that even long-banned compounds such as DDT and 666 remained in the environment in many parts of the country and may have continued to pose some human health risks from polluted air (Lifei Zhang et al. 2013), water (Luo Wang et al. 2013; Zhenwu Tang et al. 2013), and soil (Huan-Yun Yu et al. 2013; Dan Yang et al. 2013). A comprehensive study of soil contamination on over sixty thousand sites nationwide showed that a small percentage were contaminated by organochloride residues, though the areas affected were a fraction of those contaminated by heavy metals from industrial wastes (MEP 2014).

Awareness did not, however, necessarily cause farmers to use fewer pesticides, and in some cases farmers used many times the amounts that pesticide manufacturers recommended, contributing more to environmental degradation and health hazards than would even be necessary in "modern" CCCM farming (Chao Zhang et al. 2015), for several reasons. In many areas farmers had only a general knowledge of the dangers of pesticide use, and agricultural extension services did an inadequate job of informing them, so that they got most of their information from pesticide retailers (L. Fan et al. 2015, 364–67; Li Jian 2010, 68). Pesticides were not expensive, and many farmers believed that they had to use pesticides on their crops, particularly on the expanding acreage devoted to cotton (Ruijian Chen, Huang, and Qiao 2013) and to fruits and vegetables, because of the danger that insects or diseases could wipe out whole crops and cause severe economic losses (L. Fan et al. 2015; Huizhen Li, Zeng, and You 2014). In addition, since continued application of a single pesticide can lead to the evolution of disease-resistant pests, farmers sometimes reacted to diminishing effectiveness of a pesticide by applying more, adding to the possibility of resistant strains as well as to environmental pollution and food contamination (Huizhen Li, Zeng, and You 2014, 967). At the same time, industrial monocrop methods of cultivation are *dependent* on pesticide use, so that if farmers who have

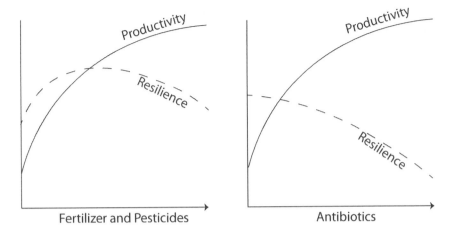

Figure 10.1. Relationships between productivity and resilience with increasing inputs of fertilizer and pesticides (*left*) and antibiotics (*right*). Fertilizers and pesticides, applied in small amounts, can increase both productivity and resilience, but as amounts increase, productivity levels off and resilience decreases. With antibiotics in animal feed, productivity also increases at low levels, but resilience drops with any level of application; productivity and resilience are in inverse relationship. Depending on risk and cost, the point at which a farmer will stop applying the inputs might vary.

abandoned traditional methods of pest control such as crop diversity then use too little of a particular pesticide, outbreaks may spread (Chao Zhang et al. 2015).

Overuse of chemical fertilizers and pesticides also illustrates the dangers of linear thinking that was so evident in the Great Leap Forward (see figure 4.2). No production input, whether labor, fertilizer, machinery, or pesticides, produces increasing marginal returns indefinitely in response to increased application. There is always a point where returns to inputs begin to diminish, where the detriments, including pollution and lost resilience, begin to outweigh the benefits of increased production (figure 10.1, left). A certain amount of chemical fertilizer was probably the quickest and most efficient way to bring China out of centuries of food insecurity (Smil 1993, 126–27; 2002, 126), and since China was adopting an industrial monocrop model, it was difficult to do without pesticides. Given these necessities, the gains in production may have initially outweighed the internal and external environmental degradation they caused. But using too much of any chemical input provided little benefit and accelerated local and regional ecosystem vulnerability, just as mobilizing labor to build waterworks in the late 1950s had been a good idea up to a point, but overdoing it brought about not just vulnerability but disaster.

Antibiotics, Crowding, and Disease

Farmers added antibiotics to feed in order to promote rapid animal growth (lowering the cost per animal raised) and to prevent the spread of diseases among large, confined herds of cattle, pigs, and chickens. Antibiotics in livestock feed differ from fertilizers or pesticides in that using antibiotics for nontherapeutic reasons, even in small quantities, can increase productivity but always diminishes resilience (figure 10.1, right). But the advent of ever larger and more concentrated livestock operations made antibiotics necessary. In 2008 almost half of the antibiotics produced in China were fed to livestock (Sharma and Schneider 2014, 32), including pigs, dairy cattle (Sharma and Zhang 2014, 23), and poultry. Some chicken farmers were reported to use as many as eighteen antibiotics at once, and predictably, officials of dragon head enterprises blamed small farmers' ignorance for overuse (C. Pi, Zhang, and Horowitz 2014, 28). Bacteria and other pathogens are notoriously capable of mutating to develop resistance to antibiotics, a potential problem for livestock as well as for humans, and antibiotic-resistant mutant genes have been found in bacteria recovered from water contaminated by waste from livestock (including pigs, cattle, chickens, and ducks) in several places in China (see B. Chen et al. 2015; W. Cheng et al. 2013; Xuelian Zhang et al. 2014). In addition, antibiotic residues in agricultural runoff are present in many sources of household water, possibly posing health risks to humans, though the level of risk does not appear to be particularly high (Ju et al. 2019).

Viral diseases, which cannot be cured by antibiotics, can be even greater problems, and they also spread more quickly in larger farms than they did previously among pigs raised by individual households. African swine fever, caused by a double-stranded DNA virus, was first detected on a family farm (with 383 pigs) in Liaoning in July 2018; the disease killed the whole herd in a few weeks (Shengqiang Ge et al. 2018, 2131; Jun Ma et al. 2020). The virus then spread quickly, and in the middle of 2019 it was estimated that perhaps 300 million pigs would be lost in the epizootic, about half the amount slaughtered in a typical year (Charles 2019). Government agencies encouraged farmers to cull their pig herds and promised compensation, but the compensation was less than farmers could get for healthy pigs in local markets. Some farmers, already operating on small margins, either quickly sold their pigs before they were infected or slaughtered and buried them to prevent further spread of the disease. More broadly, the epizootic accelerated the push to consolidate farms, on the premise that large, modern farms could afford to be more sanitary (Bradsher and Tang 2019) and could employ more scientific means of monitoring diseases. These included the facial recognition technology that also surveils Uyghurs and other minorities on the streets and in the alleys of Xinjiang (Wee and Chen 2019) and developing artificial intelligences that can identify sick pigs through unique QR codes when they display signs of fever or if they appear to be crushing each other in their crowded pens (Xiaowei Zhang 2020, 82–83). These precautions were effective in general, but six small outbreaks of swine fever were reported in early 2021, and in response China banned pig imports from Malaysia as well as "cracking down harder on fake vaccines" for the disease (Orange Wang 2021, Jane Zhang 2021).

Avian influenza or "bird flu" also broke out in concentrated poultry operations almost every year (C. Pi, Zhang, and Horowitz 2014, 29), an alarming phenomenon since, although bird-to-human transmission is rare, the disease is highly lethal when contracted (C. Pi, Zhang, and Horowitz 2014, 29; C. Zhou 2020), and China reported a human case in April 2022 (Weisheng 2022). Like the African swine fever, outbreaks of bird flu were directly connected to crowded conditions in CAFOs and will probably continue as long as these conditions are present.

Like the shrimp scientists described in chapter 9, agricultural re-searchers advocating pesticide and antibiotic use took it for granted that modern science was superior to the traditional knowledge that kept disease outbreaks under control in former times, which the scientists dismissed as nothing but ignorance and superstition. Strangely, in the era of "ecological civilization" Chinese agricultural science and practice still subscribed to the very unecological idea that human plans can con-quer nature. They no longer applied this thinking through campaigns of mass labor mobilization, as they did in the 1950s through 1970s, but through technically sophisticated but still linear science. At a time when systems science was questioning the assumption that relations in the natural world were linear (Meadows 2008, 91–94), many advocates of agricultural science failed to recognize that, as Ian Hacking (1999, 59, quoted in Y. Huang 2012, 76) has observed, "few things that work in the laboratory work very well in a thoroughly unmodified world—in a world which has not been bent toward the laboratory."

Lack of Diversity and Diminished Resilience

Reductions in both species diversity and patch diversity in industrial monocrop farms probably also lowered local agro-ecological systems' resilience to both economic and ecological disturbances (see Gunderson et al. 2002, 8–12; Cabell and Oelofse 2012, 18; and Matsushita, Yamane, and Asano 2016). The modernization of farming in three communities on the Chengdu Plain as part of the Hu-Wen era's New Socialist Countryside campaign illustrates what can happen when diversity decreases. The village of Zhanqi eliminated the traditional *linpan* dispersed housing and moved everyone into two- and four-story apartment buildings. They also consolidated their farms, leading to a much larger patch size. Though they maintained some crop diversity, they planted about 45 percent of their land in lavender to promote the local tourist trade (Tippins 2013, 128–30), rendering them vulnerable to the vagaries of changing tourist preferences. Another village, Jiang'an, maintained the traditional *linpan* settlement pattern but devoted almost all its land to ornamental trees, which farmers sold to landscaping companies building high-rise complexes in regional cities. As a result, they suffered when a disturbance hit—the housing market softened and construction slowed in 2015 (Tippins 2013, 95–96; Whiting et al. 2019, 276). Most residents of the village of Anlong participated in a partial consolidation program that moved them into something more like the villages found in other parts of China; they also devoted more than 75 percent of their fields to ornamentals and suffered similar consequences from the slumping real estate market (Whiting et al. 2019, 276). A few households in Anlong formed a community supported agriculture cooperative (see below). They maintained both their traditional housing pattern and a mix of crops, including ornamentals, but also with substantial areas devoted to grain, orchards, and vegetables (Tippins 2013, 106–7), in a conscious effort to sacrifice some productivity in the interests of resilience.

Specialization can diminish resilience even more than consolidation does. When a whole community devotes itself to a single crop, either a downturn in the market for that crop or a disease that kills the plants can effectively wipe out the local agricultural economy, rendering the population ever more dependent on migratory labor. The demise of the banana crop in "Nacun" in Guangxi, described in chapter 9, may be an example of this; the *Fusarium* fungus that causes Panama disease, which killed off the banana trees, spreads environmentally from plant to plant through contaminated water, soil, and insect vectors, and there is as yet no fungicidal or other treatment, so only quarantine and destruction of plants are effective against it (Pérez-Vicente et al. 2014, 15–19).

Figure 10.2 illustrates the effects of elements of CCCM agriculture on social-ecological system resilience and ecosystem quality.

INTERNATIONAL REPERCUSSIONS

Even if Lester Brown was wrongly alarmist in predicting that feeding China would create an international food crisis and cause prices to increase worldwide, there were still international repercussions from China's increasing appetites, particularly in the Americas and Southeast Asia, increasingly in Africa, and—significantly—in the world's oceans. The 2007 "going out" (*zou chuqu*) strategy, adopted in response to perceived food security threats (Qiu et al. 2013), encouraged agribusinesses to strengthen connections with foreign countries, not simply to provide China with imports (which continued to be a very small part of China's food supply) but as profit-making ventures (Zhan, Zhang, and He 2018, 704). These overseas expansions were not without their ecological effects.

Meat, Feed, and Deforestation in South America

The increasing numbers of animals that Chinese people ate in the early twenty-first century depended greatly on imported soybeans, mainly from the US Midwest in the early 2010s, but after 2014 also from Brazil, Argentina, Paraguay, and lowland regions of Bolivia. In addition, even though imported beef in the late 2010s accounted for only about 10 to

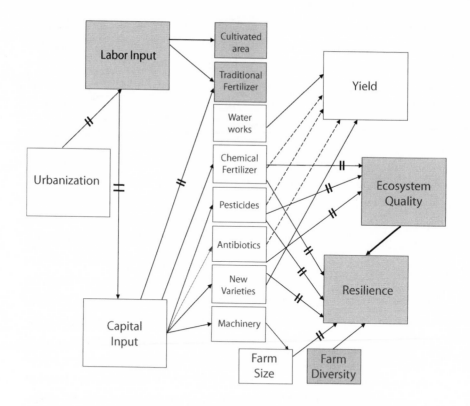

Figure 10.2. Inputs and resilience in CCCM agriculture. Gray boxes indicate factors that decreased in the early twenty-first century, while white boxes indicate factors that increased. Solid arrows indicate a positive relationship between two factors: the change in the originating factor, whether increase or decrease, has caused a corresponding increase or decrease in the affected factor. Dashed arrows mean that the relationship is curvilinear; there are positive effects at lower quantities but diminishing returns to marginal additions. Blocked arrows indicate an inverse relationship; the change in the originating factor, whether increase or decrease, has caused the opposite change in the affected factor. Thus, an increase in urbanization has contributed to a decrease in rural labor input, while a decrease in rural labor input has contributed to an increase in capital input, as capital increasingly replaced labor.

20 percent of China's consumption, Brazil, Argentina, and Uruguay together exported about six hundred thousand tons of beef to China, overtaking Australia as the largest sources of beef imports (Wang Chen 2019, 14; Trase 2020).

Most of the soy that Brazil exported to China was planted not in the Amazon but rather in the *cerrado* savanna farther east and in the long-planted parts of southern Brazil. About half of the total area in the *cerrado* that had been covered with native vegetation in 2005 was planted in soybeans by the late 2010s (Andreoni 2019, 33). Another 2.6 million hectares had been deforested in neighboring parts of Argentina, again mostly for soy production (Abrosio 2019, 34).

Beef exports to China had a greater effect on Amazon forests than did soy production. Much of the beef that Brazil exported to China was actually raised in the south of the country, where land productivity was relatively high. But because there was still domestic demand for beef, ranchers expanded into the relatively unproductive Amazon (Lazzeri 2019a, 4–5). There, deforestation reduced evapotranspiration and increased temperatures, raising the incidence of wildfires, on top of the fires deliberately set to clear the land for agriculture and ranching (Lazzeri 2019b, 7–9). This activity also contributed to increases in greenhouse gas emissions (Trase 2020; Milhorance 2020). This same pattern of deforestation for cattle ranching began to appear in lowland Bolivia as well (Jemio 2019, 19).

Worried about the persistence of African swine fever on its own farms, China also increased pork imports. In 2019 China began to import pork from ten joint-venture farms in Argentina, and in late 2020 the government signed a deal to open twenty-five more industrial pork operations there. Predictably, Argentine environmentalists worried (justifiably) about human health risks, manure pollution, greenhouse gas emissions, pressure on forests (from anticipated increases in soybean production), and exclusion of Argentina's small- and medium-scale farmers (Koop 2020).

Chinese Farms on Other Continents

Up to 2006, China had invested only US$190 million in overseas agriculture, but by 2016 that total had risen to US$3.29 billion (Xiaoyu

Jiang, Chen, and Wang 2018, 42). By 2022, Chinese firms had negotiated agricultural land deals in at least thirty-nine countries around the world (Xinhai Lu, Li, and Ke 2020, 8; Land Matrix n.d.). But contrary to journalistic narratives of "land grabs" in Africa and South America (Hongzhou Zhang 2014), investments in Africa were mostly small scale (Yangfen Chen et al. 2017, 365–68), and almost all of China's agricultural land purchases were in Southeast Asian countries, mostly in Laos, Cambodia, and Myanmar, with a lot of additional investment in Kazakhstan (Xinhai Lu, Li, and Ke 2020, 3).

Although most of the food grown on land that Chinese companies purchased or leased in Southeast Asia was sold locally, the Chinese government still saw foreign farms as a hedge against urban encroachment on farmland in China. Once they had the land, investors operating overseas could always ship food back to China if the market supported it. Also, Chinese planners hoped that technology transfer and agricultural inputs would raise yields in the host countries, thus relieving some pressure on world food markets. Finally, Chinese firms primarily grew bulk, low-priced crops on overseas land (81 percent of purchased land in Africa was planted to staples, for example), and this allowed the host countries to import higher-quality, higher-priced, labor-intensive foods, such as shrimp, vegetables, and fruits from China, boosting the market for high-return Chinese crops (Zhan, Zhang, and He 2018, 716).

It is unclear just how much ecological effect Chinese agriculture had on host countries in the early twenty-first century. For one thing, different methods for calculating total acreage, plus the opacity of the actual contracts between Chinese investors and foreign sellers, mean that estimates varied widely, from about 1.0 to 3.2 million hectares (see Yangfen Chen et al. 2017, 364; Xinhai Lu, Li, and Ke 2020, 2; and Gooch and Gale 2018, 3). For another, many deals either were never consummated or never resulted in actual use of the land; one study estimated that only about 12.8 percent of the land negotiated actually ended up being farmed (Qiu et al. 2013, 46), and many farms were planted for a while and then abandoned. In the 2010s many Chinese companies shifted from land purchases and leases to mergers with and acquisitions of foreign agribusiness concerns (Gooch and Gale 2018, 35), most notably the purchase of the US meatpacker Smithfield by Henan Shuhanghui Investment and Development.

Chinese Appetites and World Fisheries

Chinese fishing in world oceans grew rapidly in the early twenty-first century, and its effects were unequivocal. China in the 2010s produced and consumed more than 39 percent of the world's fish and other aquatic foods, and its fishing fleets continued to expand throughout the high seas, particularly on the Pacific Ocean. From the beginning of Reform until the end of the 1990s, China's leaders encouraged aggressive marine fisheries expansion, almost entirely in coastal and nearby waters, resulting in both depletion of coastal fish stocks and degradation of coastal and nearshore habitats (see Mallory 2016, 76; Hongzhou Zhang and Wu 2017, 217; and Hongzhou Zhang 2016, 69). Alarmed by this phenomenon, the State Council issued documents in 2003 advocating both aquaculture and the expansion of ocean fishing to more distant waters, aided by subsidies for boat fuel and tax allowances for depreciation (Mallory 2016, 76). This amounted to exporting the degradation of marine ecosystems. In 2010 China switched from this policy of no net growth (substituting distant for nearby fisheries) to a policy of further expansion (Hongzhou Zhang and Wu 2017, 220). By 2014 Chinese distant water fisheries were annually bringing in over two million tons of fish, about a third of that from other countries' exclusive economic zones and the rest on the high seas (Hongzhou Zhang and Wu 2017, 218). In addition, Chinese companies began in the late 2010s to build fish-processing plants in places such as Sierra Leone, raising local fears of both depletion of local fisheries and harm to marine and coastal ecosystems (Nyabiage 2021).

As they expanded outward, Chinese fishers were accused both of overfishing and of unsustainable practices such as bottom trawling (Alavi 2020), illegal mesh size, underreporting of catch, fishing in protected areas, harmful bycatch, and harvesting protected species (Hongzhou Zhang and Wu 2017, 220). The last became the object of particular concern as wealthy and middle-class appetites for tasty and prestigious seafood such as live reef fish expanded.

Shark fins were another extremely popular food in the 1990s and early 2000s. That popularity raised alarm among international conservationists, not just because of possible depletion of shark stocks but for the shocking amount of waste embodied in killing a large animal and consuming only a small part of it. Late in the 2000 aughts, however, international and domestic conservation NGOs began a public relations campaign starring basketball great Yao Ming to encourage people to avoid the delicacy. Shark fin's fall from favor was also precipitated by increasing suspicion that many of the shark fins sold in big city restaurants were fake, and although it is probable that few people could tell the difference in taste (they don't have much taste anyway), banquet hosts would lose face and restaurants would lose customers if it was discovered that they ordered or served fake, less expensive products (Hongzhou Zhang and Wu 2017, 220). This led many people to replace shark fin soup with soups or other dishes featuring sea cucumbers (Hongzhou Zhang and Wu 2017, 222–23).

Sea cucumbers, a family of mollusks that, despite their particularly unpleasant texture and lack of any taste whatsoever, are nevertheless expensive and thus attractive banquet foods. They are also considered, as are so many rare foods, to have medicinal value (Fabinyi, Barclay, and Eriksson 2017, 5). China produces some domestically but in the 2010s increasingly looked outward for an expanded supply, and Chinese traders (rather than fishers) went abroad in search of them. Some of the most desirable came from Japan, which was perfectly capable of regulating its fishery, but less so Papua New Guinea, where authorities promoted the sea cucumber fishery as a means of "development" that could earn money for coastal villagers. Concentration on the sea cucumber market not only resulted in high-value species being fished out but in villagers becoming more and more dependent on cash income from the often unsustainable fishery, as well as less and less able to produce subsistence foods for themselves (Fabinyi, Barclay, and Eriksson 2017; Barclay et al. 2019).

China in the 2010s thus entered the category of the rich countries that slowed or even reversed degradation of their own ecosystems by exporting the damage to other, usually (but not always) less wealthy areas. As the United States exported industrial pollution to China beginning in the 1990s by outsourcing manufacturing, China with its "going out" strategy, like Japan and South Korea before it (Harrell 2020), exported deforestation and agricultural pollution to South America, Africa, and fisheries around the world.

CCCM agriculture and aquaculture, along with expanding capture fisheries around the world, also brought about real and imagined fears about food safety among environmentalists and consumers. There were outright scandals, as well as more general fears about health effects of pesticide residues and rumors of unsanitary practices such as using "gutter oil" to cook street food or to sell as cooking oil, and together they raised awareness of food safety, particularly among urban consumers. This awareness prompted two reactions. From the policy side, paradoxically and probably tragically, faith in linear science and general contempt for rural people combined to spread the belief that ignorant small farmers caused breaches of food safety and that the answer was to accelerate the shift to the very CCCM model that brought about many of the threats to food safety in the first place (Sharma and Zhang 2014, 15). At the same time, small but growing groups of urban consumers came to see industrial agriculture as the problem and government regulators as the untrustworthy ones, leading to the rise of alternative food systems, including organic farming and aquaculture, community supported agriculture cooperatives, and even vegetarianism and veganism. The fact that until 2018 most urban families in China had only one child and were placing all their hopes for social mobility and old-age security in these children (Fong 2004) compounded fears about food safety and spurred the search for safe alternatives to foods grown with a variety of chemicals.

Food Scandals

In September 2008, the Chinese government announced a recall of infant milk formula powder that contained melamine, a chemical used in manufacturing plastic dinnerware, among other things. The baby formula was produced by the Sanlu Company, one of the large private dairy companies that had arisen with the recent increases in dairy consumption. About 290,000 people, mostly very young children, became ill from consuming the milk, most of them developing kidney stones, and six infants died. Urban families, who had increased their consumption of dairy products dramatically since the 1990s, cut their consumption of baby formula and other milk products (G. Qiao, Guo, and Klein 2012,

384). Sanlu was bankrupted by the scandal, the general manager was sentenced to life in prison, and two managers were executed for their role in the intentional adulteration of the product (Xiu and Klein 2010, 463; Sharma and Zhang 2014, 15).

This scandal was not simply the result of perfidy (though, as in most scandals, perfidy was a factor) but arose more fundamentally from the way the dairy industry had developed at the interface between China Proper and Central Asia since the 1990s. As government policies of "good nutrition," combined with rising urban consumer incomes and the push toward CCCM agriculture, led to an explosive expansion of the dairy industry, there was very little oversight or regulation of its complex supply chain. Most milk cows at the time were still raised in herds of twenty or fewer animals, by farmers who either milked the animals at home (by hand or with machines) or took them to milk-collecting stations established by villages and townships, where the milk was chilled and shipped to or picked up by the corporate producers, who pasteurized it and processed it for wholesale and retail markets. As demand and supply fluctuated between 2005 and 2008, people at various positions in the supply chain experienced price squeezes between costs—for feed, machinery, and other expenses—and the prices they received for their milk and thus were tempted to cut corners to make ends meet. The easiest corner to cut was to water down their milk, but watered-down milk would bring lower prices or be rejected altogether, so one solution was to add melamine, which fooled the nitrogen-based tests for milk quality into indicating that there was more protein in the milk than was actually there (Xiu and Klein 2010, 463; Gale and Hu 2009).

Who actually added the melamine was unclear. Original press reports blamed farmers, because urban people, who are mostly the ones who read (and write) the media reports, tend to be prejudiced against peasants (*nongmin*) and their "unscientific" knowledge and methods (Sharma and Zhang 2014, 15) or even their supposed lower moral and intellectual quality (*suzhi*) (H. Yan 2008). Further investigations by both Chinese and foreign authors discovered more systemic problems, however. Everyone was squeezed, there was little quality control (Pei et al. 2011, 413–15), and a consensus later developed that traders and processors were more likely than farmers to have adulterated the milk (Sharma and Zhang 2014, 15).

Paradoxically, most melamine-contaminated products came from large companies; milk from small companies and farmer cooperatives rarely contained the noxious additive (Xiu and Klein 2010, 466). Turning milk into a commodity whose production and distribution served profit rather than nutrition led to adulteration.

However, rather than questioning the appropriateness of the system, authorities moved quickly to patch it. They passed China's Food Safety Law in 2009 (Xiu and Klein 2010, 468), instituting a system of controls and inspections at all links of the supply chain. Although they at first encouraged farmer cooperatives as a way of pooling knowledge and making quality control easier (Gale and Hu 2009, 12), by the mid-2010s dairies, like other farms, were moving toward a consolidated model in which former owners of small farms became wage laborers on larger ones. By the early 2020s the dairy industry seemed destined to become another branch of China's vertically integrated CCCM agriculture; officials in Ningxia, for example, announced plans in August 2019 to have 99 percent of the region's milk produced on farms with one hundred or more cows (Xinhuanet 2019).

Adulteration with melamine was not the only milk scandal. Even earlier, in 2004, *China Daily* reported cases of watered-down or otherwise dangerous milk leading to sickness among farm children in many parts of the country, repeating the narrative that farmers were uneducated and gullible (Li Jing 2004). There were also rumors that most fluid milk was past its expiration date before it was sold or that milk powder contained excess iodine (G. Qiao, Guo, and Klein 2012, 380). Nor was it just milk and milk products. Consumers found eggs with dye added to make the yolks red, considered a sign of high quality, as well as eggshells filled with soil rather than eggs and manufactured or "faked eggs" (Alexander Lee 2005). Pork was found laced with the asthma drug clenbuterol or pumped full of water to increase weight (G. Qiao, Guo, and Klein 2012, 380), and "instant chickens," laced with feed additives, including up to eighteen antibiotics, so they would grow faster, were sold at fast-food outlets (C. Pi, Zhang, and Horowitz 2014, 15, 28).

Perhaps the most telling food scandal of the early twenty-first century was "returning gutter oil to the dining table." The term "gutter oil" (*digou you*)—"gutter" refers to streetside ditches, not roof gutters—is unappetizing in any language, and when the issue of gutter oil came to widespread public notice in the late 2000 aughts, it rightly became one of China's biggest food scandals. Most gutter oil does not come from actual gutters: the term refers to oils and fats recovered from a wide variety of kitchen waste, including leftover food from restaurants and hotel dining rooms, rotten meat, trimmed gristle or pork skin fat, reused deep frying oils or oil from hot pot broths, oil that dripped off kitchen fans or hoods, or oil wiped from woks (see Li Jingxu and Zheng 2014, 76; Zhao Z. 2014, 64; and M. Lu, Jin, and Tu 2013, 143).

As restaurants proliferated and eating out became more popular in China in the 1990s, gutter oil became an environmental problem: the excess oils that restaurants used in quantity were going to waste and clogging sewer systems. It clearly made ecological sense to collect and refine the oil for nonculinary uses such as the nascent biofuel industry or other industrial purposes (Zhao Z. 2014, 64). Municipal governments set up collection systems for restaurant owners to dispose of their used oil, but they charged a fee (Zhao Z. 2014, 64). Freelancing oil collectors, by contrast, would pay the restaurant owners for the oil, because they could flip it to refiners at a higher price (Li Jingxu and Zheng 2014, 77). The refiners, in turn, could make more money selling the refined oil back to restaurants than to the fledgling biodiesel industry, which could not afford to pay the higher prices. So everyone in the illegal business benefited. Starting in the late 1990s, the gutter oil business grew from a few local opportunists using crude equipment on a small scale to a nationwide underground industry with interregional trade and supply chains (Zhang Chengyi, Song, and Zhao 2018, 18–19).

The potential harm from food cooked in gutter oil was real: investigations found carcinogens such as aflatoxins, immunotoxic substances such as toluene aldehyde, and heavy metals, including arsenic and lead (Li Jingxu and Zheng 2014, 77). The regime began to pay attention as early as 2002, issuing regulations against reuse of gutter oil (Li Jingxu and Zheng 2014, 77), but the illegal market continued to expand. In 2010, when 100 elementary school students in Shandong were hospitalized with digestive problems traced to gutter oil in fried oil sticks (*youtiao*) in their school lunches, public attention grew, and the regime cracked down (M. Lu, Jin, and Tu 2013, 142). They issued further regulations and in early 2011 launched a three-month crackdown on violators, prosecuting 135 cases, convicting 800 offenders, and closing down over 100 "black workshops" (*hei zuofang*) (Liu Xiaohong and Wang 2013, 67; Zhang Chengyi, Song, and Zhao 2018, 18).

After the 2011 crackdown, gutter oil problems diminished, although into the late 2010s and early 2020s they were probably still happening (Zhang Chengyi, Song, and Zhao 2018, 18), to the point where in 2021 the Guangzhou city government issued regulations to make sure gutter oil was used to make biodiesel rather than "returning to the dining table" (Feng Yandan and Wu 2021). Like the melamine scandal, the gutter oil problem had its origins in skewed economic incentives: both restaurateurs and processors made more money from the process of "returning gutter oil to the dining table" than either would have made if they had purchased clean oil for cooking or sold the waste oil to biodiesel refiners or fertilizer producers. These scandals also stimulated broader reflection on public morals and the unbridled pursuit of profit. As Li Jingxu and Zheng Ruoyu (2014, 77) explain,

> The affairs of "gutter oil," "poison milk powder [melamine]," "hit and run" [an unrelated scandal at the same time] all set into motion the low road of social morality and embodied the public lack of a basis for trust. Some enterprises lacked business ethics, particularly food businesses: whether they were big corporations like Sanlu or small workshops like the gutter oil processors, they were all acting out of pursuit of profit and disregard for the safety of consumers. . . . In this kind of a society lacking in business ethics, in the eyes of businesspeople there was only money and not life. This is the reason behind our society where food safety is so fragile and causes people to have so little faith in it.

Ordinary Fears over Food Safety

Outright scandals were perhaps the most visible face of food safety concerns in early twenty-first-century China, but underlying fears about food contamination were pervasive. In particular, the increasing use of pesticides on fresh produce and contradictory beliefs and emotions about traditional markets provoked two kinds of reaction. On one side, the government attempted to cure the problem with ever purer, ever more modern cccm agriculture and science-driven management, regulation of pesticide and chemical fertilizer use, and replacing traditional markets with modern, sanitary "supermarkets." On the other, a growing urban consumer movement joined forces with peasant activists to opt out of the cccm system altogether and promote radical alternative food networks (Li Zhang and Qi 2019, 115).

Unlike specific scandals, which tended to receive huge press and social media coverage, in spite of attempts to suppress the news, and then recede from consciousness (Reyes 2009), general fears over food quality and safety became chronic among urban populations when fears over food quantity receded as the "warm and full" problem abated in the 1990s. Two surveys of urban consumers in the 2000 aughts revealed pervasive fears, including concerns about pesticides on vegetables, dangerous levels of preservatives and chemicals, bacterial contamination, meat from diseased animals, chemicals from polluted soil, and risks associated with genetically modified crops (Veeck, Yu, and Burns, 2010; Si, Regnier-Davies, and Scott 2018).

Many farmers did in fact overuse pesticides and were aware of the danger: it was quite common for farm families to grow two kinds of crops: chemically boosted ones, which would go to urban markets, and organically grown ones, fertilized only with manure, for their own household consumption, a strategy whimsically referred to as *yi jia liang zhi*, or "one family, two systems" (Li Zhang and Qi 2019, 115).[1] In one case, Guangxi farmers even raised traditional breeds of fat black pigs, fed on household scraps, for New Year and holiday slaughter, while raising faster-growing lean white pigs with hormone-laced commercial feed to sell on the market (Li Zhang and Qi 2019, 124). Consumers were well aware of this practice, thus adding to their distrust in farmers and in market foods (Kim 2014, 37–39).

Similarly, consumers were worried that, even though food safety regulations were in place, farmers, jobbers, processors, wholesalers, retailers, and restaurant cooks were all handling food in ways that made it potentially unsafe to eat (Si, Scott, and McCordic 2019; Veeck, Yu, and Burns, 2010). To try to reduce the danger of eating, some purposely bought vegetables with wormholes in them, which would indicate that farmers had not used pesticides, but they also suspected that market vendors might apply worms or dirt to mimic chemical-free food. More people bought more local food or only trusted brands, washed vegetables with rice water to remove residues, or, where space was available, planted urban gardens (Si, Regnier-Davies, and Scott 2018, 389).

Official agencies reacted to real and imagined food safety problems by promoting the "science" of ever more commercialized, ever more consolidated, ever more mechanized, but perhaps less chemicalized (CCM+/C−) production and processing. To the regime's credit, it established rational standards of regulation and certification, limiting pesticide and chemical fertilizer use and setting three levels of standards for retreating from chemicals: "green food" (reduced chemical use) in the 1990s, "hazard-free food" (a less stringent standard) in 2001, and organic food in 2005, strengthened in 2012 (S. Scott et al. 2014, 160–61). It also established food safety inspection laboratories in major cities. Although these standards were superficially designed to allow small- as well as large-scale farmers to participate, it was practically impossible for understaffed county inspection bureaus to inspect and grant permits to all small farmers or for small farmers to do the complex paperwork necessary for certification, even if their practices met organic standards (Li Zhang and Qi 2019, 114; S. Scott et al. 2014, 161, 164). As a result, large, consolidated operations were more able to receive official certification, adding to incentives for local officials to promote these big farms, even though they did not necessarily follow safer practices.

The government, with corporate support, began in 2004 to replace open markets with "supermarkets" (*nong gai chao*), in the interests of hygiene, legibility, ease of regulation, and modernity in general.[2] "Supermarkets" of all sizes—from little convenience stories to food departments in giant Walmarts to high-end stores selling mostly imported gourmet foods—grew throughout the first two decades of the century (Dinghuan Hu et al. 2004; Faber n.d.). However, open markets persisted, even into the 2020s (Q. F. Zhang and Pan 2013; Faber n.d.; S. Zhong, Crang, and Zhang 2020). The explicit effort to replace them was quickly abandoned, because many urban consumers continued to prefer food bought from retailers in either state-regulated open markets or unregulated street markets, because it was fresher (see Dinghuan Hu et al. 2004, 567; Q. F. Zhang and Pan 2013, 509; S. Zhong, Crang, and Zhang 2020, 177). At the same time, consumers did not necessarily trust the local vendors, unless they had established long-term relationships with them (S. Zhong, Crang, and Zhang 2020, 181; Faber n.d., 10). Nevertheless, for most consumers, open markets were the place to buy vegetables, seafood (preferably purchased live), and often meat, which ideally was slaughtered, butchered, sold, and cooked on the same day. Most people simply did not trust any kind of supermarket to sell meat and vegetables that were truly fresh.

At the same time, the overall supermarket business continued to grow; consumers typically bought packaged and processed foods; staples such as rice, flour, and noodles; cooking oil; sugar; and, interestingly, eggs and milk, along with snacks and drinks, in some kind of supermarket, or sometimes online (Si, Scott, and McCordic 2019, 85–86). As a result, urban shoppers often divided their shopping between the two types of markets, with the open markets for fresh food and the supermarkets for everything else. For diverse reasons, they rarely trusted either kind of outlet, but still, since they had to eat, they had to take their chances.

ALTERNATIVE FOOD NETWORKS AND OTHER RESISTANCE

Trusting neither the supply chain of the supermarkets nor the hygiene of the open markets, consumers thus faced multiple dilemmas. One way out for the affluent was to form alternative food networks by joining forces with farmers worried about their land and with environmentalists worried about the continued ecological and social sustainability of China's agricultural lands and rural communities.

It took more than food safety concerns, however, to give birth to the alternative food movement in China; it also required a crisis of rural society. The reforms instituted at the beginning of the 1980s had starkly revealed the labor surpluses in the villages and encouraged rural people to migrate and provide the cheap labor for the Chinese cities that were becoming "the world's factory." Many rural areas were left with a population of mostly old people and children. Rural and urban incomes had continued to diverge, giving rise to what critics at the time called the "three *nong* questions," namely *nongye*, or agriculture; *nongcun*, or farming villages; and *nongmin*, or peasants. In reaction, the regime implemented a series of subsidies and supports, outlined in an important white paper on rural problems in 2004. In 2006 the agricultural tax was abolished and the state announced the New Socialist

Countryside (Shehui Zhuyi Xin Nongcun) program of comprehensive rural development, which included village consolidation and relocation; new, "modern" housing; and concentrated and modernized services, including clinics, improved schools, internet, and various bureaucratic services (Looney 2012).

As the government was beginning to implement this comprehensive, top-down solution to the *nong* questions, a series of economic and social critics began to advocate a different approach, called New Rural Reconstruction (Xin Nongcun Jianshe), or NRR, based loosely on the ideas of Republican period rural reconstructionists, including the Confucian Liang Shuming and the Christian Yan Yangchu (James Yen). NRR theorists and activists advocated community-based solutions that would increase the autonomy and agency of rural communities. They promoted various forms of community organization, including cooperatives, often aided by concerned leftist urban intellectuals (Day 2008; Hale 2013). Loosely affiliated with other "peasant-based" movements in South Asia and Latin America, NRR was "united by the goals of reversing the rural-to-urban flow of resources and reviving, strengthening or creating—in a word '(re)constructing' (*jianshe* or *chongjian*) sustainable, self-sufficient communities based on cooperation among peasant households, supported by agroecological skill-sharing and alternative marketing" (Hale 2013, 3).

NRR was not exclusively a movement for organic or otherwise healthier food; many NRR cooperatives grew no organic products at all (Hale 2013, 172, 258). However, organic food did fit into the movement's ideology of a peasant livelihood not dependent on industrial inputs or government subsidies, and many NRR-affiliated communities also saw an opportunity to market chemical-free, local, and seasonal foods to urban consumers worried about food safety. There thus developed at least four kinds of symbioses between health-anxious urban consumers and community-oriented rural reformers: community supported agriculture (CSA), farmers' markets, food-buying clubs, and "urban peasant" P-Patch programs for city people to grow vegetables in the nearby countryside.

In a CSA, customers (almost always urban) pay a subscription fee to producers at the beginning of an agricultural season and receive deliveries of fresh foods, usually chosen by the producers, at regular intervals during the growing season. The earliest CSAs to gain widespread notice were the Anlong ecological farm outside Chengdu, established in 2007, and the more famous Little Donkey Farm, outside Beijing, established in 2008 and promoted by the NRR theorist Wen Tiejun (S. Scott et al. 2014, 163–64). By 2011 reporting suggested that China had more than one hundred CSAs. Most of these grew what they called "ecological food," which was local, seasonal, grown without chemical fertilizer, and free of pesticide residues (Si, Schumilas, and Scott 2015, 304), but most did not apply for or receive organic certification. For them and their customers, organic certification was at best beside the point and at worst a sham perpetrated by untrustworthy officials and farmers (S. Scott et al. 2014, 164).

Ecological farmers' markets began in the early 2010s—Beijing again took the lead with the Beijing Country Fair market, selling fresh fruits and vegetables as well as some prepared items such as tofu, rice wine, and baked goods (Si, Schumilas, and Scott 2015, 306). The market founder, who had gotten the idea from farmers' markets she had visited in New York, chose which farmers could participate, using criteria of "small or medium scale, use no pesticides or chemical fertilizers, animals are not caged and no unnecessary antibiotics are used, and farmers are willing to work with others to develop the Country Fair" (Si, Schumilas, and Scott 2015, 306). They held the market at different places throughout the city to make their food available to more potential customers (Si, Schumilas, and Scott 2015, 306).

Buying clubs got their impetus primarily from consumers rather than producers, beginning in 2004 when "a group of self-described nature lovers" banded together to purchase food in quantity from farms near Liuzhou in Guangxi (Si, Schumilas, and Scott 2015, 307). Similar buying clubs sprang up in Chengdu and elsewhere, often in connection with some of the same farms that organized CSA plans (Si, Schumilas, and Scott 2015, 307).

Finally, some cooperatives began programs for "urban peasants" (*chengshi nongmin*), or city people who could rent small plots and come out to farms on weekends to grow their own vegetables, learn from the farmers how to grow things using "ecological" methods, and also learn about the motivations behind alternative and ecological agriculture as reactions to CCCM methods (Si, Schumilas, and Scott 2015, 308).

Different groups of stakeholders in these activities often had very different motivations for participating. The farmers who organized and ran the operations were concerned with continuing economically viable small-scale farming. Many but not all of them were also determined to farm in ways that were not environmentally degrading; they were often supported by academic and nongovernmental organizations concerned with environmental sustainability. The organizers of the Anlong effort, for example, were prompted and aided both by Hong Kong–based environmental NGOs and by the Chengdu Urban Rivers Association, an NGO concerned with the effect of excessive agricultural chemicals on water quality in the city downstream from farms (Si, Schumilas, and Scott 2015; Hale 2013). In addition, some of the Anlong farmers were devout Buddhists and vegetarians, motivated by their religious beliefs both not to harm living things and to live mindful and frugal lives. Matthew Hale (2013, 226) describes their motivation as "attempt[ing] to create a certain micro-ecosystem (through intercropping certain combinations of plants, including various aromatic herbs, and introducing insect-eating animals such as frogs) that is appropriate to the area's broader ecosystem and deters the development of certain weeds and pests without killing them (including through bug zappers) or introducing potentially disruptive chemicals (including so-called 'organic' chemicals)." However, other farmers in the same network, though ecologically minded and committed to biodiversity, were more interested in economic sustainability and did not subscribe to the rather purist eco-ethical values of the Buddhists (Hale 2013).

Most customers' motivations differed from those of the farmers. Almost all customers came from the urban middle- and upper-middle classes ("ecological" foods were typically much more expensive than CCCM-grown foods), and a large number of them were mothers anxious for the safety and welfare of their only children (Si, Schumilas, and Scott 2015; Li Zhang and Qi 2019, 120). They did not trust either open market or supermarket food, nor did they trust the government's top-down attempts at regulation through complex processes of certification and inspection (see S. Scott et al. 2014; Kim 2014; and Gaudreau 2019). So they sought healthy and safe food for themselves and their children by building relationships of trust with ecological farmers. Very few of them even thought about the environmental or ecological consequences of farming with or without chemicals. They simply wanted safe and healthy food (Si, Schumilas, and Scott 2015, 304).

There were, however, exceptions. A small but perhaps growing segment of China's urban middle classes acted out of more fundamental dissatisfactions than just worries over food safety. In a series of interviews, Su Min Kim (2013) found Anlong's customers rejecting the ethics of consumerism altogether, along with the conventional food supply chains and their inherent hazards. Some saw the whole problem as something specifically Chinese: "In China there are some difficult situations; it could be better in other countries where they have more savvy standards and these standards are strictly enforced. In that way it might be okay but it is not very realistic in China. . . . When it comes to me, for me to have trust I'm going to want to know, like, I know you—like in a friendship type of relationship, I can trust in it" (Kim 2013, 46). Others seemed to be motivated by clear ecological concerns: "The couple expressed a clear preference for Xu's production [over that of a local certified organic producer], which they describe as 'more like a farm (nongjia),' where the operation follows the most basic workings of nature without any *unnecessary* commercial practices. For them, [the] Xu farm's earthy nature is more appealing" (Kim 2013, 66, original emphasis).

For these customers, as for many of the producers and for the academics and activists who work with them, ecological farming and alternative food networks served as a kind of countersymbol to what they saw as morally and environmentally untethered consumer culture. Their dissatisfaction is perhaps best summed up in the contrast between certified organic products found in many supermarkets, clearly labeled and carefully packaged in plastic, as opposed to food from a CSA delivered with no labels in a reusable shopping bag.

Ecological agriculture and alternative food networks in the late 2010s and early 2020s took up only a tiny fraction of the Chinese agricultural landscape and involved a minuscule percentage of China's 1.4 billion food consumers. But their significance reaches well beyond their size. In the first place, they really do appear to be more sustainable and resilient. Jennifer Tippins (2013) has demonstrated that the landscape of the Anlong cooperative showed several signs of greater resilience than its neighbors in Pi County, Sichuan, including plant diversity, patch diversity, and patch size. Even after the cooperative structure broke down, individual families were still growing "ecological foods" and delivering them to urban families in Chengdu (Matthew Hale, pers. comm.). In the second place, they may be the clearest indicator that China's long struggle to feed itself had, by the 2010s, come to a point where striving for quantity and efficiency was no longer necessary or sufficient: there was plenty of food available, and CCCM agriculture was beginning to hurt the prospects for medium-term sustainability.

....

That China in 2021 was feeding about two and a half times its 1949 population on very little more cultivated land, and feeding it very well, must be counted an impressive success story. The costs have of course been enormous—in the starvation of the late 1950s, the displacement of rural populations, the systematic degradation of ecosystems, and the rigidity traps of harnessing waters and chemicalizing agriculture. The shadows of degraded soils, insufficient water, and extreme dependency on rigid infrastructure and on inadequately regulated institutions loom over China's near future, but it seems possible to avoid collapse, given the turn of agricultural science to regenerative practices and the turn of consumers to organic beliefs. China feeds itself, and for that everyone should be grateful. China's food and water security face challenges, and for that everyone should be vigilant.

PART III

Cities and Industry

11. A Revolution of Steel, 1949–1961

This is a global first-class blast furnace, the first time this has happened in our country. You need to manage it well.

ZHOU ENLAI, at Baotou Iron and Steel, 1959

The Chinese Communists had come to power through a mostly rural revolution. They had little experience governing urban areas after the Nationalists massacred leaders and followers of the CCP's "workers' revolution" in the 1920s. The Communists took over some cities in the northeast in 1948 and quickly moved to restore some of the important industries that had operated there under the Japanese colonial regime, including the Anshan Iron and Steel complex in Liaoning (Li Huazhong 1990) and the Fushun collieries in Heilongjiang (Seow 2021). For the rest of the country, they needed a model for systematic industrial development. Determined not to follow their capitalist adversaries in the United States, Europe, and Japan, they turned to the only socialist model they knew—that of the Soviet Union, where Josef Stalin, the "Man of Steel," had led rapid industrialization in the 1930s.[1] China would base its industrialization not just on the example of the Man of Steel but on steel itself, along with steel's upstream connections to iron ore and coal mining and its downstream connections to heavy machinery and railroads. Every link in this chain brought about environmental degradation.

THE STALINIST STRATEGY

Economist Barry Naughton (2007, 55) has pointed out that the Chinese turn toward the Stalinist model of industrialization "wrenched [the Chinese economy] out of its traditional framework and completely reoriented [it] . . . turn[ing] their backs on China's traditional household-based, 'bottom heavy' economy . . . neglect[ing] labor-intensive sectors suitable to China's vast population, and instead pour[ing] resources into capital-intensive factories producing metals, machinery, and chemicals." Over 80 percent of investment went to heavy or strategic industry, which centered on what Astrid Kander and her colleagues (2013, 159)

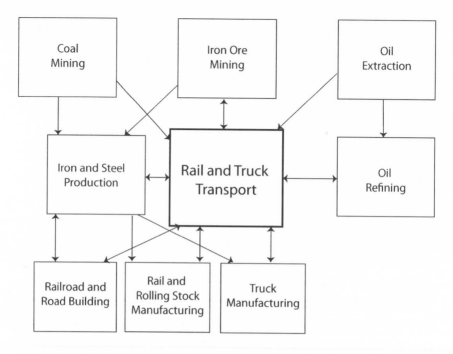

Figure 11.1. Relations between the physical elements of heavy industry built in China under the Stalinist model.

call "the steam-coal-steel development block" of industries that fueled the first industrial revolution in Europe, energized particularly by coal and including mining, steel, and railroads (see Meyskens 2020, 202; Wright 2012, 12, 27–31; and Seow 2021). They also added, only a little later, elements of the "internal combustion engine/oil block," which had fueled the second industrial revolution in Europe and America (Kander et al. 2013, 287). The relationship between the elements of China's heavy industrialization push is schematized in figure 11.1.

Chinese planners and industrialists began developing all of these elements of heavy industry in the 1950s. Transport of raw materials (coal, ore, and oil) and of industrial products was central because not all elements were always close to each other (Wright 2012, 21), but every element depends on every other, directly or indirectly. If one element falls behind, it retards the development of all the others—an important factor in explaining the mostly failed industrial strategy of the Great Leap Forward.

The Stalinist program, by pouring so much investment into heavy industry, not only had environmental consequences but also resulted in neglect of consumer goods production and lack of investment in agriculture. The labor-intensive, low-investment strategies of food production described in chapters 3 and 4 were direct results of the strategic emphasis on heavy industry. So was the scarcity of urban consumer goods and housing that plagued the country into the 1980s as planners skimped on urban construction in order to accumulate capital to invest in industry (Hou 2018, 75–78; Gaubatz 1998, 256).

Mining and Smelting

In 1949 China produced only 158,000 tons of steel (almost all of it in the northeast), but by 1953 it had already surpassed its prewar high, producing 1.35 million tons (Shi B. 1990, 124). Still, that was not nearly enough to build heavy industry and infrastructure. Consequently, the first five-year plan, extending from 1953 to 1957 and written "half in Moscow and half in Peking," set a Stalinist course of steel-and-coal-centered industrialization (Naughton 2007, 55–60). Along with expanding other industries, the plan called for quadrupling the 1952 steel output to 5.35 million tons in 1957. This meant building several large steel mills, either in existing urban areas or in new cities to be built around the smelters. Eight cities designated "heavy industrial cities"—Beijing, Baotou, Xi'an, Datong, Qiqihar, Daye, Lanzhou, and Chengdu—and thirty-one others would host heavy industries of all kinds, but central among these were iron and steel smelters (Hou 2018, 68).

The first and largest of the iron and steel projects was at Anshan, a city in Liaoning that sits in the midst of as much as a quarter of China's iron and coal reserves (Gu, Liu, and Wang 2009, 17) and that had been a site of iron mining and metallurgy for more than two millennia (Liu Xiaowen 2000, 3). Under Japanese colonial control from 1918 through 1945, Anshan was an early center of modern iron and steel production that served the Japanese empire until its defeat. The Soviet Red Army, which occupied much of Liaoning in late 1945, removed a large portion of the mining and manufacturing equipment to the Soviet Union. The Nationalist authorities attempted to revive some manufacturing activity when they occupied the area from 1946 to 1948, but when the Communists finally "liberated" Anshan in February 1948, they had to rebuild both the mining and the manufacturing almost from scratch (Liu Xiaowen 2000, 4–5). This project to rebuild and expand the Anshan Iron and Steel complex (Angang for short) became a centerpiece of the industrialization effort in the first five-year plan.[2] It also served as a model for building giant steel centers at Wuhan, where the Qing dynasty had built East Asia's first iron and steel mill in the 1890s, and at Baotou, in western Inner Mongolia.

The five major iron mines in and around Anshan (map 11.1) all began to revive production between 1948 and 1950. At the start, human labor did most of the mining, because so few machines or machine tools were available. At the Dagushan mine, the largest of the lot, workers dug ore by hand and moved it to the ore milling and sintering facilities on a narrow-gauge pushcart railway. Although each of the five mines produced just tens of thousands of tons of ore in 1949 and 1950, by the mid-1950s they were each producing more than a million tons per year, and production continued to grow into the 1980s and 1990s, as most mines changed from underground to open-pit mining and from hand labor to advanced mechanized mining techniques (Liu Xiaowen 2000, 4).

Manufacturing at Angang also grew throughout the 1950s and beyond. Steel production, begun in 1916 under Japanese management, produced about 11 million tons of raw iron, 5.5 million tons of steel, and 3.29 million tons of steel products up to 1948. Peak annual production under the Japanese puppet regime was 1.3 million tons of raw iron and 840,000 tons of steel (Li Huazhong 1990, 25), but by the time the Communists began to reconstruct the mills, they, like the mines, were in extreme disrepair, and Angang managed to produce only a little more than 100,000 tons of iron and steel in 1949.

From 1950 through 1960, production grew rapidly. During the first few years of the decade, the complex built or rebuilt six coking furnaces, three blast furnaces, eight open-hearth furnaces, and six rolling mills. By 1953, iron production had risen to ten times 1949 levels, and coke, steel, and steel products had all increased by similar amounts. During the first five-year plan, the operation expanded again, and by 1958 Angang was producing almost four million tons of iron and steel, around three-quarters of China's total production (Yu Liren 1990, 683). Responding to the mines and mills' labor demand, the urban population of Anshan shot up from 218,000 in 1949 to 391,000 in 1952 and 833,000

Map 11.1. Anshan, showing urban area, Angang steel mill, and associated mines.
Map by Lily Demet Crandall-Oral, based on Google Maps.

in 1957 (Yu Liren 1990, 677), demonstrating the urban focus of China's Soviet-derived economic strategy at the time.

The Shijingshan mill in suburban Beijing had produced some steel under Japanese occupation, but after the Founding it made only pig iron until 1958, when it produced its first steel under the new regime (Zhu G., Qiu, and Ma 2019). China's other big iron and steel operations that were part of the first five-year plan did not actually produce anything until the plan was over and the Great Leap was started—the Wuhan mill produced its first steel on 3 September 1958 (Baidu n.d. d), and Baotou in Inner Mongolia started construction in 1956 but did not finish building and produce its first blast furnace of iron until 15 October 1959, when Zhou Enlai paid a personal visit to mark the occasion, admonishing the managers to manage it well (Baotou 2015; Minwei 2017).

Metallurgy of course requires fuel, and the customary fuel is coke, made by baking coal in anaerobic conditions to volatilize impurities and leave a higher carbon content. Coal was also the primary fuel for generating electrical power, as well as for the boilers that heated northern cities in the winter and boiled the water that urban residents fetched every morning in their bundles of big thermos bottles. The Kailuan, or Kaiping, mines that fueled the industrial city of Tangshan on the Bohai coast, as well as the Fushun mines in Heilongjiang, began operating at the end of the Qing and supplied considerable amounts of coal even in the early twentieth century. Like steel, however, coal production dipped during the Civil War years of 1945–49. Since, as the 1950s slogan went, "coal is the staple grain of industry" (*mei shi gongye de liangshi*), the regime was eager to expand coal production even before the Founding, beginning when it took control of the Fushun collieries in October 1948. The Communists rapidly revived and then expanded production with the help of Soviet experts and retained Japanese engineers (Seow 2021). After 1949, production nationwide grew rapidly but did not recover its 1942 levels of around 60 million tons until about 1953. Coal production then rose to about 200 million tons at the end of the first five-year plan, in 1957 (Wright 2012, 22–23).

Moving Things Around

Steel from the mills went to the build railroad tracks, the locomotives and railcars that ran on them, and the machinery they carried to other factories. China had more than 27,000 kilometers of rail at the end of the Japanese occupation, but because of the havoc of those times and especially of the Civil War afterward, probably only around 8,000 kilometers of rail were operational in 1949, and much of the track was in poor condition (Ginsburg 1951, 470). To support industrialization, China needed to restore these lines quickly, as well as to improve and expand the network (Prybyla 1966, 269). By 1952, almost 23,000 kilometers of rail had been restored to operation, 1,400 of them double-tracked. By 1957, 26,700 kilometers were in operation, with 2,203 double-tracked (EPS China Data n.d.). China also began building more of its own railroad cars: it produced only 20 locomotives in 1952, but a total of around 900 by 1958, along with about 50,000 freight cars and around 2,000 passenger cars (EPS China Data n.d.; Prybyla 1966, 272).

China also needed roads in places where railroads were impractical, as well as trucks to carry freight over those roads. By 1956, 21 percent of China's total freight moved on its rather primitive road network, but all of the trucks that carried it were imported. China had not produced significant numbers of motor vehicles before 1949, so manufacturing trucks was a priority for the new regime, which it did with great fanfare at the First Auto Works in Changchun, built with Soviet assistance and modeled partially on the Stalin Auto Works outside Moscow (Seow 2014, 147). Ground was broken in 1953, and in October 1956 the first "Liberation" brand four-ton diesel truck rolled off the line. The First Auto Works had built three hundred of those lumbering hulks by the end of the year, and by 1957 it was turning out thousands annually. Fifteen other auto plants were operating by 1960. In the early years, most of the steel used to build the trucks was imported from the Soviet Union, and ending this dependency was a major impetus to the development of the steel industry (Seow 2014, 152). Also, like Anshan's, Changchun's urban population grew fast, doubling between 1949 and 1957 (Seow 2014, 150).

Roads were also a priority, and between 1952 and 1958 the highway network grew to four hundred thousand kilometers, including the Sichuan-Tibet highway and the Qinghai-Tibet highway, both completed in 1954 (Prybyla 1966, 273). Finally, diesel trucks need fuel, but in 1949 China had only one operational oilfield, at Yumen in western Gansu (Hou 2018, 13). Even after previously operating fields at Dushanzi in Xinjiang and Yancheng in Shaanxi were revived (Hou 2018, 22) and a new field discovered in 1956 at Karamai in northern Xinjiang (Kinzley 2018, 156), China still imported the bulk of its petroleum products throughout the 1950s. This situation spurred widespread oil exploration and culminated in the 1959 discovery of the Daqing field in Heilongjiang (Hou 2018, 26). In addition, extraction of nonferrous metals, including lithium, beryllium, and niobium-tantalum in Xinjiang (mostly exported to the Soviet Union to pay for technical assistance), combined with the expansion of the oil industry there, accelerated the development of a road network that helped solidify that Central Asian region's status as an extractive colony (Kinzley 2018, 158, 174).

Industrialization and Urbanization

All of this industrial development brought about expansion of cities and the urban population, even though the first five-year plan emphasized the development of industry rather than the development of cities (Hou

2018, 67). Total urban population more than doubled in the early years of the regime, from about 58 million in 1949 to about 131 million in 1960 (Hou 2018, 67). Under Soviet guidance, Chinese planners discarded earlier Euro-American city planning models and began to follow Soviet examples, based on the Moscow plan of 1935, which emphasized "preserving the old city core as the administrative center and channeling industrial use to the peripheral areas" (Hou 2018, 68). These plans were centered on the Soviet model of *mikrorayon*, meaning "microdistrict" or "superblock," in which a medium-sized urban area with housing blocks and attached amenities, such as stores, schools, clinics, and offices, was divided from industrial production areas by greenbelts to promote environmental health. The overall plan for factory, housing, and amenities at the First Auto Works followed this scheme (Hou 2018, 68; D. Lu 2006, 380).

At the same time, the Chinese regime, reluctant to invest much capital in anything other than heavy industry or support for industrial workers, made it clear that cities were to change from the "consumer cities of capitalism to the producer cities of socialism" (Gaubatz 1999, 1498). Planners thus slighted the consumption needs of the growing urban population, especially after 1955, when the capital constraints on industrial growth became increasingly clear. Rather than build much new housing in existing cities, for example, authorities confiscated private houses of the urban middle class, then set about dividing existing single-family homes into units for multiple families, filling in formerly open areas such as gardens and courtyards with more rooms or commercial facilities, and repurposing such buildings as temples or lineage halls to serve as multifamily housing (Gaubatz 1999, 1516; Knapp 2008). The combination of a growing urban population and the reluctance to invest in consumer goods and infrastructure led to cutting the original Soviet standard of 9.0 square meters per resident to a paltry 4.5 (Hou 2018, 76), a privation that would last until the 1980s (Y. Bian et al. 1997, 225). In addition, planners did not pay as much attention to industrial location as the previous Soviet planning advisers had. The regime often built factories along rivers, for example, drawing water from the streams and then discharging it back into the waterway without any pollution controls (Economy 2004, 47; Newman 1981). The fact that the 1979 environmental law needed to mandate the location of industry outside residential or scenic areas suggests that in the earlier years many cities had not followed their own design principles very closely (Newman 1981, 74).

Looking Back at Pollution in the 1950s

All of these interlinked efforts at rapid and massive heavy industrialization had severe and often interlinked environmental effects as well. I know of no contemporary accounts of the overall extent of environmental effects of the early industrialization efforts in the People's Republic, but we do know how this kind of heavy industrial development pollutes and degrades the environment in the absence of environmental controls. Mining iron and other ores "transforms fertile, cultivated land [or forestland that provides ecosystem services] into wasteland as mining activities generate a vast amount of solid waste which deposit at the surface and occupy a huge area" (M. Li 2006, 39). These "degraded wastelands" include stripped areas, open-pit mines, tailings dams, and waste dumps, as well as subsided lands, including gullies, sinkholes, and karst collapses, which can cause whole villages to have to be moved (see M. Li 2006; Chen H. and Ge 2003; Xiong et al. 2009). In addition, mine waste often includes both gaseous and particulate atmospheric pollution in the form of dust, ash, and chemical pollutants such as sulfur dioxide, as well as pollutants released into water, including heavy metals (Chen H. and Ge 2003, 104). And of course, iron and coal were not the only things mined in China; the nonferrous metal complex at Koktokay in Xinjiang was converted from underground to open-pit mine in 1956 (Kinzley 2018, 156), thereby exacerbating local environmental effects.

The environmental effects of coal are disastrous anywhere, and we can assume that they were as bad or worse in China in the 1950s as in other coal-producing regions, from Appalachia to the Ruhr district. Most of China's mines in the 1950s were underground, and they often caused land subsidence, sinkholes, and gullies. The small amounts of surface mining left even bigger scars on the land (see map 11.1), and slag heaps covered large areas near the mines (Wright 2012, 37–39; Zhang Yulin 2013, 88). In addition, coal production puts pressure on water supplies, diverting water from agriculture and other activities in order to control dust or wash coal and in some places contributing to desertification. And there is the air pollution—from mine dust, from dust that blows off uncovered coal cars on moving trains (Jaffe et al. 2015), and most of all from burning the coal. Coking generates both fine particulate and volatile organic air pollutants such as benzene and polycyclic organic hydrocarbons (Clean Air Council 2016). Finally, coal mining is dangerous. In the 1950s, gas explosions, cart collisions, and sometimes deadly collapses plagued China's coal mines (Seow 2021).

The process of converting iron ore into iron, in a blast furnace supplied with ore, fuel, and limestone, produces both flue dust and waste gases, including carbon monoxide, which is quickly oxidized to carbon dioxide—not a pollutant but a greenhouse gas, something that would have been of even less concern than a pollutant in the 1950s (Schueneman, High, and Biye 1963, 1). Electrostatic precipitators can be used to control dust in blast furnace exhaust, but these were just coming into use in China in the early 1980s, so dust would have been emitted without controls in the earlier period (Newman 1981, 745). Sintering plants, used to consolidate iron ore dust into a product that can more easily be fed into a blast furnace, produce airborne dust and sulfur dioxide (Schueneman, High, and Biye 1963, 2–3). Open-hearth furnaces and Bessemer converters—both of which were used in China in the 1950s to turn raw iron into steel by removing some impurities—emit sulfur dioxide, oxides of nitrogen, and small particulate matter (Schueneman, High, and Biye 1963, 3–4).

Later accounts, which unlike contemporary observations are numerous, paint a rather grim retrospective of industrial environments in the early years of industrialization, especially in terms of air and water pollution. As China was opening to the world in the early 1980s, two North American delegations visited to learn about pollution and environmental mitigation. Their reports point to some improvements that China was then making, indicating that they had not been doing those things previously, much less in the 1950s. Donald Adams's (1984, 726) account exemplifies the tone of many foreign observers of China's industrial environment in the 1980s: "Beijing is a sometimes shadowy city, its sunlight diffused by billions of dust particles—some borne by the prevailing westerly winds; some originating from the one or more million small, home coal stoves; some from the 10,000 industrial boilers; and some from local industrial enterprises. Most of the emission points are near roof level. It is fortunate that the coal used in Beijing is low in sulfur—typically 1–1.5%."

Observers also found that the entire steelmaking process from mine to mill at Anshan was producing largely uncontrolled emissions. At a copper and zinc smelter in Shenyang, sulfur dioxide and arsenic emissions were controlled when the delegation visited in 1984, but with newly installed equipment (Adams 1984, 727). Visitors to Nanjing in 1981 were told that emissions were controlled at about 10 percent of the city's factories; one observer riding on a train counted fifteen smokestacks visible in one place, "emitting colors ranging from yellow to grey to black" (Newman 1981, 745). Authorities were paying attention to pollution control measures, but they were clearly at an early stage as late as the 1980s. We can safely infer that the industrialization drive of the 1950s took place with little or no attention to concerns about pollution of water, soil, or air.

All of this heavy industry, with little concern for environmental effects, naturally had adverse effects on worker health. A retrospective study of 130,000 workers employed at Angang from 1952 through 1980, with follow-up for thirteen years afterward, hints at the problems. Researchers classified workers into white collar, blue collar nonexposed, and blue collar exposed to airborne pollutants, including dust from silica, iron, cement, welding, asbestos, coal, wood, and grinding, along with carbon monoxide, polycyclic aromatic hydrocarbons, oil mist, acid mist, benzene, and high temperatures. Exposed blue-collar workers (about 42 percent of the total sample) had significantly higher risks of mortality from tuberculosis, neoplasms, cerebrovascular diseases, and other respiratory diseases; cardiovascular disease was the only cause of mortality that was not significantly greater for exposed workers. Workers in general had lower mortality than the general population, due to the "healthy worker effect."[3] But mortality from pneumoconiosis (black lung, brown lung), was more than two and a half times as high among exposed workers as among the general population and six times as high as among white-collar workers (Hoshuyama et al. 2006).

Authorities sometimes expressed concern for the environment, especially for worker health. For example, as early as 1952, the Ministry of Railways reported, amid other concerns for worker safety, that workers in its factories in Tangshan were exposed to three times the allowable concentrations of exhaust dust and fine particulate and that tunnel drillers in the northwest often fainted (MoR [1952] 2009, 834). But it is clear that environmental degradation was far from the top of political leaders' and planners' lists of concerns as they rushed to increase China's heavy industrial capacity. As Anna Ahlers, Mette Halskov Hansen, and Rune Svarverud (2020, 5) put it, "Air pollution was, at best, a necessary evil." In addition, the Marxist-Leninist belief that pollution was a capitalist problem allowed leaders to rationalize, or sometimes even celebrate, smokestacks and their output as indications of industrial progress (Ahlers, Hansen, and Svarverud 2020, 60–61).

China was thus somewhat successful in its initial drive for heavy industrialization, building up coal, steel, rail, cement, and highways and making lesser but still substantial progress in motor vehicles and oil. We have only fragmentary information about the environmental effects, but we know that local air and water pollution were serious, with health consequences for those directly exposed, and local landscape degradation was severe in many places. In addition, cities grew rapidly, with little attention to placement of industries relative to urban residential and administrative districts. Clearly, as we would expect, the trade-offs for industrial growth often took the form of environmental injustice, exchanging local pollution and resulting illness for nationwide economic growth.

THE GREAT LEAP FORWARD IN INDUSTRY

By 1957 it was clear that, as with agriculture, the plans for industrial growth in that decade had partly succeeded but still fallen short. Reorganizing rural labor through collectivizing agriculture and sending massive teams of farmers to build waterworks not only failed to increase agricultural production enough to ensure food security, but the shortfall also slowed industrial growth. Agricultural reorganization failed to provide the extra capital in the form of agricultural products that could be sold internationally, with the receipts invested in industrial production and construction. Even reducing targets for housing space and other urban consumer goods had not produced enough savings or enough steel to fuel industrialization as fast as the regime wished.

Slow growth in steel production was the biggest bottleneck in the rush to industrial expansion (Wagner 2011a, b). It was clear by 1956 that of the large steel mills built or enlarged according to Soviet-style plans, only Anshan was producing at near the desired capacity. Wuhan would not come on line until 1958 and Baotou a year later, while Shijingshan in Beijing was still producing only pig iron, not steel. So along with the big push toward waterworks construction in the countryside in late 1957 and early 1958, the leadership decreed a massive, rapid increase in steel production, involving not only increased investment in large mines and mills but also reorganization of urban labor (Salaff 1967) to produce steel and other products at small scales.

At a certain level, this was a rational strategy (Wagner 2011b). By diverting more investment to steel production, China could unjam its development bottleneck. And as part of the leftward policy turn toward voluntarism in mid-1957, radicals in the leadership blamed China's slow industrialization on bureaucratic inertia and lack of revolutionary enthusiasm. But as with the agricultural plans of the same period, ambition and political pressure eventually led to denial of physical reality. At a Central Committee meeting in late September 1957, Mao criticized the policy of "opposing rashness" (*fan maojin*) and advocated a slogan of "More, better, faster, frugal" (Duo, kuai, hao, sheng). In terms of steel, that meant increasing production from 5.35 million tons in 1957 to 20 million tons by the end of the second five-year plan, in 1962. At the November celebration of the fortieth anniversary of the Great October Revolution in Moscow, Mao heard Nikita Khrushchev advocate "overtaking the US in 15 years," and he sinicized this line to "overtaking the UK in three five-year plans" (Mao 1957), which evolved into the Great Leap slogan, "Pass the UK and chase the US" (Chao Ying gan Mei) (Shi B. 1990, 125). The United Kingdom produced about 20 million tons of steel in 1957 and the United States, a little more than 100 million. Consequently, targets for steel production ballooned throughout 1958, from 6.2 million tons in February to 7.11 million in early May, to 8–8.5 million in late May, and to 10.7 million tons, double the previous year's output, in late August (Shi B. 1990, 128). In June, Mao shortened the time for passing the United Kingdom to two or three years (Shu 1998, 50).

The production quotas set early in the year were realistic. Angang was expanding. Construction and improvements were proceeding apace at the large iron and steel complexes in Wuhan, Shijingshan, and Baotou, as well as smaller modern mills in Taiyuan, Chongqing, Kunming, and elsewhere. These other mills were able to produce about 1.88 million tons of iron and 1.16 million tons of steel in 1958 (Shi B. 1990, 126). But this was clearly not enough to meet the rapidly expanding quotas set in midyear, so China turned to small-scale production, combining scaled-down versions of modern blast furnaces with traditional methods— blast furnaces to make iron, as well as puddling furnaces or Bessemer converters to make low-carbon steel, also known as wrought iron.[4] In Donald Wagner's (2011b) words, this was "a sensible attempt to break economic gridlock by establishing smaller industrial plants which had smaller infrastructure requirements in comparison with giant modern plants." This combination of traditional and scaled-down modern methods was able to add to China's steel production capabilities, but there was considerable waste in the process. Although the 1958 plan had called for building thirteen thousand small modern blast furnaces, only thirteen hundred remained in 1960, and about two hundred would eventually be developed into "small or medium-sized iron and steel complexes" (Riskin 1987, 217).

These plans for increasing small-scale production using practicable methods, overambitious as they were, were not the main source of waste or even the main source of environmental degradation from the Great Leap Forward in iron and steel. Much of the degradation and most of the waste came from the drive to reach the unrealistic goal of doubling 1957 production in the "big iron and steel smelting" (*da lian gangtie*) campaign. Wagner has aptly called this effort to involve the general population in steelmaking "furnaces of the masses."[5] The so-called Great Smelting campaign began in June but expanded to a nationwide mobilization in August through October (see Shi B. 1990, 128; Shu 1998, 50–51; and Y. Qian 2020). In this campaign, iron- and steelmaking was caught up in the same voluntarist fervor that had produced deep tilling, close planting, reservoir building, and sparrow bashing in the countryside. A Xinhua report from Angang modified the voluntarist agricultural slogan "The fertility of the land is as great as the courage of the people" (Ren you duoda dan, di you duoda chan) to "The output of steel is as great as the courage of the people" (Ren you duoda dan, gang you duoda chan), further linking the Great Leap mentality in agriculture and in industry.

Other voluntarist slogans appeared at the same time: "If there is grain in ideology, there will be grain; if there is steel in ideology, there will be steel" (Sixiang you liang, jiu you liang; sixiang you gang, jiu you gang) and "Laypeople can do it as soon as they learn; the day you build a furnace, it will produce steel" (Waihang yi xue jiu hui; dangtian jian lu chu gang) (Shu 1998, 51). The propaganda was screaming "Game on!"

In August, the *People's Daily* proclaimed that the way forward was by combining local and international methods (*tu yang bing ju*), and the regime began to advocate building very small blast furnaces everywhere. By the end of the month there were reportedly 240,000 of these, some with capacities as small as 0.4 cubic meter (Wagner 2011b), but in fact only 6,000 were actually operating (Shi B. 1990, 128). Still, "in the South and the North alike, old people and children alike, workers, peasants, store clerks, officials, students, soldiers, even old ladies and wearers of the red scarf—without exaggerating we can say everyone—harbored the determination to 'let the roiling molten iron drown the bandits of [insert evil imperialist country here], doing Great Smelting of Iron and Steel, even by moonlight and starlight'" (Shu 1998, 52).[6]

The campaign involved propaganda and education as well as technology: people searched classical sources for traditional methods and produced both handbooks and instructional films to show ordinary people how to make steel. Filmmaker Shi Mei stated that she could not figure out how to make steel from technical manuals, but once she witnessed ordinary people making steel, she found she could make steel too (Y. Qian 2020, 596–97).

In mid-September, leaders held a national conference call on iron and steel production, and the same day people across Henan mobilized a "great iron and steel army" of 3.6 million people, along with 400,000 vehicles, to dig ore (figure 11.2), dig coal, make coke, and smelt iron, achieving a "miracle" of 18,000 tons of iron in a day. Small Bessemer converters in the Hexi District in Tianjin produced their first small-scale furnace-load of steel on 15 September, helping to publicize the slogan of "small, local, masses" (*xiao, tu, qun*) as a way to make iron and steel (Shi B. 1990, 129). The Tianjin offices of the Xinhua News Agency built a small crucible smelter and assigned five or six people to smelt day and night. Doctors and nurses in local hospitals used their spare time to chop wood for the furnaces and collect scraps of iron to be melted.

For fuel, if they had coal, they used coal; if they had wood, they made charcoal (Shu 1998, 52).

In Fengshui Commune in Zibo, Shandong, "big smelting" began right after winter wheat was planted in the early fall. One group of peasants went to a nearby iron ore deposit to dig by hand and with simple machines. Workers from the commune canteen took food to them in the daytime, and they slept in tents right at the site. After they had dug for three days, the brigade organized a transportation team to haul the ore, with students carrying it on their backs in flour sacks. At other mines, they used horse carts or bicycles to haul the ore, or hand carts with one person pulling and another pushing. They took the ore to a converted brick kiln, which was now the "headquarters factory," with an office, meeting room, broadcast booth, canteen, barbershop, and bathhouse. Nearby were a carpenter shop and drying kiln. To the south was a brick kiln and to the west, almost a hundred newly constructed local-style furnaces. To the northeast lay the coking plant, with eight big furnaces; to the southeast, a cement kiln; to the east, under construction, a modern (*yang*) blast furnace; and to the north, workers' dormitories. Pulverized ore could be fed into the semisubterranean furnaces directly, but bigger chunks had to be broken up, so schoolchildren were given allotments of about twenty-five kilos to take home overnight and smash with hammers. Both the blast furnaces (figure 11.3) and the kilns used to make the brick and tile to construct those furnaces needed fuel, so leaders instructed villagers to cut down whatever trees were available, including those in ancestral cemeteries. And even with local people digging ore, they still did not have enough raw materials, so each family had to contribute its woks, griddles, and other iron implements to go into the furnace, a process they referred to as "smelting iron from iron" (*yi tie lian tie*) (Jin C. 2009).

In September 1957 more than 50 million people across the nation participated in the big smelting effort, and between then and the end of the year the total involved increased to about 90 million (Shi B. 1990, 129). In December the Ministry of Metallurgy announced that the nation had met the target of doubling 1957 production and in fact exceeded it a bit, producing 10.73 million tons of iron and steel. However, there were problems. Much of the iron contained too much sulfur or phosphorus, and although it might be suitable for making simple tools like knives or

Figure 11.2. Masses digging and gathering iron ore, 1958. Photo by Joseph Needam, reprinted courtesy of Needham Research Institute.

Figure 11.3. Masses operating small blast furnaces, 1958. Photo by Joseph Needham, reprinted courtesy of Needham Research Institute.

axes, it was useless for making steel (Shi B. 1990, 130; Shu 1998, 52). Of the 10.73 million tons produced in 1958, only about 3 to 4 million tons was produced by the Great Smelting campaign, which turned out to have been a colossal waste of labor time and local resources (Shi B. 1990, 132). Still, China's output of usable steel did increase, fulfilling the less utopian quotas that the central planners had proposed early in the year.

Aside from waste, however, the citizen smelting campaign had much more serious effects on environment and livelihood. The most drastic of these stemmed from the need for fuel to feed blast furnaces and other steelmaking apparatus; whether the furnaces were fueled with coal or charcoal, the environmental effects were severe. The effects of coal mining and coking, detailed above, continued during the Great Leap

Figure 11.4. An illicit charcoal kiln in the hills of Liangshan, Sichuan, 2007, probably resembling those used to provide charcoal for local furnaces in the Great Leap Forward. Photo by the author.

but expanded in scope and severity. Coal production more than tripled during the Great Leap, from 130 million tons in 1957 to 425 million in 1960 (although it fell again with the collapse of the Great Leap), and the quality of coal dropped, canceling out improvements in furnace and boiler efficiency. With huge increases in output, many mines were unable to undertake such rudimentary measures as coal washing, further contributing to thermal inefficiency and thus to the pollution emitted per unit of industrial product. Mine accidents also increased (Wright 2012, 22–23; Seow 2021).

The amount of coke produced in "primitive rural ovens" with no pollution controls expanded from 2.8 to 40 million tons between 1957 and 1960 (Zhang Z. 1999, 64–65), contributing to rural air pollution (Ahlers, Hansen, and Svarverud 2020, 65–66). But even the huge increases in coal and coke production did not provide enough coal in many localities to fuel the modern furnaces, let alone the furnaces of the masses, so they had to be fed with charcoal. Charcoal of course is made from wood, and, like the coking process, making charcoal is a source of pollution, as it emits copious smoke (figure 11.4) Sourcing the wood was a bigger problem. Forests were cut all over the country, resulting in a probable loss of about a quarter to a third of China's already meager remaining forests (Robbins and Harrell 2014, 382–84).

URBAN TRANSFORMATIONS
AND THE GREAT LEAP IN INDUSTRY

The Great Leap also brought about big changes in the urban environment. Urban population had grown between 4 and 9 percent every year from 1950 through 1957 except for 1955, and in 1958, the first year of the Great Leap, it grew 7 percent (K. Chan and Xu 1985, 603). In addition to rapid growth, attempts to form urban communes, begun in August 1958, would alter the spatial and political organization of the cities. The earliest plans called for reorganizing the manufacturing-residential-urban amenity microdistricts established under Soviet guidance, integrating them with nearby agricultural communities to form autarkic units. This never actually happened, however, and suburban agricultural communes remained separate from the urban workers' communities. The urban communes, however, did establish low-tech and low-capital "satellite factories" to manufacture consumer goods. Other communes were created in residential areas not directly tied to industrial enterprises (Salaff 1967, 87–88).

Zhengzhou before 1949 was a small city devoid of manufacturing, but with industrialization's advent it grew rapidly (from 250,000 residents in 1950 to 700,000 in 1960) and was thus an ideal place to try creating new urban forms (Salaff 1967, 90). Zhengzhou's Red Flag Commune became a national model. It began when the neighborhood residents' committee of Mosque Street mobilized local women from twenty-one households to set up handicraft workshops to make shoes, clothes, and paper bags. They were hampered, however, by conflicts between housework (considered "unproductive," following Engels and orthodox Marxist-Leninist theory) and "productive" work, and they soon began to set up nurseries and communal canteens. Mao visited in August and suggested that they were creating a people's commune, which soon grew to include one of Zhengzhou's largest new enterprises, the Spinning and Weaving Machinery Factory, along with all its attached residential areas and the local "stores, post offices, banks, restaurants, coal godowns, vegetable supply stations, bean curd [tofu] establishments, tobacco and wine factories, tailors, laundries, barbers, cobblers, watch repairers, crop warehouses, etc., together with two agricultural productive teams and a goat-milking place" (*Guangming Daily*, 16 October 1958, quoted in Salaff 1967, 93). The touted success of the Red Flag Commune led to plans to communize or cellularize the entire city, "creat[ing] a totality of urban living—bringing together industry, agriculture, commerce, culture, education and defence" (Salaff 1967, 94) and relocating residents of traditional areas in the old city to the housing areas of the new manufacturing and residential compounds.

This utopian model never took over all of Zhengzhou, much less all the cities of China, stalled as it was by the rural crisis and horrific famine of 1959–61. But it had lasting effects. The closely integrated community of residence, work, and services that was tried out in a few of the communes was key to the development of the classical "unit" (*danwei*) organization of cities that persisted through high socialist times and well into the Reform era. This form of cellular organization represented a real change in the scale of social and political interaction and also required women to join the "productive" workforce on purportedly equal terms with men.[7] Cellularization also accelerated urban pollution and environmental degradation in general. Gone were the greenbelts of

Soviet urban design, replaced by tightly packed, walled compounds that placed workers and cadres in close proximity to mostly uncontrolled urban emissions, further concentrating the effects of industrial development in local communities while the benefits from their products remained diffused.

Probably partly because newly established and expanded industries were recruiting previously rural workers and converting former farmland to factory *danwei* and partly because people were fleeing starvation in rural areas to try access some of the grain that authorities had confiscated from their villages, urban population increased by a whopping 15.4 percent, or sixteen million people, in 1959, overwhelming the best-laid utopian community plans. This surge led to attempts to reduce the urban population after 1960 and to keep urban growth rates very slow through the end of the high socialist period. China had established its household or population registration (*hukou*) system in 1951 and extended it to rural areas in 1955, but before the Great Leap it was used primarily to monitor population rather than to control migration (K. Chan 2009b, 200; K. Chan and Zhang 1999). In fact, the Household Registration Law of 1958 was an attempt to prevent unplanned migration, but it was overwhelmed by the refugee crisis of 1959 and did not have any real effect on constraining urbanization until 1961 (K. Chan and Xu 1985, 603).

Cities in Central Asia grew as well. In the mining and oil-producing regions of northern Xinjiang, new cities arose, populated mostly by Han migrants from China Proper. Their residents and workers expanded the refinery capacity at the older oilfield in Dushanzi, developed the nonferrous metal complex at Koktokay, and, most significantly, built the new oil cities. Karamay, in the north, took shape rapidly after the oilfield began production in 1956 and continued its growth into the Great Leap years, primarily according to Soviet urban planning models (Kinzley 2018, 161). Korla, the first oil city in southern Xinjiang, grew from a small Uyghur town of thirty thousand residents at the time of the Founding to a predominantly Han city of more than one hundred thousand in the 1960s. Han people from China Proper were moved there to build an oil industry and to develop the agricultural fields and canals that supported the city (Cliff 2016, 30).

....

The Great Leap Forward thus had four major effects on city-scale social-ecological systems. First, it led to major changes in the physical, spatial structure of the cities, culminating in the cellular model that prevailed through the rest of the high socialist period and well into the Reform era. Second, it concentrated the environmental effects and injustices of industrial pollution even more locally than previously, as people lived right next to where they worked—and not necessarily upwind. Third, the rush to rapidly expand industry led to an increase in the ratio of pollution to output. Any environmental controls on large-scale polluting industries were neglected in the rush to expand. Many of the ill-advised, small-scale industrialization efforts, particularly the Great Smelting campaign, produced more pollution per unit of output in unfiltered coke ovens and inefficient furnaces than was produced by larger, more energy-efficient factories. Inferior grades of coal, containing too much sulfur to begin with, were washed improperly or not at all, leading to greater emissions of ash and gaseous pollutants. Worker safety, including protection from pollution, was neglected or honored in the breach. All of these increased the numerator of the pollution-to-output ratio, while the unusable nature of much of the product decreased the denominator. Fourth and finally, the relationship between city and countryside was changed; although agricultural failures of the Great Leap resulted in some capital investment in agriculture, the exploitative nature of the urban-rural relationship endured and probably worsened.

12. Normal Socialist Industry, 1962–1980

Two fists—agriculture, national defense industry, one butt—basic industry, we have to put in place.

MAO ZEDONG to party leaders, May 1964

By 1962, when the famine and the ensuing disorganization were over and China's population was again fed (if insecurely), the Chinese leadership began to pursue a less harebrained but not necessarily less radical course in industry—curtailing investment, construction, and population growth in large cities in the eastern and northeastern parts of China Proper (K. Chan 2009b, 200) and instead developing industry in rural areas and in new cities built from the ground up in places remote from the coast and the major urban centers. There were three parts to this strategy: adopting the self-contained work unit, or *danwei*, model both in existing industrial centers and in the newly built cities; building new cities as part of the inland "Third Front" industrialization program; and encouraging industry in small cities and towns in both China Proper and Chinese Central Asia. This basic program began in the years of political relaxation between 1961 and 1965 but lasted through the Cultural Revolution, whose influence on the biophysical environment was minimal other than leading to declining production in some areas because of political activity between mid-1966 and early 1969.

THE *DANWEI* MODEL

During the Great Leap and earlier, walls had come down—the city walls of Beijing and other major centers were razed to permit redevelopment and improve traffic flow (Gaubatz 1998, 264), but after the Great Leap walls of a different kind went up. Self-contained work unit cells, or *danwei*, lined China's urban streets and sidewalks with miles and miles of whitewashed walls topped with barbed wire or broken glass. Facing one street was a gate, left open in the daytime but with a traffic barrier, framed by one or more vertical signs—black for industrial, educational, medical, and administrative units, red for party units—and secured by a person in a guardhouse who would let members and frequent visitors through but ask unrecognized visitors, "Ni na'er?" (Where [meaning what *danwei*] are you [from]?) and not let the outsider in without a satisfactory answer or an okay from someone inside. One of the first things a kindly foreign affairs officer told me in the 1980s when I began visiting China for research was "Jin renhe danwei de damen bixu yao dismount" (Whenever you enter the main gate of any *danwei*, you have to get off your bike).

The *danwei* became, in Piper Gaubatz's (1998, 257) words, "the basic building block of the Chinese city in the Maoist period." Inside those white walls was everything that people needed for everyday life. There were residential areas, offices, social services, child-care facilities, public toilets and bathhouses (since only high-ranking officials and senior managers had private facilities), stores selling everyday dry goods, meeting rooms and auditoria, retirees' clubs, sports courts and fields, and, if the *danwei* was a large one, also perhaps schools and an inpatient clinic. Factory *danwei* of course included the production facilities as well (see Bjorklund 1986; Lü and Perry 1997; Gaubatz 1998, 1999; and D. Lu 2006). With all necessary urban functions available within the walls, people would have little need to travel outside their *danwei*, so cities could skimp on public transportation and prohibit private ownership of motor vehicles altogether, again curtailing consumption in the interests of accumulating capital to increase production. Big streets had bicycle lanes, but in the 1960s and 1970s even bicycle traffic was light (Gaubatz 1998, 257). Cities were planned and built this way to increase the percentage of total urban investment allotted to industrial production; skimping on housing and urban transportation served a function similar to earlier skimping on agriculture (Naughton 2007, 56). Within industry itself, however, efficiency was not as much of a concern, and Chinese industries continued to be some of the most energy-intensive in the world (Smil 1993, 127–28). The environmental consequences of efficiently directing investment toward inefficient industry could be severe, to the point where shoddy or poorly maintained construction could exacerbate damage from natural disasters, such as the cataclysmic Tangshan earthquake of 1976.

Not all *danwei* could provide all these amenities inside their walled compounds, particularly in older urban districts where pre-Founding housing and offices remained usable, so some *danwei* constructed or refurbished existing housing in nearby neighborhoods (Bjorklund 1986, 22), but work units actually *controlled* almost all housing, whether it was located within their main compounds or elsewhere.

The ideal place to construct the full-scale *danwei* compound was not in an existing city, where there might not be room, but either on the urban margins or near the resources the *danwei* was extracting, which would allow all the components to be built from scratch and located in one place. A sawmill compound on the outskirts of a small city in northern Fujian, built and expanded from 1958 through the 1970s (S. Zhou, forthcoming), provides a representative example. The entire mill complex (map 12.1) actually contained two separate enterprises, one owned by the state, employing mostly male workers with state salaries and pensions, and the other a collective, employing mostly female family members (*jiashu*) without the benefits of state employment (S. Zhou, forthcoming). Because the two enterprises shared both space and services, however, residents considered them a single *danwei*. The mill took in logs from a series of logging camps in the nearby hills and shipped out finished products on the Yingtan-Xiamen railroad, the only intercity railroad line in Fujian at the time the mill was built (Wikipedia 2021b). The mill had internal spur lines connecting to the main railroad. Liberation Road, which ran east-west through the center of the mill complex, served truck traffic in and out and ran to the center of the local city. Within the compound were a large number of production facilities. There was a sawmill proper, plus workshops making not only finished products—plywood, particle board, railroad ties, and wooden crates—but also intermediate products such as wood fiber and steel baling mesh for wrapping bundles of plywood. A cinder block factory used residues from the mill's own coal-fired power plant. There were various warehouses and yards that provided storage space for raw logs, finished products, and machinery, along with at least two motor pool garages.

All these components, along with administrative offices for each *danwei*, were things one might find in or near a sawmill anywhere in the world, but this mill was also a living space. Workers' apartment blocks were scattered throughout the mill grounds and in a separate living area

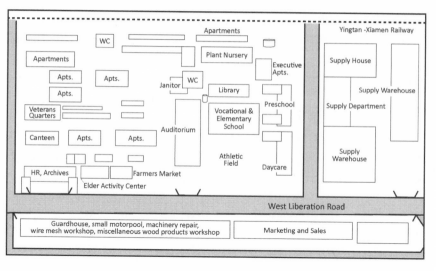

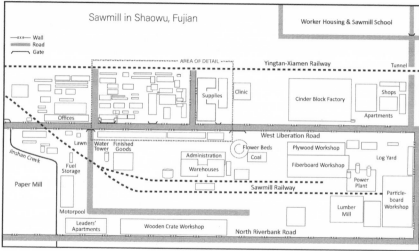

Map 12.1. The sawmill complex in Shaowu, Fujian. Drawn by Lily Demet Crandall-Oral from a reconstruction by Zhou Shaoping and Zhou Shuxuan.

(*shenghuo qu*) across another road to the south. There were separate, smaller apartment blocks with slightly nicer accommodations for administrators and for workers who were military veterans. There was a childcare facility, a kindergarten, an elementary school, a secondary school, and a vocational school. Public baths and a canteen were available, as well as a clinic, a library, a farmers' market where local peasants came to sell their produce, a series of small convenience shops, and a senior activity center. There was even a plant nursery that provided ornamental greenery to line the paths, yards, and lawns between the buildings.

This quintessential form of the comprehensive *danwei* remained through the 1980s and into the 1990s in some cases, but it could not outlast the more complete transformation to bureaucratic capitalism and the rapid expansion of the economy in the twentieth century; Dr. Zhou Shuxuan and her father, Zhou Shaoping, had to draw map 12.1 from memory. But it represents the apotheosis of Communist ideas of a total community of work, family, and recreation.

OIL AND A NEW MODEL OF URBANISM

Determined to continue its emphasis on investing in heavy industries, restricting urban growth, and investing only minimal amounts in agriculture, China still faced a hydrocarbon bottleneck. There was plenty of coal, which was being mined at ever greater rates (Wright 2012, 22–23). Still, China was short of petroleum and even in 1958 produced only 2.26 million tons (16.6 million barrels) of crude oil domestically (EPS China Data n.d.), importing over half of the oil it consumed, mostly from the Soviet Union (Hou 2018). It was thus great news for Chinese industry (and its international accounts) when oil was discovered at a site called Datong, in Heilongjiang, about 125 kilometers west of Harbin. However, because there was already a larger Datong in north Shanxi and because the petroleum discovery occurred right before the big celebration of the tenth anniversary of the founding of the People's Republic, the oilfield was named Daqing, or "Big Celebration." The development of this extraordinarily rich oilfield began the explosive growth of China's oil industry (from 2.26 million tons in 1958 to 15.99 million tons in 1968 to 1 billon tons in 1978), starting China on the road from net importer to net exporter of oil, a status that would persist until 1993 (Hou 2018, 168). Perhaps even more significantly, Daqing became a model for the new style of city that was touted for the rest of the high socialist period.

The original plans for Daqing called not for a city at all but rather for a "new socialist mining district" populated by families with "the husband in industry and the wife in agriculture." The most radical version of this new plan envisioned "one oil well and one family house surrounded by two *mu* of farmland," complemented by refineries and other more concentrated production operations (Hou 2018, 87). This ideal was embodied in the slogan "Industry and agriculture integrated, city and countryside integrated, conducive to production, convenient for living" (Gongnong jiehe, chengxiang jiehe, youli shengchan, fangbian shenghuo) (Hou 2018, 88).

Meanwhile, the workers who opened the oilfield and built the refinery had to be housed, and the quickest way to do this was to adopt the local peasant housing style called *gandalei*, a semisubterranean house with tamped-earth walls and roofs sealed with locally abundant bitumen and hay (Hou 2018, 92–94). Some of these were built in small clusters around the oil wells themselves, while larger *gandalei* settlements were built near the factories or offices where the residents worked (Hou 2018, 112) (figures 12.1 and 12.2). The only multistory buildings built during the first decade of Daqing housed the design institute and the hospital, while the workers in those units lived in nearby *gandalei* as well (Hou 2018, 113). Interspersed among all these facilities was farmland that fed the workforce, and everything was connected by a newly laid out road grid. The only direct connection of Daqing to the outside world was a line of the Japanese-built South Manchurian Railway, which fortuitously ran right through the 140-kilometer-long oilfield. This frugal, industrial-agricultural kind of rural-ish model was enshrined in one of the most famous slogans of the Cultural Revolution: "In industry, emulate Daqing; in agriculture, emulate Dazhai" (Gongye xue Daqing; nongye xue Dazhai).[1]

The original design for Daqing was dictated partly by necessity: China needed the oil desperately, especially after its break with the Soviet Union. Building a conventional city would have been slower and more costly, taking money away from foreign exchange that would have been needed even more desperately had the oil from Daqing and another large oilfield, Shengli, near the mouth of the Yellow River in Shandong, not come on line quickly. At the same time, this model reflected both Maoist ideals of local and national self-reliance and the stinginess of the leadership toward cities in general and urban amenities such as housing in particular. All through the 1970s, there was little investment in housing, leaving "urban areas in a dilapidated state" (Hou 2018, 182).

"Emulating Daqing," in frugality if not always in semisubterranean mud houses, yielded great results—it provided much of the oil needed

Figure 12.1. Gandalei settlement in Daqing, 1977. From Hou 2018. Reprinted courtesy of the President and Fellows of Harvard College.

Figure 12.2. Oil workers building *gandalei* houses, 1970s. From Hou 2018. Reprinted courtesy of the President and Fellows of Harvard College.

for the "second industrial revolution" to happen right on the heels of the first in China. However, it was disastrous for the resilience of the built environment, rendering it even more vulnerable than before to acts of nature such as earthquakes, locally concentrating negative effects of industrial intensification while the benefits of economic growth from Daqing and the other oilfields spread out over the whole country.

Daqing, though the most famous and in fact the most productive oilfield discovered in China in the Great Leap period and afterward, was by no means alone. Other oilfields sprang up across the country, allowing China to expand its oil production almost five-hundred-fold in the twenty years from 1958 to 1978. The Shengli field was one of the most successful; in addition to producing much-needed oil, it became the site of China Petroleum University. It also gained notoriety for a half-century-long process of soil contamination that threatened groundwater supplies and released dust into the air from the residue left after light hydrocarbons vaporized (Shi De-qing et al. 2007). But many of the most important oilfields were discovered in Central Asia; these included the Karamay field in northern Xinjiang, which was discovered in 1956 but greatly expanded during the Great Leap and immediately afterward. In 1958, Karamay's 357 thousand tons of oil accounted for about 16 percent of China's oil production, and just two years later, in 1960, its output had risen to 1.66 million tons (Kinzley 2018, 172).

THE THIRD FRONT INDUSTRIALIZATION DRIVE

Until the mid-1960s, most of China's industry remained solidly in China Proper; the only major exceptions were in resource extraction industries—oil and mining complexes in Xinjiang and the big steel complex at Baotou in Inner Mongolia. Amid the growing three-way Cold War tensions of the mid-1960s, however, China's leadership became afraid that either the imperialist United States or the revisionist, "social-imperialist" Soviet Union might attack China, perhaps with nuclear weapons, and that its heavy industries were vulnerable. China had built its first atom bomb in 1964, but its nuclear arsenal was still embryonic. The solution was to relocate some industries out of the way of possible attacks and to build new ones in remote, mountainous areas in the interior, far out of their presumed enemies' military reach. The slogan "Prepare for war, prepare for famine, for the people" (Bei zhan, bei huang, wei ren-min) (Shapiro 2001, 145) summed up the policies of the time. If the first front of industrialization was in the northeast and along the coast and the second in such interior cities as Lanzhou and Wuhan, the remote, mountainous areas in both China Proper and Zomia were to be the third; Mao's geographical description was "Two fists—agriculture, national defense industry, one butt—basic industry" (Liangge quantou—nongye, guofang gongye, yige pigu—jichu gongye) (Mao 1964b). If agriculture was one fist and defense was the other, they needed to be anchored by a firmly seated butt of basic industry, meaning steel and heavy machinery (Shapiro 2001, 145). Building the "butt" in the mountainous interior was encapsulated in the slogans "in the mountains, dispersed, and hidden" (kao shan, fensan, yinbi) or, more simply, "mountains, dispersed, caves" (shan, san, dong) (Shapiro 2001, 147).

The Third Front (Meyskens 2020) encompassed three regional projects. The first, beginning in 1965, was built in the southwest, much of it in Zomian parts of Sichuan, Yunnan, and Guizhou, and centered on the steel complex at Panzhihua on the Sichuan-Yunnan border and the coal complex at Liupanshui in Guizhou. A series of defense-related industries, including aluminum, arose around Chongqing. In addition to the mining and industrial complexes themselves, the project involved building a huge transport network to get the raw materials and steel to more advanced manufacturing facilities near Chengdu and Chongqing. So railroad lines were built from Kunming to Guiyang and, most famously, from Chengdu to Kunming by way of Panzhihua. These lines spanned 1,083 kilometers, fully a third of them in tunnels (Naughton 1988, 356–58).

The second project, centered on machine building, covered an area of western Hubei, southwestern Henan, and southernmost Shaanxi. Started in 1969, its most famous factory was the Second Auto Works at Shiyan in northwestern Hubei, with production lines modeled on those of the First Works, in Changchun. The layout of that entire machine-building danwei, however, followed a more Daqing-like model, with thirty-four branch factories dispersed throughout mountain valleys, each with its own air raid shelter in addition to the usual urban services (Naughton 1988, 207).

The project in the northwest, which overlapped in time with the other two, intensified a drive, begun in the 1950s, to industrialize the edges of Central Asia in Gansu and northeastern Qinghai. The Third Front added nuclear industries, mostly to build weapons of mass destruction. (China's first nuclear power plant for civilian use did not come on line until 1991 [WNN 2019].) The project also included a steel complex at Jiuquan in northwestern Gansu and the Liujiaxia hydropower plant on the upper Yellow River, which had begun construction during the Great Leap but was finished as part of the Third Front (Meyskens 2020, 217). Machinery enterprises moved from Harbin and Changchun to Longxi and Tianshui in Gansu, and from Beijing to Ningxia (see Naughton 1988, 360; Bramall 2007, 14–15; and Meyskens 2020).

In the 1980s, Panzhihua, the most famous and most successful of the cities built from scratch as part of the Third Front, extended along the southern bend of the Jinsha River and its nearby tributaries. At the foot of the grand staircase in the Bingcaogang commercial and administrative district stood a marble statue of two dancing human figures, representing the "flowers of vanadium and titanium" (*fan tai zhi hua*) (figure 12.3).[2] Vanadium and titanium, used in steel alloys and for lightweight metal construction of things such as airplanes, were the resources, along with steel and coal, that made the Panzhihua steel complex the priority project for the first phase of the Third Front and indeed a metonym for the Third Front as a whole. Not only was Panzhihua audacious and, unlike many other Third Front projects, quite successful, but also, in spite of its success, it illustrates the economic and environmental problems and trade-offs associated with the entire industrialization drive. It was also kept secret from the rest of the world, reflecting the military purpose of the Third Front; when it first appeared on published maps, it was called Dukou and was not officially named Panzhihua until 1987, when it was opened to foreign visitors.

Chinese planners chose Panzhihua for the steel project because of its remote location and the area's rich stores of magnetite iron ore. There was nothing much there before—a few villages of Han farmers in the valley and, higher up in the mountains, Zomian farmers speaking various languages and with various degrees of acculturation to Han norms. Into this remote landscape the regime moved battalions of construction workers from eastern Sichuan, along with steelworkers and engineers from Angang and elsewhere; they built coking and sintering plants,

Figure 12.3. Flowers of Vanadium and Titanium statue in downtown Panzhihua, Sichuan, 1988. Photo by the author.

along with a smelter, in a relatively flat area called Nongnongping and began mining the ore by gradually taking down a mountain just to the north (figure 12.4). Huge earthmovers shoved great chunks of the mountain down a well several hundred meters deep and into railroad cars waiting in a tunnel on a spur track off the main Chengdu-Kunming line, to take them a few kilometers to the steelworks at Nongnongping. The mill produced its first iron in 1970 (Shapiro 2001, 154), just when the railroad through to Chengdu was finished (Naughton 1998, 357), and the following year the mill produced its first steel. Railroad cars running on parallel tracks on the spur could carry steel back to the main line, bound for the machine-building factories in Sichuan.

Panzhihua was an archetypical *danwei* city (albeit with fewer walls, perhaps because of its steep topography), a company town with each of its four dominant companies based in one of the "slices" (*pian qu*) along

Figure 12.4. Mountaintop mining of magnetite ore, Panzhihua, Sichuan, 1988.
Photo by the author.

the Jinsha River into which the city was divided. "Eldest brother," the richest and most powerful, was Pangang, the steel company; its base in Nongnongping sported not only the steel mill but housing, stores, theaters (I attended an African pop music performance at one of them), clinics, several primary and secondary schools, and a major hospital. The general manager of Pangang was the second most powerful person in town after the municipal party secretary; the mayor was a distant third. "Second brother" was Pankuang, a mining company with its base a short distance downstream at Guaziping (I gave a lecture at a high school there). Although Panzhihua originally imported coal from Guizhou, later on the coal deposits at Geliping, upstream to the west, were opened, and the railroad spur allowed "brother number three," Panmei, a coal company, to ship coal directly to Nongnongping. Upstream from Geliping the Jinsha ran clear; downstream it became a dump for a variety of industries. Finally, Metallurgical Company 19 (Shijiu Ye) was the "youngest brother," in charge, despite its name, of building facilities for the other three.

Given the perceived military urgency of the Third Front, it is not surprising that Panzhihua planners and managers paid little heed to environmental concerns. In its deep, narrow valley, the city could easily trap air pollution, particularly in the hot summer months when there was little wind, and although original plans called for pollution controls, few or none were installed in the early years. Similarly, from the coal company upstream to the steel mills to the ore tailings, all sorts of pollution flowed into the tributaries and the mainstream of the Jinsha, turning the river yellow on occasion and also causing outbreaks of pollution-related diseases. Soil contamination was also severe around the industrial sites (Shapiro 2001, 154).

The environmental degradation at Panzhihua exemplifies the widespread effects of the Third Front industrialization drive. In China as elsewhere, mining and steel industries always pollute the air, water, and soil, but the particular emphases of the Third Front exacerbated the effects in many ways. The perceived military urgency did what it has done in so many places, from building factories on the sites of former gardens in London during World War II (Laakkonen 2020) to US manufacture and testing of atomic weapons during and after that war (B. Johnston and Barker 2008). Even though planners were fully aware of possible environmental consequences, governing authorities did not consider them important enough to do anything about them. Building industries "in mountains, dispersed, in caves" caused additional problems, including flash floods, mudslides, water shortages, and land subsidence in caves enlarged to hold factories, as well as endemic diseases from lack of ventilation (Meyskens 2020, 204). In addition, building factories in places that required new transport infrastructure and where transport costs were high, even after the roads and rail lines were in place, increased the energy intensity of the projects. Given that energy was almost entirely produced from fossil fuels, the pollution intensity also increased. Building industries in caves raised the monetary and energy costs even further (Meyskens 2020, 206). Finally, many Third Front projects—perhaps as many as half of them—failed. Although Pangang was still a going concern in the early 2020s, with a second large steel mill near Xichang to the north, many Third Front projects were abandoned after the reforms, thus increasing the energy and pollution intensity of the whole effort (Shapiro 2001, 154; Meyskens 2020, 204–5).

Unlike the Great Leap Forward, the Third Front caused no nationwide famine or other acute human disaster. Like the Great Leap, however, it involved a huge amount of waste, including both polluting waste products of inefficient production and the waste of money and effort. Still, it would be a mistake to condemn every aspect and every project of the Third Front as total folly. Who knows how useful it would have been if the Americans or the Soviets had actually attacked, which did seem like a real possibility in the mid-1960s? Third Front projects expanded China's transport network and helped integrate peripheral areas into the national economy and polity and into the web of relations shown earlier (see figure 11.1). Panzhihua did produce a lot of steel and valuable alloys, and the Second Auto Works continued to turn out Dongfeng trucks (its headquarters moved to Wuhan in 2006). In 2020, when it no longer produced the old Dongfengs, it was number 100 on the Fortune Global 500 list of the world's largest corporations (Fortune 2020). Still, there were trade-offs. As the successful Third Front projects contributed to the heavy-industrial complex and thus benefited the national economy (and perhaps the national defense), they also brought pollution and environmental degradation to local areas, illustrating the environmentally unjust relationship between national-scale growth and local-scale pollution so characteristic of heavy industrial development. It exploited the resources of the interior (much of it belonging to Zomia and Central Asia) for the benefit primarily of the urban areas in China Proper, setting up a classical kind of environmental injustice characteristic of extractive colonialism (Greaves 2018; Atiles-Osoria 2014).

The Third Front was not the only innovative industrialization project of the late years of high socialism. Just as important was the development, beginning in 1970, of the "five small industries" (iron and steel, chemical fertilizer, machine building, coal mining, and cement). As chapter 13 elaborates, at first these small industries were owned by county governments, but they increasingly devolved and expanded to commune and brigade levels after 1972; commune-run industries grew by 30 percent and brigade industries by 17.4 percent between 1971 and 1978. For example, in Zhejiang over four hundred agricultural machinery factories and eighty small-scale fertilizer plants were begun during the 1970s (Bramall 2007, 20). These industries, although limited both in extent and in products, set the stage for the rapid growth of a much wider variety of rural industries (later known as township and village enterprises, or TVEs) in the early Reform period, with all the economic growth and environmental degradation and pollution that they brought about.

EVALUATING INDUSTRY AND ENVIRONMENT IN THE HIGH SOCIALIST PERIOD

We know that China's industries, particularly the heavy industries that were the focus of the Stalinist industrialization strategy, grew rapidly during the period of the planned economy, but at what cost? Was industrialization in the high socialist period a "war against nature," as Judith Shapiro's (2001) pioneering work so catchily describes it? Some of the rhetoric of the time—the battles, brigades and battalions, heroic struggles, heroes and martyrs, sacrifice for the nation—certainly portrayed a metaphorical war. At the same time, the environmental results suggest another explanation in terms of social-ecological systems. I return to Brian Walker and David Salt's (2006) admonition that whenever we emphasize one variable in a system at the expense of all others, the resilience of the system decreases. In this case, emphasizing a single variable—industrial growth, undertaken in the interests of both economic development and national security—involved neglecting environmental health (ironically, at the same time that China was making great progress in public health programs to eliminate infectious diseases [Liang Huigang et al. 2020, 25; Banister 1987, 84–85]), neglecting building safety, neglecting long-term productivity of landscapes that could never be agricultural again once they had been mined, neglecting the ecological buffering effects of forests cut down for fuel or valleys dug out for factories. This led to decreased resilience, increased disasters, and increased environmental injustice. Perhaps it is more accurate, if less dramatic, to say that industrial development in the high socialist period was not a war against nature but rather a war against poverty and against national weakness and humiliation and that nature was the "collateral damage" of that war.

In retrospect, we might also ask how much the environmental effects of state socialism and the Stalinist "big push" industrial strategy differed from those of industrial capitalism at the same state of capital accumulation. My provisional answer is "not very." The push for development, outlined in chapter 2, had the same goals in socialist China as in capitalist Japan or capitalist North America, caused the same kinds of environmental degradation and loss of system resilience, and perpetrated environmental injustice and extractive colonialism, even if the victims of China's colonial extraction were almost exclusively its own citizens and its own water, air, and soil. In this sense, the difference between a high socialist and a bureaucratic capitalist economy, or China's transition from one to the other in the 1980s and 1990s, was less important in ecological than in ideological or economic terms. The trade-offs between economy and environment were real in nations around the world throughout the whole period of postwar economic growth, whatever the ideologies of the ruling regimes.

One way in which state socialism did contribute directly to loss of ecosystem resilience was that the state purposely shortchanged urban infrastructure in order to direct investment toward heavy industry. This not only made city living in some ways more difficult than country living—peasants, once freed from the ecosystem collapses and starvation-inducing requisitions of the Great Leap, could at least build roomy if crude houses and grow what they needed—but also decreased the resilience of cities to major disturbances, such as the great Tangshan earthquake of 1976. Tangshan, on the Bohai Gulf northeast of Tianjin, was an industrial center focused on mining and related industries, including steel and cement. It was built on soft soils and mostly out of brick and precast concrete, with little attention to earthquake-resistant building codes. The region was known to be seismically active; several other major quakes had occurred within a few hundred kilometers in recent decades. When the ground shook in Tangshan, the consequences were cataclysmic. Liu Huixian et al. (2002, prologue) graphically describe the disaster:

> At 4:00 a.m. on July 28, 1976 the city of Tangshan, China ceased to exist. A magnitude 7.8 earthquake was generated by a fault that passed through the city and caused 85% of the buildings to collapse or to be so seriously damaged as to be unusable, and the death toll was enormous. The earthquake caused the failures of the electric power system, the water supply system, the sewer system, the

telephone and telegraph systems, and radio communications; and the large coal mines and the industries dependent on coal were devastated. The railway and highway bridges collapsed so that the city was isolated from the external world. Before the earthquake Tangshan had 1,000,000 inhabitants and it has been estimated that about one half were killed.[3] Although the building code had seismic design requirements, Tangshan was in a zone requiring no earthquake design.

The nearby Kailuan coal mines experienced flooding and shaft collapse (Liu Huixian et al. 2002, 20), seven trains derailed (Liu Huixian et al. 2002, 18), and craters formed in nearby villages. Molten iron solidified in blast furnaces, which then had to be blown up to dispose of them (Liu Huixian et al. 2002, 23). A dam holding back the Miyun Reservoir near Beijing, 150 kilometers away, partially failed (Liu Huixian et al. 2002, 68–69). The earthquake was felt at distances of up to 1,100 kilometers, and severe damage extended to Tianjin, 70 kilometers distant. When I visited Tianjin four years later, in 1980, thousands of people were still living on the sidewalks of the broad, colonial-Soviet style city boulevards, in shoddily built red-brick "earthquake shelters," which would all have collapsed if even a much smaller quake had occurred. The official line in the months and years after the great quake blamed the shoddy condition of the cities on "leftist thinking of the Cultural Revolution," but in fact such thinking reached all the way back to the mid-1950s, when planners halved the living space recommended in the Soviet plans in order to save on investment.

....

China's ideology suffered a sea change in the late 1970s, perhaps foretold by environmental disasters. The word *beng*, meaning landslide or avalanche—often the result of an earthquake—was a traditional euphemism for the death of an emperor, and thus earthquakes were popularly believed to presage the deaths of eminent men. In the first half of 1976, when there were major quakes centered at Horinger County in Inner Mongolia and at Longling in western Yunnan, two of the most eminent leaders of the Communist revolution passed from the scene: Zhou Enlai in January and Zhu De in early July. Then, on 28 July, the Great Quake, one of the most severe disasters in world history, leveled Tangshan and caused damage for hundreds of kilometers around.

Mao Zedong died on 9 September, just forty-three days after the Tangshan quake. In late 1978, his successors began dismantling and replacing his model for revolutionary development. Part of this change involved industrializing the countryside, building export-oriented consumer goods industries, expanding the cities, and greatly enlarging the heavy industrial base at the same time. I examine these developments and their environmental effects in the chapters that follow.

13. Factory to the World, 1984–2015

> Only development is a solid truth.
>
> DENG XIAOPING, during a factory tour in Guangdong, 1992

By the late 1970s, the Chinese economy was stuck in a kind of Stalinist rut (Naughton 2007, 79–83). Despite having increased agricultural output just fast enough to keep famine at bay and despite continuing to produce ever more coal and steel, 84 percent of the population were still farmers (Banister 1987, 297). Cities were dominated by administration and big *danwei* producing mostly heavy-industrial goods; they were drab places with few vehicles, understocked state stores, and slow growth. One could walk quite a distance in a Chinese city center without encountering a restaurant. Deng Xiaoping had advocated reforming the economy for several years, and on 18–22 December 1978 he orchestrated a meeting— the Third Plenum of the Eleventh Party Central Committee—that ratified his ideas of pursuing a different course of development, decollectivizing the economy, and relying more on market mechanisms. The changes, which came to be called Reform and Opening, had huge effects on China's social-ecological systems: within a few years the countryside would fill up with dirty factories and inefficient small mines, while the cities would hum with new, light industry in factories staffed by hundreds of millions of rural migrants. The big, dirty heavy industries inherited from the high socialist era would grow even bigger and a little less dirty.

The economic rationale for these changes was simple: China's economy was woefully inefficient. Agricultural productivity per unit of land was moderate by world standards—higher than anywhere else in Asia besides Japan and on a par with Canada (World Bank n.d. b), but per unit of labor it was abysmal—a lot fewer people could have done the work that the great masses of peasants were doing (Naughton 2007, 242). Low labor productivity, seen from another angle, was massive rural unemployment and a huge surplus of labor. But the collective agricultural system, with its emphasis on staple production and its fixed quotas and prices, offered very little incentive to try anything new or to work harder. At the same time, although heavy industry had continued to grow, the value added of the product was low, and the amount of energy consumed revealed a shameful level of waste growing twice as fast as GDP from 1952 though 1980 (Levine, Zhou, and Price 2009), when it was about six times the world average (see Kambara 1992, 608; Tverberg 2011; and Naughton 2007, 336). Because China's heavy industry depended on mining, processing, and burning coal, its extremely low energy efficiency multiplied the environmental harm of industrial processes, unmitigated by any kind of pollution controls. Urban consumer living standards had not increased since the 1950s; urban infrastructure was everywhere creaky and in some places dangerous. Reform meant not only increasing the volume of all kinds of industrial production but also using labor and resources more efficiently.

CHINA'S RURAL INDUSTRIAL REVOLUTION

China's rural industrialization in the early Reform era was built on the foundation of the "five small industries," which had begun spreading out from county seats to communes and brigades in the mid-1970s. By the time the Reform began, there were about a million and a half small commune and brigade enterprises, employing around twenty-eight million full- or part-time workers. Notably, these enterprises included fifteen hundred chemical fertilizer plants, which made about half of the ammonia produced nationally, and around twenty-one hundred "local" (medium-sized) and thirty-five hundred "rural" (small) cement plants, together producing more than 80 percent of China's total output. These commune and brigade enterprises mostly supplied producer goods to agriculture, with only 18 percent of their income from agricultural processing (Zhang Z. 1999, 66–67; S. Zhao and Wong 2002, 261). In the 1970s a small number of rural factories, mostly in Guangdong, began to produce goods for export (Zhang Z. 1999, 81–82), foreshadowing China's transformation in the 1980s and 1990s into the "factory to the world."

While the commune and brigade enterprises of the late high socialist period provided a material base upon which to build the rapid rural industrialization in the 1980s and 1990s, their embeddedness in the planned economy also limited their possibilities for growth and innovation. Plans dictated what and how much they could produce, where they could sell it, and how they could finance themselves (Mingchuan Yang 1994, 158–59; S. Zhao and Wong 2002, 261).

The programs approved at the Third Plenum were designed to eliminate these obstacles to rural industry's growth. As agriculture was decollectivized and land contracted to individual households, farm families quickly realized that they had surplus labor beyond what it took to farm their small plots (Tilt 2010, 38). Although some underemployed farmers could migrate to the cities to work, the regime still restricted rural-to-urban movement, meaning that many surplus farm laborers were available to work in local industries (Z. Liang, Chen, and Gao 2002, 2176). At the same time, the state also raised prices of farm products and provided bank loans, seed money, and tax holidays that enabled communes and brigades to expand their industrial enterprises and establish new ones. They began by expanding industries that directly served agriculture, such as food processing, energy (including small hydropower and biogas), mining, metallurgy, building materials, construction, transportation, and other services (Zhang Z. 1999, 71).

After agriculture was decollectivized and communes and brigades reassumed their pre-1958 designations as townships and villages, the newly christened township and village enterprises (*xiangcun qiye*), or TVEs, boomed throughout the mid-1980s, increasing the value of their output by more than 20 percent per year between 1982 and 1988. Economic adjustment policies slowed their growth for the next few years, but it resumed after 1991 at an even greater annual rate of 40 percent. From the beginning of the Reform to the turn of the century, the rural industrial economy grew more than twice as fast as the Chinese economy as a whole (Zhang Z. 1999, 85), a big part of the reason why the world could celebrate China's "economic miracle."

As rural industries grew in number and size, they also became more diverse and included many small rural factories and workshops established as satellites of larger urban consumer goods factories. By 1993, approximately 23 million TVEs were operating in China, employing about 112 million people. The TVE average of fewer than 5 employees means that many of them were very small, but still the average village had more than 20 TVEs (Z. Liang, Chen, and Gao 2002, 2181). Total TVE industry output of ¥1.6 trillion (around US$200 billion) was 75 percent of rural income by 1995, meaning agriculture supplied less than a quarter of the income of rural families on average (Zhan 2015, 413). By then, manufacturing had replaced resource extraction as the primary rural industry: the largest sectors were, in order, machinery, building materials, textiles, chemicals, food processing, metallurgy, apparel, and metal products, each of which generated more than ¥100 billion gross value. Foreign investments in rural industry also grew, amounting to US$30 billion in 2004, invested in at least 40,000 individual TVEs (Zhan 2015, 83) and accounting for about one-third of China's total exports (M. Wang et al. 2007, 649).

As demands on local governments increased and central subsidies decreased, county and township governments, as well as village committees, began to depend more and more on TVE revenue to fund both further economic growth and the social services that the central authorities and local residents came first to demand and then to expect, including schools, clinics, roads, and communications infrastructure. In addition, local officials' career advancement came to depend more and more on expanding their local economies. Bureaucracy and industry thus came to depend on each other—industrial firms depended on bureaucrats for permission and lenient treatment, while bureaucrats depended on industries for revenue. The environmental impact of the TVEs also increased, as investing in cleanup ran against the interests of both the bureaucrats and the factories (Tilt 2010, 39–40; M. Wang et al. 2007, 650–55).

The TVE boom, like all booms, eventually slowed. Most TVEs were privatized between the late 1980s and the mid-1990s (Zhan 2015, 420–22), and the newly privatized factories faced changes in the nature of the market. Domestic and international demand for low-end products slowed, rural industries lost competitiveness because of their generally inefficient use of materials and energy, and the pollution they generated became less and less acceptable to both local people and regulatory agencies (see M. Wang et al. 2007, 650; Tilt 2010, 44; and Zhang Z. 1999, 87). In addition, rapid urbanization after 2005 allowed local governments to replace industrial revenue with income from land transfer to developers. This led many rural enterprises, particularly factories, to close down in the 2010s, their sites then repurposed as urban housing or industrial parks occupied by large enterprises (Zhan 2015, 425–31). By the late 2010s, plans for economic revitalization of the countryside hardly mentioned manufacturing or mining, encouraging instead large-scale agriculture, agricultural processing, tourism, and the information economy, all of them both locally adapted and purportedly "green" (*lüse*) (Guowuyuan 2019).

Rural Industry and the Environment

In the 1980s and 1990s, when TVEs provided a road out of poverty for both rural residents and local officials' and when those officials' careers were often made or broken by whether they were able to expand local economies, environmental quality could easily become a secondary consideration or no consideration at all. Consequently, TVEs contributed mightily to pollution of water, air, and soil and in some places had serious deleterious effects on public health.

By the late 1990s, more than 80 percent of China's rivers and more than 75 percent of its lakes were polluted or contaminated. A 2002 report showed that water quality in 70 percent of the monitored river sections (mostly in populated areas) was grade IV or grade V (M. Wang et al. 2007, 648), defined in national standards as "mainly suitable for general industrial purposes and recreational uses that do not involve direct human contact with water" and "mainly suitable for agricultural uses and general scenic purposes," respectively—in other words, very polluted (grade IV) and even more polluted (grade V) (Changhua Wu et al. 1999, 253). Although urban residential and large industrial discharges accounted for much of the pollution, TVEs discharged about 15 percent of the total wastewater and half the industrial wastewater discharged in China, and unlike urban wastewater, essentially none of the water dumped by TVEs was treated at all (M. Wang et al. 2007, 651).

In addition to almost never being treated, much TVE wastewater flowed from industries whose waste products were particularly polluting, including pulp milling and paper production, chemical manufacturing, textile dyeing, metal casting, brickmaking, and cement production. These industries grew explosively in the era of TVE expansion: rural paper making increased 26-fold, cement 40-fold, and bricks 7-fold. All of these generated water pollutants, including acids and bases, nitrogen, phosphates, phenols, cyanide, lead, cadmium, mercury, and bichromate (M. Wang et al. 2007, 651; Changhua Wu et al. 1999, 251). A 1989 study found that TVE discharge densities of heavy metals, cyanides, and phenols were 2.2, 3.3, and 9 times greater, respectively, than those of comparable urban industries (Changhua Wu et al. 1999, 251). The State Environmental Protection Administration reported in 2002 that over half the population drank water contaminated with chemical and biological wastes (M. Wang et al. 2007, 651). Although most urban families would not have dreamed of drinking unboiled water and many rural families also boiled their water for tea, I personally observed rural people dipping water for drinking straight out of streams and reservoirs, and when I questioned them, they would say, "We're used to it here." And anyway, boiling only eliminates biological contaminants, not chemicals.

TVE factories polluted the soil and air as well as the water: the most prominent sources of rural air pollution were coal processing, cement, brickmaking, and ceramics (Andrews-Speed et al. 2003, 188). Adding to the problems from TVEs, authorities concerned about urban air pollution in the 1980s began to advocate moving urban factories to more rural areas, specifically to export their air pollution (Ahlers, Hansen, and Svarverud 2020, 92). In addition, a combination of incentive structures for TVEs and a shortage of regulatory personnel ensured that rural air pollution would be laxly regulated if at all.

In 1994, when I spent a week in Futian Township in the outskirts of Panzhihua, there were already places where the air smelled foul, but I remember nevertheless a young truck driver telling me that "our air here is really fresh." By the time Bryan Tilt conducted extensive fieldwork in Futian in 2002–3, there was no room for such wishful thinking. Although most of Futian's TVEs had disappeared in the interim due to declining profits, three of the largest and most polluting survived as privatized enterprises: a zinc smelter, a coking plant, and a coal-washing facility (Tilt 2010, 45).

The zinc smelter used ore from a neighboring township and "consisted of six brick furnaces connected by a rudimentary ventilation system that drew coal smoke through a series of pipes and vented it through a smokestack. . . . The smelter consumed coal at an astounding rate . . . approximately ten to fifteen tons of raw coal . . . for every ton of pure zinc produced" (Tilt 2010, 46–47); this compares with a little over one ton in a modern smelter (Van Genderen et al. 2016, 1588). Tilt (2010, 47) visited the smelter one day:

> Wearing soiled clothing, their faces and hands blackened with soot, the men shoveled piles of raw coal into the furnace chutes. Other workers placed semirefined zinc ore in cylinder-shaped ceramic crucibles and hoisted them into the furnaces with metal tongs. After a period lasting from a few hours to more than a day, depending on the amount and quality of the ore, the men would carefully remove the cylinders and pour the molten zinc into rectangular molds. The hillside surrounding the smelter was littered with coal piles, empty ceramic cylinders, and unusable zinc slag. The men worked slowly and methodically, without masks or eye protection, despite the thick, sulfurous smoke pouring out from the sides and top of each furnace.

In the coking plant, workers manually shoveled coal into leaky brick ovens they lit from beneath. They then allowed the coal to smolder for several days (Tilt 2010, 48). After interviewing coking workers, Tilt (2010, 65) writes,

> I recall little beyond the cursory details in those early field notes, but I do remember feeling short of breath and nauseated. . . . We were also taken aback by some of the sensory details that a longer period of research allowed us to observe: the constant, lung-burning haze in the air; the factory laborers in soot-blackened clothes; the fine, gray dust covering vegetation, rooftops, laundry, and anything else left outside for more than a few hours; the endless, serpentine rows of coal trucks plodding along steep mountain roads; the local stream that often looked like flowing sludge.

All these pollution-producing manufacturing industries needed raw materials and energy, of course, and the growth of TVE manufacturing would not have been possible without simultaneous growth in TVE mining, sometimes called artisanal mining or simply small-scale mining. In 2006, 94 percent of China's mines were artisanal and small mines (ASMs), employing 65 percent of the nation's miners and producing about 2.7 billion tons of metallic and nonmetallic minerals, or about 53 percent of the national total (L. Shen, Dai, and Gunson 2009, 150–51). About half of the miners employed in these small mines dug coal—there were about 75,000 or 80,000 township and village coal mines, ranging widely in capacity from 100,000 tons down to just a few hundred tons per year and producing around 45 percent of China's coal—around 650 million tons in the mid-1990s (Andrews-Speed et al. 2003, 185–86). Compared to larger, mostly state-owned mines, these ASMs were characterized by "large numbers of illegal operations; irrational locations; low recovery rates; poor safety; and substantial environmental damage" (Andrews-Speed et al. 2003, 186). Their labor-intensive methods (often digging by hand or with small equipment) resulted in inefficient use of labor and other resources (Wright 2012, 93). In addition to legal township and village mines, there were also large numbers of illegal small mines or "black holes" (hei kouzi), which were even less susceptible to regulation; Fenxi County in Shanxi alone was reported to have more than 1,000 of these in 2005 (Wright 2012, 93; Zhang Yulin 2013, 90).

Environmental damage from artisanal coal mines, particularly in the 1980s and 1990s, was widespread, diverse, and severe. It included "destruction of arable and grazing land through accelerated erosion of topsoils, landslides, collapse of old workings, dumping of tailings, lowering of water tables, contamination of soils by dust from mines, increased levels of sediment load and flooding in adjacent rivers, and disturbance of local water tables leading either to flooding of land or to a shortage of water" (Andrews-Speed et al. 2003, 186–87).

Damage from coal mining was most severe in Shanxi, the province with the most coal mines. At the beginning of the twenty-first century, coal mining in Shanxi affected a total of almost 20,000 square kilometers or 13 percent of the province, of which about a quarter was already mined out. Of the total area affected by mining, there was detectable surface subsidence in 2,700 square kilometers, increasing by about 94 square kilometers per year; other, undefined "geological disasters" affected a total of 6,400 square kilometers. About 1,900 natural villages were affected, with a population of a little over 2 million people. Over 20,000 square kilometers also experienced problems with water resources—over 1,300 wells went dry, 4 springs dried up, and 7 were reduced to a trickle, leaving nearly 5 million people and over half a million large livestock short of water (Zhang Yulin 2013, 88). In addition to the environmental effects, accidents in those small mines also killed thousands of miners and other workers nationally each year (L. Shen, Dai, and Gunson 2009, 150); in Shanxi the yearly toll was probably around 500, though there are no accurate statistics (Zhang Yulin 2013, 88). Mining accidents continued to plague China decades later, with 573 miners killed in 2020 (J. Lau 2021).

Coal, of course, was not the only mineral extracted by TVEs. Metal mines had problems of their own. Laiwu Municipality, Shandong, which has some of the richest deposits of iron ore in China, experienced major mine collapses and floods, including an iron mine flood in 1998 that killed 20 people. In Laiwu's karst landscapes, naturally riddled with caves and then mined, overextraction of water by mining operations caused at least 180 faults to open, resulting in 2,000 homes having to be moved (Chen H. and Ge 2003, 104–5).

Problems with lead mining were even more serious. Most of China's lead and zinc mines were ASMs, exploiting resources spread over a wide swath of the country (J. Lu and Lora-Wainwright 2014, 192). A village in western Hunan, one of the hotspots of lead mining, started "limited artisanal mining by villagers" in the 1950s, and large-scale state-owned mines operated from the early 1960s to the late 1970s. Small-scale mining grew again as part of the general TVE boom in the 1980s and 1990s: poor villagers struck deals with state-owned mines to allow them to dig low-quality lead deposits of little interest to the big mines (J. Lu and Lora-Wainwright 2014, 193). Although some village elites got rich, the environmental consequences were predictable:

> Particularly in areas of the village situated downstream from extraction and processing, such as the sub-village of Fengcun, water pollution became a severe problem. Villagers recalled that at that time the water in the local stream turned black and smelly, they would develop itchy skin rashes if they came into contact with it, and prawns and fish died. . . . [Villagers'] response to mounting evidence of pollution was a series of petitions in the 1980s, including two petitions to the central government in Beijing, but there were no visible outcomes. At this time of boom, villagers increasingly

focused on mining as a livelihood strategy and regarded it as a path toward a better life, largely conceived in terms of financial wealth. (J. Lu and Lora-Wainwright 2014, 194)

There was more: in 2006, nearly half the village farmland was affected—some buried in tailings, some polluted, some dried up because of groundwater extraction. Lead content in soils exceeded national standards by a factor of 5, and lead in rice harvested from some fields exceeded the allowable level by about 120 times. Soils also contained high levels of cadmium, mercury, and arsenic. The mines extracted groundwater, resulting in reductions and pollution of drinking water supplies, and tests of the river water put its quality into the notorious category V, or really dirty. Local people predictably experienced high blood levels of lead and other heavy metals (J. Lu and Lora-Wainwright 2014, 195). A lead-mining and smelting area in distant central Fujian had similar difficulties: villages near mines, separation facilities, and smelters showed levels of lead in both soil and air that were more than twice as high as those in nearby control villages (Sihao Lin et al. 2011).

Why Were/Are Rural Industries So Polluting?

Almost all extractive and manufacturing industries pollute, but China's mining and manufacturing TVEs, especially in the 1980s and 1990s, produced especially high amounts of pollution per unit of energy input and per unit of output and also kept up their level of pollution in spite of multifaceted efforts to clean them up, especially after the late 1980s. Why were TVEs so bad for the environment for so long? The answer seems to lie in the primacy of development in CCP ideology and legitimation, TVEs' importance to local economies, the costs of making them more efficient, their role in promoting the careers of local officials, and the difficulty of enforcing environmental regulations upon them.

From the beginning of Deng Xiaoping's rule, the whole country took to heart his assertion that "only development is a solid truth." So many of the problems that China faced in the early Reform era stemmed from its poverty, including its weak positions in international diplomacy, military affairs, and trade; its inability to ensure that the population would be "warm and full" and have access to housing and a reasonable standard of living; and even its inability to clean up the environmental messes created during the high socialist period. In addition, rural people quite understandably wanted more, materially, for themselves—they were tired of manual labor, of the lack of access to modern goods and services, and of rudimentary education that lagged far behind that in China's cities. In this predicament, neither the leaders nor the common people considered environmental sustainability or eliminating pollution to be pressing concerns. Almost everyone was aware of the adverse health, environmental, and aesthetic consequences of rapid, uncontrolled, polluting industrial development. But they did not care much about it—many of them saw what a local agricultural official in Panzhihua explained in a clever wordplay as a contradiction between *wenbao* 温饱, or "warm and full," and *huanbao* 环保, or "environmental protection" (Tilt and Xiao 2007, 132–33; Tilt 2010, 142–43). They saw environmental degradation as a necessary evil, a reasonable price to pay for the benefits of, as Deng himself put it, "getting rich."

Even with their development obsession, Chinese leaders were not going to solve the problems of rural poverty and labor surplus completely if they just let people migrate to urban areas, leaving behind only those needed to staff a more labor-efficient agriculture. By restricting rural-to-urban migration, they hoped to avoid the developmental Hobson's choice between unaffordable urban services and unconscionable slums. Under the slogan "Leave the fields but not the countryside; enter the factory but not the city" (Li tu bu li xiang; jin chang bu jin cheng), TVEs provided employment for rural people as well as goods for the economy. Agricultural productivity did grow in this period, and more profitable cash crop agriculture could employ some people who had previously grown only grain with their production teams. But most of the income growth in rural areas through the 1980s and into the 1990s came from TVE employment. Temporary labor migration was allowed beginning in the early 1980s, but its contribution to rural incomes did not overtake that of TVE employment until migration became a flood in the 1990s, reaching about two hundred million in 2006, and wages for migrant laborers began to rise at the same time (K. Chan 2009a).

If TVEs were to remain profitable, however, they had to remain competitive with larger, better capitalized state-owned industries. These big factories benefited from the "soft budget constraint" (Kornai 1986; X.

Dong and Putterman 2003), meaning that governments would bail them out if they went into the red, whereas TVEs were not only on their own but had to generate revenue for local governments. Efficient equipment and environmental safeguards were expensive, and if TVEs invested in these they would lose what little competitive edge they had (M. Wang et al. 2007, 650; Wright 2012, 111). Labor, and to an extent raw materials, were abundant, while capital was scarce (Naughton 2007, 265–76). The only rational course for a local government (or, later on, the owner of a privatized TVE) was to use as much coal, clay, ore, and other relatively cheap raw materials or energy as necessary but minimize investment in expensive machinery or pollution abatement equipment, or in worker and miner safety (Wright 2012, 111). Low energy efficiency, in turn, meant more pollution, especially since most energy came from coal and most factories burned coal with no environmental mitigation (Tilt 2010, 69).

For local governments, low-capital, inefficient, dirty mines and factories were not only desirable but necessary (Tilt 2007, 916). In the early years, when TVEs really were run by township governments and village committees, their profits often compensated governments for decreasing subsidies from higher up. They were one of the main sources of revenue to fund bureaucrats' salaries, pay local labor, purchase construction materials to build roads, schools, and public buildings, and even to invest in expanding existing TVEs and establishing new ones. As TVEs were increasingly privatized in the 1990s, taxes, fees, and often unofficial payoffs replaced profits as the form of revenue from rural industries to governments, but the sources—the mines and factories—remained the same. There was also a further incentive for local officials: the all-important cadre evaluation system, which determined cadres' prospects for raises and promotions. The system emphasized economic growth above all other indicators except maintaining public order, and although environmental protection was sometimes on the list of criteria, it was not one of the standards that officials were absolutely required to meet (Whiting 2001, 100–118). Compounding the effects of quota-based cadre evaluation were short terms of office (typically three to four years), which forced cadres to show their "administrative accomplishments" (*zhengji*) in a hurry, leading to corner cutting and showy projects that they could accomplish quickly, instead of longer-term, more fundamental environmental improvements (Eaton and Kostka 2014).

In the time of the TVEs' greatest growth and flourishing, there were in fact strict environmental standards on the books, and every level from the county on up had an environmental protection bureau (EPB) charged with enforcing those standards. However, the EPBs were faced with impossible tasks. Not only were they chronically short-staffed, often with only two or three inspectors for a county that had hundreds or even thousands of TVE factories and mines (Tilt 2007, 923; M. Wang 2007, 650–53), but the logic of local profitability as the fuel of development worked against them. If they were to demand cleanup, they would be demanding that TVE owners—local governments or private entrepreneurs—cut into their profits and thus endanger the continued existence of the factory or mine and with it the incomes of workers and owners, as well as the career prospects of local officials (Tilt 2007, 925). As a result, local governments often colluded with inspectors to inform enterprise managers that they should shut down when the inspectors were planning to visit (M. Wang et al. 2007, 650, 655). Even when pollution fees (a form of Pigovian taxes) were mandated, they were not always paid (Sinkule and Ortulano 1995, 115), and when fines were levied for pollution that exceeded standards, it was sometimes still cheaper to pay the fine than to invest in pollution mitigation (M. Wang et al. 2007, 652).

Given their importance to development, the local economy, and local government budgets, the initial solution to polluting rural industries was not to clean them up; everyone concerned (except perhaps for the long-suffering environmental protection officials) recognized the problems but accepted them as trade-offs for the economic boom, but neither was the solution to ignore the problems altogether. Rather, those who had been harmed by environmental degradation often quit trying to stop the polluting activities, resigned themselves to pollution as a "fact of life," and demanded compensation of other kinds, usually money (Lora-Wainwright 2017, 83–86). This kind of deal in effect gave TVEs and other industries and mines a sort of license to pollute, often one that was not particularly expensive, thus further exacerbating the problem of polluting rural industry.

In the longer run, however, TVE pollution decreased beginning in the 1990s for a variety of reasons. Perhaps most important, their products became less competitive. The market for low-end goods produced by crude technologies began to decrease as early as the mid-1990s, as consumer demand switched to higher-quality products (Zhang Z. 1999, 87; S. Zhao and Wong 2002, 258). For those who did remain in business, there was economic pressure to increase efficiency, and they did so (Chen Wei-hong, Zhang, and Liu 2009, 11).

At the same time, both local residents and consumers came to place greater value on a cleaner environment, so enforcement became easier. Many highly polluting TVEs were shut down (Zhang Z. 1999, 87), some of them in a concerted drive that closed tens of thousands of small, inefficient, and unsafe mines between 1999 and 2006. Part of this effort was due to pressure from large, more modern state-owned mines that saw the TVE mines as unfair competitors, though workers and local officials often resisted, still worried about the *wenbao-huanbao* contradiction and the effect on local government revenues. Quite naturally, this led to direct bribery and other forms of corruption. Nevertheless, the net result was a reduction in the output of small mines (those that produced less than thirty thousand tons per year) from a peak of more than six hundred million tons in 1995 to less than one hundred million in 2008 (Wright 2012, 110–35), and given that large mines were more able to install controls on discharges into air and water, closing small mines probably also resulted in a decrease in the amount of pollution produced *per ton of coal*. But since China's coal production and consumption continued to expand until at least 2014, the problems of air pollution and greenhouse gas emissions also increased.

THE GROWTH OF MODERN INDUSTRY AND ITS ENVIRONMENTAL EFFECTS

The usual narrative of China's rise from a poor to an upper-middle-income country centers on light industry, on factories that employed hundreds of millions of rural migrants (mostly women working long hours for low pay) in manufacturing consumer goods first for export and then increasingly for the domestic market (C. Lee 1995). In fact, from 1980 to 1995 these consumer industries largely drove urban industrial growth. During this period, the fastest-growing industries, all of which grew by more than 19 percent per year in monetary terms, mostly manufactured labor-intensive products: electronics and communication equipment, furniture, plastic products, wood and cane products, synthetic fibers, electric machinery, pharmaceuticals, leather and fur products, metal products, and clothing. Among fast-growing industries, only nonmetal mining stands out as an energy-intensive "heavy industry." Mainstays of the earlier high socialist industrialization, such as metal mining, metallurgy, coal, and coal gas, also grew but at slower though still remarkable annual rates of 7 to 15 percent (Naughton 2007, 331).

Detailed figures, however, tell a more complex story. Table 13.1 shows the increases in China's consumer goods industries from 1980 to 2016 or 2018 (depending on available data), while table 13.2 shows the equivalent increases in heavy industry, including mining, metals, chemicals, cement, and glass.

These tables show that China's fastest-growing industries from the 1980s through the early 2000 aughts were indeed the labor-intensive consumer goods sectors, and only a few of the fastest-growing industries depended heavily on increased supplies of energy. Clothing and fabrics both grew, because they served export markets and because people added variety to their wardrobes, but clothing grew faster because people stopped making their own clothes. The same thing can be said for leather shoes, which replaced cloth shoes (often homemade as well) as everyday good-weather wear for the majority of the population. These industries had rather fewer environmental consequences than heavier consumer goods such as refrigerators, washing machines, and motorcycles, which depended on growing steel output.

After 2000, however, China's accelerating economic growth and material prosperity came to depend increasingly on urban and transportation infrastructure and also on heavier, environmentally more impactful durable consumer goods, setting in motion a return to heavy-industry growth similar in some ways to that of the high socialist period. The difference between the time before 1980 and the time after 2000, however, lies in energy and raw materials intensity. During the high socialist period, energy demand grew about twice as fast as gross domestic product, reflecting the Stalinist emphasis on heavy industry and the lack of budgetary constraints on what the state considered to be vital industries (Smil 1993, 127; Levine, Zhou, and Price 2009, 19). After 1980, however, the energy intensity of the Chinese economy plummeted as strong incentives developed to conserve; from 1980 to 2000, GDP grew about three times as fast as energy consumption, but because production was

Table 13.1. China's consumer goods industry, 1980–2010s

	1980	2000	2000/ 1980[a]	2016–18	2018/ 2000[a]	2018–21 % OF WORLD TOTAL
ELECTRONICS [b] (billion US$)	4.5	120	26.6	350	2.92	25
CLOTH (billion meters)	13.4	27.7	2.06	90.6	3.27	52
CLOTHING (billion pieces)	0.94	20.9	22.2	29.8	1.34	
SHOES (million pairs)	157	1468	9.3	4618	3.14	54
BEER (billion liters)	0.69	22.3	32	45.1	2.02	23.5
NEWSPRINT (million tons)	0.37	1.45	3.91	2.91	2	19
PLASTIC PRODUCTS (million tons)	1.14	10.35	9.07	77.12	8.50	31[c]
HOME REFRIGERATORS (millions)	0.24	14.2	59	76.2	5.3	39.5

Note: Newest data available, different years for different products.

[a] "2000/1980" and "2018/2000" indicate ratios of the later to the earlier date or, alternatively, the factor of increase from the earlier date to the later (e.g., China produced 9.3 times as many shoes in 2000 as it did in 1980).

[b] Different sources use different measures; included to indicate order-of-magnitude changes.

[c] Percentage of world's plastics (as reported by Plastics Europe 2020), a slightly larger denominator than plastic products (as reported by the State Statistical Bureau).

Sources for tables 13.1 and 13.2: Data from National Bureau of Statistics (through EPS data), in Chao Feng, Huang, and Wang 2019; Wen 2017, 16–29; N. Li, Ma, and Chen 2017, 1840; Jin-Hua Xu et al. 2014, 592; BizVibe n.d.; mayor box n.d.; Kirin 2019; Plastics Europe 2020; Holst 2020; Berg and Lindqvist 2019; Guoji Taiyang 2020; Chao Chen and Reniers 2020; Xun 2020; Shaoqing Huang and Qiao 2005; Pecht et al. 2018; Moko Technology 2022; World Footwear 2022.

Table 13.2. China's heavy industry, 1980, 2000, and 2018

	1980	2000	2000/ 1980[a]	2016–18	2018/ 2000[a]	2018–21 % OF WORLD TOTAL
IRON ORE (million tons)	112	223	1.99	375	1.59	16
STEEL (million tons)	37	128	3.46	807	6.3	51
NONFERROUS METALS[b] (million tons)		7.8		51.56	6.6	56[c]
COAL (million tons)	620	1355	2.18	3693	2.72	46
CHEMICALS (billion yuan)						40
CEMENT (million tons)	79	597	7.55	2410	4.03	59
FLAT GLASS (million cases)	24	183	7.6	927	5.06	~60

Note: Newest data available, different years for different products

[a] 2000/1980 and 2018–19/2000 indicate ratios of the later to the earlier date or, alternatively, the factor of increase from the earlier date to the later. For example, China produced 1.99 times as much iron ore in 2000 as it did in 1980.

[b] Primary nonferrous metals include copper, aluminum, lead, zinc, tin, and nickel.

[c] Aluminum only. China is also the world's leading producer of lead, tin, and zinc.

growing so fast, total energy consumption slightly more than doubled (Levine, Zhou, and Price 2009, 20). Energy efficiency grew in the 1990s and the 2000 aughts mainly as a result of the shift from meeting quotas in the plan to making profits in the market. All enterprises, whether state owned or private, had incentives to cut costs, including the cost of raw materials; these incentives mandated improving the efficiency of industrial processes, to the point where investment in efficiency became a significant share of total energy investment between 1981 and 1998 (Levine, Zhou, and Price 2009, 21).

The imperative to save on energy and raw materials, and thus on costs, had several results. First, the oldest and least efficient heavy industries, mostly state-owned enterprises, particularly in what has become known as China's Rust Belt in the northeast, went out of business, resulting in a severe decline in the region's share of industrial output and in massive worker layoffs and involuntary early retirements (Naughton 2007, 333; Zhang P. 2008, 112). Second, both new plants and those older ones that managed to survive the shakeout reacted to strong incentives to conserve. Third, as a result of the first two, the energy intensity of Chinese industry in general, which had already fallen dramatically, from 230,000 tons of coal equivalent per billion *yuan* (at constant 2015 prices) in 1980 to 80,000 tons in 2000, decreased further, to under 60,000 tons in 2017 (Lewis 2020, 49), mainly because of continued energy efficiency gains in steel, cement, glass, and other heavy manufacturing. For example, in the short period between 2005 and 2010, the energy used to produce a ton of cement declined by 28 percent, from 137 to 98 kilograms of coal equivalent (N. Li, Ma, and Chen 2017, 1841). Improvements in steel were even more dramatic, with the energy inputs almost halving, from 1,123 to 619 kilos of coal per ton of steel produced between 1996 and 2010, and declining further to 572 kilos in 2014 (Feng He et al. 2013, 205; Chao Feng et al. 2018, 837–41). However, this level of energy consumption was still above the levels in Japan or Germany, indicating that further improvements were possible and necessary (Chao Feng et al. 2018, 844).

All these improvements in energy efficiency, however, were happening at the same time industrial production was growing by leaps and bounds. Thus, part of the growth in air pollution, water pollution, and greenhouse gas (GHG) emissions happened simply because energy efficiency improvements could not keep up with industrial growth. In the mid-2010s, several studies estimated that the iron and steel industry contributed about 9 or 10 percent of China's total GHG emissions (J. Shen et al. 2018; R. Jing et al. 2014) and the cement industry, about 13 percent (Jin-Hua Xu et al. 2016).

In addition, like China's food and wood products industries, China's steel industry had an ever-increasing impact on world environments through the early twenty-first century. In 2017–18, China produced about half the world's steel but only about a sixth of its iron ore (see table 13.2), so the output of China's iron mines supplied only about a third of the raw material needed for iron and steel production (Xun 2020). Recycled scrap provided about another quarter (Fubao 2021). This meant that China relied on imported ore for almost half its steel production, and it imported more than a billion tons of iron ore in both 2016 and 2017, which was more than 40 percent of total world iron mine production. In 2020 about 60 percent of these imports came from Australia (Su-lin Tan 2021), which absorbed the environmental costs of iron mining.

Even though China imported only about a quarter of its zinc ore (Xun 2020), the effects of outsourcing environmental degradation with economic development were very evident in China's zinc industry. In 2016 China closed most small zinc mines and a few larger ones because of the environmental problems so vividly described in the passages by Bryan Tilt, so China began to import both more zinc concentrates for its smelters and more finished zinc (Xun 2020). Some of this imported zinc came from mines with majority Chinese ownership, such as one in Sumatra, where nearby villagers complained to the World Bank about the possibilities of pollution from a tailings dam, which, if it failed, would release tons of toxic mud and potentially wash the village away (Simangunsong 2021).

The Chemical Industry

China's chemical industry was another egregious polluter during its rapid growth in the late twentieth and early twenty-first centuries. Already in 2007, China produced 54 percent of the world's dyes (J. Tian et al. 2012, 264), and by 2011 China's chemical industry as a whole had become the world's largest (Chao Chen and Reniers 2020, 1), accounting for 40 percent of the world's total chemical sales by 2017. Like other industries in the late twentieth century and the first decade of the twenty-first, it "prioritized growth over environmental quality" (Hong et al. 2019).

Much of the growth in China's chemical industry after the 1990s took place in more than twelve hundred chemical industry parks (Wei Liu et al. 2020, 1), established in the suburbs of many Chinese cities as part of the urbanization process. Chemical industry parks produced both chronic, point source pollution and severe, acute pollution. The latter occurred increasingly frequently as the industry and the parks grew in the 1990s and early 2000s (Meng et al. 2014, 2217), even as chemical production diversified to "fine chemicals," or specialty products used in one or a few industries (Hong et al. 2019; J. Tian et al. 2012). In the Shangyu industrial area in northeastern Zhejiang, for example, researchers found that 35 percent of the sulfur used between 2006 and 2011 to make various kinds of specialty chemicals (mostly dyes) ended up in liquid, solid, and gaseous wastes rather than in the end products of the chemical plants. Most of the wastes were quantities of calcium sulfate deposited in landfills, but leaks of atmospheric sulfur dioxide resulted both from burning coal to fuel the processes and from sulfuric acid manufacturing processes that were part of the chemical syntheses (J. Tian et al. 2012, 262–66). A study of nine chemical plants in the Binhai industrial area in Tianjin, which produced everything from dyes to pesticides to pharmaceuticals, showed that they emitted a total of sixty-five different polluting compounds, including end products such as pesticides and drugs as well as intermediary by-products of the manufacture of dyes and pharmaceuticals. Although most of these pollutants were supposed to be removed in the wastewater treatment plants downstream from the factories, in fact eleven pollutants were found *only* in the effluent from the treatment plants, indicating that they might have formed in the treatment process itself (Wei Liu et al. 2020, 5–6).

Environmental pollution from petrochemical processes was especially pervasive. Petrochemical factories generated both chronic pollution— they were the primary source of surface water and sediment pollution by polycyclic aromatic hydrocarbons in the Chang delta (T. Jia et al. 2021)— and acute pollution outbreaks from increasingly frequent accidents (Shao et al. 2013, 1611). In the aforementioned seaside Binhai industrial area in Tianjin, accidents were caused by leaks during production and transportation, oil leakage, and improper operation of pollution treatment equipment (Shao et al. 2013, 1619). As in almost every industrial or industrializing country, pollution from chemical processes and accidents differentially affected the poor and marginalized (Chiu 2020). In Nanjing, long a center of the petrochemical industry, authorities moved over a hundred petrochemical plants to two periurban industrial parks in areas populated by "socially disadvantaged groups, including retired workers, villagers, and migrant workers" (Mah and Wang 2019, 1967; Meng et al. 2014), creating classical instances of environmental injustice.

The most iconic petrochemical accident and spill occurred in Jilin

City, Jilin, in the early afternoon of 13 November 2005 (Sina 2005). Ten days prior to the spill, the Jilin Petrochemical Works of the China Petroleum Corporation had received an award as "a company friendly to environmental protection" (Zhang Yue and Shen 2005). But when a blocked duct in an aniline production facility was not dealt with properly, serial explosions and a fire immediately killed five people and injured twenty-three. The accident also released into the nearby Sungari River highly dangerous organic compounds, including benzene, nitrobenzene, anilines, and dimethyl benzene (Sina 2005; Wang Jing 2005; Yang X. 2005). According to initial accounts—typical for the unfree Chinese press—officials responded quickly to seal off the spillway to the river, but in fact they acted too late, and by the next morning the air was full of the chemical smell of bitter almond, and concentrations of dangerous benzene compounds in the river reached one hundred times the allowable levels (Wang Jing 2005).

By the afternoon of 20 November, a week after the original explosions, the pollutants floating downstream had reached the Jilin-Heilongjiang border, and it was predicted that they would reach the major city of Harbin on 24 or 25 November (Yang X. 2005). Sensing the approaching chemical danger and perhaps also the twin political dangers of admitting an environmental problem or doing nothing about it, the Harbin government announced on 21 November that water supplies would be cut off for about four days starting the next day "for repairs to equipment" (Yu Jintao, Zhang, and Shen 2005). The city government provided some substitute supplies taken from groundwater, urged residents not to hoard water, and asked work units with surplus water to donate it to hospitals (it is not clear how that was to work). Not surprisingly, within a half hour of the announcement the hoarding began. Harbin grocery stores were running low on bottled water, sausages, and the dark bread that reflected Russian influence in the region. One reporter covering the story stopped by a store that afternoon and managed to grab a case of Coca-Cola and then waited in line for two hours to pay. The price of bottled water tripled (Yu Jintao, Zhang, and Shen 2005).

Eventually dilution, the usual pollution solution, took over. Major tributaries flowed into the Sungari downstream from Harbin, municipal water was turned back on after the slick passed Harbin, and national and provincial leaders had a chance to demonstrate their extreme concern through quick visits and sanctimonious statements—when people in Harbin questioned the quality of groundwater that the municipal government would provide during the shutoff, Governor Zhang Zuoji promised to "drink the first mouthful" (*People's Daily* 2005). But the Sungari flows into the Amur, which forms the border between Heilongjiang and the Russian Far East, so the pollution became an international problem as emergency agencies in the city of Khabarovsk (only a few kilometers from the Chinese border) worried about the effects there (Zhang Yue and Shen 2005). A lot of foreign media picked up the story, couching it in the tired terms of how polluted China was and how authorities tried to hide the fact (see, e.g., *The Guardian* 2005).

More seriously, the Sungari chemical spill accelerated concerns not only about the environmental effects of China's rapid industrialization in the Reform era but also about the legacy of even less environmentally conscious industrial development in earlier times. The Jilin Petrochemical Works were first built according to Soviet plans between 1954 and 1957, in the Longtan District of Jilin. It was originally planned to be built in Jimusi, near the Soviet border, but delicate diplomatic concerns—Soviet planners apparently knew the factory posed potential environmental dangers—changed the site to Jilin, far from the border. Environmental problems, apparently related to chronic benzene discharges from the factory, started early. As the number of chemical plants in the district grew to more than one hundred in the 1980s, Minamata disease (mercury poisoning) from the same industrial complex spread widely (Zhang Yue and Shen 2005).

In the relatively open atmosphere for discussing environmental issues during the Hu-Wen administration (between 2002 and 2012), the Chinese press began to ask questions such as "Why can we have polluting factories upstream of populations? How can governments and populations establish dialogues [on environmental questions]? Why is our only form of development 'on the basis of money'"? (Zhang Yue and Shen 2005).

Popular protest against industrial pollution, particularly from chemical and metallurgical factories, continued to grow through the early twenty-first century, despite officials' insistence that economic growth had to remain the first priority. In Shifang, Sichuan, protesters against a proposed molybdenum and copper processing plant stormed the offices of the local party committee and were dispersed by police using tear gas and flash-bang grenades (Yan Dingfei 2014). Initial analyses interpreted this as a typical NIMBY protest by otherwise powerless local people who would bear the brunt of industrial projects that were supported by local officials eager for the revenue and promotion evaluations such projects would generate (Tang H. 2012). Later, revisionists doubted this explanation, pointing out that opposition to the plant came not so much from local residents—many of whom were also eager for the economic opportunities—but from competing interests, including small chemical industries afraid of being outcompeted and local food producers afraid their sales would be hurt by their proximity to a big chemical plant (Yan Dingfei 2014). Similar protests broke out in such far-flung locations as Dalian in the northeast (*China Daily* 2011) and Xiamen in the coastal southeast (Tang H. 2012).

....

Whatever the reasons for individual protests, it became clear around the beginning of the 2010s that environmental protection was not simply a matter of local fixes, installing equipment here and there, or holding local officials responsible (everyone from peasants to national leaders to the foreign press was eager to blame local officials for almost everything bad that happened in China around that time). In reaction to the increasingly visible pollution problems and also to the discourse around them, the state would adopt environmental protection, or more precisely eco-developmentalism, as a basic principle of development policy. Slogans like "ecological civilization," originally promoted in the early 2000 aughts by what were then fringe figures such as Deputy Environment Minister Pan Yue (2003, 2006) became mainstream, as indicated by the publication of an editorial in *People's Daily*, under the pseudonym Ren Zhongping, indicating an official position of the party's propaganda and promoting ecological civilization as a core value of the Xi Jinping era. Even more significant was Premier Li Keqiang's "declaration of war against pollution" at the second meeting of the Twelfth National People's Congress in March 2014 (*People's Daily* 2014).

This conscious turn toward taking sustainability seriously, as opposed to just sloganeering about it (as authorities had done since the 1990s), had a bifurcated effect. On the one hand, Chinese leaders continued to believe that expanding the economy and raising living standards were absolutely necessary if they were to maintain their legitimacy and political power. And fundamental to political power was power in the literal sense—increasing the energy supply to power economic growth. At the same time, in the spirit of sustainable development and ecological civilization, the state also took serious measures to curb the worst environmental damages, with impressive results in some areas, as for example the air pollution that consumed Chinese cities at the beginning of the century and into the 2010s. That story must wait, however, until we examine the growth of the cities themselves.

14. Building an Urban Continent, 1980–2022

> Urbanisation has put enormous pressure on both the natural and built environments. —QI YE et al., *China's New Urbanisation Opportunity*, 2021

Even though accelerated rural industrialization after 1980 absorbed huge amounts of surplus farm labor under the slogan "Leave the land but don't leave the countryside," this was only half of the transformation. The other half—rural-to-urban migration, along with the explosive economic growth that began then—radically transformed China's cities. They evolved from relatively small, regimented islands of bureaucracy and heavy industry into growing metropolises ahum with factories making clothing, toys, tools, and every conceivable kind of gizmo, widget, and doohickey that foreign consumers might want to buy at a price reflecting the cheapness of China's migrant labor at that time. As this industrialization created more wealth, cities evolved further, into glittering, high-rise, traffic-congested, inequality-plagued tributes to consumerism and national pride (figure 14.1). According to official statistics, they contained 65 percent of China's population in 2021 (Statista 2022b).

This urban transformation had multiple effects on China's environment. First, the cities, expanding to build factories, house workers, and accommodate burgeoning numbers of migrants, occupied more and more of China's most productive agricultural land and sucked up more and more of China's limited water supply. Second, air and water pollution increased, coming from three sources: the factories themselves, the extractive industries that supplied the raw materials for these factories, and the urban residents who produced more waste and recycled less. Third, the increases in urbanized land caused secondary effects such as land subsidence and microclimate changes.

POPULATING THE CITIES

For most of the high socialist period, very few rural people were allowed to move to the cities. During the Great Leap Forward the cities

Figure 14.1. A typical urban scene in Beijing: (top) 1992 and (bottom) 2000. Photos courtesy of Daniel Benjamin Abramson.

grew too fast for the available services or the budget, and tens of millions of people were forced to go back to villages. To prevent future unplanned growth, authorities repurposed the *hukou* household registration system, changing its primary function from monitoring people's movements to controlling them. Only a tiny minority of rural people who attended colleges or got jobs in urban *danwei*, plus a few whose villages were absorbed by new heavy industries, could convert their *hukou* and become urban residents. It was almost impossible for people without urban registration to live in cities, partly because local officials in their home villages had to watch them very closely (for a while, peasants planning to visit a city even temporarily were advised to have a letter from authorities, a supply of ration coupons, and some cash, none of which they would have used in everyday life). If they did manage to sneak into the city, it was difficult to eat, since it was impossible to buy food with just money and no ration coupons, and rural people had very little money anyway (Mobo Gao 1999, 177). The only way for a rural resident to purchase food would be to share the ration coupons of urban relatives (few rural people had any urban relatives) or purchase food coupons on the black market, which required money that the rural people didn't have. This way, the regime could severely restrict the numbers of urban people to whom it owed salaries and benefits, from housing to medical care to pensions to the right of children to inherit jobs in their parents' units, and it could then keep the cities free of people it could not afford to care for.

All this changed when, "after some experimentation, as China latched onto a labor intensive, export-oriented growth strategy in the mid-1980s, rural labor was allowed *en masse* to the cities to fill industry's labor demand, which later became a major state industrialization strategy" (K. Chan 2012, 188). Migration started slowly in the late 1970s, but by 1982 about 46 million rural migrants had moved to China's urban areas, a figure that had increased to 130 million by 2000 and to more than 170 million in the late 2010s (K. Chan 2012, 190; Chen Jinyong 2023). Most of these people were employed in one of three kinds of low-skilled, low-wage jobs. Construction of China's rapidly growing and transforming cities was a favorite of young and middle-aged men (and some tough women) who were physically strong and could tolerate rough conditions, long hours, and unpredictable schedules. Women and teenage boys tended to favor boring, repetitive factory jobs that paid less but had more reliable—if longer—hours (Pun 2005; C. K. Lee 2007), or else jobs in service industries, as retail sales, hotels, and especially restaurants all boomed when consumer goods became available and urban families acquired more disposable cash income. Some rural women also became domestic servants for prosperous urban families (H. Yan 2008). In addition, some rural migrants became petty entrepreneurs, producing clothing, opening their own small restaurants, peddling various goods at formal and informal markets on urban streets, and occasionally fulfilling Deng Xiaoping's admonition to "get rich" (Li Zhang 2002).

Despite the effective caste system that denied them the benefits of formal urban residency, rural people continued to swarm into the cities, and the urban population grew dramatically even as China's total population began to level off in response to the planned-birth policy. Although it is difficult to classify population discretely and unambiguously into urban and rural, Kam Wing Chan's meticulous research indicates that China's de facto urban population (including both legal residents and people with rural *hukou* living in cities) more than doubled, from about 214 million in 1982 to about 459 million in 2001, and then doubled again, to 848 million in 2019. Of these, about 621 million had official urban residence and 227 million were living in cities without urban *hukou*. Of the latter, around 170 million were temporary rural migrants, or in Chinese parlance "peasant workers" (*nongmin gong*) or "rural migrants," also known in English as the "floating population" (K. Chan 2001, 131; 2012, 190; Chen Jinyong 2023, 145–46).

Besides increasing by over 650 million (twice the total population of the United States) in four decades, the urban population also got much richer and thus consumed many more resources of all kinds, particularly those related to housing and transportation. From an unbelievably cramped 6.7 square meters per person in 1978, the average per capita floor area in urban housing reached 38 square meters in 2018 (*People's Daily* 2018), meaning that as the urban population quadrupled during the same period, residential floor space increased by a factor of twenty. China had 2.4 billion square meters of floor space completed or under construction in 2000; in 2009 that figure had grown to 13.4 billion and, in 2019, to 18.4 billion. Although most of the additional space was in multistory buildings (much of it in high-rise towers), this growth still contributed to the enlarged footprint of urban housing and at any rate consumed massive amounts of concrete, steel, and glass. In addition, private automobile ownership went from essentially zero at the beginning of the Reform period to over 243 million in 2020 (Statista 2022c), putting huge strains on the street network. Consumption in general increased greatly, as reflected in an elevenfold increase in urban consumption expenditures, adjusted for inflation, from 1978 to 2018, about two-thirds of that occurring after 2000. Even in the short period from 2013 to 2018, per capita urban disposable income increased 40 percent, from about ¥18,000 to almost ¥31,000 (SSB data, from EPS China Data n.d.).

In sum, China's urban population not only quadrupled but increased its consumption standards exponentially between 1980 and 2020. However much strain this immense quantitative increase in urban resource use put on China's urban (and, indirectly, also rural) ecosystems, the qualitative changes are arguably even more significant. In metamorphosing from clusters of industrial and bureaucratic *danwei* to wealthy, modern, functionally differentiated cities, China's urban areas underwent a fundamental transformation in size, form, layout, resource use, environmental impacts, and social-ecological system resilience.

BUILDING THE CITIES

Chinese cities, like cities elsewhere in rapidly industrializing countries, grew in a characteristic double pattern (Angel 2012, 164). Most urban growth consisted of expansion around city peripheries, mostly into former farmland. This kind of outward expansion continued into the 2020s. At the same time, previously existing urban areas, where infrastructure stagnated and often decayed during the high socialist period, underwent massive urban renewal; this filling-in process began slowly in the 1980s but rapidly intensified in the 1990s (Gaubatz 1999, 1415) and continued through the first decades of the twenty-first century.

Chinese planners in the early Reform period redesigned their cities not only to accommodate the rapidly increasing urban population but also to help boost economic growth by building light industry that produced consumer goods. In the process they transformed urban areas from "production cities to . . . livable cities that provided residents with a high quality of life" (Yin et al. 2011, 615). The regime promoted this urban remodel in three ways. First, it required municipal governments to adopt and periodically revise master plans that divided both the existing urban area and the area of projected expansion into different kinds of development and redevelopment zones (Jiang G. et al. 2017, 260). Second, it reformed property laws so that both state *danwei* and private developers could develop urban land for commercial, industrial, or residential purposes (F. Deng and Huang 2004, 225). Third, it gradually phased out the system in which work units assigned rent-free housing to state employees. Reform policies in the early 1990s allowed people to purchase their former work-unit housing at low prices (Y. P. Wang 2011, 20–21). These reforms proved unsuccessful, so after 1998 the government moved to a fully marketized housing system of purchase and rents, not unlike those in traditionally capitalist countries but with subsidies for low- and middle-income families (Y. P. Wang 2011, 22–23). Along with this marketization there developed a hierarchy of housing locations—the farther a neighborhood from the traditional city center, the lower the purchase price or rent per square meter (Y. P. Wang 2011, 34).

Expanding Urban Footprints

Urban footprints had expanded somewhat during the planned economy period, as *danwei* blocks rose up on the peripheries of existing cities and whole new cities arose, but city footprints expanded much faster after 1980, especially as urban population growth accelerated to more than twenty million a year after 1990. Most of this growth occurred as land at the edge of already built-up areas became urbanized (known in China as "pancake growth"), with smaller amounts in new satellite cities removed from the main urban agglomeration (known as "leapfrog growth"). Urban infilling of previously vacant land within the old urban cores was a third, albeit minor, mode of expansion, which also continued from the planned economy period (Min Xu et al. 2016, 8; Q. He et al. 2017, 731). As a result, the total urban built-up area grew to approximately 200,000 square kilometers by 2020, probably representing a fourfold increase since 1981, and occupying about 2 percent of China's total area (Min Xu et al. 2016: 1; Q. He et al. 2017, 732–33, Shanchuan Wang 2018; Guanyan Baogao Wang 2022). The increase in floor space per capita meant that overall urban population density declined despite the increasing height of buildings, a trend that China shared with other rapidly industrializing and urbanizing countries (Angel 2012, 175).

Some of this periurban growth was organic, or "bottom-up growth," accommodating the tens of millions of temporary migrants who began streaming into cities in the 1980s and 1990s. Many factory workers lived in crowded dormitories within the factory compounds (Peled 2005), a situation reminiscent of the old *danwei*, while construction workers, who moved around a lot, typically lived in ramshackle sheds on demolition or building sites, cooking over makeshift stoves made of oil drums and using public bath and toilet facilities (Sniadecki 2008). For many migrants, however, especially those involved in petty trading or running small businesses, the only choice was to rent housing from farmers (or former farmers) living on urban peripheries (see Li Zhang 2002; Yue, Liu, and Fan 2013, 364–67; and L. Tian and Yao 2018, 77). Villagers often took advantage of the migrants' housing needs, first renting out rooms in their own houses (Siu 2007) and later building one-story brick row houses on deserted land to rent out to the migrants. For some migrant communities, such as the Wenzhou clothing manufacturers and merchants in Beijing, this was an unsafe and unsatisfactory situation, so they began to lease land from village committees to build their own walled compounds, known as "big yards" (*da yuan*). In some cases, whole suburban neighborhoods turned into colonies of migrants from particular areas, usually specializing in particular trades and informally named after the residents' province of origin. By 1994 the outskirts of Beijing had Zhejiang, Henan, and Anhui Villages, plus two separate Xinjiang Villages (Li Zhang 2002, 14–15). The municipal government demolished these yards in 1995, but by 1998 members of the Wenzhou migrant community had taken over abandoned factory *danwei* compounds and converted them into new yards (Li Zhang 2002, 200). Among these urban villages were some devoted to recycling and reprocessing urban waste (J. Goldstein 2021, 190–206).

In other areas, particularly in the migrant-magnet light industrial districts of the Pearl River Delta in Guangdong, former farming villages formed land-share cooperatives that engaged in a sort of localized planning by designating land to be retained for agriculture, developed for migrant housing, or leased out to industrial concerns, resulting in a patchy not-quite-urban but no longer rural landscape (L. Tian

2015, 120–25). Sometimes these villages were gradually surrounded by expanded urban areas driven by city planning and became "villages within the city" (*cheng zhong cun*) where villagers, no longer farming or farming only remnant bits of land, redeveloped their land as rental housing, while retaining the rural *hukou* that allowed them to keep their land rights (Siu 2007, 333; Zacharias, Hu, and Huang 2013, 1). In the early 2000s, the city of Guangzhou (one part of the Pearl River Delta metropolis) had 138 of these urban enclaves; nearby Zhuhai had 26 (Siu 2007, 334). In 2010 urbanized villages covered as much as 20 percent of Guangzhou's urban footprint and may have housed as much as a third of its population (Zacharias, Hu, and Huang 2013, 4). Most of them, however, lacked both modern infrastructure and security (Siu 2007, 334; Yuan Gao, Shahab, and Ahmadpoor 2020). Gradually, some urban villages built basic service infrastructure as part of very partial inclusion in municipal plans (Zacharias, Hu, and Huang 2013, 2–3). At the same time, though, they were very densely populated—buildings, mostly medium-height ones, covered much of the land, and there was very little open space (Yuan Gao, Shahab, and Ahmadpoor 2020, 11). Municipal governments redeveloped a few of these districts to make them more like twenty-first-century housing estates, with buildings as tall as thirty stories and more green space between the buildings (Zacharias, Hu, and Huang 2013, 4).

Fiscal reform and state policy also contributed to the urbanization—often unsystematic—of periurban areas. After 1994, when fiscal reforms made local and provincial governments more responsible for financing their development programs and other activities (L. Tian 2015, 119), governments from the provincial to the township level needed new sources of revenue and often found it in land conversion. Essentially, the tax and land policies instituted in the 1990s allowed municipal, district, and township governments to expropriate collective land (held by village committees) and convert it to state-owned land. The governments compensated the villages for this loss and then flipped the land at a profit to industries and residential developers for the rights to build on it (Yue, Liu, and Fan 2013, 368; Jinlong Gao et al. 2014, 464; L. Tian and Yao 2018, 73; L. Tian et al. 2017, 427). The resulting mix of residential, industrial, commercial, and other uses often foiled the efforts of systematic planners, thereby contributing greatly to the unplanned urban growth usually condemned as "sprawl" (see Gibson, Li, and Boe-Gibson. 2014, 7851; Yue, Liu, and Fan 2013, 368; and Yin et al. 2011, 615).

Rebuilding Existing Cities

Eventually, as urban peripheries grew and demands for urban services and amenities increased, the old socialist cities grew increasingly obsolete and dysfunctional. Beginning in the 1990s, rapid urban renewal in these older areas proceeded parallel to periurban expansion. Whole neighborhoods, consisting of prerevolutionary housing, multifunctional *danwei* compounds, or a mixture of both, were demolished (Sniadecki 2008) and redeveloped. Overall, residential populations in the old city cores declined (Yue, Liu, and Fan 2013, 364). A large proportion of the courtyard-style housing in the famous *hutong* lanes of Beijing was still intact in 1989, but in the next thirty years most of what remained was demolished and replaced by modern buildings (see Abramson 2007, 136–38; Knapp 2008; and R. Mao 2018). Many of the remaining lanes were made into tourist attractions, and spaces surrounding buildings considered worthy of historic preservation were often cleared and left open to create sightlines to the monuments (Abramson 2007, 146–49). Likewise, the characteristic two-story, wooden-front houses that remained in parts of Chengdu as late as the late 1990s were practically gone by the 2010s, and comparable kinds of low-rise housing in other cities, from Quanzhou on the southeast coast to Kunming in the southwest interior, disappeared as well. Residents of old-style housing were resettled in new, high-rise residential parks. A few of the more expensive ones were built within the preexisting city footprint, but most of them rose in the expanding urban peripheries.

As modern buildings replaced the old houses, the pedestrian alleys that had connected them gave way to a modern traffic grid. The grid was anchored by multilane arterial streets consisting of a central median, a couple of auto lanes in each direction, a bicycle-and-scooter lane separated from the main traffic by a low fence or sometimes a narrow planted strip, and a wide sidewalk, usually of patterned paver stones (Abramson 2007, 150–51; Abramson 2008) (figure 14.2). Connecting these boulevards were two-lane side streets, often reduced to a single functional lane by parked vehicles. In large, flat cities, the whole network was tied together by a series of concentric ring roads, numbered from the center outward. The outer ones were often built as elevated highways with on- and off-ramps. As cities expanded, they added new rings. By the 2010s, Chengdu, for example, had three ring roads and an "around the city expressway" plus an expressway to its airport, formerly nestled amid farmland but newly surrounded by urban developments. Beijing at the same time was completing its fifth ring road and then its sixth. Cities built in mountainous regions, such as Chongqing or Panzhihua, however, did not adopt the concentric ring model but conformed their highways to the natural topography.

Polluting factories were also moved out of city centers and into new peripheral developments (L. Tian and Yao 2018, 72, 77). The spaces previously occupied by prerevolutionary neighborhoods and high-socialist *danwei* were mostly rebuilt with several new-style buildings—new offices for the still existing work units, whose employees no longer lived in *danwei*-owned housing but rather in the new residential compounds; the residential compounds themselves (the more luxurious ones at least); fancy commercial streets and huge indoor shopping malls (Jiang G. et al. 2017, 263–65); hotels rated by numbers of stars; and many multifunctional commercial buildings housing restaurants and nightclubs, retail stores, and offices. Some large *danwei*, primarily universities and colleges, maintained and modernized their campuses. Most old and new campuses still included student housing, but faculty and staff living quarters gave way to academic and service buildings as their former occupants bought nicer, off-campus housing.

Planners recognized that even people living in spacious apartments with modern conveniences still needed open space, and most parks and squares continued to be places of myriad outdoor activities, from viewing art exhibitions to strolling among the bamboo or along river banks to tea drinking to old people's disco dancing, calisthenics, and *taiji* (tai-chi) fist or sword practice to evanescent water calligraphy destined to evaporate with the next rays of full sunshine (Zito 2011; Sniadecki 2012) (figure 14.3).

Historic buildings, abundant in most Chinese cities, were preserved and sometimes refurbished, and some former *danwei* districts were rebuilt as "imitation ancient streets" (*fang gu jie*), where consumers and tourists could indulge their appetites for food, tea, liquor, games, high-

Figure 14.2. Baishiqiao Road in Beijing, 2000. Photo courtesy of Daniel Benjamin Abramson.

Figure 14.3. Calligraphy teacher Liu Lanbo demonstrating technique to his students, Tuanjiehu Park, Beijing, 2007. Photo courtesy of Angela Zito.

end clothing, or a variety of "imitation ancient" curios and handicrafts. Outdoor food markets remained as well, though in the 1990s many of these were enclosed and roofed over while still composed of stalls rented to individual vendors (Faber n.d.). The notorious Huanan Seafood Market in Wuhan, where the SARS-CoV-2 virus was first detected, was such an indoor market, consisting of individual-proprietor stalls. Beginning in the 2000 aughts, there was a push to move food shopping into indoor "supermarkets."

New Urban Spaces

The typical residential housing compound (*xiaoqu*), whether in the inner city or the suburbs, was surrounded by a whitewashed wall, usually topped with broken glass or barbed wire, not unlike the wall of the old socialist *danwei* compound. Residents could enter by scanning their smartphones at one or more turnstile-equipped entrances; visitors had to announce themselves and be greeted at the entrance by a resident. Inside were paved paths, lined with more or less carefully tended evergreen shrubs and trees, leading to the entrances of the several apartment towers within the compound, where elevators to the upper floors had replaced the grimy concrete stairways of the old *danwei* walk-ups. There were typically some services within the compound—a place to dispose of garbage and, in the higher-end developments, to separate and recycle it (Schmitt 2016, 200–212) (figure 14.4); a small grocery store selling fresh meat and produce along with dry foods; perhaps a gym or recreation room; and benches where people could sit outside and relax. The apartments themselves contained full kitchens and bath facilities, plus a small balcony to grow herbs or flowers, stash extra stuff, and hang laundry under cover, since washers were ubiquitous but not dryers. What distinguished the twenty-first-century compound from the twentieth-century *danwei*, in addition to the much more spacious and luxurious housing space, was that it was entirely a space of consumption. It lacked major services such as schools or clinics, and the residents did not work there.

In the 2000 aughts, a spatial hierarchy began to develop among compounds, based on their locations and amenities. Compounds closer to the city center ranked higher on the whole, with more spacious units, better construction materials, more and better-kept green space, and better

Figure 14.4. Waste separation bins at a residential complex in Chengdu.
Photo courtesy of Edwin Schmitt.

security arrangements to keep crime and "people who do not belong"
outside the gates. Some of the most luxurious housing of all was not
in inner-city high-rises but in suburban developments of townhouses
or stand-alone houses called "villas" (*bieshu*), including in at least one
prominent case a large development on the shores of Dian Lake near
Kunming that was ultimately torn down because it had been built ille-
gally in a conservation zone (Li Zhang 2010, 109–19; Xie 2021b).

The final components of the rapidly growing urban peripheries were
relocated industries and university campuses. Some large and polluting
factories formerly located closer to city centers moved to more distant
exurban locations: the Capital Iron and Steel works, built in the 1950s
at Shijingshan in what had since become Beijing's urban core, moved
to a more distant site in 2011 as it modernized its production facilities.
Shanghai converted the former Pudong Iron and Steel complex to the
grounds for the Expo 2010 world's fair, moving the steel mill to the
Baoshan district (*China Daily* 2005), where steel had been produced since
1938. In 2020 work began to move one of its large mills to a yet more
distant site, in order to convert the Baoshan site to a high-tech "smart
city" (Yang Jian and Song 2020). Also, as early as the late 1990s many

large and medium-sized cities established "development zones" (*kaifa qu*), industrial parks where they provided utilities and infrastructure for industrial and commercial operations (L. Tian 2015).

College campuses also moved. The only ones that remained in old urban areas had either been there since pre-Founding times or were built in the early years of the People's Republic. In the early twenty-first century some of them moved out of the city altogether, decamping for the suburbs. The prestigious Zhejiang University sold its downtown campus near West Lake to developers for ¥2.4 billion and used part of the proceeds to build a new suburban campus (Yue, Liu, and Fan 2013, 367). Both of Sichuan University's early twentieth-century downtown Chengdu campuses were gradually surrounded, first by little stores and restaurants in the 1980s and 1990s, then in the 2000 aughts by residential compounds, and in the 2010s by gaudy high-rise hotels with multicolored flashing lights. The university also renovated or replaced most of its old buildings, retaining a few with historical interest. At the same time, the university also built an entirely new campus twelve kilometers away, near the Shuangliu Airport, meaning that, given the likelihood of traffic jams, administrative documents had to travel an hour or more by courier to collect necessary signatures. Southwest Minzu University, whose downtown Chengdu campus was built in the 1950s, did the same thing, concentrating most of its undergraduate programs at a new campus built in 2008 in the airport district (Banfill 2021, 26).[1] Many new universities and colleges, founded as part of the education boom of the twenty-first century, were built in the periurban areas.

Different levels of government developed different kinds of land uses on periurban lands owned by either state agencies or village collectives: municipal and county governments developed state-owned land for residential compounds, industrial parks, and university campuses. Township governments developed mixed state- and collectively owned land for both rural and urban housing plus township industrial clusters. Village committees developed collectively owned land for village housing (some of which was quite modern and multistoried) and sometimes continued using the land for rural industries (L. Tian 2015, 125).

Although the urban footprint grew mostly by expanding at urban peripheries, beginning in 2000 central ministries formulated a national plan to develop up to four hundred "satellite towns" or "new towns," intended to be secondary centers of existing metropolises, separated from the primary centers by agricultural land. These new towns were ostensibly planned and built to alleviate pressure on the old cities, often with the aim of sustainability, ecological development, or even "low-carbon" development (Visser 2016, 44). They had varied degrees of success. For example, Lingang Marine Eco-City, built beginning in the 2000 aughts on fill in a formerly rural part of the municipality of Shanghai, was planned for an eventual population of eight hundred thousand, but by 2012 it had become a failed "ghost city" where some partially finished buildings were abandoned and other planned structures had never been started (Visser 2016, 45–46).

Whatever the ostensible environmental principles that motivated the planning and construction of Lingang and other eco-cities (Sze 2015), financially they were another manifestation of the congruent interests of local officials, who needed land conversion revenue to balance their budgets, and real estate developers, who pursued profits. In addition, local officials' typically short terms in office incentivized them to pursue short-term goals of economic growth (Eaton and Kostka 2014). As a result, as Robin Visser (2016, 43) has explained, "rather than actually developing a knowledge economy or green economy, such land-related projects are strategies used by Chinese sub-national authorities, in collusion with transnational corporations, to generate local finance and real estate development." Because such projects enriched governments at all levels through land transactions and thus encouraged speculation, and because real estate was one of the few avenues for investment (Rithmire 2017, 124), these projects often led to oversupply and thus to repeated real estate bubbles, first in the early 1990s and later on after 2008 (Rithmire 2017, 128, 140), and when demand faltered, quite naturally many of them ended up half empty or even half built.

Getting around the Big Cities

With home and work now separated for almost everyone in China's large cities, there needed to be a way to get from one to the other, particularly for people who lived in new compounds in the suburbs and had jobs in the inner cities. The network of surface streets and elevated highways functioned to connect city and suburbs to an extent, and public buses went just about everywhere frequently. However, with the exponential growth of car ownership, streets and expressways were subject to monumental traffic jams, so that it often took an hour or more just to get from home to work or back again. To try to control these tie-ups, municipal officials in many cities restricted traffic on certain days to vehicles with certain license plate numbers, often prohibited small, slow vehicles from traveling during certain busy hours, and sometimes instituted lotteries for plates with city numbers. There are even reports of people paying as much as ¥160,000 for a sham marriage to someone who had a coveted plate, then transferring the plate and divorcing (Zao 2019). Still, despite all the evasion, these controls both loosened traffic just a bit and perhaps contributed to the decline in air pollution in the late 2010s.

Even with these restrictions, there were just too many vehicles on the streets, so planners embarked on massive programs of subway construction. Beijing's first two lines were started as early as 1969 and operational in 1971, but no more lines opened until 2002, and new lines continued to be added in the following decades. By 2019, 33 Chinese cities had subway systems (*China Daily* 2019). Beijing had 17 operational lines crisscrossing the ever-expanding city, with a length 581 kilometers, serving 405 stations and a daily traffic of more than 10 million riders, with a further 9 lines under construction and 13 more planned, which would expand the length of the network to 985 kilometers (Jintian Toutiao 2019; Baidu n.d. e) (map 14.1).[2] Other cities soon followed, often building lines with astounding speed. Chengdu's first line was not completed until 2010, but by 2020 it had 12 working lines with 350 kilometers of track serving 215 stops, and even more were under construction (Wikipedia n.d.). On 20 October 2020, just over 4.5 million riders used the Chengdu subway (Chengdurail 2017–20).

Small and Medium-Sized Cities

Large cities such as Beijing, Hangzhou, Chengdu, and the Pearl River Delta Metropolitan Region were not the only urban centers that grew substantially beginning in the 1980s. By 2017, China had 152 cities with urban populations greater than 1 million (G. Wan et al. 2017, 4), 50 of them with populations over 2 million. But a large number of China's urbanites lived in small and medium-sized cities (the former with pop-

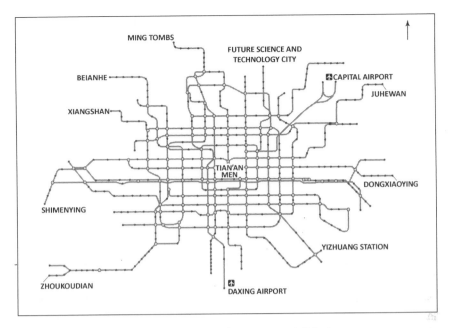

Map 14.1. Beijing subway system. Drawn by Lily Demet Crandall-Oral from public system maps.

ulation below 200,000 and the latter with between 200,000 and 1 million), which also grew substantially beginning in the 1980s. These cities differed considerably from the big metropolises. For example, Xichang, the prefectural capital of Liangshan Yi Autonomous Prefecture in Sichuan, had only about 180,000 people in 1999 (Bamo Erha, pers. comm.); it had no buildings taller than a few stories, no streets wider than four lanes, and traffic was almost completely unregulated: the sound of horns reverberated day and night at a time when honking had already been prohibited in Beijing and other megacities. Air pollution was not noticeable; night skies were dark and clear enough to see stars even from the city center. By 2018, the city population was approaching a million, and cars on wide, landscaped boulevards whizzed honkless past elegant public squares, modern gymnasia, and high-rise apartment towers.

Cities in Zomia and Central Asia

Zomia, by definition, was until recently not a land of cities, and what cities traditionally existed in the lowlands (such as Xichang) had always been Han intrusions into the territory of other peoples. Higher in the Zomian mountains of Liangshan, the county seat of Yanyuan, which in prerevolutionary times was nothing but a small garrison and a source of widely traded salt (Zeng Y. 1988), underwent an even more fundamental transition. In 1993 it was still a relic of the high socialist era. Its twenty thousand or so residents lived among a small maze of shabby government offices, shabby schools, one shabby hotel, and some streets with small shops and food stalls. There were so few vehicles that people socialized in the middle of the street. The few paved streets quickly turned to dirt (or mud, in season) a few hundred meters outside the town center. Almost no buildings had indoor plumbing; the one hotel in town had installed it since my previous visit in 1991. By 2018, the population had expanded to one hundred thousand or more, hotels were multistory, government offices bordered on the luxurious, restaurants bordered on the elegant, schools were modern, clean, and efficient, and there were enough wide, paved streets to make taxi driving a profitable business. Xichang and Yanyuan, both located in the Zomian region, were not much different from similar-sized cities in China Proper, having participated in the enlargement and physical transformation of China's overall urban landscape.

Elsewhere in Zomia, new cities were built according to some of the same ostensibly ecological principles as satellite towns around Shanghai or other major metropolises. The old socialist county seat town of Beichuan, in a valley about twenty kilometers into the mountains that border the Chengdu Plain on the north, was utterly destroyed by the 2008 Wenchuan earthquake. Like other areas in the quake zone, Beichuan was presented with an opportunity for rebuilding, not only according to principles of eco-development but also to celebrate the culture of the Qiang minority people who lived there. However, planners decided to leave the twisted ruins of the old town as a memorial to those lost in the disaster (figure 14.5, top), so they transferred two townships from lowland An County to Beichuan's administration and built a new county seat for Beichuan there. When I visited in 2011, the new town had an impressive public square built to incorporate decorative elements from Qiang architecture, a market with people in Qiang dress selling corn cakes and other "minority flavored" snacks, and blocks of four-story "Qiang-style" apartments that were less than half occupied (figure 14.5, bottom).

Figure 14.5. After the 2008 earthquake in Sichuan: (top) ruins of the city of Beichuan preserved as a monument; (bottom) the newly rebuilt, "minority-themed" Beichuan, 2011. Photos by the author.

Central Asia, unlike Zomia, has a millennia-old urban civilization, and even as late as the 1990s many of its cities still had predominant or at least significant populations of indigenous ethnic groups. With the growth of the Reform-era economy, however, as peasant workers expanded the cities of China Proper, long-distance and long-term Han migrants expanded the cities of Central Asia. Lhasa, for example, was Tibet's only real city until the 1980s, but it had been the administrative capital of Tibet since the seventeenth century, and immediately before the Chinese invasion in 1951 it probably had somewhere between thirty thousand and sixty thousand residents, all of them Tibetan. Its modest three-square-kilometer urban area centered on the Barkor circumambulation route and the Jokhang Temple and contained administrative buildings, stores, and workshops in addition to private residences. Tibet had no other urban places that would qualify as cities (Yeh 2013, 199–200). Lhasa grew modestly in the 1960s and 1970s with the addition of a small administrative presence, but in the 1990s Han immigrants began streaming in. The regime invested heavily in Chinese-style buildings, streets, and other urban infrastructure, so that by the 2010s the urban population was several hundred thousand (perhaps half Han and half Tibetan), living in a built-up area of over seventy-eight square kilometers. To accomplish this transformation, both old and new urban housing, as well as much of the housing in the villages absorbed into the expanding city, was demolished and replaced with standard, Chinese-style apartment blocks and villas, often with superficially Tibetan design features such as brightly colored door lintels or window frames (Yeh 2013, 210–27). In addition to Lhasa, Tibet in 2019 had five other cities, each with a population of under two hundred thousand, and small towns sprang up all over the landscape (Yeh 2013, 204).

Xinjiang, by contrast, had had its own widespread urban civilization for centuries before the consolidation of Chinese rule in the late nineteenth century, but city dwellers were primarily Turkic Muslims, and their cities were built in a distinctive Central Asian style. They primarily functioned as stations along the trade routes known as the Silk Road and as modest market centers for surrounding oasis agricultural populations. As the Qing dynasty consolidated its rule over parts of Xinjiang (Perdue 2005), many cities there acquired a hybrid character—a walled city built according to Chinese administrative geography sat side-by-side with walled or open Muslim quarters that exhibited a more purely Central Asian character (Gaubatz 2002, 243–45). After 1950, the PRC regime sponsored massive Han migration into the region (Bellér-Hann 2013, 175–76), and Xinjiang's urban population grew from 530,000 in 1949 to 7.47 million in 2005 (Mansur and Rahman 2007, 118), but not every city grew in the same way. In the regional capital, Ürümchi, and in some other preexisting oasis cities, traditional Uyghur neighborhoods, or *mehelle* (Dautcher 2009), were first surrounded by expanded modern Chinese construction like that in the cities of China Proper. After 2009 *mehelle* were often subjected to urban renewal in the form of demolition and replacement by Chinese-style housing blocks (Byler 2018, 236), a process of "socialist creative destruction" carried out by the "bulldozer state" (Bellér-Hann 2013, 179). New cities that sprang up in oil-producing regions such as Korla in the central-south of Xinjiang and Karamay and Shihezi in the north, where there were no "silk road cities" to renovate, were barely distinguishable from new cities in China Proper (Cliff 2009, 2016).

In Kashgar, a traditional city in the far southwest, "modernization" took a similar but more complex course. Kashgar was already flooded by Han migrants in the early 2000s, but even in 2009, when there was urban unrest in Ürümchi as a result of the murder of Uyghur workers in Guangdong, Kashgar's "old town" pre-Founding Silk Road city was still largely intact. However, as part of an attempt to ameliorate ethnic conflict and soften the effects of colonial repression with development, the government not only established a special economic zone to attract investments from Chinese companies, it also embarked on a renewal of the old town. It invested almost half a billion dollars in the Uyghur Historical and Cultural Preservation Project, which would "make older buildings earthquake resistant, improve sanitary standards, alleviate poverty, and even promote Uyghur culture" (Steenberg and Rippa 2018, 282), the latter by making the area into a tourist attraction. The project "almost entirely demolished and rebuilt [the old town area] in what authorities have labeled 'ancient Islamic architecture'" (Steenberg and Rippa 2018, 282) (figure 14.6). Residents could remain if they could afford to rebuild their own houses in accordance with government standards, but those who could not afford it were moved to typical 2010s high-rise apartment blocks in periurban areas (Steenberg and Rippa 2018, 284).

Figure 14.6. Streets in Kashgar's old town area: (top) in 2013, before urban renewal; (bottom) after urban renewal, 2016. Photos courtesy of Alessandro Rippa.

Then, after 2014, when the Xinjiang government introduced the "people's war on terror" (M. Anderson 2017; Byler 2018), security checkpoints, surveillance cameras, and police stations completed the landscape of Kashgar and other formerly traditional Uyghur cities in early twenty-first-century Xinjiang.

CONNECTING THE CITIES

In the early 1990s, the only practical way to get from Chengdu to Xichang, the capital of Liangshan Prefecture, was to ride a night train on the Chengdu-Kunming line, a journey of ten to twelve hours, much of it spent in tunnels as the train passed through rugged mountain topography. By 2008, driving was a reasonable proposition: two full days on two-lane paved roads, careening around corners and crossing the center line to pass slow Jiefang and Dongfeng trucks as well as lumbering local buses, a few small *bengbeng che* (diesel tractors), and lots of sheep and goats. When one reached the Anning River valley north of Xichang, one could pay a toll and whiz the last fifty kilometers or so on a limited-access expressway, which had taken more than a decade to build. Then in 2012 the world was changed. The expressway was completed all the way from Chengdu to Xichang, and it was an engineering wonder. It passed not so much over or around the mountains as *through* them. The longest tunnel extended 10,008 meters, and to gain altitude around Liziping, the road exited a tunnel, crossed a valley on a high trestle, and entered another mountain, where it turned more than a full circle on a helical route *inside* the mountain, to emerge on another trestle before entering the tunnel that passed through the next mountain (figure 14.7). Instead of two full days, a drive from Chengdu to Xichang, obeying all speed limits, took only four hours.

Although spectacular, the Chengdu-Xichang Expressway was but one relatively short link in the over 3,000-kilometer Beijing-Kunming Expressway (Jing-Kun Gaosu), which in turn accounted for only about one-fortieth of China's expressway network. Beginning in the late 1980s, China sought to connect its cities and thus embarked on a huge upgrade of not only its urban transport systems but also its intercity travel networks. While China did not open its first expressway—the 18.5-kilometer stretch from downtown Shanghai to Jiading (Guowuyuan 2016)—until

Figure 14.7. A trestle between tunnels on the Chengdu-Xichang Expressway, 2012. Photo by the author.

1998, by 2015 China had the world's longest expressway network, stretching 123,500 kilometers (Yanqun Yang et al. 2019, 1). These expressways connected not only the large metropolitan areas in China Proper but even parts of Zomia and Central Asia. Many more expressways were still under construction, particularly in mountainous areas, as China entered the 2020s.

The expressway-building campaign went hand in hand with China's boom in private car ownership, and it seemed for a while that China was following the US path toward mobility, paying little attention to the environmental much less the climate costs of highways and car travel. But in 2004, not long after the expressway campaign got into full swing, Chinese planners decided to go the Japanese route in addition to the American one and began building a high-speed rail network (Lawrence, Bullock, and Liu 2019, 9).

In 2008 the first high-speed line opened. It covered the short distance from Beijing to Tianjin at an average speed of 240 kilometers per hour, though the trains could actually go as fast as 350 kilometers per hour. By 2012, the 2,105-kilometer Beijing-Guangzhou line opened, covering the distance in about eight hours.[3] By 2020 China had around 30,000 kilometers of high-speed rail, more than the rest of the world put together; plans were also afoot to construct an additional 8,000 kilometers in the next five years. One could travel by bullet train from Qiqihar in western Heilongjiang to Beihai on the Guangxi coast, from Shanghai to Ürümchi, from Kunming in the southwest to Hangzhou on the coast (Lawrence, Bullock, and Liu 2019, 17) (map 14.2). In 2019 China's high-speed trains carried about 2.3 billion passengers, who racked up about 800 billion passenger-kilometers (Statista 2022d). Meanwhile, China announced in 2022 that railways (albeit not high-speed ones) now encircled the Taklimakan Desert of southern Xinjiang (Kate Zhang 2022).

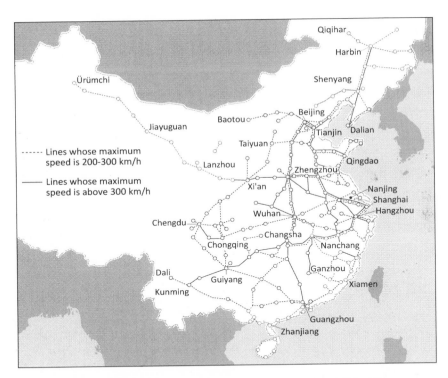

Map 14.2. High-speed rail network, 2019. Adapted by Lily Demet Crandall-Oral from Martha Lawrence, Richard Bullock, and Ziming Liu. *China's High-Speed Rail Development*. License: Creative Commons Attribution CC BY 3.0 IGO. Views and opinions expressed in the adaptation are the sole responsibility of the author of this book and are not endorsed by the World Bank.

The final piece of the transport infrastructure was the airline system. In the 1950s, with Soviet help, major cities all built airports, or in some cases enlarged existing airports, often an hour's drive or more away from city centers over narrow roads. As late as the 1980s, only high-ranking cadres, senior academics, or foreigners could travel by airplane; tickets were scarce and usually secured by *danwei* bureaucrats at downtown offices; tickets were not sold at airports, and there were no round-trip tickets. Since China had never allowed smoking on airplanes, the un-heated, high-ceilinged passenger waiting rooms were redolent with the exhalations of officials and professors loading up on nicotine before the long flights. When the flight attendants gave the safety lecture before takeoff, passengers craned their necks to see the demonstration and also listened attentively. In 1980 Chinese domestic flights carried 2.5 million passengers (World Bank n.d. a); since many of these were round-trips and repeat trips, probably less than one Chinese in a thousand had ever been near an airplane.

In 1990 air travel was still rather unusual; only 17 million passengers (World Bank n.d. a) flew a total of 23 billion passenger-kilometers (S. Yao and Yang 2008, 1). By 2005, however, the number of passengers had octupled, to 136 million, flying more than 200 billion passenger-kilometers (S. Yao and Yang 2008, 1) as China updated its fleet and expanded its airports. The showpieces were the huge, ultramodern third terminal at Capital Airport in Beijing, built for the 2008 Summer Olympic Games, and the wholly new Pudong Airport in Shanghai, connected to the city center by one of the world's only magnetic levitation trains (which covers the distance of 30 kilometers in eight minutes, at an astonishing top speed of 430 kilometers per hour). China also opened an average of ten to fifteen new airports per year in the 2010s (Chao 2015). Some of these were built in difficult mountainous terrain: Panzhihua's, opened in 2003 and enlarged in 2013, sat at an elevation of 1,902 meters (Baidu 2020b); Jiuzhaigou's, in northwestern Sichuan, opened the same year at an elevation of 3,447 meters (Baidu 2020a).

The big city airports, however, were soon bursting at the seams and underwent repeated rounds of enlarging and reconstruction as China's yearly passenger numbers grew to 611 million in 2018 (World Bank n.d. a). Even with its world-famous third terminal, the Beijing Capital Airport soon grew too small and cramped to serve the metropolis. In September

2020, General Secretary Xi Jinping inaugurated the new Beijing Daxing International Airport, featuring the "world's largest terminal building," designed by Zaha Hadid, as its first flight—an Airbus A380 jumbo jet—took off for Guangzhou.

....

At the beginning of the Reform period, China was a primarily rural nation where it was difficult to get from one place to another and only members of a privileged elite ever voluntarily traveled very far from home. By 2020 China had become a mostly urban nation where almost everyone traveled, from migrants on crowded trains looking for work to middle-class families out for road trips in their cars to jet-setters riding the world's fastest trains or flying between its hundreds of sparkling modern airports.

None of this modernization—industrial growth, energy growth, urbanization, or travel—came without environmental costs, however. China progressively eliminated ecological buffers, becoming more dependent on, and more tightly locked into, the infrastructure of housing, industry, commerce, and transport. As a result, its social-ecological systems not only lost resilience but suffered from increasing and often dangerous air, water, and soil pollution, along with depletion of many of the resources that fed the remarkable infrastructural transformation. The next chapter examines the ecosystemic consequences of China's grand transition (*ju bian*) to an urban country.

15. Sludge Abides, 1990–2022

China . . . abruptly closed its ports to "foreign trash" in 2018, and today urban waste managers throughout the world are still scrambling to figure out what to do with your yogurt container.

JOSHUA GOLDSTEIN, *Remains of the Everyday*, 2021

In addition to growing by orders of magnitude, Chinese cities became much less self-sufficient after 1980. They had previously been relatively closed systems, and their residents had not consumed much. Supplies came primarily from nearby hinterlands, a tendency encouraged by the autarkic, cellular ideals embodied in the Communist Party's economic policies. In Chongqing you got Hilltown Beer, and in Chengdu, a mere 325 kilometers away, you got Blue Sword Beer, both produced locally. Similarly, cities disposed of their waste locally. Night soil went to fertilize suburban farms; garbage was either composted for fertilizer or land-filled nearby. Recycled materials went to local factories, and anything that could be reused was either repaired at home or collected by people with bicycle carts and resold at local shops and markets (J. Goldstein 2021, 55–57). As the cities and the transport networks grew, however, cities took in more and more goods from wider and wider networks, and smaller and smaller proportions of urban wastes were reused or recycled, until the waste streams, the landfills, and the suburban dumps swelled to the point where the state had to begin prioritizing the waste problem.

In addition to growing consumption, growth of urban and suburban industries after 1980 contributed to the growing national waste heap. As industrialization proceeded, manufacturing overtook resource extraction as the most important industrial sector, and since most manufacturing was urban, more and more of the environmental degradation took place in the cities. The growth of urban buildings and transport infrastructure relied on energy-intensive industries such as steel, cement, glass, and chemicals, and both the fuel burned in the production process and the by-products of production itself contributed to pollution of air and water, along with increases in greenhouse gas emissions.

Expansion of the cities also contributed to urban problems other than waste. Expanding cities took over farmland and other ecologically important lands that previously buffered food and water supplies against climatic and other disturbances. Also, because cities in China consumed more resources per capita than rural populations (Naughton 2007, 133), they strained supplies of water, metals, and minerals, competing with rural users and exporting air and water pollution to the places where those materials were extracted and processed. Urban expansion and increases in consumption often reinforced each other. As urban incomes and living standards rose, consumers used more water, and withdrawals of groundwater caused land under cities to subside. Heat emitted from urban buildings and vehicles altered local microclimates, leading to even higher resource use, including electricity to run air conditioners.

All these problems arising from rapid urban expansion came to a head late in the 2000 aughts, when authorities realized that such a course of development was unsustainable. They thus instituted a series of "eco-developmental" policies (Esarey et al. 2020) attempting, with various degrees of success, to set the country on a more sustainable path of urbanization. These policies brought remarkable results in a short period of time, but planners and engineers still faced overwhelming challenges as the cities continued to grow and the people continued to get richer.

EXPANDING URBAN FOOTPRINTS AND LOSS OF RESILIENCE

As urban areas expanded tenfold in the four decades after 1980, they took over great expanses of farmland and ecological buffer land, such as wetlands and even lakes. Losing the lands' former food production and ecosystem services potentially endangered food supply and increased the chances of floods, while urban resource use caused land to subside and temperatures to rise. Leaving aside increasing resource use and rising standards of living, the mere fact that cities grew geographically had huge environmental consequences.

Farmland Conversion and Food Security

Between the Founding and 1980, the small amount of urban expansion was not much threat to agriculture. Most cities sat in flat, lowland areas, and waste from urban toilets and pigpens went not into sewers but into shit buckets to be carted or shoulder-poled out to suburban farmlands, increasing their productivity and ensuring supplies of vegetables and other high-priced, perishable crops for urban populations (Skinner 1977b, 288; J. Goldstein 2021, 34–41). But when cities began to expand rapidly in the Reform era, they took up not just farmland but the most productive farmland. Periurban expansion claimed around three million hectares of farmland between 1995 and 2005, or about 2.5 percent of China's total farmed area. Urban expansion thus theoretically threatened the food supply, but food production continued to grow because of increasing yields, and grain harvests continued at record levels through 2020 (Statista 2022a).

Nevertheless, urban expansion took up more farmland than it needed to. Municipal governments could earn revenue by expropriating agricultural land at low prices and selling rights at a profit to developers, who in turn speculated on profits from selling or leasing space in the new buildings (Rithmire 2017, 129). When speculation led to oversupply, a lot of land was taken over but not built on, or built on but not occupied. Farmland preservation policies seem to have been too weak to impede the incentives for urban development, so municipal governments and developers simply ignored the zoning lines drawn in municipal plans, as happened in Hangzhou from 1995 to 2005 (Yue, Liu, and Fan 2013, 367). Kuancheng District, a rapidly urbanizing area on the outskirts of Changchun, lost about four thousand hectares of farmland, or 28.5 percent of its cultivated area, between 2004 and 2014. Of the total agricultural land lost, about a third was converted to a regional park, a quarter to industrial land, and a tenth to urban settlement. But strikingly, fully 20 percent of the land that was farmed in 2004 was unutilized in 2014, perhaps waiting for a developer (W. Li et al. 2017, 218). Similar tracts of bare former farmland emerged around Shanghai (Yin et al. 2011, 615–16) and Chengdu (personal observation).

Partly as a result of continued urban expansion, from 2020 through 2022 the Chinese government continued to express concerns about food security, even as yearly grain harvests set records while population growth had slowed to almost zero—Xi Jinping stressed an old proverb in a May 2020 speech: "Grain in the hand, no worry in the mind" (Shou zhong you liang, xin zhong bu huang) (Xinhua 2020b), and in the fall the

State Council issued regulations limiting planting of nongrain crops, implicitly blaming greedy or selfish farmers for the perceived harm done by urban expansion (Orange Wang 2020).

Removal of Ecosystem Services and Urban Flooding

Expanding cities did not just take over land formerly used to grow things; they also took over areas that had previously provided ecosystem services, particularly flood prevention. Growing cities reclaimed or constrained natural bodies of water, disconnected those that remained, and paved over or built on empty or vegetated land. This eliminated important buffers against extreme weather disturbances and caused serious increases in the frequency and severity of urban floods (Z. Zheng et al. 2016, 1149; Yong Jiang, Zevenbergen, and Fu 2017, 521), echoing on a smaller scale the filling in of major lakes, which was causing regional-scale flooding. Cities in floodplains or in low-lying coastal areas were particularly vulnerable. Wuhan, for example, a lake-and-river city built at the confluence of the Han and Chang Rivers, had always been subject to flooding, but historically the city's lakes and wetlands had buffered its populated areas against river surges caused by heavy rainfall events. Modern infrastructure projects had already increased flood susceptibility by 1931 (Z. Ye 2015; Courtney 2018b), but after 1980, over 90 percent of the city's three hundred lakes had been filled in, creating 229 square kilometers of new urban construction land but effectively blocking the channels through which excess stormwater had previously flowed, thereby decreasing the resilience of the urban system to heavy rainstorms (Z. Zheng et al. 2016, 1150; N. Du, Ottens, and Sliuzas 2010, 176–78).

Concrete streets, squares, and sidewalks, along with concrete, glass, and steel buildings, are all impervious to water, so when open land, forests, or agricultural land gives way to expressways and skyscrapers (or even alleys and shopping malls), more water runs off and drainage systems with diminished capacity must deal with more water. For example, between 1965 and 2015 the proportion of Nanjing's urban area covered by water decreased from 20 percent to 2 percent, leading to an estimate that a three-year rainstorm would produce twice the runoff volume of a normal year (Z. Zheng et al. 2016, 1150). On a nationwide scale, 62 percent of 351 cities surveyed by the Ministry of Housing and Urban-Rural Development experienced flooding just in the three years from 2008 through 2010 (Yong Jiang, Zevenbergen, and Fu 2017, 523); 184 cities experienced urban flooding in 2012, as did 234 cities in 2013 (Huan Liu, Jia, and Niu 2017, 1).

Recognizing that expanding urban footprints increased flooding and water waste generally, in 2013 the newly elevated CCP General Secretary, Xi Jinping, gave his official imprimatur to the concept of the "sponge city" (*haimian chengshi*) as a way to deal with heavy rainfall. Rather than building everything out of concrete and other impermeable materials, cities needed to protect their ecological environments by adopting a six-point program promoting rainwater infiltration, rainwater stagnation, rainwater storage, rainwater purification, rainwater utilization, and rainwater discharge.[1] The concept also called for cities to use more permeable materials in buildings, roads, and public open spaces; remediate natural features, such as wetlands, urban lakes, and river channels that were blocked or filled in during the previous decades of full-speed urban construction; and build adequate drainage systems (Huan Liu, Jia, and Niu 2017, 2–3).

Some of this recommended remediation happened. Beijing, for instance, reported that by following these measures it saved 183 million cubic meters of rainwater in 2014 alone and that heavy rain events in 2016 brought no significant flooding, in contrast to what had happened in a similar storm just two years earlier (Huan Liu, Jia, and Niu 2017, 4). But the effort to build sponge cities, like other attempts at solving environmental problems, was hampered by some of the same problems of governance. In order to show results fast enough to get credit for "administrative achievements," officials often rushed to build "physically visible projects on the ground in a short period" (Yong Jiang, Zevenbergen, and Fu 2017, 526). Although some cities built impressive pervious pavement areas and urban wetland parks, including restored or artificial wetlands and artificial lakes, too often authorities relied on new drainage systems—a fix to fix the fix that had left most surfaces impervious and did little to unblock natural water flow systems—or they built the wetland parks hastily so that they, too, could count as administrative achievements (Song, Albert, and Prominski 2020). Divided and uncoordinated administrative responsibility did not help, nor did relying on "traditional engineering oriented" approaches, which made it difficult to adjust and adapt flood mitigation measures in the light of experiences gained or to adapt appropriately to different conditions in different cities (Yong Jiang, Zevenbergen, and Fu 2017, 526). As with so many of China's eco-developmental projects, the sponge city effort was not so much two steps forward and one back as one step forward when there could have been two—progress, but short of the ideal.

Land Subsidence in Urban Areas

Thirsty cities do not just take water away from farms. Many of them also take water from aquifers directly underneath or nearby, and this causes land subsidence, which has destabilized urban land and buildings in areas as different as the Chang River delta, the North China Plain, and river valley cities such as Xi'an and Taiyuan. By 2004, researchers had identified forty-five major cities with land subsidence, nine of them where the land had subsided a meter or more (R. Hu et al. 2004, 66). In the Chang delta, land subsidence was detected as early as 1921, when fast-growing Shanghai was China's largest coastal city and treaty port (J. Wu et al. 2007, 1731; T. Yang et al. 2020, 831). Subsidence worsened with increasing groundwater withdrawals in the 1950s, to the point where authorities began large-scale water injection into Shanghai's aquifers in 1966. They increased the amounts of water injected throughout Cultural Revolution times and thus slowed the rate of subsidence for the next few decades (J. Wu et al. 2007, 1733; T. Yang et al. 2020, 833–34). After 1990, however, as Shanghai grew from a mere megacity of eleven million inhabitants to a truly gargantuan conurbation of well over twenty million, groundwater withdrawals increased enough to start land subsidence in parts of the city once again (J. Wu et al. 2007, 1734; T. Yang et al. 2020, 833).

Combined residential, industrial, and agricultural water usage contributed to widespread and severe land subsidence in many parts of the North China Plain. In Beijing and Tianjin, subsidence was noticed in the early twentieth century, but beginning in the early 1960s agricultural overexploitation of groundwater resources across the entire North China Plain severely depleted the complex aquifers there (S. Foster et al. 2004; H. Guo et al. 2015). Before the cities expanded in the Reform era, farms had drawn most of the water, but as the cities grew, they took larger and larger shares—urban households with "modern" standards of living, including daily showers and frequent clothes-washing, use more water per capita than their rural counterparts (L. Fan et al. 2017, 127). Subsidence was particularly serious in areas close to the coast, including such cities as Tianjin, Beijing, Cangzhou, Hengshui, Tangshan, and Dezhou (H. Guo et al. 2015, 1416). Parts of Tianjin had sunk by as much as 2.8 meters by 2004 (R. Hu et al. 2004, 69). And because the soft mud geology that underlies much of the plain is distorted when the water is sucked out of it, the subsidence is irreversible even when the aquifers are artificially recharged (H. Guo et al. 2015, 1415–17). In addition, in coastal areas the combination of land subsidence and sea level rise due to climate change exacerbated floods in Tianjin and elsewhere (H. Guo et al. 2015, 1417).

Changing Urban Microclimates

Big cities even change local climates. As cities and transport networks expand, the buildings and impervious surfaces that replace vegetation absorb visible light and radiate infrared, raising air temperatures, sometimes dramatically. Already in the 1950s there were urban heat islands in the center of Shanghai, and they gradually spread outward through the 1960s and 1970s (Juan-juan Li et al. 2009, 413). But like so many other urban environmental problems, heat islands expanded faster as the cities grew even more rapidly after the 1980s. In spring 1997, there was an astonishing yearly average 6.4 degrees C difference between surface temperatures in Shanghai's urban core and its rural areas. By 2008 that average difference had shrunk to a "mere" 2 degrees, because the rural areas had partially urbanized and thus gotten hotter in the intervening decade, although summer temperature differentials remained above 3 degrees C (Ying-ying Li, Zhang, and Kainz 2012, 132). There were also notable heat islands and heat corridors along the ever-expanding expressway network and in suburban industrial parks, port areas, and newly built satellite towns (Ying-ying Li, Zhang, and Kainz 2012, 132). Studies of land cover change demonstrated close positive correlations between surface temperatures and population density, as well as inverse correlations between temperatures and vegetative cover (Ying-ying Li, Zhang, and Kainz 2012, 136–37).

Similar urban heat islands appeared in the spring and summer across many major urban areas, including the large urban agglomerations of Jing-Jin-Ji (Beijing, Tianjin, and Hubei), the Chang River Delta, the Pearl River Delta (Yonghong Liu et al. 2018), and Xi'an (Meiling Gao et al. 2017). But surprisingly, in the Jing-Jin-Ji area—where winters are cold, dry, and sunny—barren, frozen farmlands absorbed more solar radiation than did the urban expanses of concrete and steel, turning Beijing and other large cities into urban *cold* islands. In other words, urban areas of Beijing and Tianjin were about 3 degrees C hotter in summer and about 1 degree colder in winter than the nearby countryside (Yonghong Liu et al. 2018, 479–81).

URBAN CONSUMERS AND RESOURCE DEPLETION

City dwellers in China have consumed more than their rural cousins at least since the late Qing; *hukou* restrictions on migration kept the consumption gap large during the high socialist period, allowing authorities to grant generous benefits to urban workers, while rural collectives received few subsidies. Urban incomes were about two and a half times as large as rural incomes at the start of the reforms (Naughton 2007, 133). However, the cities at that time were still relatively small—urban residents did not exceed 20 percent of the national population until 1983 (Naughton 2007, 127), and even though their consumption levels exceeded those of farmers, they were low by world standards. Urban consumption thus placed only moderate demands on resources such as water, construction materials, and energy; in 1978, when only 20 percent of China's population was urban, city dwellers consumed roughly a third more resources than an entire country consuming at the rural level would have.

Urban consumption began to grow noticeably during the early years of the Reform. Such amenities as running water, bath and toilet facilities, twenty-four-hour electricity, and especially expanded residential floor space gradually began to spread downward from the privileged cadre classes to the masses of urban residents. The combination of a 45 percent increase in total population, an elevenfold increase in national consumption standards, a fairly constant 2.5-to-1 ratio of urban to rural incomes, and an increase in the urban proportion of the total population from 20 percent in 1980 to 65 percent in 2022 (EPS China Data n.d.) meant that total consumption grew by a factor of about twenty-six, and about 40 percent of that growth was directly due to the growth of cities.[2] This has led some analysts to conclude that by the early 2020s the Chinese urban middle class was the biggest driver of consumption worldwide and by extension the biggest threat to global sustainability (Kharas and Dooley 2020). These exponential increases in consumption put parallel strains on resources, including water, iron ore, nonferrous metals, rubber, wood, and various kinds of energy resources.

PRESSURE ON RESOURCES

In 1980 China used a total of about 443 cubic kilometers (or gigatons) of water. Of this, about 394 cubic kilometers (88 percent) went to agriculture, as a result of increases in irrigation (figure 15.1). An additional 45 cubic kilometers (10 percent) went to industrial uses, and only 6 cubic kilometers (1.5 percent) was used by households (C. Bao and Fang 2012, 533).[3] By 2008, the growth of cities had changed those proportions dramatically: total usage had increased by 30 percent, but the amount used in agriculture had actually declined by about 5 percent, with all of the increase in water usage coming from households and industry. As a result, the percentage used in agriculture had declined to 64 percent, while industry now consumed 24 percent and households 12 percent of a higher total volume used (C. Bao and Fang 2012, 533). By 2019, efficiency campaigns had somewhat curtailed both agricultural and industrial water usage, which by then constituted 61 percent and 20 percent of the total, respectively, while domestic use, at 87 cubic kilometers, had increased to 14.5 percent (S. Wong 2020).

Despite some improvements, much of this water was still used inefficiently; one estimate states that China used ten to twenty times as much water per unit of GDP as "advanced nations" (C. Bao and Fang 2012, 547). Inefficient urban water use was a particular problem in northwestern and North China, where water is scarce: cities in the Hai River basin, which includes most of the large cities of North China (e.g.., Beijing and Tianjin), used approximately 126 percent of available regional water resources in 2008; that is, they imported water from other regions even before the completion of the South-to-North Water Transfer Project (C.

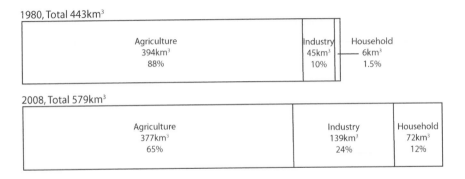

1980, Total 443km³

| Agriculture 394km³ 88% | Industry 45km³ 10% | Household 6km³ 1.5% |

2008, Total 579km³

| Agriculture 377km³ 65% | Industry 139km³ 24% | Household 72km³ 12% |

Figure 15.1. Water use in China, 1980 and 2008.

Bao and Fang 2012, 537). Paradoxically, in the 2010s people living in more "advanced" (wealthier) cities both used more water than their "backward" counterparts and used it more efficiently. This is because more affluent households were more likely to have water heaters, washing machines, and other water-guzzling appliances, and thus take more showers, wash clothes more often, and simply draw more water from household taps (L. Fan et al. 2017, 131). At the same time, however, those people in affluent cities were also more likely to adopt water-saving measures such as recycling wastewater and minimizing evaporation (Chu, Wang, and Wang 2015, 4–7). As a result, although total residential water usage increased almost linearly from 1980 to 2015 as the cities grew, per capita usage peaked around 2000 and then declined (L. Fan et al. 2017, 127).

Cities are built mostly out of minerals, including sand (for concrete and glass), clay (for bricks and tiles), cement (for concrete), and ores (most notably iron ore for steel but also copper for electrical wiring, as well as aluminum, lead, zinc, and other metals). In addition, manufacturing and operating the ever-growing numbers of vehicles on urban streets consumed a lot of steel and glass, as well as rubber and petroleum. Table 15.1 shows the growth in production of various city-building minerals from 1979 to 2018.

Just in terms of materials used to make buildings and to build and propel vehicles, China produced forty-five times as much glass, twenty-seven times as much steel, and more than thirty times as much cement in 2018 as it had thirty-nine years earlier, almost seventy-six times as many tires, and thirteen times as much gasoline in 2018 as in 1979.

Table 15.1. Increases in production of materials needed for cities and vehicles, 1979–2018

	1979	1988	1997	2006	2015	2018	2018/ 1979[a]
GLASS (million tons)	1.04	3.65	8.3	23.25	39.3	47	45.1
BRICKS (billion)						541	
TILES (billion)						12.8	
CEMENT (million tons)	73	210	512	1,236	2,359	2,236	30.6
IRON ORE (million tons)	118	168	267	588	1,381	801	6.8
STEEL (million tons)	34.5	59.4	108.9	419.1	803.8	929.0	26.9
GASOLINE (million tons)	10.7	18.6	35.2	55.9		139.6	13.0
TIRES (million)	11.7	29.9	96	436	928	886	75.7 [a]

a. 2018/1979 indicates ratios of the later to the earlier date or, alternatively, the factor of increase from the earlier date to the later. For example, China produced 45.1 times as much glass in 2000 as it did in 1979.

Source: EPS China Data n.d.

Wood is not a primary construction material in China; only people in the forested regions of Zomia built their houses mostly out of wood. Elsewhere, construction of both traditional and high-rise modern housing uses wood only as trim, not as framing. Nevertheless, China cut almost three hundred million cubic meters of standing timber in 2017 and imported about forty-five million tons of logs and another twenty-two million tons of cut lumber (Clever 2017, 2, 5). Most of the logs, both domestically harvested and imported, went to produce plywood and other panels used in furniture and wood flooring (Zhang Jianlong 2018, 72). These wood products supplied both the burgeoning domestic market (mostly to provision urban residential and commercial buildings) and a sizable furniture export market. Due to mostly judicious policies of reforestation and forest conservation, however, China's forest cover still managed to increase, from about 12 percent in 1980 (Robbins and Harrell 2014, 384) to 23 percent in 2017 (Zhang Jianlong 2018, 2). Part of the reason why, however, was that by importing logs and lumber, China exported deforestation (Clever 2017, 5). In 2018 China imported 40 percent of the roundwood it used, up from 28 percent in 2009. Particularly heavily affected were conifer forests in New Zealand, Russia, the United States, and Australia, along with the hardwood forests in Africa, Brazil, Russia, and the Solomon Islands (Zhang Jiaran, Yang, and Yi 2019, 133–34).

As China's urban continent developed, its patterns of energy consumption were paradoxical. Except for a short period in the 2000 aughts, the energy efficiency of its economy increased steadily from the start of the Reform era to the 2020s. Despite these gains in efficiency, however, China used considerably more energy at the end of this period than it had at the beginning, largely because standards of living and resource use increased so fast. I treat this paradox in quantitative detail in chapter 16.

CLEANING THE CITIES

In preindustrial times, Chinese city dwellers took small amounts of food, water, minerals, and energy from their immediate surroundings, consumed them, recycled some of the nutrients and other usable components, and discharged small amounts of pollutants into the air and water (J. Goldstein 2021, 27–28). With the growth of urban consumption, more and more resources and energy left the biophysical environment and entered human bodies and the social and built environments. At the same time, a smaller and smaller percentage of those resources cycled directly back to the biophysical environment. They either remained in the cities as construction materials in the buildings and roads, stayed in human bodies as tissues and toxins, or found their way into disposal facilities such as landfills and incinerators, from which they eventually did return to the biophysical environment, albeit in unusable and sometimes toxic or at least environmentally harmful forms such as air pollutants, water pollutants, solid waste, and greenhouse gases (J. Goldstein 2021, 85–88). Beginning around the turn of the century authorities at all levels realized that this model was unsustainable, and began to promote comprehensive cleanups of air, water, and urban waste. By 2020, Chinese urban ecosystems were still depleting resources and generating waste, but not as much. It is thus no wonder that the idea of the "circular economy" (*xunhuan jingji*) became part of the whole turn to eco-developmentalism and the push to develop an "ecological civilization." The logic behind the circular economy was not only to use fewer resources but also to change the proportion of used resources that were reused or recycled as opposed to discarded, thus eventually reducing the overall quantity of resources used.

In 2009 the law on circular economy promotion (*xunhuan jingji cujin fa*) defined "circular economy" as "a generic term for the reducing, reusing and recycling activities conducted in the process of production, circulation and consumption" and established procedures and guidelines for those "three Rs" (X. Chen, Geng, and Fujita 2010, 718). The law declared that "developing a circular economy is an important strategy for the economic and social development of the state" and that "the development of a circular economy shall be propelled by the government, led by the market, effected by enterprises and participated in by the public" (Standing Committee 2008). China's need for such a law showed just how modern and noncircular its economy had become since the Founding, particularly since the Reform. By the late 1990s, people were pretty much throwing everything away. Urban residents were particularly prodigal with the plethora of newly available stuff, and urban governments' capacity to dispose of that waste responsibly was grossly inadequate.

Solid Waste

In the high socialist period and the first years of the Reform era, cities still produced relatively little waste. Materials were scarce, for one thing, and so people consumed frugally and reused or recycled what they could (A. Zhang 2017, 81; J. Goldstein 2021, 65–67). Most of what waste the cities did discard, however, was simply dumped into some convenient, low-lying place outside town. In Wuhan, dumps even filled in parts of lakes and contributed to flood vulnerability, and many cities became surrounded by garbage hills (J. Wei, Herbell, and Zhang 1997, 573, 576). Multiplied by the increasing urban population and the rapidly rising consumption levels, the lack of disposal facilities created a garbage crisis. Whereas China had collected about 25 million tons of urban household garbage in 1978 (Shidong Ge and Zhao 2017, 2), the total increased to 118 million tons in 2000 (Liu Wenhua and Xu 2019, 94). In 2004 China passed the United States as the world's biggest garbage generator (Hua Wang et al. 2011).

The first solution was to build sanitary landfills. By 1996, cities such as Wuhan, Yichang, Hangzhou, and others had initiated large projects to build or improve landfills (J. Wei, Herbell, and Zhang 1997, 579). But urban garbage continued to pile up, and for a while landfill construction could not stay ahead of the need for space. From 2000 through 2003, the percentage of urban garbage that received "harmless treatment" (*wu hai chuli*)—meaning incinerated, composted, or landfilled (as opposed to just dumped)—actually decreased. Landfill capacity started to catch up between 2005 and 2010, so that the percentage of treated garbage increased to 78 percent (Liu Wenhua and Xu 2019, 94). But solid waste volume began to increase again after 2010, despite construction of a lot more landfills—their number increased to 580 in 2014 (Z. Han et al. 2016) and to 654 in 2017 (Liu Wenhua and Xu 2019, 109). Beginning in the 2000 aughts, even small places like county seats in Yunnan needed landfills, and the government instituted a program to establish them (Hua Wang et al. 2011). As late as 2017, even as the responsible ministries were pushing other solutions, about 57 percent of China's garbage being "harmlessly treated" was in landfills.

Landfills, however, created their own environmental problems, particularly when liquid, decomposed material contaminated nearby groundwater. A survey of studies of thirty-two landfills across the country, done in the early 2010s, assessed groundwater quality near landfills as "very bad," revealing the presence of at least ninety-six groundwater pollutants, including a variety of organic chemicals, heavy metals, alkali metals, inorganic salts, and high chemical and biological oxygen demand (Z. Han et al. 2016, 1259). The study also showed that, because decomposition is a slow process, the highest levels of pollution occurred between five and twenty years after the garbage was landfilled (Z. Han et al. 2016, 1262). This meant that materials landfilled in 2015 could still be polluting local groundwater in 2035 before they were sufficiently diluted. Groundwater pollution extended as far as one kilometer from the landfills, though the most serious pollution was confined within two hundred meters. In light of this, a 2008 requirement to surround landfills with at least a ten-meter buffer zone planted with trees (X. Chen, Geng, and Fujita 2010, 717) seemed inadequate.

In addition to industrial chemicals and bacteria, pharmaceuticals in landfills posed a problem for nearby groundwater and reservoirs. In the early 2010s in Guangzhou, where 90 percent of the metropolis's eighteen thousand tons of daily municipal solid waste went into landfills, researchers sampled groundwater from nearby village wells and from surface water in two reservoirs 2.8 and 4.6 kilometers downstream from the landfills, one of which had closed in 2002 and one of which was still accepting waste at the time of the 2013 study. The researchers used more distant wells and upstream reservoirs as controls. In water near the landfills, they found antibiotics, azole antifungal creams, aspirin and ibuprofen, and a variety of endocrine-disrupting ingredients from skin-care products. Many of these products posed risks to algae in the reservoirs and in some cases even to fish (X. Peng et al. 2014). In addition to polluting the local waters, organic decay in landfills also contributed methane to the greenhouse gas inventory (L. Wei et al. 2020, 5).

Given the limited capacity of landfills and the fact that they continue to pollute for up to two decades after they are closed, authorities knew by the early 2000 aughts that even the best equipped and best managed landfills were not going to be a sustainable waste solution, so they turned to incineration. Municipal authorities, taking advantage of central government subsidies, quickly jumped on the bandwagon, building incinerators everywhere, often creating more capacity than they needed. By 2006, incineration capacity had sextupled its 2001 mark (X. Chen, Geng, and Fujita 2010, 721), and in 2017 eighty-four million tons, or 40 percent, of the total "harmlessly treated" garbage was going up in not entirely benign smoke (see A. Zhang 2017, 81; Shapiro-Bengtsen 2020; Liu Wenhua and Xu 2017, 110).

Like landfills with their downstream populations, incinerators have their unhappy downwinders. In 2009 the Guangzhou municipal government revealed plans to build a large incinerator in densely populated and somewhat upscale Panyu District, south of the central city. Middle-class NIMBY protestors, fearing that the incinerators would release heavy metals, dioxins, and other pollutants that would threaten their air, water, and soil quality, staged a successful protest. The Panyu incinerator was not built, thus perpetuating Guangzhou's reliance on landfills and prompting the government to build six more incinerators in poorer areas—a clear instance of class-based environmental injustice (Jiang Yifan 2019; Shapiro-Bengtsen 2020; Wang Chen 2020; Wu Yixiu 2019).

It is possible to reduce pollution from incinerators, but they only work well for what in Chinese is called "dry trash" (*gan laji*), so the best way to make incinerators work is to have urban residents sort their garbage according to whether it is "wet" (*shi laji*), dry, or recyclable. Beijing had tried to have its citizens sort waste into excrement, garbage, and recyclables as early as 1956, but the effort was hard to manage or police (J. Goldstein 2021, 80–81). The central government tried again in 2001, attempting to implement waste separation in eight large trial cities, not only to take the pressure off landfills and incinerators but also to facilitate composting wet waste and recycling whatever it was possible to recycle. But there was a big problem. Waste separation campaigns and guidelines had almost no effect because cities neglected to build infrastructure to collect the waste. Many cities simply put out bins for sorted waste without providing for pickup or processing. Itinerant waste-pickers pulled out most of what was recyclable, no matter what bin it was in, and other discarded materials just stayed there. As a result, the bins at apartment complexes were always full or only open at certain hours, and most residents, realizing the situation, did not bother to sort garbage (see J. Goldstein 2021, 208–10; Wu Yixiu 2019; A. Zhang 2019; and Schmitt 2016, 200–213).

Finally, led by Shanghai, in 2019 the authorities began to get serious. In July they rolled out compulsory waste-sorting in apartment complexes throughout the city, with fines for violating the rules (Wu Yixiu 2019; Jiang Yifan 2019; J. Goldstein 2021, 247–50). A media propaganda campaign accompanied the rollout of the program, and it included such things as WeChat messages showing urban cops (*chengguan*) issuing a citation to a restaurant server who, despite pleading the contrary, was seen mixing dry and wet waste; it was one of 623 such citations issued on the first day the regulations went into effect (Ms. Pi 2019). The authorities planned to have similar programs in place in 46 provincial capitals and other large cities in 2020 and in all cities at the prefectural level and above by 2025. They also pledged to invest ¥21 billion (around

US$3 billion) in the program (Qiao Xuefeng 2019), and in 2022 they announced plans to replace most large-city landfills with incinerators by 2025 (Kong Dechen 2022).

Recycling

Recycling was even more complex. Amid the austerity of urban life in the high socialist period, some of the traditional spirit of frugality remained, as Amy Zhang (2017, 81) colorfully relates:

> During the Maoist era, state-run recycling depots were filled with citizens lined up to sell everything from old blankets to empty toothpaste containers. Itinerant migrants roamed back alleys singing and calling out for chicken and duck feathers. In an era of material scarcity, the objects of daily life were repurposed for immediate reuse. In each of these instances, however, reprocessing began with fine-grained sorting focused on separating and disaggregating objects. Recycling, refurbishment, and the transformation of matter focused on the integrity of materials—steel was melted to forge more steel, paper pulp hardened to produce more paper.

But in China as elsewhere, with the first bloom of prosperity came the throwaway ethic, and by 2000 municipal governments were doing about as well with recycling as they were with other forms of waste sorting—that is, not well at all. Campaigns to recycle fell as flat as those to separate wet from dry garbage (A. Zhang 2019), as migrant workers continued to collect recyclables to sell to reprocessors. China was also becoming the destination for recyclable materials from other, then-wealthier countries. Even though China produced 15 percent of the world's plastic in 2009 and 28 percent in 2016, it relied on foreign imports to supply extensive reprocessing industries for everything from paper to plastic to electronic waste. Throughout the early twenty-first century, Beijing grew a ringworm-like infestation of increasingly specialized scrap markets, expanding one ring road at a time as urban construction took up the next ring inward (J. Goldstein 2021, 191–96 (map 15.1).

These scrap markets, in turn, fed sorted waste to specialized reprocessing areas. Paper went to Manchang in Baoding Municipality; ferrous metals went to Tangshan, where they were fed into electric-arc

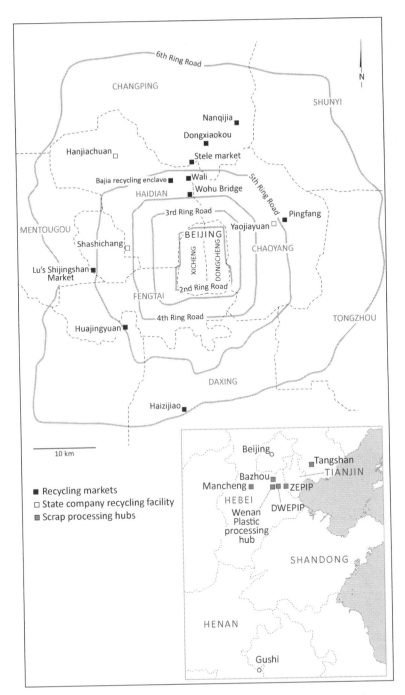

Map 15.1. Representative waste-processing sites in and around Beijing in
the 2010s. Map by Bill Nelson from Joshua Goldstein, *Remains of the Everyday:
A Century of Recycling in Beijing*, 190. Used by permission.

furnaces for reprocessing into usable steel; glass went to Handan, near the Hebei-Henan border. Plastics went two hours south to Wen'an, which became the world's largest plastic reprocessing hub; it took in plastic not only from Beijing and other northern Chinese cities but also from overseas. Somewhere between one and two hundred thousand local people with rural *hukou* crowded sites in several townships in the county, melting down toxic plastics into reusable form and paying for it in greatly increased rates of hypertension, liver hypertrophy, and other ailments. Runoff from detergent baths used to clean dirty plastic polluted local rivers. Food-delivery containers made of plastics recycled in Wen'an reportedly caused health problems among Beijing customers (J. Goldstein 2021, 199–201). Melting insulation off of recycled copper wire, much of it imported, similarly toxified the environment at Ziya near Tianjin and at several sites in the Pearl River Delta in the south (J. Goldstein 2021, 203).

By the mid-2010s, the quantities of domestic and foreign recyclables were overwhelming China's capacity to reprocess them. In addition to growing amounts of domestic waste, by 2017 China was importing an estimated 70 percent of the world's exported waste plastic and 30 percent of its exported recycled paper. Overwhelmed and hoping to be able to reprocess its own growing burden of recyclable wastes (if it could get citizens to sort them out), in mid-2017 China issued a virtual ban on waste imports, effective January 2018, sending recyclers in other countries into tailspins but at least buying time for its own waste-management efforts (Early 2017, 2020).

In 2009 China also imported large amounts of used and discarded electronics (K. Zhang Schnoor, and Zeng 2012, 10862), though it is impossible to know how much, since a considerable amount of the trade was illegal or unregulated (Lepawsky, Goldstein, and Schulz 2015; Huisman et al. 2015). At the same time, domestically generated volumes of discarded electronics continued to expand at least through 2018, as urban consumers' appetites for newer phones, faster computers, and bigger-screen televisions led them to discard the old ones, thus keeping recyclers busy (Jia Fu et al. 2020, 1).

Evidence points to the fact that the informal sector repaired, refurbished, and resold a large proportion of these discarded electronic materials (Lepawsky, Goldstein, and Schulz 2015). What they could not repair, they took apart to recover the materials, in so-called e-waste recycling. Informal recyclers in periurban and rural areas, particularly in Guangdong and Zhejiang, continued to dominate the industry, processing at least 60 percent of both domestic and imported wastes (Xianbing Liu, Tanaka, and Matsui 2006, 92). E-waste contains valuable materials, such as precious and semiprecious metals, at much greater concentrations than they occur in nature, and thus e-waste recycling conserves both mineral resources and energy (K. Zhang, Schnoor, and Zeng 2012, 10863). At the same time, these informal operations polluted air, soil, and sediment with organic compounds and heavy metals, as well as exposing their own workers, who often toiled with no safety precautions at all, to health hazards from the waste products (K. Zhang, Schnoor, and Zeng 2012, 10863–65; Xianbing Liu, Tanaka, and Matsui 2006, 95–96).

Since recycling valuable materials was clearly preferable to dumping them, beginning in 2001 successive environmental commissions and ministries attempted to cut pollution and its health effects by prohibiting informal reprocessing and recycling workshops and by encouraging and subsidizing large commercial recyclers that could minimize waste, environmental pollution, and health hazards. However, it didn't work exactly as intended. Informal brokers provided more convenient outlets—they maintained extensive networks of reprocessors and materials recyclers and could also resell recyclable materials to the larger, supposedly safer operations, which still depended on the informal sector for much collecting and processing (J. Goldstein 2021, 219–20). In addition, some licensed reprocessors gamed the system, using ruses such as displaying counterfeit television sets to inspectors in order to increase the count of discarded sets for which they were entitled to receive a subsidy (K. Zhang, Schnoor, and Zeng 2020, 2). Thus, the informal networks and workshops persisted.

Sewage and Sludge ("Biosolids")

Cities, of course, discharge not only solid but also liquid waste, in the form of municipal sewage. In premodern cities, as indicated above, there were no sewage systems; human and animal waste was recycled for fertilizer on suburban farms, and wastewater simply drained into rivers and other bodies of water. China's first underground sewers were built in modern cities beginning with Shanghai's International Settlement in the 1920s, when foreign residents demanded flush toilets. Since there were no treatment plants, sewers simply discharged raw sewage into rivers and coastal waters, contributing to the pollution there. In the 1950s, most new urban developments separated sewage from rainwater, but there was still no treatment (C. Jiang 2011, 17). Urban wastewater was often mixed with industrial and mining wastewater, and less than 10 percent of all wastewater was treated at all. A lot of urban wastewater was simply used to irrigate fields, a practice that had become widespread during the Great Leap Forward (Xia Z. and Li 1985). China built its first sewage treatment plants in the late 1970s, but treatment did not keep up with the volumes generated. Problems such as eutrophication of nearby lakes finally forced governments to face the problem of sewage pollution (Jianru Fu 2020, 2). Even when treatment plants were built, mostly in comparatively prosperous eastern cities, they were inadequate. In 1996, for example, the four sewage treatment plants in Changzhou on the lower Chang River could treat only about a third of the city's daily wastewater discharge, with the rest flowing directly into the river. Authorities hoped that the huge volume of water in the river would prevent any large-scale ecosystem effects (Yang Yaliang 1996, 87). As the saying goes, the solution to pollution is dilution, but it has to be diluted enough.

The wastewater problem was not easy to solve, given the rapid pace of urbanization. The total volume of discharged sewage continued to increase, from 21 billion tons in 2000 to 46 billion tons in 2012, with concomitant rises in nitrates and chemical oxygen demand (XiaoHong Zhang et al. 2015, 1010). This required a crash program, and in fact China implemented one, increasing the number of wastewater treatment plants from fewer than 200 in 1997 to 1,199 in 2009, 3,508 in 2013, and 5,476 in 2019, boosting treatment capacity by a factor of about 16 between 2000 and 2019 (Q. H. Zhang et al. 2016, 12–13; L. Wei et al. 2020, 2). Goals for 2020 were to treat 95 percent of the wastewater in large cities, 85 percent in county seats, and 70 percent in townships (Jianru Fu 2020). Most of the treatment removes a fair amount of the nitrogen and phosphorus in the water, although not enough for the discharge to rise above grade IV water, defined as "mainly suitable for general industrial purposes and recreational uses that do not involve direct human contact with water"

(Q. H. Zhang et al. 2016, 18). Even if those goals were not met, moving from treating essentially zero to almost all wastewater in the space of thirty years was a remarkable achievement, although many systems still leaked badly (Y. Cao et al. 2019).

Like so many other environmental fixes, however, the sewage treatment fix created problems, requiring fixes to fix the fix. Wastewater treatment leaves behind solids, called sewage sludge in most places but also known euphemistically as "biosolids" in the United States (EPA n.d.). Sludge can be treated to make it relatively harmless, and it can be used as fertilizer, especially on nonfood crops (thereby alleviating the fears of squeamish modern urbanites). In the early twenty-first century, however, more than half of the sludge remaining after primary sewage treatment in China simply went to what one article euphemistically calls "land application" (L. Wei et al. 2020). This meant dumping it on farm fields, often not even spreading it as fertilizer—farmers were using mostly chemical fertilizers by that time—but just piling it up and paying kickbacks to local cadres. In 2013 the Chinese investigative news organ *Caixin* published a gross-out exposé revealing that several private waste-disposal companies in Beijing were taking unprocessed sludge and dumping it in nearby rural areas in the middle of the night. At first they dumped it into gravel pits and abandoned mines, but later they hauled it to villages that charged them tipping fees and used the sludge to fertilize melon crops. The villagers stated that they never ate these melons themselves but shipped them to markets, where they were sold to supposedly unsuspecting urban buyers (Cui and Liu 2013). In earlier centuries, in fact as late as the high socialist period, all of this sludge would have been fertilizer and no one would have complained.

The worst abuses of sludge appeared to have lessened in the late 2010s, as more and more of it was processed before disposal. "Land application" declined from almost 60 percent in 2010 to less than 25 percent in 2017 but rose again to around 30 percent in 2019, much of that as "strategic land application," apparently referring to application of treated sludge. Meanwhile, in 2019 more than a quarter of sludge was incinerated, while another 20 percent went to sanitary landfills. Sludge incineration, like solid waste incineration, removes pathogens and heavy metals from the ash but of course releases some of them into the air, while disposal in sanitary landfills can lead to leachate ending up in the water and release of greenhouse gases into the atmosphere (L. Wei et al. 2020, 5). Paradoxically, large cities with their high consumption standards generate more sludge per capita, but small towns are less likely to be able to afford to build treatment plants to treat sludge or even wastewater.

....

"Sludge abides" is a fitting motto for the ecology of urbanization in China over the four decades from 1980 to 2020. The overall nadir for the urban environment was probably around 2013 or 2014, when "air-po-calypses" and midnight sludge-dumping aptly captured the zeitgeist. This promoted a barrage of initiatives that strengthened China's turn to eco-developmentalism under the banner of ecological civilization. Mitigation of air pollution, water pollution, urban flooding, heat islands, land subsidence, traffic congestion, and other problems was impressive, and China's cities were much safer, healthier, more livable, and more sustainable in the early 2020s than they had been just a decade before. As rapid urban construction had eliminated many ecological buffers, especially since the 1980s, the newly eco-developmental regime attempted, with some success, to replace some of them with infrastructural buffers and even restore some of the ecological ones. It could have done better, though, had it not faced problems in environmental governance that were as difficult to solve as the worst biophysical aspects of pollution. While recognizing the problems and facing them head-on, the authorities continued to employ approaches that reduced the effectiveness of the proposed solutions. These approaches included linear, engineering-based solutions to ecosystem problems, top-down management by decree, emphasis on short-term results that would look good on officials' résumés, and a tendency to favor superficial appearances over fundamental structural improvements. There were pervious parks in the early 2020s, but urban floods were more frequent than before massive urban growth. People used less water, but water tables were not necessarily recuperating. Grain yields were at record highs, but officials still worried about food insecurity, as there was less good farmland than before. Most wastewater was treated, but sludge, real and metaphorical, continued to abide.

16. Paradoxes of Eco-Development, 1998–2022

There is no such thing as a free lunch.

—Saying from the 1930s, first use unknown

China's energy system was greener than ever as this book went to press in 2023, and the air in China's cities was much cleaner than it had been just a few years earlier. At the same time, China was using more energy than ever before, coal still accounted for more than half this energy, and China still contributed the largest share of the world's greenhouse gas emissions. Its cities also remained some of the most polluted in the world. Whether the glass was half full (Mathews and Huang 2021) or half empty (Standaert 2021) depends on whether we look at ratios or at absolute numbers. China was experiencing a paradox of environment and development, doing better in relative terms while still doing badly in absolute terms.

GROWING AND GREENING THE ENERGY SUPPLY

China's energy use became vastly more efficient from the 1980s to the late 2010s (figure 16.1, top), but growth in total energy production and consumption continued to outpace efficiency gains into the 2020s: consumption increased every year except for the pandemic year of 2020, rising from about 600 million tons of coal equivalent (Mtce) in 1980 to 1,400 in 2000, 4,800 in 2018 (Lewis 2020, 49; Sönnichsen 2020a), and was forecast to peak at around 5,600 Mtce in about 2035 (Muyu Xu and Chen 2020) (figure 16.1, bottom).

China's energy mix also got a bit greener. In 1996 China consumed about 1,350 Mtce of primary energy, almost all of it from fossil fuels—94 percent if we leave out the considerable amounts of biomass consumed by rural households, which never entered official figures. Specifically, coal accounted for 72 percent of total energy; oil, 20 percent; gas, 2 percent; and "other," including hydroelectricity and renewables, 6 percent (figure

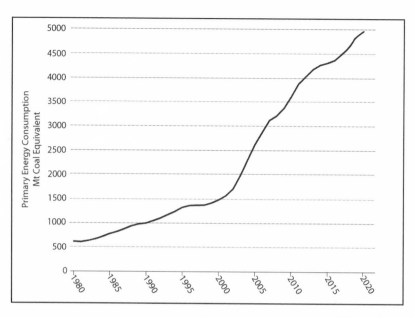

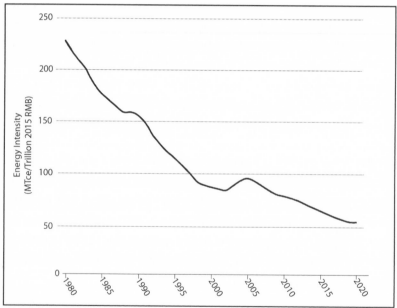

Figure 16.1. Energy consumption and energy intensity of China's economy, 1980–2020. Adapted and updated from Lewis 2020, 49, with additional data from EPS China Data n.d.

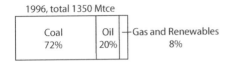

1996, total 1350 Mtce

| Coal 72% | Oil 20% | Gas and Renewables 8% |

2019, total 4500 Mtce

Wind and Solar 5%

| Coal 58% | Oil 20% | Hydro 8% | Gas 8% | Nuclear 2% |

Figure 16.2. China's total energy consumption, 1996 and 2019.

16.2). In 2019, when the total more than tripled to about 4,500 Mtce, the mix was coal, 58 percent; oil, 20 percent; hydroelectricity, 8 percent; gas, 8 percent; renewables, including wind and solar, 4.7 percent, and nuclear, 2 percent (see Downs 2000, 5; Shi Y. 2020; Sönnichsen 2020b) (figure 16.2). In short, the percentage of coal in China's energy mix declined sharply, but coal was still the largest source of energy. The share of natural gas grew, but less than anticipated, and the shares of "other," including hydro, wind, solar, and nuclear, grew by a factor of about two and a half.

China became the world's largest emitter of greenhouse gases in 2013, and its 2021 emissions of 9.9 gigatons of CO_2 equivalent were more than twice those of the United States, the second largest emitter. Although its per capita emissions of about 7 tons were still less than half those of such monumentally wasteful economies as Australia, the United States, and Canada, they already exceeded those of some other wealthy countries, such as Italy and France (UCS 2022). In 2020 General Secretary Xi Jinping promised "peak carbon" before 2030, but even if he were able to deliver on that promise, until then greenhouse gas emissions would continue to rise, exacerbating the worldwide climate crisis.

The share of fossil fuels in China's energy mix decreased only because energy generated from other sources grew faster than fossil fuel energy, but fossil fuel energy still grew. From 2000 to 2013, China almost tripled its coal-produced energy, and from 2011 to 2020, China consumed more coal than the rest of the world combined, fluctuating around 4.6 billion tons after 2013 (China Power 2020). Petroleum consumption tripled from 5 million barrels per day in 2000 to almost 15 million in 2020, two-thirds of it imported (EIA 2020, 4), and natural gas consumption increased from 28 billion cubic meters in 2000, virtually all domestically produced, to 314 billion cubic meters, about half of it imported, in 2019 (EIA 2020, 8). In sum, in the first two decades of the twenty-first century, even as China was making ever-greater efforts to decarbonize its energy mix, its total consumption of all three major fossil fuels increased many times over.

In late 2020, in conjunction with the publication of the fourteenth five-year plan, when General Secretary Xi pledged "peak carbon" by 2030, he also called for "carbon neutrality" by 2060, the first time the regime had committed to such ambitious general targets (Shi Y. 2020). At the UN Climate Action Summit in December, Xi strengthened China's nationally determined contributions in the Paris Agreement of 2015. While China had pledged in Paris to reduce the carbon intensity of its economy by 60 to 65 percent from 2005 levels by 2030, Xi raised the pledge to "over 65 percent," having already achieved a 48 percent reduction by 2019. Similarly, the General Secretary raised the 2030 goal for the nonfossil share of the energy mix from 20 to 25 percent, which would require upping the share of electric power above its current 40 percent (EIA 2020, 2), electrifying much of the energy, buildings, and transportation sectors, and replacing thermoelectric power (coal and natural gas) with other sources, including hydro, wind, and solar.

Electrification has benefits even when the electricity is generated from fossil fuels. Electric vehicles essentially emit zero air pollutants on heavily trafficked urban streets and intercity roads, and in 2019 China had over 3.8 million electric cars, almost half the world's total, as well as hundreds of millions of electric bicycles and tricycles and a few hundred thousand electric buses (IEA 2020). Electrifying home heating and cooking has similar effects on air quality (W. Peng et al. 2018). However, replacing fossil fuels with electricity for heating and transportation requires generating more electricity, and as of 2018 China's electricity generating capacity of about 1.9 terawatts consisted of 59 percent fossil fuels (mostly coal), 19 percent hydropower, 10 percent wind, 9 percent solar, 2 percent nuclear, and 1 percent bioenergy (figure 16.3). Because of curtailment of hydro and other renewable sources, fossil fuels still generated about 69 percent of the 6,712 terawatt-hours of total electric power nationwide. China was still building coal-fired plants in the early 2020s, while at the same time retiring some of the older, more polluting,

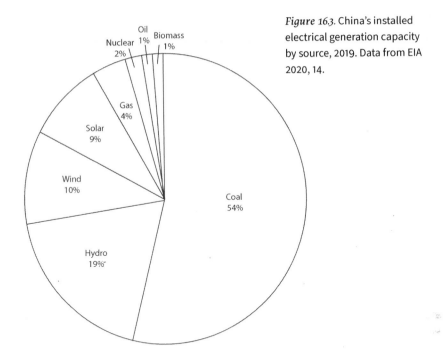

Figure 16.3. China's installed electrical generation capacity by source, 2019. Data from EIA 2020, 14.

and less efficient plants and replacing them with natural gas, which generates only about 55 percent as much CO_2 as coal per unit of energy produced. And as long as most electricity comes from fossil fuels, running electric cars or heating buildings with electricity still emits carbon dioxide. In other words, to gain the full benefit of electrification, electricity must be generated by nonfossil sources (W. Peng et al. 2018, 512). In addition, there are co-benefits to "greening" the power generation mix, including reductions in pollution-related morbidity and mortality (W. Peng et al. 515–16).

The Hydropower Boom

As early as the 1990s, Chinese planners recognized the problems of continuing to generate most of the country's electricity by burning coal, but the only substitutes available at the time were nuclear power and hydroelectricity (Wright 2012, 34). Although there were calls over the years to rely more on nuclear power, authorities were cautious about expanding the nuclear sector and instead turned to dams (J. Roney 2013). But building big dams is an ecological trade-off. Dams produce only a small fraction of the life-cycle greenhouse gas emissions of thermoelectric plants, but they have other severe negative social and environmental effects, including population displacement, loss of agricultural productivity, biodiversity loss, microclimatic alterations, hydrological changes both up- and downstream, and local water pollution.

Partly because of these known environmental hazards of dams and partly, one suspects, out of a desire to protect the vested interests of various bureaucracies and corporations, a sharp dispute referred to as the "struggle between water and fire" (*shui huo zhi zheng*) erupted in late 2010. In response to the twelfth five-year plan's goal of increasing newly installed hydroelectric power generation from 50 to 80 gigawatts by 2015, Ling Jiang, vice-director of the Pollution Control Office in what was then the Ministry of Environmental Protection, declared at a meeting on 12 December 2010 that "to a certain degree, pollution caused by hydroelectric power can be even greater than that caused by thermoelectric power." At the same meeting, he referred to specific environmental problems such as lake eutrophication caused by hydroelectric dams. The next day Zhang Boting, deputy secretary general of the Hydroelectric Power Engineering Association, exploded in an opinion piece under the heading "This environmental official ought to understand common scientific knowledge and fundamental logic." Ling was, he said, "saying nothing new, and is repeating some false environmental rumors to smear hydroelectricity" (*Shidai Zhoubao* 2010).

As Chen Yongjie (2011) pointed out in a column in *Beijing Youth News* two weeks later, this water-fire fight had little effect on the headlong rush to hydro other than as a reminder to take environmental effects into account when planning dams. At that time, greenhouse gas emissions (as opposed to environmental pollution) from thermoelectric power were not even part of the dispute. So hydropower development continued apace, stopping only in the extreme case of the Nu, which was called "China's Last Wild River." Although hydropower in China goes back to 1912, the small numbers of dams constructed before 1949 had a total installed capacity of only 163 megawatts. At the end of the high socialist period, in 1978, this capacity had grown to 18.67 gigawatts. With accelerated construction of dams, mostly in the southwest, by 2000 China had 79.4 gigawatts of installed hydropower, accounting for 25 percent of China's total generating capacity and providing about 17 percent of its power generation. After that came the real hydro boom—hydropower more than quadrupled between 2000 and 2017, to an installed capacity of 341 gigawatts, about 27 percent of the world's total hydro generation. At the same time, hydro's percentage of China's total power supply grew only a tiny bit, to 18 percent, because fossil fuel generation grew just as fast. About half of China's generating stations were very small (less than 500 kilowatts), but there were also 142 "large" stations with more than 300 megawatts in generating capacity (X. Sun et al. 2019, 568, 577, map 9.1). There could be no clearer demonstration of the difficulty of greening while growing. And in spite of the prospect of more land inundated, more species threatened, and more people displaced, China in 2018 had plans to increase installed hydropower capacity to 430 gigawatts by 2030 and 510 gigawatts by 2050 (X. Sun et al. 2019, 578).

China also became a leader in pumped-storage hydropower stations, which allow power suppliers to smooth out the peaks and valleys of electric power demand by using some of their power to pump water upstream during periods of low demand when it otherwise might be curtailed, and then using the pumped-up water to supply extra power at times of high demand. In June 2018, China had twenty-eight gigawatts of pumped storage power capacity and was building infrastructure to produce another forty-four gigawatts (X. Sun et al. 2019, 569).

Shying Away from Nuclear Power

Although nuclear power, unlike wind or solar power, was available as an affordable and feasible substitute for fossil fuels in the late twentieth century, it definitely took a back seat to hydropower, and in fact in 2019 it accounted for only 2 percent of China's total energy, far less than the newer renewable technologies. China's first nuclear plant came on line in Guangzhou in 1991, but as late as 2015 installed nuclear power capacity was only about twenty gigawatts; a reassessment after the 2011 Fukushima disaster probably contributed to the delay (WNN 2021). Concerns about decarbonization spurred growth after 2015, so that capacity doubled to forty-six gigawatts in 2020, with a further eleven gigawatts in generating capacity under construction (EIA 2020, 14; WNN 2021).

Still, planners in the mid-2010s were already balancing the advantages of a "low carbon" power source against questions of safety, radiation exposure, waste disposal, and plant lifetime (Chen Jie, Zhou, and Zhou 2016). Although nuclear power was not originally planned to exceed 3 percent of China's total power generation, this changed with the fourteenth five-year plan, released in 2021, which set a goal of seventy gigawatts of installed nuclear power by 2025 (Xie 2021a), partly drawing on China's recent success in driving down the cost and the construction time of new plants (Ng 2021). Still, this would be only about one-fifth the capacity of China's hydroelectric dams.

"Clean" Renewables

Rather than rely primarily on nuclear power, however, China embarked in the early 2000s on the world's largest program to expand "clean renewable" (nonhydro) electric power generation, including wind and solar power, as a key part of its energy strategy in the eleventh five-year plan and in the Renewable Energy Law, both published in 2005 (Lewis 2020, 50). Wind power was the first to take off, followed by solar a few years later.

When the Renewable Energy Law came into force in 2005, China had about 1 gigawatt of installed wind power capacity, making up 17 percent of the total nonfossil capacity, while the rest was entirely hydro. Windmills went up quickly, though, and by 2010 wind accounted for about 45 gigawatts, or roughly 20 percent of nonfossil capacity, again the remainder being hydro. Installed wind capacity continued to grow, to 184 gigawatts in 2019 and 270 in 2021, China having installed over 45 additional gigawatts in capacity annually in both 2020 and 2021 (IEA 2020, 23). Still, China's wind generating industry continued to be plagued by curtailment problems caused by poor siting (sometimes wind farms were built in places where there wasn't as much wind as anticipated), insufficient transmission capacity, and problems with turbine functioning (Hayashi, Huentler, and Lewis 2018). As a result, curtailment in 2015 amounted to as much as 15 percent, a huge loss of potential compared to curtailment rates in other countries. The government moved quickly thereafter to reduce curtailment, but it had limited success (Lewis 2017). As a result, even though in 2019 China led the world in installed capacity and had increased its efficiency, there were still improvements to be made—China generated slightly less than 2,000 kilowatt-hours per kilowatt of installed capacity, compared with 2,900 kilowatt-hours in the United States. In addition, China's wind- and solar-generated electricity was only 5.9 percent of its total energy production (Ritchie and Roser 2022), about the same percentage as Canada but less than the United States and less than a third the proportion in Spain, Germany, or the United Kingdom (IEA 2020, 23).

Solar energy development in China did not begin with large-scale photoelectric installations but when scientists at Tsinghua University in the early 1990s developed a new, efficient "evacuated tube" solar water heater design, which was commercialized a few years later. These units were ubiquitous in rural China in the 2000 aughts, as far flung as the winding roads of Chinese Zomia, and by 2014 China had over eighty-five million solar water heating systems, most of them installed in rural households (Urban, Geall, and Wang 2016, 538).

Solar electric water heaters took a little pressure off the grid and avoided a bit of the particulate pollution that would otherwise have emanated from coal boilers and stoves, as did the manure-to-energy biogas systems, developed in the 1970s and adopted by as many as forty-three million rural households between 1990 and 2010 (Xinyuan Jiang, Sommer, and Christensen 2011; Maloney 2019). But it was not until the 2010s that photovoltaic (PV) power actually became a major contributor to China's electricity generation mix. Although China led the world in the manufacture of silicon photovoltaic cells and panels throughout the early twenty-first century, producing two-thirds of the world output in 2013 (Urban, Geall, and Wang 2016, 535), most of its production was for export, and in 2013 China had less installed PV capacity than Italy or Germany and only slightly more than Japan or the notoriously fossil fuel–loving United States (Urban, Geall, and Wang 2016, 535). This began to change, however, around 2014 with Premier Li Keqiang's declaration of war on pollution and China's increasing commitment to reducing growth of greenhouse gas emissions. As with wind power, after 2015 the regime's goals for solar power development were cautious. The thirteenth five-year plan set a goal of 100 gigawatts of installed capacity by 2020, but in fact capacity was already 205.2 gigawatts and growing in 2019, and the 224 terawatt-hours of electricity generated was 32.9 percent of the world solar total (IEA 2021, 25). Huge fields of solar panels replaced wide swaths of relatively unproductive dry farmland along Zomian roads, joining rooftop water heaters as visual signs of China's turn to the power of the sun. China also began building large-scale floating solar farms, including the world's largest, with an installed capacity of 150 megawatts, built in Huainan, Anhui, on lakes created by earlier coal mine collapses (Sina 2017; Lao Sun 2020).

Decarbonization, however, has its environmental costs. The Ministry of Water Resources issued a "guiding opinion" (*zhidao yijian*) to halt overwater wind and solar projects in the interest of water quality, and in response at least one major floating solar project in Jiangsu was ordered to dismantle (China Dialogue 2022). The longer-term results of this directive were unclear when this book went to press, but concerns with water quality joined those about the environmental effects of mining and the land occupied by utility-scale solar projects as environmental trade-offs in large-scale decarbonization projects (Wyatt and Kristian 2021).

Building generation capacity, however, was not the only challenge for energy planners and engineers: solar, wind, and hydroelectric power all had problems getting the power from where it was generated to where it was used. Large amounts of utility-scale solar and wind power were generated in Central Asia, and hydropower came almost entirely from Zomia. Even coal was concentrated in Inner Mongolia and northern Shanxi. That meant that much of the burden of supplying energy fell on the electric grid, and China became the world leader in ultrahigh-voltage transmission lines, which brought power from distant wind and water turbines or solar fields to the power-hungry cities (Ye Ruolin and Yuan 2021). However, technical problems and jurisdictional conflicts meant that such lines did not always transmit power at full capacity (Zhu Yan 2020), so cities such as Changsha experienced power shortages even while the ultrahigh-voltage lines were operating at partial capacity and some of the generating capacity was being curtailed (Ye Ruolin and Yuan 2021). Thus, effective overcapacity and high curtailment rates accounted for the low ratio of electricity produced to installed capacity (EIA 2020, 13). In the solar industry, the mismatch between generation and transmission led to an emphasis on building-integrated photovoltaics (i.e., rooftop or "backyard" solar) beginning in 2018 (Lewis 2020, 51), but similar solutions were not available for hydro or wind power.[1]

China's investment in dam building and "clean renewable" development accounted for 40 percent of the worldwide total, but coal still provided 58 percent of total primary energy use and 54 percent of electric power generation. We can thus either condemn China as the world's largest fossil fuel consumer, a major polluter, and the main obstacle to mitigating climate change, or we can laud China as the world leader in renewable and sustainable energy development. In fact, of course, it was both, and one reason for this paradox is because, as General Secretary Xi proudly announced in February 2021, China had lifted 850 million people out of absolute poverty (M. Lau 2021), damning the environmental and climate consequences. It did so not by redistributing resources (the last time that happened was during the Land Reform, 1947–51) but by increasing the supply of food, housing, consumer goods, and infrastructural amenities. All of this required increasing the supply of energy. At first China did this with shocking inefficiency, but after the turn to Reform, efficiency increased dramatically—just not nearly as dramatically as total energy consumption. China's leaders recognized this paradox, and, being unwilling for political reasons to stop economic growth, they increased their efforts to make that growth less energy dependent and to make energy less carbon dependent. Whether they would, in fact, reach "carbon neutrality" by 2060 and whether that would be too late to forestall the worst effects of global warming was impossible to know in the early 2020s.

THE SUCCESSFUL ATTACK ON AIR POLLUTION

Just as it had set China on the road to becoming the world's largest greenhouse gas emitter (and eventually the world's largest GHG mitigator), the Stalinist heavy-industrialization strategy pursued before 1980 had also fouled the air in much of China. Still, the huge Reform-era expansion in industrial capacity of all kinds exacerbated the problem, even as the regime recognized that the problem was real. What Anna Ahlers, Mette Halskov Hansen, and Rune Svarverud (2020, 79) describe as "the largely unmonitored growth of generally local, small-scale factories that belched out unregulated emissions" led to "incredible amounts of solid and gas pollutants—pollutants that not only blackened China's skies, but also diffused into the global atmosphere at unprecedented rates."

The air in late twentieth- and early twenty-first-century Chinese cities contained a multitude of pollutants from a multitude of sources. Ultra-small suspended particulate matter ($PM_{2.5}$) arose from coal combustion, diesel and gasoline exhaust, dust, biomass aerosols (including cooking and barbecuing as well as agricultural burning), cigarette smoke, and other sources, but different sources predominated at different times and places. In Beijing in the 2000 aughts, a time of severe pollution, coal and biomass (from urban and rural heating fuels, respectively) were the primary sources in January and February. In April, winds blew from the northwest and west, bringing dust from the loess plateau in Northwest China and both dust and organic matter from nearby parched and still fallow agricultural lands. Neither heating nor dust storms were prevalent in the summer, when vehicle exhaust and cooking were the primary culprits. In October, farmers burned their fields after the fall harvest, and the biomass aerosols became the primary source of carbon in suspended particulate (M. Zheng et al. 2005, 3973–75; Q. Wang et al. 2009). Other cities, including Qingdao and Shanghai, showed similar patterns, with coal burning a primary pollutant in the winter months, but not at other times, and dust as an important contributor (Z. G. Guo et al. 2003; Yilun Jiang et al. 2009).

Even though China implemented its first air pollution control law as early as 1987 (Ahlers, Hansen, and Svarverud 2020, 90), pollutant concentrations in China's urban air continued to increase throughout the following decades. The first air pollutant to come to wide national and international concern was sulfur dioxide (SO_2), produced primarily by burning hydrocarbons with high sulfur content. SO_2 was the prime cause of acid rain, whose effect on water quality and plant growth in southern and eastern China had been the subject of considerable research since the early 1980s. In the north, by contrast, high particulate matter concentrations (which tended to be alkaline) neutralized much of the emitted SO_2 (Ahlers, Hansen, and Svarverud 2020, 86–89).

By the time of the 2008 Olympic Games in Beijing, however, $PM_{2.5}$ had become the flagship pollutant in both research and popular environmental discourse in China. The Ministry of Ecology and Environment's air pollution data website is called "Historical Statistics of $PM_{2.5}$" ($PM_{2.5}$ Lishi shuju), even though it presents detailed daily statistics for six different pollutants (MEE n.d. b). Despite Premier Zhu Rongji's 1999 comment that "if I were to work in your Beijing, it would shorten my life at least 5 years" (Economy [2004] 2010, 71), little was done, and pollution continued to worsen through the 2000 aughts.

At the time of the 2008 Olympics, the Beijing government managed to show the world that it could blue up the skies for special events, something it continued to do every year in March, when the "Two Sessions" of the People's Congress and the People's Consultative Conference brought representatives from all over the country to the capital. But these temporary reprieves were no match for the severe smog events that happened every winter and culminated in the "air-pocalypses" of 2012–13 through 2014–15, when air quality indexes routinely varied between "unhealthy" and "hazardous" (or "heavy pollution" [zhòngdu wuran] and "severe pollution" [yanzhong wuran] in Chinese parlance). Public concern was raised not only by what people saw and breathed when they stepped outside but also by a pas de deux that started when the US embassy began tweeting out air quality measurements for its own personnel and other expatriates in 2008 and local citizens soon began to follow the tweets.[2] The Beijing municipal government reacted by posting less worrisome ratings—not phony, but conveniently measured in less polluted parts of the municipality. But the authorities soon realized that the citizenry would trust the American ratings more anyway, and so they started posting more realistic levels in 2012 and 2013 (Bradsher 2012; Barboza 2012). In addition, in 2015 investigative journalist Chai Jing released a documentary, Qiongding zhi xia (Under the dome), on air pollution and its health effects; it was viewed more than two hundred million times before it was suppressed and probably at least as many times afterward on underground sites or by using VPNs (Chai 2015; Gardner 2015). China became known as the world locus of air pollution, with frequent attention in foreign media, especially during the worst occurrences.

Thus, by the early 2010s, everyone was taking air pollution seriously, and regulators began more serious efforts to control it. Daily statistics of six pollutants in more than two hundred cities, compiled and posted online by the Ministry of Environmental Protection (later the Ministry of Ecology and Environment) beginning in December 2013, enable us to see the results of these efforts clearly and quantitatively. I have tabulated the results for eight cities—Beijing, Tangshan, Baotou, Panzhihua, Nan-

Table 16.1. Monthly air quality index (AQI) averages and concentrations of selected air pollutants in 2014 (top) and 2021–2022 (bottom) for selected cities in China

2014	AQI[a]	AQI Max[b]	$PM_{2.5}$[c]	SO_2[c]	NO_2[c]	Ozone[d]
Beijing	125	402	94	51	65	170
Tangshan	179	378	139	155	71	150
Baotou	126	236	76	113	57	148
Panzhihua	101	131	74	61	45	89
Nanjing	166	347	129	46	73	128
Guangzhou	120	207	90	22	74	133
Chengdu	222	462	179	32	73	122
Ürümchi	151	289	114	50	67	102
2021–2022	AQI[a]	AQI Max[b]	$PM_{2.5}$[c]	SO_2[c]	NO_2[c]	Ozone[d]
Beijing	68	198	45	3	27	121
Tangshan	82	204	56	11	44	134
Baotou	82	183	56	19	45	139
Panzhihua	62	107	43	22	41	97
Nanjing	82	137	58	5	44	102
Guangzhou	55	100	31	7	23	121
Chengdu	91	158	67	5	41	135
Ürümchi	145	288	110	6	62	128

Note: All pollutants except ozone measured in January; ozone measured in July.

[a] Monthly average air quality index, January 2014 and 2022.

[b] Highest air quality index for any day in January 2014 and 2022.

[c] Micrograms/cubic meter (µg/m³), January 2014 and 2022.

[d] µg/m³, July 2014 and 2021.

Source: Daily data from MEE n.d. b.

jing, Guangzhou, Chengdu, and Ürümchi—for January, April, July, and October from 2014 to 2020 (MEE n.d. b). These statistics show a definite trend of improvement (table 16.1).

Sulfur dioxide (SO_2) levels decreased first and fastest. In 2011 the state issued the China IV gasoline standard, which mandated a maximum of 50 parts per million (ppm) sulfur in gasoline. The 2013 State Council Air Pollution Control Action Plan tightened the screws much further, instructing localities to phase in the China V motor fuel standard, which allowed a maximum of 10 ppm of sulfur (Pei Li, Lu, and Wang, 2020, 4). The Environmental Protection Law followed in 2014 and the Air Pollution Prevention and Control Law, in 2015 (L. Zheng and Na 2020). These measures, perhaps with a boost from license plate–linked driving restrictions implemented in some cities in order to lessen traffic (Viard and Fu 2015; Z. Chen, Zhang, and Chen 2020), dramatically reduced SO_2 concentrations in the air of almost all major cities in China. For example, in Beijing from 2014 to 2022, January SO_2 concentrations declined from 51 to 3 $\mu g/m^3$, partly due to fuel standards and partly due to burning less and higher-quality coal, and July concentrations, almost entirely due to motor vehicle exhaust, declined from 9 in 2014 to 2 $\mu g/m^3$ in 2021.[3] SO_2 in highly polluted industrial Tangshan declined from 155 to 11 $\mu g/m^3$ in January and from 43 to 5 $\mu g/m^3$ in July, while Ürümchi's SO_2 went from 50 to 6 $\mu g/m^3$ in January and from 24 to 7 $\mu g/m^3$ in July, despite the fact that other pollutants there experienced little if any decline during the six-year period (MEE n.d. b). By 2020, China's atmospheric SO_2 levels were comparable to those in the United States (EPA 2017). China still had air quality problems, but acid rain was no longer a major one.

$PM_{2.5}$, the iconic pollutant, was somewhat harder to reduce, but China still made progress; the Air Pollution Control Action Plan and the laws that followed had real results, "driven above all by reductions in coal use and emissions of power plants and heavy industry factories" (Greenpeace 2018, 1). From 2013 to 2018, yearly average levels of $PM_{2.5}$ fell almost every year in every provincial capital (FORHEAD 2018, 16). However, declines varied widely from place to place, depending on local topography and weather, strictness of environmental regulations, and the mix of fuels burned. In the largest cities—including Beijing, Nanjing, Chengdu, and Guangzhou, $PM_{2.5}$ concentrations declined precipitously from 2014 to 2020, by more than half at most seasons. Chengdu, for example, recorded a monthly average of 179 µg/m^3 in January 2014, seventeen times the World Health Organization (WHO) guideline for the annual average and seven times the guideline for the daily maximum, while the comparable concentration in 2022 was 67 µg/m^3, or still polluted but close to WHO's "interim target" (WHO 2006, 9–11). Concentrations in July declined from 40 to 22 µg/m^3 and, in October, from 64 to 24 µg/m^3. Pollution levels in smaller and politically less important cities also declined but more slowly and from higher beginning levels. Tangshan's January 2014 concentration was 139 µg/m^3, less than Beijing's at the time, and by 2020 it had declined to 56 µg/m^3, now a little higher than Beijing's. From a very high 106 in October 2014, Tangshan's PM$_{2.5}$ level declined, but was steady in the low 50s or upper 40s from 2018 through 2021. By contrast, Ürümchi's January PM$_{2.5}$ levels did not decline at all from 2014 to 2022, perhaps because the overwhelming majority of its households and businesses continued to heat and cook with coal, although the municipal government did make some effort to replace goal with natural gas. The fact that Ürümchi sits in a basin between mountain ranges, like Los Angeles or Taipei, may have made reducing pollution levels more difficult (MEE n.d. b; Darren Byler, pers. comm.).

Chinese cities made similar progress against carbon monoxide and oxides of nitrogen, but one pollutant resisted efforts to control it: ozone. Ground-level ozone is primarily a product of photochemical processes; pollutants already released into the atmosphere, mostly by motor vehicle exhaust, become oxidized. Unlike particulate matter, ozone is primarily a warm-weather problem, as exemplified by Los Angeles's summertime smog alerts of the 1950s and 1960s. When particle pollution in China was most severe—in the 2000 aughts and the early 2010s—ozone was not considered a major problem; particulate matter was much more visible and much higher by world standards. However, no place in China had managed to bring down ozone levels by 2020 (Tong et al. 2020). In all the cities I tabulated, ozone levels in spring and summer 2021 exceeded national air quality standards and the WHO guidelines (WHO 2006, 14) at values of between 100 and 160 µg/m^3, or 50–80 ppb (MEE n.d. b). These levels were only slightly lower than the Los Angeles levels of about 100 ppb, although LA levels had been much higher in its smog emergencies of the 1950s and 1960s (CARB 2018). China is not alone in its inability to control ozone pollution; Japan, South Korea, and Taiwan had similar lack of success in the 2000 aughts and the 2010s (Harrell 2020).

How did China achieve these improvements in air quality? First, the industries that produced the greatest amounts of pollution—steel, cement, and glass—were regulated and restructured. In 2017 China still produced about half the world's steel, much of it in the Beijing-Tianjin-Hebei region, even though its mills were not operating at full capacity (Mele and Magazzino 2020). The region also produced large amounts of cement and glass. This over-capacity production presented an opportunity to close small, inefficient, and polluting mills, upgrade the remaining mills to higher-efficiency processes and better pollution-control equipment (including better dust collection), and ban high-sulfur coal and petroleum (FORHEAD 2018, 46).

Implementing fuel standards was at least as important as closing polluting factories. As private motor vehicle ownership exploded in China, from near zero in 1980 to around 20 million in 2000, estimated PM emissions from cars increased even faster, to around 550,000 tons. But after the first fuel standards were implemented in 2002, followed by progressively stricter standards in 2006, 2011, and 2014, vehicle PM emissions leveled off even as vehicle ownership increased exponentially once more, to an estimated 243 million cars in 2020 (Statista 2022c). The gasoline IV standard of less than 50 ppm of sulfur in fuel, implemented in 2011, had a particularly striking effect on particulate matter, not to mention its contribution to the virtually complete solution of the problems of SO_2 and its contribution to acid rain (Pei Li, Lu, and Wang 2020).

In addition to motor fuel standards, replacing coal with other fuels for industrial and residential uses also had dramatic effects, though improvements were greater in urban than in rural areas. Liaoning, for example, a province heavily dependent on coal for a variety of uses, implemented a pilot policy of replacing coal with gas and electric power in 2015, reducing wintertime coal consumption from monthly highs of around 4 million tons in 2005 to around 1.5 to 2 million tons in 2017–18, with concomitant reductions in emissions of carbon dioxide, sulfur dioxide, and oxides of nitrogen (Shou et al. 2020, 34358, 34366). Coal-fired boilers were replaced as rusty factories and run-down housing gave way to service industries and new residential compounds (Chang Su,

Madani, and Palm 2018), but rural cooking and heating with both coal and biomass fuels were more difficult to address, and particulate matter often wafted toward nearby cities on prevailing winds (see Qili Dai et al. 2019, 66; Ying Zhou et al. 2020, 5; Tong et al. 2020, 4). Rural ambient air pollutant levels have rarely been studied, but it seems clear that some areas remain as polluted as Futian was when Bryan Tilt visited in the 2000 aughts (Yun et al. 2020, 3; Tilt 2010). In addition, indoor air pollution from burning biomass in rural households remained a significant health hazard even in the 2020s (Yue Li et al. 2020).

GOVERNANCE AND ENVIRONMENTAL MITIGATION

China's air quality improved largely due to good environmental governance, spurred on by the realization that dirty development was not sustainable either environmentally or politically—the citizenry and the semi-independent media (including social media) demanded cleaner air as clearly as did the plans for sustainable growth. No longer did national leaders make the late twentieth-century argument that China, as a "developing country," had a license to pollute now, clean up later. At the same time, however, further improvement ran into obstacles in the Chinese governance system, which we can call the three Cs: coordination, corruption, and cadre evaluation. All three Cs were not only obstacles to cleaner air in general but also resulted in environmental injustice, placing the economic, ecological, and health burdens of economic growth on the backs of people living in poorer and less powerful areas.

Coordination problems stem from the fact that, despite China's authoritarian ruling party, provincial and local governments have considerable leeway to try out different ways of meeting policy directives issued by the party center and the central government. As pollution gained increasing public notice in the 1990s and afterward, many local governments issued regulations that made it difficult for polluting industries to continue operating in growing major cities. Consequently, industrial enterprises moved to more distant locations or to other jurisdictions where regulation and monitoring were laxer, creating "pollution shelters" (T. Feng et al. 2020, 1) or "pollution havens" (Ahlers, Hansen, and Svarverud 2020, 92) that resulted in higher pollution levels in the late 2010s in much of Hebei, for example, than in Beijing and Tianjin

(Greenpeace 2018, 1; FORHEAD 2018, 38; T. Feng et al. 2020, 6). Wealthy, highly educated populations in the major cities in effect exported their pollution to poorer and more sparsely populated areas, where local authorities had less control and less leverage against polluting enterprises operated by powerful central or provincial state–owned corporations.

Coal gasification plants built in Inner Mongolia and Shanxi in the mid-2010s clearly illustrate the practice of exporting environmental degradation. Recognizing the contribution of coal-fired power plants and vehicle exhaust to increasing air pollution in Beijing and other cities, as well as the pollution and highway clogging caused by the masses of coal trucks hauling fuel out of the coalfields to the urban power plants, China's coal and energy companies accelerated construction of coal-based synthetic natural gas plants. The US Department of Energy cataloged 46 gasification plants built between 1974 and 2007, but by 2015, 72 more were in operation and 60 under construction. Of these 132 plants, 62 were in the northwest, mostly near the coalfields in Shanxi, Xinjiang, and Inner Mongolia, where pollution from generation would dissipate in relatively sparsely populated areas, thus helping reduce air pollution in the big cities (NETL 2015; E. Wong 2014). Fuel gas from these plants would be piped not only to nearby areas but in some cases as far away as southeastern China (Y. Ding et al. 2013, 446), or it would be used to generate electricity that would travel to major cities over long-distance high-voltage lines. Not only did these projects pollute the air in less powerful and minority-inhabited regions of Central Asia, they also increased China's net output of greenhouse gases: the sequential processes of coal gasification and burning the cleaner natural gas in the urban plants would approximately double the life-cycle GHG emissions per unit of energy produced (Y. Ding et al. 2013, 451). Similarly, using the coal gas for vehicles fueled with liquid natural gas (LNG) would emit about twice as much life-cycle GHG as either a gasoline-fueled vehicle or an electric vehicle powered by electricity from a coal-fired plant (Y. Ding et al. 2013, 451–52). Coal gasification plants located in arid regions (almost all such plants were in dry areas) also consumed vast amounts of scarce water (Greenpeace 2012, 12). In recent years, China has pioneered underground coal gasification technology, which presumably reduces the local air pollution but does nothing to address the life-cycle greenhouse gas emissions.

Such lack of coordination results in environmental injustice. People living in poorer and less powerful areas bear the burden generated by wealthy areas' ability to clean up their air. Corruption and the perverse incentives of the cadre evaluation system raise similar issues. Environmental cleanup imposes short-term costs on local governments and economies. Most local governments derive much of their revenue from taxes and fees paid by industry, and many industries cannot afford to invest in expensive equipment to reduce pollution. There are thus incentives to cheat, but cheating requires the collusion of local governments, and many of them gladly collude, because the industries fund much of their budgetary needs. Inspections by the Ministry of Environmental Protection (L. Zheng and Na 2020, 2) found hundreds of factories and over thirteen thousand construction projects out of compliance with environmental regulations and standards in Hebei alone (FORHEAD 2018, 43). These inspections resulted in temporary reductions in $PM_{2.5}$ emissions in inspected provinces compared to those not inspected, but in many places pollution had risen again thirty days after the inspections (L. Zheng and Na 2020, 2) and repeated inspections had longer-lasting effects in wealthier areas (L. Zheng and Na 2020, 3), revealing pervasive environmental injustice. Reductions in air pollution brought about by anticorruption campaigns were also mostly temporary (M. Zhou, Wang, and Chen 2020).

The cadre evaluation system presents further obstacles to pollution control, similar to those it created in the heyday of the polluting township and village enterprises (TVEs). Environmental criteria were included in the point system for cadre evaluation beginning in 2005, but it was much later that environmental protection was elevated to the status of "one ticket denial" (*yi piao foujue*), which prevents cadres from being promoted if they do not meet a specific target. Officials therefore had originally placed less emphasis on environmental goals than on those of public safety and economic development. In addition, since the evaluation system is based on achieving specific targets, it becomes easy, in Jie Gao's (2015) words, to "game the system" and comply with the letter of the regulations while ignoring their original purpose. Through subterfuges such as blue skies for special events or warnings to factory managers that the inspectors would be coming Wednesday so they had better turn off the furnaces on Monday and Tuesday to clear the air, cadres can meet specific targets while not really contributing to the long-term improvement of air quality or other environmental measures. In addition, officials at higher levels have some discretion in setting the specific targets for cadres at subordinate levels in their own jurisdictions and thus can set environmental targets so that they are easier to meet (Heberer and Trappel 2013, 1050).

These governance problems aside, China's urban air improved spectacularly from the winter air-pocalypses and year-round foul air of the mid-2010s to the only moderately severe pollution of the early 2020s. This improvement tends to validate the environmental Kuznets curve (EKC), which states that environments in general will degrade during earlier stages of industrialization but then improve as mitigation technologies become more affordable and both public opinion and the state begin to value a clean environment over increases in the material standard of living.

Air pollution, however, is more susceptible to the EKC than many other kinds of environmental degradation—cleaning the air is relatively cheap and relatively fast (Dinda 2004; Harrell 2020). China's experience in the late 2010s shows that other kinds of pollution are harder to remedy. Water pollution from China's industries, urban wastes, and chemical agriculture remained far greater problems in the early 2020s than the still-polluted but cleaner air. Soil contamination from industries and agricultural chemicals will take even longer and be more difficult to remediate. Soil cleanup is intrinsically hard—there is no quick fix, and, unlike air pollution, which dissipates with the next windstorm, or crud in the river, much of which eventually gets diluted in the sea, the chemicals in the soil don't go anywhere (Harrell 2020, 257–61).

....

China thus stood at a crossroads in the late 2010s and early 2020s. On the one hand, everyone from the General Secretary down to the poorest farmers and migrant laborers was aware of the environmental crisis— terms such as "environmental protection" (*huanjing baohu*), "green energy" (*lüse nengyuan*), and "ecological civilization" (*shengtai wenming*) were ordinary language by that time, and everyone knew that environmental problems had to be addressed. On the other hand, China might have eliminated absolute poverty, but in the speech in February 2021 when he announced this, the General Secretary also pointed out that the next milepost, "moderate prosperity" or *xiaokang*, was already on the horizon (M. Lau 2021). That meant further economic growth and continuing environmental challenges.

CONCLUSION Chinese History as Ecological History

> Seek truth from facts.
>
> —Chinese Communist reform slogan

This customary four-character phrase, originally used in *History of the Han Dynasty* (Hanshu), has become a touchstone of Chinese Communist philosophy. Mao Zedong first used it in 1936, but Deng Xiaoping popularized it more widely at the beginning of the Reform period. It implies not just respect for facts but using facts to inform theory and policy. Combining theory from human ecology and the facts laid out in this book can lead us to an informed appraisal of the current condition and likely future of China's ecosystems.

As I write in 2022, China has tightened the Malthus-and-Boserup ratchet to an unprecedented degree; Chinese people experience plenty as never before, but they rely increasingly on institutional and infrastructural buffers, locking in all scales of ecosystems and thus rendering them increasingly vulnerable to disturbances. Mainstream media outlets such as the *New York Times* have started to take notice of this very fact, without using the theoretical framework I employ in this book (see Myers et al. 2021). Whether the nation slips back into a new misery—more affluent for sure than at the Founding but also at least as vulnerable to environmental disasters—is the question the nation now faces. No one in China denies that the nation's previous trajectory of development, proceeding with little if any regard for environmental resilience, was at best shortsighted and at worst a disastrous mistake. Whether or not such a predicament was inevitable, the regime and the populace now need to fix it. At the same time, the legitimacy of the CCP regime continues to depend on ongoing economic growth or, more narrowly stated, on continued increases in material consumption and social services for both urban and rural people. Thus, the regime faces a delicate balancing act: implementing environmental cleanups, restoring ecosystems, and strengthening institutional and infrastructural buffers while still promoting rapid increases in living standards.

Awareness of this predicament came gradually, but over the two decades starting in the final few years of the twentieth century, China turned from developmentalism, in which it committed to economic growth above all else, and moved toward eco-developmentalism (Haddad and Harrell 2020). The leadership and most of the people came to recognize that economic growth was not in direct contradiction with environmental protection or ecosystem resilience. They realized instead that policy that neglected environmental considerations would limit economic growth—the nation would consume the very resources and ruin the very ecosystems that it depended on to continue growing. Also, leaders recognized that providing a livable environment was becoming more and more central to retaining the support of wide sectors of the population.

The impetus to move toward eco-developmentalism thus came both from public dissatisfaction and from scientific worries about the direction of the country's ecosystems. The turn started out with specific programs. Controls on industrial emissions at the beginning of the Reform period led the way, though they were often ineffective before the turn of the century. Next came the "returning" programs—farmland to forest, pasture to grassland—at the very end of the twentieth century, in response particularly to the 1998 floods in the middle Chang River region and in the northeast. Soon specific programs evolved into general slogans and policies. At the end of the Jiang-Zhu administration (2002–3), "sustainable development" was everywhere, even when it meant something other than ecosystem resilience, such as economic growth rates that could be sustained or any other development programs that local officials might want to implement (Li Yongxiang 2005). During the Hu-Wen administration, both the "scientific perspective on development" (*kexue fazhan guan*) (Guizhen He et al. 2012, 33; Qian G. 2012) and "constructing the new socialist countryside" (*xin shehui zhuyi nongcun jianshe*) (Looney 2012, 2015, 2020, chap. 4) incorporated elements of sustainability into their principles and policies. The most important slogan, "ecological civilization" (*shengtai wenming*), began to appear at this time (Pan 2006), though it only gained full prominence as part of Xi Jinping's "China dream" (*Zhongguo meng*) (Ren Z. 2013).

Much of China's turn in policy was a response to public pressure, as was typical of countries reaching a certain stage of material development. Environmental protests were ubiquitous, ranging from opposition to dam building (Mertha 2008) to protests against petrochemical plants (Yan Dingfei 2014) to complaints about urban air pollution (Chai 2015). And in the early 2000s, at least, the regime was extraordinarily tolerant of protests and exposés about pollution and other environmental problems. At a time when advocating for human rights, ethnic autonomy, or what the party dismissed as "Western-style 'democracy'" could land one in jail, and when newspapers and websites published little if anything critical of policies in these areas, environmental protests, muckraking, and disputes filled the airwaves, the press, and the internet, and environmental NGOs proliferated (J. Dai and Spires 2020). It was as if the regime were covertly encouraging this kind of debate and critique in order to support its own increasingly eco-developmentalist policies.

With the turn toward environment and ecology, China put aside its earlier rhetoric that condemned environmentalism as an excuse for the rich countries of the "West" to keep down the "developing countries," including China (see figure 2.1). Propagandists stopped using the line "Hey, you guys polluted on your way to prosperity, and now you want us to bear the burden of cleaning up after a mess that you created in the first place" (the line was of course true). Instead, China sought to be seen as a world environmental leader, especially through its huge progress in developing and deploying renewable energy (Lewis 2013; Mathews and Huang 2021).

Alongside the rhetoric, China also gave more power after 2000 to those agencies charged with enforcing environmental regulations and rehabilitating ecosystems. The primary environmental agency of the central government had started out in 1979 as the understaffed (only 30 employees) and underfunded State Environmental Protection Office (Jahiel 1998, 768). Even after its upgrade to the State Environmental Protection Agency in the late 1980s, it did not fill its employment quota of 300-plus officials (Jahiel 1998, 768). The eleventh five-year plan (2006–10) began to emphasize environmental protection much more seriously, and the agency was elevated in status to the Ministry of Environmental Protection in 2009 and then the Ministry of Ecology and Environment (MEE) in 2018. In 2021 MEE employed 478 people in fifteen central departments with mandates ranging from solid wastes and chemicals to climate change to nature conservation (MEE 2021), as well as another 240 people in six regional inspection bureaus. The ministry also operated seven watershed inspection offices jointly with the Ministry of Water Resources (Baidu 2021). In addition, governments at all levels down to the county have environmental protection bureaus, and their enforcement powers were greatly strengthened by the Environmental Protection Law of 2014 (X. Wang 2019, 11).

Laws and regulations themselves have also been strengthened. In the Reform period the Chinese government had promulgated increasingly strong environmental laws, beginning with the Environmental Protection Law of 1979 (revised in 1989) and a suite of comprehensive laws in the 1990s (Economy [2004] 2010, 101–5). However, enforcement lagged for a long time, especially at county levels and in rural areas, primarily because environmental protection branches of governments at all levels lacked both institutional strength and sufficient staff (Jahiel 1998, 767–72; Tilt 2010), but changes began with the eleventh five-year plan and the increasing power and staffing of environmental offices. In 2014 environmental rules were updated under a new and more comprehensive version of the Environmental Protection Law, as well as a set of laws and regulations on pollutants, natural resource protection, and clean production/circular economy (X. Wang 2019, 6). Very importantly, beginning with the eleventh five-year plan the evaluation systems for cadres and factories were revised to make environmental protection, in some cases at least, a "one ticket denial" item that would prohibit promotion of cadres or awards to factories that failed to rise to its standards (X. Wang 2019, 12; Zeng Zhiyang 2012; Wu You 2017).

THE CONDITION OF CHINA'S ECOSYSTEMS

So what has recently increased attention to pollution abatement and ecosystem restoration actually accomplished? It is possible to see China's turn toward eco-developmentalism cynically, as an authoritarian regime making excuses for its depredations, and certainly the regime is still capable of such prevarication. But I think we need to give the leadership the benefit of the doubt (Chotiner 2021) and at least credit them with good intentions, even if the results are uneven. It is also possible to look at Chinese eco-developmentalism in the context of a new cold war, with advocates of authoritarian rule pitted against those favoring democratic governance, using China's recent successes and failures as a test of whether "authoritarian environmentalist" (Yifei Li and Shapiro 2020, 9–21; Kostka and Zhang 2018) regimes can act more quickly and decisively than messy electoral democracies. Here I take a third tack on China's eco-developmentalism, examining positive and negative results in specific areas and drawing lessons from them for our study of China and of ecological history.

Encouraging Trends

The environmental Kuznets curve model predicts a curvilinear relationship between environmental quality and economic growth (Grossman and Krueger 1994; Panayotou 1993; Kaika and Zervas 2013; Dinda 2004). In the early stages of industrial development, nations have neither the technology nor the funds to place environmental curbs on industrial pollution or ecosystem degradation, and there is little pressure to do so, since citizens are eager to free themselves from what is now defined as poverty. When nations reach a certain level of economic development, however, citizens begin to place a higher value on environmental quality and ecosystem health, technologies of cleaner industry are available and affordable, and ruling regimes begin to see environmental cleanup as central to continued economic prosperity as well as an important facet of legitimacy. Thus, citizens and their rulers, whether authoritarian or democratic (Haddad and Harrell 2020), develop a common interest in ecosystem health and environmental cleanup. Some recent environmental trends in China, analyzed in more detail in previous chapters, support the EKC hypothesis:

- Population growth has slowed, and population will soon begin to decline. China's 2020 census showed the lowest rate of population growth in any intercensal interval since reliable censuses started in 1953. Between 1970 and 2020, crude birthrates declined from 39.5 to 11.9 (O'Neill 2021a) and total fertility from 6.3 to 1.7 (O'Neill 2021b). With a boost from the COVID-19 pandemic, population will already have reached its maximum of 1.41 billion by the time this book is printed (UNPD 2022a, b, c). Although China's maldistribution of wealth probably contributes as much or more to some kinds of consumption, and particularly to GHG emissions, any further growth in China's population would put pressure on food supplies as well as periurban lands, and it looks like this will not happen.

- China has become a leader in green or clean energy. China's growth in hydroelectric, solar, and wind power generation since 2000 has been spectacular: its hydropower in 2019 accounted for 27 percent of the world total; wind, 28 percent; and photoelectric, 31.9 percent. Whereas hydro, wind, and solar accounted for only 17 percent of China's electricity generation in 2010, by 2020 this figure had reached 27 percent—slightly below the world average of 29 percent but gaining fast and projected to increase substantially in the next decade. Since renewables account for more than 40 percent of China's installed capacity, better storage and a more efficient grid could allow the share of renewables in generation to grow even faster (Mathews and Huang 2021). The trajectory so far has prevented large amounts of both air pollution and greenhouse gas emissions, and projections for the near future point to this trend continuing.

- Air pollution has been cut considerably. That the Chinese and foreign media published alarming reports on the burst of air pollution in northern Chinese cities in March 2021, caused primarily by PM_{10} from dust storms, shows that these events are increasingly rare. In 2014–15 dangerous levels were reached almost every month in almost every city (MEE n.d. b). Many factors, including the turn to green energy and increasingly strict controls on vehicle exhaust, are responsible for this remarkable turnaround.

- Forested area has more than doubled. Aggressive programs of reforestation have doubled forest cover since the low point in around 1960 (Harrell 2020, 256). There are legitimate questions about how forest cover and standing biomass are measured, and poor species and site selection, along with perverse incentives for local governments, have kept the quality of the replanted forests lower than it would otherwise have been. Still, reforestation has been a huge success.

- Organic and sustainable agriculture has emerged. The main trend in China's agriculture is toward the CCCM model—consolidated, chemicalized, commercialized, and mechanized, with predictable effects on soil and water. However, it is not happening without resistance, and both consumer worries about food safety and environmentalist worries about pollution are driving a small but active movement to farm more regeneratively. We can expect this trend to grow because of both these kinds of pressure.

The environmental Kuznets curve thus works for a few kinds of environmental degradation, including air pollution and deforestation. These problems are relatively noncontroversial—there is broad public support for both, and they are relatively easy to solve. Air pollution is visible, transient, and easy to remedy, and forests grow again when they are replanted (Harrell 2020). Other kinds of environmental problems are trickier.

Persistent Worries

Despite the encouraging trends listed above, the state of Chinese ecosystems is still parlous as I write in 2022. The worst degradations have mostly been halted, but with the exceptions of air pollution and deforestation, they have not been rolled back. More generally, Chinese ecosystems are if anything more dependent on infrastructural and institutional buffers than ever before—much of the progress against degradation has come at the cost of increased technological lock-in and ever more rigid rigidity traps, making the ecosystems even less resilient to natural and human-caused disasters. In many areas, degradation has not even slowed down.

- Waters are still very polluted. Water is harder to clean up than air is. Air pollution is both more transient—it stops with the smokestacks and disperses with the wind—and more visible—everyone in a smoggy city sees it and breathes it. There are also alternatives to air-polluting practices, such as cleaner fuels and cleaner generation. Water pollution is a more difficult problem—it is hidden from many people's everyday life and thus not the subject of continued popular unrest, there are fewer alternatives to the processes that produce it, and the pollutants are slower to disperse. Consequently, improvement in river water quality was slow, and there was no overall improvement in lakes. The State Statistical Bureau numbers for 2019 showed that, of the seven major river basins monitored, only in the Chang and Pearl watersheds was less than 15 percent of river length classified as grade IV or below. The Yellow River basin, with 27 percent of its water "dirty," and the Liao River basin, with 43 percent, were in a bit better shape but still not good (EPS China Data n.d.). Lakes fared worse. National data for 2017 showed that of forty-one large lakes monitored, only ten had water quality of grade III or above, and from 2013 to 2017 water quality had deteriorated in eleven, stayed the same in twenty-four, and improved in only six (SSB 2014, 2018b, 20). No lake was classified as heterotrophic (clear); fifteen were mesotrophic (ten of those were in Tibet), twenty-two were lightly eutrophic, and four moderately eutrophic. Between 2013 and 2017, three had become less eutrophic, twenty-nine stayed the same, and nine had gotten worse (SSB 2014, 2018b, 20).

- Soil contamination persists. If water pollution is less tractable than air pollution, soil pollution or soil contamination is even less so. Contaminants stay in the soil even after the sources of contamination are shut off. An eight-year survey by the Ministries of Environmental Protection and Soil Resources showed that 17 percent of China's farmlands were contaminated by heavy metals, particularly cadmium, and that 8 percent were so severely affected that they should not be used for growing food (MEP and MOA 2014). Removing these contaminants is slow and expensive and can take valuable agricultural land out of production. Soil remediation, even where it is possible, is too slow and too costly for the environmental Kuznets curve to be applicable. After this problem attracted national attention in the mid-2010s, it largely disappeared from Chinese discourse, presumably because there is little that can be done quickly, so it detracts from the "green" image that China is trying to project to the world.

- Biodiversity keeps diminishing. Extinction is forever; no environmentalist crusade or scientific program can bring a species back. However, while extinction draws the headlines, species decline because of habitat decline is more widespread. Some restoration programs have been successful, particularly internationally prominent ones to build up populations of charismatic megafauna such as the golden monkey, the giant panda, and the Tibetan antelope (Gao Baiyu 2020b). Less charismatic species get less attention from media or from planners, and while hasty programs of reforestation have produced admirable gains in land cover and standing biomass while retarding erosion and sequestering carbon, the monocultural and single-cohort plantation forests are much less resilient to fires and particularly to insect damage. The environmental Kuznets curve can apply to protected species, particularly charismatic ones, if conservation happens in time, but it does not work when change is irreversible.

- Desertification is proceeding. Severe air pollution in spring 2021 was not primarily due to exhaust from coal-fired power plants or high-sulfur gasoline-burning vehicles but rather to dust storms originating in the deserts of the central and eastern parts of Central Asia. These storms occurred in spite of the intensive afforestation efforts of the Three North Shelterbelt Project, also called the Great Green Wall. Misguided policies of agricultural expansion before 1998 had left these areas vulnerable to increases in extreme weather due to climate change. Progress happened in some areas, but sandstorms in 2009, 2014, and 2021 showed that progress was fragile (Z. Kong et al. 2021, 12).

- Floods, floods, floods. I awoke one morning in July 2021 to headlines of floods in central China, but a quick search of the stories revealed that this was not a repeat of the flood disasters in the Chang River basin in 1998 or 2020 but rather the result of torrential rains in the Yellow River and Huai River watersheds on the western parts of the North China Plain (Baptista et al. 2021; Hassan 2021). As extreme weather events appear to become ever more common with climate change, engineering and constructing environments designed to provide water supplies to ever thirstier populations have not made these environments any more resilient against these weather events.

- Industrial agriculture now dominates. If the citizen-led movements for organic and regenerative agriculture are an encouraging sign of a possible future, for now the general trend is in the other direction, with agriculture continuing to consolidate into the CCCM model. Agrarian radicals have been predicting the collapse of agriculture in the rich countries for decades, and although, despite erosion and soil loss, it has not happened, it still could. Ecologically, there is now a race between soil damage and water pollution caused by chemicals and concentrated animal feeding operations and the realization that agriculture must be more regenerative. Only time will tell the outcome.

- Greenhouse gas emissions continue to increase, despite a short-term dip during the worst of the COVID-19 epidemic in 2020 and 2021. Xi Jinping and other Chinese leaders have been touting the national goals of "peak carbon" by 2030 and "carbon neutrality" by 2060, but meanwhile the greenhouse gases China emits will be in the atmosphere much longer than that. Inasmuch as they contribute to increasingly frequent extreme events such as the April 2021 dust storms or the July 2021 floods, they are increasing the size of disturbances because China's ecosystems have not built up resilience against them.

- Environmental injustice remains rampant. Even as China "goes green" (Yifei Li and Shapiro 2020), its leadership seems extraordinarily insensitive about who bears the burden of increasing pollution or the burden of the trade-offs that are part of the greening. From Mongolian herders choked by the exhaust from coal trucks, to Sichuanese farmers who live near polluting factories, to urban migrants who live near waste incinerators, it is the poor, the powerless, and the ethnic minorities who suffer most.

- China is exporting environmental degradation. One reason China has been able to do such an admirable job of reforestation is that it now imports a lot of its wood from poor and resource-dependent countries. One reason Chinese people can now enjoy such rich and varied diets is that Chinese fishing vessels ply the world oceans and Chinese pigs feast on soybeans from deforested environments in South America. Some of China's restoration of its own environments is good for the world, just as it is good for the China's poor. Some of it, however, merely deflects the cost to people elsewhere who are least able to bear it.

- New things come up. Lance Gunderson (2003, 37) has described a type of ecological surprise he christens "true novelty . . . something truly unique and new or not previously experienced by humans." Since about 2016 China has become a center of cryptocurrency mining, a process that uses huge amounts of electric power. In 2019 China accounted for about 75 percent of world electricity consumption in this dubious process, though its share fell to "only" 49 percent in April 2021 (CBECI 2021). Bitcoin miners used mostly surplus hydroelectricity from Zomian dams in the summer, shifting to coal power from Central Asia in the winter when Zomian rivers ran low (CBECI 2021; Coco Feng 2021). These peripheral regions have thus fallen victim to a "truly novel" form of environmental injustice stemming from this strange manifestation of world financial capitalism. Cryptocurrency mining comprised only about 1 percent of China's total electricity consumption (around 75 terawatt-hours), so recent crackdowns (Coco Feng 2021; Areddy 2021) will not make a huge dent in China's total energy consumption.[1] But beyond the numbers, Bitcoin mining illustrates the principle that we don't know what the next disturbance might be, and insofar as China's ecosystems are increasingly dependent on institutional and infrastructural buffers that have been designed to provide specific but not necessarily general resilience, China seems unprepared for the inevitable "true novelties." The effects of Russia's invasion of Ukraine in February 2022, another true novelty, are as yet unclear.

Given these lists of encouragements, worries, and uncertainties, it seems foolish to prognosticate either doom or salvation. As in most other countries experiencing development, industrialization, urbanization, and rising standards of material consumption, the Chinese regime and its citizenry have realized that this development has come at a cost, and they are determined to do something about it. Much is already lost and irretrievable, but much else can still be repaired and regained.

LESSONS FOR THE STUDY OF ECOLOGICAL HISTORY

An extended and thorough (but still far from comprehensive) study of the ecological history of China ought to teach us some lessons about things other than the particulars of China's predicament in the present and its possibilities for the near future. Since the Chinese economy represents about 14 percent of world GDP (Global Economy n.d.), China's development has a significant effect on global ecosystems. Aside from its direct effect, however, study of China's ecosystems should help us think about how to study ecological and environmental history generally.

China's Environmental History Has Little to Do with Its Authoritarian Politics

It is clear to philosophers of science that science is never and has never been apolitical—science is something that scientists do, and scientists are perforce members of political, social, and economic communities, with personal, career, and financial interests in their work.[2] But there is extreme danger in taking this insight too far and letting politics move from influencing science, which is inevitable, to dictating science, which is disastrous. When politics either dictates or ignores science, the result is what Bruno Latour (2018) calls "epistemological delirium": science threatens ideology to the point where the only ideological choice is to ignore, denigrate, or at the very best spin science to conform to ideology. As hostility grows between China and the United States, and to a lesser extent between China and Europe or China and Japan, critics of the authoritarian Chinese regime use their political beliefs to evaluate the state of China's natural systems and of course find those systems teetering on the brink of immediate collapse. At the same time, members and supporters of the CCP regime spin the science the other direction and see China's authoritarian governance as the last, best hope for humanity's ability to live on an increasingly stressed planet. Of course air pollutants, floods, dams, and newly planted trees are unaware of the politics that led to their existence. That the Chinese regime oppresses Central Asian people (Roberts 2020; Byler 2018; Zenz 2018), ruthlessly suppresses political dissent, and facilitates the exploitation of its migrant populations (K. Chan 2021) or, on the contrary, that China has experienced lower overall COVID-19 death rates than any other country says little about the condition of its soils, water, or flood control infrastructure, even if the regime's callousness toward the welfare of its own citizenry may enable some of its environmental policies (Salimjan 2021).

Using one's personal political preferences or one's national geopolitical interests to predict the likelihood of a flood or evaluate the recovery of a panda population will lead to a misunderstanding of that likelihood or that population and quite possibly lead to the next flood or the next population crash. So while we must always recognize that environmental policy and action are inherently political, we must never use our political preferences to evaluate the state of the environment. When we put aside our personal politics, we can see more clearly both the encouragements and the worries detailed above.

One might be tempted to compare the ecological performance of the CCP authoritarian regime with that of the US, Canadian, or German electoral democratic regime to ask whether authoritarian or democratic regimes do better at environmental governance (Yifei Li and Shapiro 2020, 8–13). My answer lies in the Chinese saying "big similarity, small difference" (*da tong xiao yi*). China can build a high-speed railroad or a solar power plant much faster than the United States can; US citizens can bring the plight of the victims of environmental injustice to public attention more easily than their counterparts in China. But on the whole, the grand historical process of development, environmental degradation, and partial remediation seems to proceed about the same no matter the form of governance. China's environmental trajectory has been remarkably similar to those of its neighbors in Taiwan, South Korea, and Japan, even though Japan has been a bureaucratic regime with electoral democracy ever since General MacArthur headed the occupation, and Taiwan and South Korea moved from authoritarian to democratic government in the late 1980s and 1990s. China has simply followed the same path later because its economic growth came later (Haddad and Harrell 2020; Harrell 2020).

The Environmental Kuznets Curve Is a Useful Model, but Only for Limited Cases

EKC theorists were overly ambitious when they posed a curvilinear relationship between economic growth and the *general* health of the environment. China's general turn to eco-developmentalism, as well as the rise of environmental activists and alternative food movements, validate the EKC theorists' observation that attention to ecological and environmental problems grows after economic growth has reached a certain point. But the ability to turn concern into effective remediation varies from one environmental problem to another. Both the originators of the EKC model (Panayotou 1993) and later scholars who have applied it empirically (J. He and Wang 2012) have taken air pollution and deforestation as two of their primary case studies. China, like other countries, has gone a long way toward mitigating these two specific environmental scourges, but other equally problematic processes and policies do not disappear or even diminish with development, despite the best intentions of governments and activists.

There are several reasons why EKC predictions do not work for all problems. Some environmental harms result automatically from development and thus present an inevitable trade-off. These include the recent replumbing of the East Asian continent and the increased use of chemical fertilizers, pesticides, and mechanized irrigation to solve the "warm and full" problem in the late twentieth century. Other forms of degradation and lost resilience are the result of prioritizing development over environment and might have been avoidable had people cared to avoid them. It is possible to remediate air pollution and still allow more cars on the road; all one has to do is "clean" the fuel standards, as China has belatedly done. Restricting harmful fishing practices would negatively impact the diet variety of urban dwellers and thus restrict the regime's ability to provide bread and circuses, but the cost would be small. Still other changes happen at time scales much longer than the course of economic development (Harrell 2020, 246). Soil takes centuries to remediate without prohibitively costly interventions. Climate change will take thousands of years to reverse even if we "get to net zero" or "achieve carbon neutrality" by midcentury, and species extinction is forever. In cases like soil pollution, extinction, or climate change, the EKC can only help us understand or predict a *deceleration* of environmental degradation, not its reversal. One can mobilize public opinion to save the panda but not the dodo.

The Jury Is Still Out on the Possibility of Green Growth

Economists have recently advocated a "decoupling" of economic growth and various kinds of environmental degradation in a kind of tangential offshoot of the environmental Kuznets curve (Y. Kong et al. 2021; F. Dong et al. 2021). As dominant industries change from resource extraction to manufacturing to services, each stage is assumed to lower the ratios of inputs to outputs and of negative externalities to output value. The current Chinese discourse on "ecological civilization" is a variation on this theme that preserves the Marxist teleology of the historical progression of modes of production. But China is not yet at the stage where the declining ratio of inputs to outputs or even the declining *ratio* of negative externalities to economic outputs has led to an actual decline in the *volume* of negative externalities. As we cannot predict the course of China's environment over the next few decades, we cannot predict whether the changing content of economic growth will allow its detrimental effects to decline. Some environmental radicals continue to espouse the twentieth-century view that capitalism (or alternatively, any industrialized economy) inevitably consumes the source of its own growth; they advocate a restructuring of economic inequality through a drastic reduction in overall consumption as well as redistribution. Degrowth, in their opinion, is the only salvation of the planet (Demaria et al. 2013, 196–98). Advocates for green growth, on the other hand, think that growth can be measured in terms of quality of life, which itself does not depend on material consumption beyond a certain, perhaps rather low, minimum standard of comfort and security (Kallis et al. 2018). You need your next meal, but you really don't need your extra car, let alone your private jet.

The evidence for or against the possibility of green growth in China is equivocal. Certainly there has been relative decoupling, as exemplified by China's dramatic gains in the energy efficiency of its economy, in the efficiency of urban water use, and in the proportion of renewables in electricity generation. But there is as yet no absolute decoupling; economic growth has continued to depend on increasing resource use and diminished social-ecological system resilience. Many think, mostly on the basis of theoretical assumptions, that absolute decoupling is impossible (Kallis et al. 2018, 296); China in the next three or four decades will provide an empirical test case.

Resilience Is Knowable Only in Retrospect

Given the multiple trade-offs between production and resilience narrated in this book, along with the increasing rigidity traps of dependence on infrastructural and institutional buffers, it is tempting to pontificate that collapse is inevitable. In this view, declining ecosystem resilience along with increased frequency of extreme weather events will ultimately lead to a turn of the adaptive cycle toward a collapse and then perhaps a reorganization phase. But at the same time, research on ecosystem resilience leads us to more modesty in our prognostications. At the end of the Great Famine in 1962, few people (other than faith-based Marxists and Rostovians), would have predicted that China could have more than doubled its population and increased its material standards of consumption by tens of times between then and the present. There have also been definite signs of diminished resilience. These include the Great Leap Forward itself, along with its ensuing famine on a national scale, as well as such local-scale phenomena as the Tangshan and Wenchuan earthquakes, the Banqiao Dam collapse, and the central Chang River basin floods of 1954, 1998, and 2020 and the lower Yellow and Huai River basin floods of 2021. But in all those cases the systems at various scales recovered or probably will recover quickly, though the growth phases of their succeeding adaptive cycles were different from their predecessors. Predictions of doom on the basis of supposed *overall* diminished resilience are often belied by recovery after all sorts of disturbances. Because some dams did not collapse while Banqiao did, does this mean that Chinese dams in general were resilient, or did it mean that the particular patterns of rainfall in Zhumadian would have overwhelmed the best dams China had built at the time? Would so many Qiang schoolchildren have died in the Wenchuan earthquake if schools had been better constructed? Will accelerated climate change in the next few decades overwhelm the Chinese agricultural system and lead to general famine, or will only some crops fail and others be sufficient to compensate? There are sometimes indicators of diminished resilience before a sudden transition (Scheffer et al. 2012; Barnosky et al. 2012), and in some cases it seems obvious in retrospect that a system or a structure was vulnerable. But the fact that the Soviet Union collapsed when no one had expected it a few years earlier and that the Chinese Communist Party just celebrated its one-hundredth anniversary should make us humble in the face of the complexity of social-ecological systems.

....

As noted in the preface, this book started out as a rather audacious but still small-scale attempt to improve our understanding of recent Chinese history through using an ecological lens rather than a conventional one of politics and economics. Obviously, none of these lenses is independent of any of the others, but domination of Chinese historiography by elite politics, popular rebellion, media studies, or even local culture leaves out an important vantage point. The ecological view has a long and honored history that began when the story of Great Yu taming the waters entered the classical canon. Its most notable early incarnation in Europhone literature came when Karl Wittfogel put forth his idea of "hydraulic society" in *Oriental Despotism* (1957), but Wittfogel's ideas were mostly discredited because of his ideological anticommunist starting point and the ways he stretched his data to make his argument. But China's rise to wealth and power has depended on a radical reengineering of its ecosystems, and that ascent also affects natural resources around the globe. In other words, the ecological results of China's rise affect not just world ecology but world economics and politics. An ecosystem-centered account such as this one can help us understand both why Chinese history is important to the world and why we must study China's ecology to understand its history and its future.

GLOSSARY OF CHINESE TERMS, SAYINGS, AND SLOGANS

"Bai hua qi fang; bai jia zheng ming" 百花齐放，百家争鸣 • A hundred flowers all bloom; voices of a hundred schools [of thought] compete

baise wuran 白色污染 • "white pollution" from plastic mulch

baochan dao hu 包产到户 • contracting production to households

Bei Da Huang 北大荒 • Great Northern Wilderness

"Bei zhan, bei huang, wei renmin" 备战，备荒，为人民 • Prepare for war, prepare for famine, for the people

bengbeng che 蹦蹦车 • "putt-putt" diesel riding tractor

"Bian dongtian wei chuntian, zhiyao sixiang budong, di jiu budong" 变冬天为春天，只要思想不冻，地就不冻 • To change winter to spring, if our thought isn't frozen, the land won't be frozen

bieshu 别墅 • "villa," detached single-family house

Bingtuan 兵团 • [Xinjiang] Production and Construction Corps

"Bu pa zuobudao, zhi pa xiangbudao" 不怕做不到，只怕想不到 • Don't be afraid of what you can't do, only be afraid of what you can't imagine

budao weng 不倒翁 • a weighted doll that rolls back upright when pushed over

"Cang liang yu di, cang liang yu ji" 藏粮于地，藏粮于技 • Store staples in soil, store staples in skill

Chang Jiang san xian 长江三鲜 • three savories of the Chang River

chao gang lu 炒钢炉 • "steel stir-frying furnace," puddling furnace

"Chao Ying gan Mei" 超英赶美 • Pass the UK and chase the US

chaoshi 超市 • "supermarket," any kind of indoor store that sells groceries

cheng zhong cun 城中村 • village within the city

chengshi nongmin 城市农民 • "urban peasants"

chiu-ts'ai (jiucai) 韭菜 • garlic chives

chongjian 重建 • reconstruct, reconstruction

"Chu si hai" 除四害 • Eliminate the four pests

chuji nongye hezuoshe 初级农业合作社 • beginning-level agricultural producers' cooperative

"Chun zhong yi pian po, qiu shou yi luokuang" 春种一片坡，秋收一箩筐 • Plant a whole hillside in the spring, harvest one basketful in the fall

da hu 大户 • "big firms," large-scale farmers

da lian gangtie 大炼钢铁 • big iron and steel smelting

da tong xiao yi 大同小异 • big similarity, small difference

da yuan 大院 • "big yard," urban migrant housing compound

danwei 单位 • unit, particularly a work, residential, or work-and-residential unit or its compound digou you 地沟油 • "gutter oil," re-used or unsanitary oil used for cooking

dimo 地膜 • "ground membrane," plastic agricultural mulch

Dixiashui Yundong 地下水运动 • Groundwater Campaign

Dongfeng Qu 东风渠 • East Wind Canal

doufu 豆腐 • tofu

"Duo, kuai, hao, sheng" 多快好省 • More, better, faster, frugal

fan maojin 反冒进 • opposing rashness

fan tai zhi hua 钒钛之花 • flowers of vanadium and titanium

fang gu jie 仿古街 • imitation ancient street

fazhan 发展 • development

"Fazhan cai shi ying daoli" 发展才是硬道理 • Only development is a solid truth

Fenghuang Shan Gu Jianzhu Qun 凤凰山古建筑群 • Phoenix Mountain Historic Structures Complex

fengshui 风水 • feng shui, geomancy

fu jian qu 复建区 • reconstruction district

fudan 负担 • burden

Gaige Kaifang 改革开放 • Reform and Opening

gan laji 干垃圾 • dry trash

gandalei 干打垒 • semisubterranean, mud-walled building

gangxing wending 刚性稳定 • rigid stability

gaoji nongye hezuoshe 高级农业合作社 • higher-level agricultural producers' cooperative

geming 革命 • revolution

Genzhi Huanghe Shuizai he Kaifa Huanghe Shuili de Zonghe Jihua 根治黄河水灾和开发黄河水利的综合计划 • General Plan to Fundamentally Control Yellow River Flood Disasters and Develop Yellow River Waterworks

"Gongnong jiehe, chengxiang jiehe, youli shengchan, fangbian shenghuo" 工农结合, 城乡结合, 有利生产, 方便生活 • Industry and agriculture integrated, city and countryside integrated, conducive to production, convenient for living

"Gongye xue Daqing; nongye xue Dazhai" 工业学大庆, 农业学大寨 • In industry, emulate Daqing; in agriculture, emulate Dazhai

guai yu 怪鱼 • weird fish

guo zhi dazhe 国之大者 • major national issue

guojia shuiwang 国家水网 • national water network

haimian chengshi 海绵城市 • sponge city

hei kouzi 黑口子 • "black holes," illegal mines

hei zuofang 黑作坊 • "black" (illegal or underground) workshop

huajiao 花椒 • Sichuan peppercorn

huanbao 环保 • environmental protection (short form of *huanjing baohu*)

huang ye bing 黄叶病 • yellow-leaf or Panama disease, or *Fusillarium* wilt (of banana trees)

huanjing baohu 环境保护 • environmental protection

huanjing wenhua 环境文化 · environmental culture

hukou 户口 · household registration (system or status)

hutong 胡同 · Beijing residential lane

jia 家 · home, house, family, lineage, clan, school of thought

"Jianku fendou, yongyu kaituo, gu quan daju, wusi fengxian" 艰苦奋斗，勇于开拓，顾全大局，无私奉献 · Bitter struggle, courageous reclamation, attention to the whole situation, selfless contribution

jianshe 建设 · construct, construction

jiashu 家属 · family members, family dependent workers without benefits

jiating zeren zhi 家庭责任制 · household responsibility system

Jiefang 解放 · Liberation

jiejue wenbao wenti 解决温饱问题 · solve the warm and full problem

Jihua Shengyu Zhengce 计划生育政策 · Planned Birth Policy (also called "One-Child Policy")

jin 斤 · half a kilogram

Jing-Kun Gaosu 京昆高速 · Beijing-Kunming Expressway

ju bian 巨变 · big transition

kaifa qu 开发区 · (industrial) development zone

kang 炕 · sleeping platform with heating flues

"Kao shan, fensan, yinbi" 靠山, 分散, 隐蔽 · In the mountains, dispersed, and hidden

kexue 科学 · science

kexue fazhan guan 科学发展观 · scientific perspective on development

"Li tu bu li xiang; jin chang bu jin cheng" 离土不离乡, 进厂不进城 · Leave the fields but not the countryside; enter the factory but not the city

"Liangge quantou—nongye, guofang gongye, yige pigu—jichu gongye" 两个拳头—农业, 国防工业；一个屁股—基础工业 · Two fists—agriculture and defense industry; one butt—basic industry

liangshi 粮食 · starchy staples (often incompletely translated as "grain")

linpan 林盘 · dispersed settlement pattern of the Chengdu Plain

longtou qiye 龙头企业 · "dragon-head enterprises" (domestically owned agribusinesses)

"Lü shui qing shan shi jin shan yin shan" 绿水青山是金山银山 · Green waters and blue mountains are gold and silver mountains

lunzheng zu 论证组 · argumentation panel

lüse 绿色 · "green" (said of food, power, industry, or development that is kinder to the environment)

lüse nengyuan 绿色能源 · green energy

"Mei shi gongye de liangshi" 煤是工业的粮食 · Coal is the staple grain of industry

Minzhu Gaige 民主改革 · "Democratic Reforms" in ethnic minority regions, 1956–58

mu 亩 · unit of land area equivalent to 1/15 hectare or 666.7 square meters

muzhu 牧主 · herdlords, owners of large numbers of livestock

Nan Shui Bei Diao 南水北调 ▪ South-to-North Water Transfer [Project]

"Nanfang shui duo, beifang shui shao; ru you keneng, jie dian shui lai ye shi keyide" 南方水多，北方水少. 如有可能，借点水来也是可以的 ▪ There is a lot of water in the south and not much water in the north; if it's possible, it would be okay to borrow a bit of water

"Ni na'er?" 你哪儿? ▪ Where [what *danwei*] are you [from]?

nong gai chao 农改超 ▪ converting from open markets to "supermarkets" (indoor grocery stores)

nongcun 农村 ▪ farming village, countryside

nongjia 农家 ▪ farmhouse, rural household

nongjia fei 农家肥 ▪ farmhouse fertilizer, usually human and animal manure mixed with other organic wastes

nongken diqu 农垦地区 ▪ agricultural reclamation district

nongmin 农民 ▪ peasants

nongmin gong 农民工 ▪ peasant migrant workers

nongye 农业 ▪ agriculture

Nongye Zouchuqu Zhanlüe Guihua 农业走出去战略规划 ▪ Strategic Plan for Agricultural Going Out (i.e., for agricultural projects in foreign countries)

pian qu 片区 ▪ "slice," urban district

PM$_{2.5}$ lishi shuju 历史数据 ▪ PM$_{2.5}$ historical statistics

"Qing huang bu jie" 青黄不接 ▪ The green not meeting the yellow [Last year's crop doesn't last until this year's crop is in]

Qiongding zhi xia 穹顶之下 ▪ "Under the Dome," 2015 air pollution documentary

"Quebao guwu jiben ziji, kouliang juedui anquan" 确保谷物基本自给, 粮食绝对安全 ▪ Guarantee basic self-sufficiency in grain and absolute staple security

"Ren ding sheng tian" 人定胜天 ▪ Human plans overcome nature

"Ren duo, liliang da" 人多力量大 ▪ More people, greater strength

"Ren you duoda dan, di rengran you xian" 人有多大胆，地仍然有限 ▪ No matter how great the courage of the people, the earth still has its limits

"Ren you duoda dan, di you duoda chan" 人有多大胆，地有多大产 ▪ The fertility of the earth is as great as the courage of the people

"Ren you duoda dan, gang you duoda chan" 人有多大胆，钢有多大产 ▪ The output of steel is as great as the courage of the people

Renmin Qu 人民渠 ▪ People's Canal

"San nian ziran zaihai" 三年自然灾害 ▪ Three years of natural disasters

"Shan, san, dong" 山, 散, 洞 ▪ [In the] mountains, dispersed, [in] caves

"She xiao li, hai da mou; ji jin gong, yi yuan huan" 射小利，害大谋；急近功, 遗远患 ▪ Aim for small advantage and ruin a great plan; fret over short-term success and leave behind long-term disaster

Shehui zhuyi xin nongcun 社会主义新农村 ▪ New socialist countryside

shen fan mi zhi 深翻密植 ▪ deep tilling and dense planting

shen geng 深耕 ▪ deep plowing

shenghuo qu 生活区 ▪ living [residential] district

shengtai wenming 生态文明 ▪ ecological civilization

shi laji 湿垃圾 ▪ wet trash, garbage

"Shishi qiu shi" 实事求是 ▪ Seek truth from facts

"Shou zhong you liang, xin zhong bu huang" 手中有粮, 心中不慌 ▪ Grain in the hand, no worry in the mind

shui huo zhi zheng 水火之争 ▪ struggle between water and fire (hydropower and thermal power)

si da jia yu 四大家鱼 ▪ four great domestic fish

"Sixiang you liang, jiu you liang; sixiang you gang, jiu you gang" 思想有粮，就有粮，思想有钢，就有钢 ▪ If there is grain in ideology, there will be grain; if there is steel in ideology, there will be steel

suzhi 素质 ▪ moral and intellectual quality [of a person or population]

taiji 太极 ▪ tai-chi

Tian Fu 天府 ▪ Heavenly Precinct

"Tianqi xianzai dou suizhe zhengce zou" 天气现在都随着政策走 ▪ Weather now always follows policy

Tianran Lin Baohu Gongcheng 天然林保护工程 ▪ Natural Forest Protection Project

ting 厅 ▪ main room or guest room

tu 土 ▪ soil, land, local, native, folk

tu yang bing ju 土洋并举 ▪ using the native (or "folk") and the foreign together

Tui Geng Huan Lin 退耕还林 ▪ Returning Farmland to Forest ("Sloping Land Conversion Program" or "Grain for Green") program

Tui Mu Huan Cao 退牧还草 ▪ Returning Grassland to Pasture (program)

"Waihang yi xue jiu hui; dangtian jian lu chu gang" 外行一学就会, 当天建炉出钢 ▪ Laypeople can do it as soon as they learn; the day you build a furnace, it will produce steel

wan, xi, shao 晚稀少 ▪ late, spaced, and few (births) population control policy

wei hai zao tian 围海造田 ▪ enclose the lake and create rice fields

wenbao 温饱 ▪ warm and full

wu hai chuli 无害处理 ▪ harmless treatment

xiangcun qiye 乡村企业 ▪ township and village enterprises (TVEs)

xiao, tu, qun 小土群 ▪ small, local, [of the] masses

xiaokang 小康 ▪ moderate prosperity

"Xiaomie danji dao" 消灭单季稻 ▪ Wipe out single-cropped rice

xiaoqu 小区 ▪ "little district," contemporary gated housing community

Xibu Da Kaifa 西部大开发 ▪ Great Western Development

xin liangshi anquan guan 新粮食安全关 ▪ new staple security vision

Xin Nongcun Jianshe 新农村建设 ▪ New Rural Reconstruction

Xingxiu Nongtian Shuili Jianshe Yundong 兴修农田水利建设运动 ▪ Campaign to Build Agricultural Water Conservancy xunhuan jingji 循环经济 ▪ circular economy

Xunhuan Jingji Cujin Fa 循环经济促进法 ▪ Circular Economy Promotion Law

yang 洋 ▪ foreign, cosmopolitan, modern (literally, "ocean")

yanzhong wuran 严重污染 ▪ severe pollution

yi dao qie 一刀切 ▪ "cutting with one knife," treating all cases uniformly

yi gong dai zhang 以工代账 ▪ using labor in place of funds

yi guo liang zhi 一国两制 ▪ one country, two systems

yi jia liang zhi 一家两制 ▪ one family, two systems

"Yi liang wei gang, qiyu sao guang" 以粮为纲，其余扫光 ▪ Emphasize staple production; sweep away everything else

"Yi liang wei gang, quanmian fazhan" 以粮为纲，全面发展 ▪ Emphasize staple production in overall development

yi piao foujue 一票否决 ▪ "one ticket denial," a requirement or quota that can block a cadre's promotion if not fulfilled

yi tie lian tie 以铁炼铁 ▪ smelting iron from iron

"Yige bu shao, liangge zhenghao, sange duoliao" 一个不少，两个正好，三个多了 ▪ One isn't too few, two are just right, three are too many

"Yin Han ji Wei" 引汉济渭 ▪ Taking [water from] the Han to fill the Wei

"Yin Jiang bu Han" 引江补汉 ▪ Taking [water from] the [Chang] Jiang to replenish the Han

"Yin Jiang ji Han" 引江济汉 ▪ Taking [water from] the [Chang] Jiang to fill the Han

you dian dao mian 由点到面 ▪ from the point to the field (from a small-scale experiment to wide-scale implementation)

you hong you zhuan 又红又专 ▪ both red and expert

youtiao 油条 ▪ dough fritters (a common breakfast food)

yuan 元 ▪ unit of Chinese currency, also called *renminbi* 人民币 or "people's currency" (2022 exchange value around seven to the US dollar)

yuejin 跃进 ▪ leap forward

zhengji 政绩 ▪ administrative accomplishments

zhidao yijian 指导意见 ▪ guiding opinion

zhongdu wuran 重度污染 ▪ heavy pollution

Zhongguo Meng 中国梦 ▪ China Dream

zhuanye hu 专业户 ▪ specialized households

ziliu di 自留地 ▪ land retained for themselves, private plot

zou chuqu 走出去 ▪ going out

NOTES

INTRODUCTION

1. The Chinese word for the Revolution (Geming) refers not to the change of government in 1949 but to the whole process of transformation that followed the change of regime. The change of regime itself is called Liberation (Jiefang). However, because it wasn't really a liberation, I use the neutral term Founding (Jianguo), which is also becoming common in the People's Republic.
2. Standard works by unofficial or foreign writers include those by Becker (1996), Yang J. (2008), and Dikötter (2011). For official histories, see People's Education Press (2006, chaps. 6 and 7, 98–113; 2019, 180). I am grateful to Qumo Asa for letting me have her used textbooks and to Bamo Ayi for sending along her daughter's hand-me-downs.
3. Notable exceptions to this continuation of the previous borders were the Republic's claims of sovereignty over Mongolia and Taiwan, neither of which it actually controlled until it took over Taiwan in 1945, but then it lost China four years later.
4. Thanks to Ed Grumbine, who challenged me to show how China was an ecosystem, since it is not ecologically uniform and its borders are not ecologically determined.
5. China's government also adapted a de facto policy of total fire suppression after 1950. See the article by Hayes (2021).

1. A TOUR OF CHINA'S SOCIAL-ECOLOGICAL SYSTEMS

1. Much of Manchuria (the northeastern provinces of Heilongjiang, Jilin, and Liaoning) was excluded from the intensive agro-ecology of Chinese agrarian civilization for political or military reasons until late in the Qing period. Ecologically, however, Manchuria shares the climatic and topographic features of China Proper, so I include it in that zone.
2. Ecologically, China Proper includes flat parts of Taiwan. This implies nothing about Taiwan's political status. During the Ming and Qing, relatively flat areas within Zomia region were also converted to this kind of agro-ecology, so I use some examples from there.
3. In much of Sichuan, there were no villages at all; farmhouses were scattered throughout the fields (Skinner 2017).
4. Ethnographic descriptions are provided in works by Hsieh (1995), Sturgeon (2005), and Hathaway (2013).

2. DEVELOPMENT, REVOLUTION, AND SCIENCE

1. This "economic miracle" has been analyzed in works by Amsden (1979), S. Ho (1978), and Gold (1986).

2. The noncommunist leftist thinkers espousing such views as dependency theory include Frank (1969), Leys (1977), and Lippit (1978).

3. I disagree with the common idea that the Chinese party-state pursued revolution under Mao (Zweig 1989, 192, quoted in Schmalzer 2016, 26) and then switched to development with Deng Xiaoping's revisionist economic reforms beginning in the late 1970s. What changed was the *means* to development, from class struggle to capitalist investment, not the *goal* itself, which had always been development (Schwartz 1973).

4. Ecological Marxists, including some in China (Schmitt 2016, 82–83), have found passages in Marx and Engels, notably in Marx's idea of the "metabolic rift" (J. Foster 1999; Schneider 2015), to support their proposition that capitalists are compelled to use natural resources as economically and efficiently as possible and thus deplete the material basis of their own economic existence (Mingione and Spence 1993). But Marxist-Leninist *regimes* have paid almost no attention to this critique.

5. The story that Marx offered to dedicate volume 2 of *Capital* to Darwin turns out, however, to have been based on a misunderstanding (see Carter n.d.).

6. The term "immoral economy" is derived from Scott's (1976) "moral economy." Esherick (1995, 62, 68–70) and Bianco (1971, 87–107) explain the appeal of Communist revolution in similar terms.

7. Shouzhang County 寿张县 lost its administrative existence in 1964, when the southern part was transferred to Taiqian County, Henan, and the northern part to Yanggu County, Shandong (Weiji Baike 2017).

8. Useful accounts of the famine can be found in works by Dikötter (2011), Yang Jisheng (2008), and Bramall (2011).

3. FEEDING A STARVING NATION, 1949–1957

1. Relevant long-term narratives of the Qing empire can be found in works by Elvin (1998, 2004), Marks (2012), and Ling (1983).

2. Accounts of specific areas of Zomia can be found in works by Giersch (2006), Herman (1993, 2007), Weinstein (2013), and Whittaker (2008).

3. Accounts of historic population growth can be found in works by Harrell (1995a, 6–16), J. Lee and Wang (1999, 27–28), and Lavely and Wong (1998), but for a revisionist view see the article by K. G. Deng (2004).

4. Regarding carbohydrate staples, the term *liangshi* is customarily translated into English as "grain." However, Chinese *liangshi* statistics have always included sweet potatoes and potatoes, so accurate translation is a problem. I use "staples" because staples, or foods that form the foundation of peasant diets in China, are nearly always carbohydrate-rich foods, either grains or tubers. See the article by Field and Kilpatrick (1978, 369–70).

5. It is almost impossible to find accurate figures for China's cultivated land area before the 1990s. The regime systematically and intentionally underreported cultivated land data before 1995, and reconstructions by indirect methods have large margins of error. See the relevant article by Smil (1999b).

6. Detailed analyses of this complex and often confusing process can be found in works by Shue (1980, 41–96), P. Huang (1990, 165–95), Bramall (2000), and Kung, Wu, and Wu (2012).

7. The formula for how many work points a laborer of a particular age and sex was allocated varied from place to place, even within counties, as did the ratio between in-kind and in-cash payments, as well as the share that was paid per capita and the share that was paid according to work points (Harrell et al. 2011, 30–32).

8. Numbers vary considerably; official statistics cited by Smil (1993) and the Nongye Bu (2009) place the 1957 total at about four million hectares lower than Perkins's (1973) figures. This amounts to about 3 percent of China's total agricultural lands. Nickum (1998, 884) suggests that this discrepancy may be at least partly due to different reporting systems of the Ministry of Agriculture or the State Statistical Bureau versus those of the Ministry of Water Resources. However, the overall trend is clear—irrigated area expanded by about half between the Founding and the start of the Great Leap Forward.

9. The Wei River 卫河 is a tributary of the Hai River and runs parallel to the "hanging" Yellow River to its north. It is not to be confused with the larger and better-known Wei River渭河, a major tributary that flows into the Yellow River before it enters the North China Plain.

10. Official figures (Nongye Bu 2009, 14) show an increase from 288 to 308 kilograms per capita (7 percent increase), whereas Kueh (2006, 705) gives the figure of 220 to 256 kilograms (16 percent increase).

4. THREE YEARS OF ~~NATURAL~~ DISASTERS, 1958–1961

1. The Chinese Communist Party–led regime in the autonomous region of Inner Mongolia was first called the Inner Mongolia Autonomous Government; it was changed to the Inner Mongolia Autonomous Region in December 1949, after the establishment of the People's Republic of China.

2. I ran the numbers. If the 25 billion cubic meters is accurate, the road would indeed stretch to the moon. To move this much dirt, a hundred million people would have had to move 250 cubic meters each, which would be feasible but require dogged exertion, at the rate of 3–5 cubic meters of earth per day.

3. Local examples are presented by Wu Q. (1998, 139); E. Friedman, Pickowicz, and Selden (1990, 217); and Yang J. (2008, 88).

4. *Shen geng* and *shen fan* are conventionally translated as "deep plowing." However, local accounts show that a lot of this "deep turning over" was done with shovels and hoes as well as plows, so "deep tilling" seems a better translation.

5. Standard accounts include those by Becker (1996), Smil (1999), Yang J. (2008), and Dikötter (2011).

6. Mgebbu Vihly, Mgebbu Ashy, and Hxiesse Vuga, interviewed by the author in June 2001.

7. These memories were collected by Dr. He Wenhai.

5. NORMAL SOCIALIST AGRICULTURE, 1962–1978

1. The potentiometric surface is the level at which water in a confined aquifer would find its equilibrium in the absence of confinement, since withdrawal from a confined space decreases water pressure.

2. There is a common misconception among political activists opposed to large-scale, commercial agriculture that the Green Revolution arose from the development of genetically modified organisms (GMOs). GMOs are a separate, later development, after the bump in food production (and many of the social ills) brought about by the Green Revolution. Both may represent a particular scientific mind-set of technology over ecology, but they are different phenomena.

3. See the article by H. Bian (2022) for a critique of this narrative.

6. SOLVING THE WARM AND FULL PROBLEM, 1978–1998

1. It is difficult to use the term "rocket scientist" in a nonsardonic way these days. However, Song Jian, chair of the commission that recommended the Planned Birth Policy, was an actual rocket scientist (Greenhalgh 2003). The Planned Birth Policy is commonly and erroneously known as the One-Child Policy.

2. After decollectivization, the production teams (*shengchan dui*) were renamed "villager small groups" (*cunmin xiaozu*). However, in the late 2010s, thirty-five years later, local people still called them "teams."

7. EVERY LAST DROP, 1998–2022

1. There was, however, evidence of long-term accumulation of sediment in some places, especially around Chongqing (Zong and Chen 2000, 176).

2. In a word-for-word translation, Tui Geng Huan Lin means "retreat plowing return forest." The program name has been translated in a bewildering variety of ways, most prominently as either "Sloping Land Conversion Program" or "Grain for Green." "Returning Farmland to Forest" states clearly what the program is about.

8. DAMMED IF YOU DO, 1993–2021

1. How much of this climate variability is cyclical on a decadal-or-so scale and how much is linear climate change is very difficult to determine (David Shankman, pers. comm.). See also works by X. Lai et al. (2014); Mei et al. (2015); and R. Wan, Die, and Shankman (2018).

2. Investigators do not attribute either the presumed extinction of the *baiji* (Yangtze River dolphin) or the endangerment of the Yangtze finless porpoise directly to the effects of dam building. See the article by Turley et al. (2007).

9. A TOXIC CORNUCOPIA, 2000–2022

1. Data come from surveys done in 2002–3 by Ross Nadal and Ma Fagen, as well as a repeat study done by John Chaffee and He Wenhai in 2015.

2. Pi County was reclassified administratively as urban, becoming Pidu District in 2016.

3. In the English edition of the 2019 white paper, the terms *liang* (staples), *liangshi* (staples), *gu* (grain), *guwu* (grain), *liangshi anquan* (staple security), and *guwu ziji* (grain self-sufficiency) are all translated as "food," "food security," "food self-suffi-

ciency," etc. Translators may have chosen these terms to harmonize with the international discourse on food security.

4. There are thousands upon thousands of such articles (and even more in Chinese, often with the same content), making it impossible and pointless to cite any not directly used in my analysis.

10. BIG AG AND ITS ECOSYSTEM EFFECTS, 2002–2022

1. This saying is a satirical reference to China's touted governance strategy for Hong Kong and Macao: "one country, two systems" (*yi guo liang zhi*).

2. I avoid the term "wet market," which apparently originated with European and American expatriates in China and appears not to be a translation of any term in Chinese. The Chinese term *chaoshi*, a direct calque of the English "supermarket," refers to any food or grocery store where customers pick items directly from shelves rather than being served by market personnel, as in traditional food markets. The term includes very small establishments that would be called convenience stores in the English-speaking world.

11. A REVOLUTION OF STEEL, 1949–1961

1. Josef Vissarionovich Djugazhvili took the nom de guerre "Stalin" from the Russian word *stal*, meaning steel.

2. The name Angang is composed of *an*, for Anshan, and *gang*, for steel. Similarly, the Baotou complex is called Baogang, and the Capital Steel complex in Beijing is called Shougang (*shou* is short for *shoudu*, or capital).

3. The healthy worker effect refers to the fact that even highly exposed workers often have lower mortality than the general population, which includes many people who are not healthy enough to work, and if studies are not controlled for age (which this study was not), the general population may include a lot of very old people who will have high mortality just because of age. Thus, the proper reference group for exposed blue-collar workers is nonexposed blue-collar workers, who differ from exposed workers only in exposure, not in age, baseline health, or type of work.

4. In Chinese, a puddling furnace is known as *chao gang lu*, or "steel-stir-frying furnace," which aptly describes the process.

5. These "furnaces of the masses" are usually called "backyard steel furnaces" in English, but they were typically centralized in communes or brigades, not in anyone's own domicile, since most Chinese houses don't have backyards.

6. The wearing of red scarves by elementary school students signified membership in the Young Pioneers, the first step toward joining the Communist Youth League and eventually the Communist Party.

7. In fact, women's participation in "productive" work brought them lower salaries, fewer benefits, and less prestige than their male counterparts received (see Y. Tian 2019; and S. Zhou, forthcoming).

12. NORMAL SOCIALIST INDUSTRY, 1962–1980

1. The usual translations are "learn from Daqing" and "learn from Dazhai," but *xue* 学 in this case clearly means "emulate" or "copy" or "do as *x* does."
2. Unreferenced observations about Panzhihua come from personal observations, photographs, and field notes from my three-month sojourn (January through March 1988) as the first foreign resident there after it was opened to visitors in 1987.
3. Casualty estimates vary widely, from the predictably low official estimate of 242,000 to a maximum of 655,000 reported by the *South China Morning Post* on the basis of a reputed "top secret" state document (Liu Huixian et al. 2002, 17). Even 242,000 is an enormous loss of life.

13. FACTORY TO THE WORLD, 1984–2015

No notes.

14. BUILDING AN URBAN CONTINENT, 1980–2022

1. Southwest Minzu University was formerly known as Southwest University for Nationalities and before that, Southwest Nationalities Institute.
2. By way of comparison, in 2020 the New York subway system was slightly larger, with 1,070 kilometers of tracks serving 472 stations, but its daily ridership was 5.5 million, about half of Beijing's (MTA 2020). Other very large cities such as Seoul, Tokyo, Shanghai, Moscow, and London have systems of comparable size.
3. The Beijing-Guangzhou line is about one hundred kilometers shorter than the rail route from New York City to Dallas, Texas, which takes over forty-seven hours to travel.

15. SLUDGE ABIDES, 1990–2022

1. "Rainwater stagnation" in this context does not mean letting rainwater stagnate into pools harboring amoebae and other parasites but rather allowing water to seep into the ground slowly instead of running off rapidly.
2. Because part of the increase in urban consumption expenditures went to services, the twenty-six-fold figure is certainly high for material resources, but for purposes of understanding the effect of urbanization on resource consumption, this exercise is useful.
3. The residential usage proportion is certainly an underestimate, as it omits much of the water used by rural households in water-rich southern farming areas. Those rural residents draw water from household wells or nearby streams, so that usage probably never enters statisticians' calculations, although in 1990 an estimated one-fourth of the rural population had access to tap water (Smil 1993, 48).

16. PARADOXES OF ECO-DEVELOPMENT, 1998–2022

1. In 2021 China announced a commitment to develop hydrogen-based energy, one of six "key industries" for focused development, although it was still a minuscule factor in total energy production. How "green" China's hydrogen will turn out to be is unclear at the time of this writing (Collins 2021a, b).

2. Twitter was not blocked in China until a year later, in 2009.

3. China measures all air pollutants in micrograms per cubic meter ($\mu g/m^3$), while the United States measures them in parts per billion (ppb). The ratio of these two measures depends on the pollutant and the air temperature; at a room temperature of 20 degrees C, two $\mu g/m^3$ of ozone very conveniently equals one part per billion.

CONCLUSION

1. There is some doubt that the attempt to ban Bitcoin mining has been effective; see the assessment by CDAP (2022).

2. Bruno Latour (2002), Roberto González (2001), H. Holden Thorp (2020), and Audra Wolf (2020) each provide a slightly different take on the relationship between science and politics.

REFERENCES

Abel, Thomas. 2007. "World-Systems as Complex Human Ecosystems." In *The World System and the Earth System*, edited by Alf Hornborg and Carole Crumley, 56–73. Walnut Creek, CA: Left Coast Press.

Abramson, Daniel Benjamin. 2007. "The Aesthetics of City-Scale Preservation Policy in Beijing." *Planning Perspectives* 22:129–66.

———. 2008. "Haussmann and Le Corbusier in China: Land Control and the Design of Streets in Urban Redevelopment." *Journal of Urban Design* 13 (2): 231–56.

Abrosio, Martín de. 2019. "Stark Images Show Soy-Linked Deforestation in Argentina." In *Tracking China's Soy and Beef Imprint on South America*, edited by Isabel Hilton, 34–36. Diálogo Chino, 5 December 2019. https://dialogochino.net/32059-tracking-chinas-soy-and-beef-imprint-on-latin-america/.

Adams, Donald F. 1984. "Air Quality Technical Information Exchange to the People's Republic of China." *Journal of the Air Pollution Control Association* 34 (7): 726–28.

Ahlers, Anna L., Mette Halskov Hansen, and Rune Svarverud. 2020. *The Great Smog of China: A Short Event History of Air Pollution.* Ann Arbor, MI: Association for Asian Studies.

Ai Sixiang 爱思想. 2011. "Sanxia gongcheng lunzheng 20 nian shanhui 9 wei zhuanjia jujue qianzi" 三峡工程论证20年闪回9位专家拒绝签字 [Shedding light after 20 years on the Sanxia project debates: 9 experts did not sign their names]. 25 May 2011. www.aisixiang.com/data/40893.html.

Alavi, Heshmat. 2020. "As Criticism Grows against China." Twitter, 9 April 2020. https://twitter.com/HeshmatAlavi/status/1248267738591776769.

Altieri, Miguel. 1995. *Agroecology: The Science of Sustainable Agriculture.* 2nd ed. Boulder, CO: Westview Press.

Amsden, Alice H. 1979. "Taiwan's Economic History: Étatisme and a Challenge to Dependency Theory." *Modern China* 5 (3): 341–80.

Anderson, E. N., Jr. 1990. *The Food of China.* New Haven: Yale University Press.

Anderson, Michael. 2017. "Genealogical Analysis of Discourse on Ethnic Minority Protests and Its Manifestation and Reinforcement in News Media and State-Sponsored Art." PhD diss., University of Washington, Seattle.

Andreoni, Manuela. 2019. "US-China Trade War Raises Fears of Deforestation in Brazil." In *Tracking China's Soy and Beef Imprint on South America*, edited by Isabel Hilton, 30–33. Diálogo Chino, 5 December 2019. https://dialogochino.net/32059-tracking-chinas-soy-and-beef-imprint-on-latin-america/.

Andrews-Speed, Philip, Minying Yang, Lei Shen, and Shelley Cao. 2003. "The Regulation of China's Township and Village Coal Mines: A Study of Complexity and Ineffectiveness." *Journal of Cleaner Production* 11:185–96.

Angel, Schlomo. 2012. *Planet of Cities.* Cambridge, MA: Lincoln Institute of Land Policy.

Apinya Wipatayotin. 2019. "Dam Tests Spark Crisis." *Bangkok Post*, 20 July 2019.

Areddy, James T. 2021. "China Reconsiders Its Central Role in Bitcoin Mining." *Wall Street Journal*, 5 June 2021.

Atiles-Osoria, José M. 2014. "Environmental Colonialism, Criminalization and Resistance: Puerto Rican Mobilizations for Environmental Justice in the 21st Century." *Revista Crítica de Ciências Sociais* 6:3–21.

Backus, Charles. 1981. *The Nan-chao Kingdom and T'ang China's Southwestern Frontier.* New York: Cambridge University Press.

Baidu 百度. 2014. "Mingdai 'qianyan qiao' chong chu shui mian, gudai jianzhu qiyi huanfa shengji!" 明代"千眼桥"重出水面，古代建筑奇迹焕发生机! [Ming dynasty "Thousand Eyes Bridge" surfaces again, a curious relic of ancient construction comes back to life!]. 18 January 2014. https://baijiahao.baidu.com/s?id=15895525349 78663612&wfr=spider&for=pc.

Baidu Encyclopedia 百度百科. 2019. "Bei Da Huang" 北大荒. [Great Northern Wilderness]. https://baike.baidu.com/item/北大荒/72832.

———. 2020a. "Jiuzhai-Huanglong Jichang" 九寨黄龙机场 [Jiuzhai-Huanglong Airport]. https://baike.baidu.com/item/九寨黄龙机场/12521782.

———. 2020b. "Panzhihua Bao'anying Jichang" 攀枝花保安营机场 [Panzhihua Bao'anying Airport]. https://baike.baidu.com/item/攀枝花保安营机场/8190025?.

———. 2021. "Zhonghua Renmin Gongheguo Shengtai Huanjing Bu" 中华人民共和国生态环境部 [PRC Ministry of Ecology and Environment]. https://baike.baidu.com/item/中华人民共和国生态环境部/22428850#5.

———. n.d. a. "Danjiankou Shuiku" 丹江口水库 [Danjiangkou Reservoir]. Accessed December 2022. https://baike.baidu.com/item/丹江口水库.

———. n.d. b. "Yinhan Ji Wei Gongcheng" 引汉济渭工程 [Taking the Han to Fill the Wei Project]. Accessed July 2022. https://baike.baidu.com/item/引汉济渭工程.

———. n.d. c. "Qiandao Hu Yinshui Gongcheng" 千岛湖引水工程 [Qiandao Lake Water Transfer Project]. Accessed December 2022. https://baike.baidu.com/item/千岛湖引水工程/10935578.

———. n.d. d. "Wuhan Gangtie (Jituan) Gongsi" 武汉钢铁 (集团) 公司 [Wu Han Iron and Steel Company (Group)]. Accessed December 2022. https://baike.baidu.com/item/武汉钢铁（集团）公司/727864.

———. n.d. e. "Beijing Ditie" 北京地铁 [Beijing Subway]. Accessed October 2020 https://baike.baidu.com/item/北京地铁/408485.

———. n.d. f. "Yin Jiang ji Han Gongcheng" 引江济汉工程 [Taking the Yangtze to Fill the Han Project]. Accessed December 2022. https://baike.baidu.com/item/引江济汉工程.

———. n.d. g. "Yin Jiang bu Han Gongcheng" 引江补汉工程 [Taking the Yangtze to Replenish the Han Project]. Accessed December 2022. https://baike.baidu.com/item/引江补汉工程/60058999.

Baker, Hugh D. R. 1968. *Sheung Shui: A Chinese Lineage Village.* Stanford, CA: Stanford University Press.

Bambaradeniya, C. M. B., J. P. Edirisinghe, D. N. De Silva, C. V. S. Gunatilleke, K. B. Ranawana, and S. Wijekoon. 2004. "Biodiversity Associated with an Irrigated Rice Agro-ecosystem in Sri Lanka." *Biodiversity and Conservation* 13 (9): 1715–53.

Banfill, Kaitlin. 2021. "Clans and Classmates: Kinship, Migration, and Education in Southwest China." PhD diss., Emory University.

Banister, Judith. 1987. *China's Changing Population.* Stanford, CA: Stanford University Press.

Bao, Chao, and Chuang-lin Fang. 2012. "Water Resources Flows Related to Urbanization in China: Challenges and Perspectives for Water Management and Urban Development." *Water Resources Management* 26:531–52.

Bao Ge 鲍戈. 2002. "Yi jiu jiu ba nian Zhongguo hongzai zhenxiang" 一九九八年中国洪灾真相 [A true picture of the 1998 flood disaster in China]. *Zhongguo Baodao Zhoukan* 中国报道周刊, 17 May 2002.

Bao Maohong. 2004. "Environmental History in China." *Environment and History* 10 (4): 475–99.

Baotou. 2015. "Photos Document Baotou's History." BaotouChina. Accessed July 2022. www.chinadaily.com.cn/m/innermongolia/baotou/2017-06/14/content_29738645_2.htm.

Baptista, Eduardo, Wendy Wu, Guo Rui, Cissy Zhou, and Mimi Lao. 2021. "China Floods: Zhengzhou Tries to Get Back on Its Feet after Heavy Rains Displace over 1.2 Million People." *South China Morning Post*, 21 July 2021.

Barboza, David. 2012. "China to Release More Data on Air Pollution in Beijing." *New York Times*, 6 January 2012.

Barclay, Kate, Michael Fabinyi, Jeff Kinch, and Simon Foale. 2019. "Governability of High-Value Fisheries in Low-Income Contexts: A Case Study of the Sea Cucumber Fishery in Papua New Guinea." *Human Ecology* 47:381–96.

Barfield, Thomas. 1989. *The Perilous Frontier: Nomadic Empires and China*. Cambridge, MA: Blackwell.

Barnosky, Anthony D., Elizabeth A. Hadly, Jordi Bascompte, Eric L. Berlow, James H. Brown, Makael Fortelius, Wayne M. Getz, et al. 2012. "Approaching a State Shift in Earth's Biosphere." *Nature* 486:52–58.

Baum, Richard D. 1964. "'Red and Expert': The Politico-ideological Foundations of China's Great Leap Forward." *Asian Survey* 4 (9): 1048–57.

BBC Chinese. 2020. "Zhongguo hongshui: Zhishao 141 ren siwang huo shizong, Poyang Hu shuiwei zhi lishi zui gao" 中国洪水：至少141人死亡或失踪，鄱阳湖水位至历史最高 [China floods: At least 141 people dead or missing; Poyang Lake water level reaches historic high]. 13 July 2020. www.bbc.com/zhongwen/simp/chinese-news-53388985.

Becker, Jasper. 1996. *Hungry Ghosts: Mao's Secret Famine*. New York: Henry Holt.

Bellér-Hann, Ildikó. 2013. "The Bulldozer State: Chinese Socialist Development in Xinjiang." In *Ethnographies of the State in Central Asia: Performing Politics*, edited by Madeleine Reeves, Johan Rasanayagam, and Judith Beyer, 173–97. Bloomington: Indiana University Press.

Bennett, E. M., G. S. Cumming, and G. D. Peterson. 2005. "A Systems Model Approach to Determining Resilience Surrogates for Case Studies." *Ecosystems* 8:945–57.

Bentley, Jeanine. 2014. "Trends in U.S. Per Capita Consumption of Dairy Products, 1970–2012." https://ageconsearch.umn.edu/record/210933?ln=en.

Berg, Peter, and Oskar Lingqvist. 2019. "Pulp, Paper, and Packaging in the Next Decade: Transformational Change." McKinsey & Company. www.mckinsey.com/~/media/McKinsey/Industries/Paper%20and%20Forest%20Products/Our%20Insights/Pulp%20paper%20and%20packaging%20in%20the%20next%20decade%20Transformational%20change/Pulp-paper-and-packaging-in-the-next-decade-Transformational-change-2019-vF.pdf.

Berkes, Fikret. 1999. *Sacred Ecology: Traditional Ecological Knowledge and Resource Management*. Philadelphia: Taylor and Francis.

Berkoff, Jeremy. 2003. "China: The South-North Water Transfer Project—Is It Justified?" *Water Policy* 5 (1): 1–28.

Bi, Naishuang, Zhongqiang Sun, Houjie Wang, Xiao Wu, Yongyong Fan, Congliang Xu, and Zuosheng Yang. 2019. "Response of Channel Scouring and Deposition to the Regulation of Large Reservoirs: A Case Study of the Lower Reaches of the Yellow River (Huanghe)." *Journal of Hydrology* 568:972–84.

Bian, He. 2022. "Science and Really Existing Socialism in Maoist China: A Review of Recent Works." *Historical Studies in the Natural Sciences* 52 (2): 265–75.

Bian, Yanjie, John R. Logan, Hanlong Lu, Yunkang Pan, and Ying Guan. 1997. "Work Units and Housing Reform in Two Chinese Cities." In *Danwei: The Changing Chinese Workplace in Historical and Comparative Perspective*, edited by Xiaobo Lü and Elizabeth Perry, 223–50. Armonk, NY: M. E. Sharpe.

Bianco, Lucien. 1971. *Origins of the Chinese Revolution: 1915–1949*. Stanford, CA: Stanford University Press.

BizVibe. n.d. Global Textile Industry Factsheet 2020. Accessed February 2021. https://blog.bizvibe.com/blog/top-10-largest-textile-producing-countries.

BJNews. 2022. "Henan Ruzhou: Gongyuan chou hushui bu 'guai yu'" 河南汝州—公园抽湖水捕 "怪鱼" [Ruzhou, Henan: Park drains lake water to catch "weird fish"]. Twitter, 25 August 2022. https://twitter.com/BJNewsOfficial/status/1562666080853688321. Also see *Beijing News* video, 25 August 2022, www.youtube.com/watch?v=SEYFq1 GQ9tc.

Bjorklund, E. M. 1986. "The *Danwei*: Socio-Spatial Characteristics of Work Units in China's Urban Society." *Economic Geography* 62 (1): 19–29.

Blaikie, Piers, and Harold Brookfield. 1987. *Land Degradation and Society*. London: Routledge.

Blecher, Mark. 1979. "Consensual Politics in Rural Chinese Communities: The Mass Line in Theory and Practice." *Modern China* 5 (1): 105–25.

Blumenfield, Tami. 2010. "Scenes from Yongning: Media Creation in China's Na Villages." PhD diss., University of Washington, Seattle.

Boland, Alana, 1998. "The Three Gorges Debate and Scientific Decision-Making in China." *China Information* 13 (1): 25–42.

Borthwick, Alastair. 2005. "Is the Lower Yellow River Sustainable?" *SOUE News*, issue 4. www.soue.org.uk/souenews/issue4/yellowriver.html.

Boserup, Ester. 1965. *The Conditions of Agricultural Growth: The Economics of Agrarian Change under Population Pressure*. London: George Allen and Unwin.

Bradsher, Keith. 2012. "China Asks Other Nations Not to Release Its Air Data." *New York Times*, 5 June 2012.

Bradsher, Keith, and Ailin Tang. 2019. "China Responds Slowly, and a Pig Disease Becomes a Lethal Epidemic." *New York Times*, 17 December 2019.

Bramall, Chris. 1995. "Origins of the Agricultural 'Miracle': Some Evidence from Sichuan." *China Quarterly* 143:731–55.

———. 2000. "Inequality, Land Reform and Agricultural Growth in China, 1952–55: A Preliminary Treatment." *Journal of Peasant Studies* 27 (3): 30–54.

———. 2007. *The Industrialization of Rural China*. Oxford: Oxford University Press.

———. 2011. "Agency and Famine in China's Sichuan Province, 1958–1962." *China Quarterly* 208:990–1008.

Brandt, Jodi S., Tobias Kuemmerle, Haomin Li, Guopeng Ren, Jianguo Zhu, and Volker C. Radeloff. 2012. "Using Landsat Imagery to Map Forest Change in Southwest China in Response to the National Logging Ban and Ecotourism Development." *Remote Sensing of Environment* 121:359–69.

Brazelton, Mary Augusta. 2019. *Mass Vaccination: Citizens' Bodies and State Power in Modern China*. Ithaca: Cornell University Press.

Brown, Lester. 1995. *Who Will Feed China: A Wake-Up Call for a Small Planet*. New York: Norton.

Brown, Melissa J. 2017. "Dutiful Help: Masking Rural Women's Economic Contributions." In *Transforming Patriarchy: Chinese Families in the Twenty-First Century*, edited by Gonçalo Santos and Stevan Harrell, 39–58. Seattle: University of Washington Press.

Buck, John Lossing. 1937. *Land Utilization in China*. Chicago: University of Chicago Press.

Bulag, Uradyn E. 2002. *The Mongols at China's Edge: History and the Politics of National Unity*. Boulder, CO: Rowman and Littlefield.

Burnham, James, Jeb Barzen, Anna M. Pidgeon, Baoteng Sun, Jiandong Wu, Guanhua Liu, and Hongxing Hang. 2017. "Novel Foraging by Wintering Siberian Cranes *Leucogeranus leucogeranus* at China's Poyang Lake Indicates Broader Changes in the Ecosystem and Raises New Challenges for a Critically Endangered Species." *Bird Conservation International* 27 (2): 204–23.

Byler, Darren T. 2018. "Spirit Breaking: Uyghur Dispossession, Culture Work, and Terror Capitalism in a Chinese Global City." PhD diss., University of Washington, Seattle.

Cabell, Joshua F., and Myles Oelofse. 2012. "An Indicator Framework for Assessing Agroecosystem Resilience." *Ecology and Society* 17 (1): 18. http://dx.doi.org/10.5751/ES-04666-170118.

Cai, Hua. 2001. *A Society without Fathers or Husbands: The Na of China*. Cambridge, MA: MIT Press.

Cai, Lingen. 1988. "Efficient Conjunctive Use of Surface and Groundwater in the People's Victory Canal." In *Efficiency in Irrigation: The Conjunctive Use of Surface and Groundwater Resources*, edited by Gerald T. O'Meara, 84–87. Washington, DC: World Bank.

Cao, Shixiong. 2008. "Why Large-Scale Afforestation Efforts in China Have Failed to Solve the Desertification Problem." *Environmental Science and Technology* 15 (March): 1826–31.

Cao, Shixiong, Xiuqing Wang, Yuezhen Song, Li Chen, and Qi Feng. 2010. "Impacts of the Natural Forest Conservation Program on the Livelihoods of Residents in Northwestern China: Perceptions of Residents Affected by the Program." *Ecological Economics* 69 (7): 1454–62.

Cao, Shuyou, Xingnian Liu, and Huang Er. 2010. "Dujiangyan Irrigation System: A World Cultural Heritage Corresponding to Concepts of Modern Hydraulic Science." *Journal of Hydro-Environmental Research* 4 (1): 3–13.

Cao, Y. S., J. G. Tang, M. Henze, X. P. Yang, Y. P. Gan, J. Li, H. Kroiss, et al. 2019. "The Leakage of Sewer Systems and the Impact on the 'Black and Odorous Water Bodies' and WWTPs in China." *Water Science and Technology* 79 (2): 334–41.

CARB (California Air Resources Board). 2018. "Fifty Year Air Quality Trends and

Health Benefits." Slide presentation. https://ww3.arb.ca.gov/board/books/2018 /020818/18-1-2pres.pdf.

Carpenter, Stephen R., and Kathryn L. Cottingham. 2002. "Resilience and the Resto-ration of Lakes." In *Resilience and the Behavior of Large-Scale Ecosystems*, edited by L. H. Gunderson and L. Pritchard, 51–70. Washington, DC: Island Press.

Carson, Rachel. 1962. *Silent Spring*. Boston: Houghton Mifflin.

Carter, Richard. n.d. "Didn't Karl Marx Offer to Dedicate *Das Kapital* to Darwin?" Friends of Charles Darwin. Accessed November 2022. http://friendsofdarwin.com /articles/marx-capital/.

Cary, Annette. 2015. "Decades-Long Hanford Downwinder Lawsuit Settles." *Tri-City Herald* (Kennewick, WA), 7 October 2015. www.tri-cityherald.com/news/local /hanford/article38165607.html.

CBECI (Cambridge Bitcoin Electricity Consumption Index). 2021. Bitcoin Mining Map. https://cbeci.org/mining_map.

CCPCC (Zhonggong Zhongyang, Guowuyuan 中共中央，国务院 Chinese Commu-nist Party Central Committee and State Council). 2008. "Zhonggong Zhongyang, Guowuyuan guanyu 2009 nian cujin nongye wending fazhan nongmin chixu zeng-shou de ruogan yijian" 中共中央，国务院关于2009年促进农业稳定发展农民持续增收的若干意见 [Some opinions of the CCP Central Committee and the State Council on promoting steady development of agriculture and continuing to raise the in-comes of farmers]. www.gov.cn/gongbao/content/2009/content_1220471.htm.

CCTV (China Central Television [Zhongguo Zhongyang Dianshitai]). 2019. "'Haishui xidiao' jinnang miaoji haishi chiren shuomeng?" "海水西调" 锦囊妙计还是痴人说梦? ["Transferring ocean water westward": A clever scheme or a mad-man's dream?] *Jinri tan* 今日谭 [Today's conversations], 5 November 2019. http:// news.cntv.cn/special/view/10/1117/.

CDAP (Cambridge Digital Assets Programme). 2022. "Bitcoin Mining—An (Un)Surpris-ing Resurgence?" Judge Business School, Cambridge University, 17 May 2022. www. jbs.cam.ac.uk/insight/2022/bitcoin-mining-new-data-reveal-a-surprising -resurgence/.

Cerny, Astrid. 2008. "In Search of Greener Pastures: Sustainable Development for Kazak Pastoralists in Xinjiang, China." PhD diss., University of Washington, Seattle.

Chai Jing. 2015. *Qiongding zhi xia* 穹顶之下 [Under the dome]. Documentary film. Ac-cessed December 2022. www.youtube.com/watch?v=A2vPyUbZMv4.

Chan, Anita, Richard Madsen, and Jonathan Unger. 1984. *Chen Village: The Recent His-tory of a Peasant Community in Mao's China*. Berkeley: University of California Press.

Chan, Kam Wing. 2001. "Recent Migration in China: Patterns, Trends, and Policies." *Asian Perspective* 25 (4): 127–55.

———. 2009a. "Introduction: Population, Migration, and the Lewis Turning Point in China." In *The China Population and Labor Yearbook, Volume I: The Approaching Lewis Turning Point and Its Policy Implications*, edited by Cai Fang and Du Yang. Leiden: Brill.

———. 2009b. "The Chinese *Hukou* System at 50." *Eurasian Geography and Economics* 50 (2): 197–221.

———. 2012. "Migration and Development in China: Trends, Geography, and Current Issues." *Migration and Development* 1 (2): 187–205.

———. 2021. "What the 2020 Chinese Census Tells Us about Progress in Hukou Reform." *China Brief* 21 (15), 30 July 2021. Jamestown Foundation. https://jamestown .org/program/what-the-2020-chinese-census-tells-us-about-progress -in-hukou-reform/.

Chan, Kam Wing, and Xueqiang Xu. 1985. "Urban Population Growth and Urbanization in China since 1949: Reconstructing a Baseline." *China Quarterly* 104:583–613.

Chan, Kam Wing, and Li Zhang. 1999. "The *Hukou* System and Rural-Urban Migration in China: Processes and Changes." *China Quarterly* 160:818–55.

Chang, Kwang-chih. 1968. *The Archaeology of Ancient China.* Rev. and enlarged ed. New Haven: Yale University Press.

Chao, Francis. 2015. "China's Airport Growth Plan." Aviation Pros, 28 June 2015. www .aviationpros.com/ground-handling/ground-handlers-service-providers/article /12076081/chinas-airport-growth-plan.

Charles, Dan. 2019. "Swine Fever Is Killing Vast Numbers of Pigs in China." *Morning Edition,* NPR, 15 August 2019. www.npr.org/sections/thesalt/2019/08/15 /751090633/swine-fever-is-killing-vast-numbers-of-pigs-in-china.

Chen, Biao, Lijun Hao, Xinyan Guo, Na Wang, and Boping Ye. 2015. "Prevalence of Antibiotic Resistance Genes of Wastewater and Surface Water in Livestock Farms of Jiangsu Province, China." *Environmental Science and Pollution Research* 22:13950–59.

Chen, Chao, and Genserik Reniers. 2020. "Chemical Industry in China: The Current Status, Safety Problems, and Pathways for Future Sustainable Development." *Safety Science* 128: article 104741.

Chen, Cheng, Hannes J. König, Bettina Matzdorf, and Lin Zhen. 2015. "The Institutional Challenges of Payment for Ecosystem Service Program in China: A Review of the Effectiveness and Implementation of Sloping Land Conversion Program." *Sustainability* 7 (5): 5564–91.

Chen Guiquan 陈桂权. 2014. "Sichuan dongshuitian de lishi bianqian" 四川冬水田的历史变迁 [Historical changes in Sichuan's winter wet-rice fields]. *Gujin Nongye* 古今农业 2014 (1): 83–91.

Chen Huanzhen 陈焕珍 and Ge Baona 葛宝娜. 2003. "Laiwu shi kuangchan ziyuan kaifa de shengtai pingjia ji duice" 莱芜市矿产资源开发的生态评价及对策 [Ecological assessment and countermeasures for mineral resource exploitation in Laiwu Municipality]. *Guotu Ziyuan Jishu Guanli* 国土资源技术管理 2005 (5): 103–7.

Chen, Jack. 1973. *Red Flag Canal* (videocassette). Albany: Bureau of Mass Communications, New York State Education Department.

Chen Jie 陈杰, Zhou Tao 周涛, and Zhou Lan-yu 周蓝宇. 2016. *Woguo neilu hedianzhan wuge wenti de yanjiu* 我国内陆核电站五个问题的研究 [Research on five questions about nuclear plants in core areas of our country). *Huabei Dianli Daxue Xuebao (Shehui Kexue Ban)* 华北电力大学学报 (社会科学版). 2016 (2): 1–4.

Chen Jinlin 陈锦林 and Bo Jiaju 薄家驹, eds. 1993. *Dujiangyan Zhi* 都江堰志 [Gazetteer of Dujiangyan]. Chengdu: Sichuan Cishu Chubanshe.

Chen Jinyong 陈金永 (Kam Wing Chan). 2023. *Da guo chengmin: Chengzhenhua yu huji gaige* 大国城民：城镇化与户籍改革[Becoming urban citizens: Urbanization and China's hukou reform]. Beijing: Peking University Press.

Chen, Ruijian, Jikun Huang, and Fangbin Qiao. 2013. "Farmers' Knowledge on Pest

Management and Pesticide Use in Bt Cotton Production in China." *China Economic Review* 27 (C): 15–24.

Chen, Stephen. 2021. "Chinese Engineers Drilling World's Longest Tunnel in Xinjiang Desert Hit a Wall—of Rushing Water." *South China Morning Post*, 18 December 2021.

———. 2022. "China's Robot-Built 3D-printed Dam Ready in 2 Years: Scientists." *South China Morning Post*, 8 May 2022.

Chen, Wang. 2019. "Chinese Consumers Ignore Calls to Eat Less Beef." In *Tracking China's Soy and Beef Imprint on South America*, edited by Isabel Hilton, 13–16. Diálogo Chino, 5 December 2019. https://dialogochino.net/32059-tracking-chinas-soy-and-beef-imprint-on-latin-america/.

Chen Wei-hong, Zhang Rui-de, and Liu Gui-chuan. 2009. "Development Model of Rural Industrialization in China." *Asian Agricultural Research* 亚洲农业研究1 (7): 8–12.

Chen, Xudong, Yong Geng, and Tsuyoshi Fujita. 2010. "An Overview of Municipal Solid Waste Management in China." *Waste Management* 30 (4): 716–24.

Chen, Yangfen, Xiande Li, Lijuan Wang, and Shihai Wang. 2017. "Is China Different from Other Investors in Global Land Acquisition? Some Observations from Existing Deals in China's Going Global Strategy." *Land Use Policy* 60 (1): 362–72.

Chen Yongjie 陈永杰. 2011. "Shuidian wuran chaoguo huodian? Shuidian shifou pohuai huanjing zhenglun yi duonian" 水电污染超过火电？水电是否破坏环境争论已多年 [Does hydroelectric pollution exceed thermoelectric? The debate over whether hydroelectricity harms the environment has already gone on for many years]. *Beijing Qingnian Bao* 北京青年报, 11 January 2011.

Chen, Zhongfei, Xiaoyu Zhang, and Fanglin Chen. 2020. "Have Driving Restrictions Reduced Air Pollution: Evidence from Prefecture-Level Cities of China." *Environmental Science and Pollution Research* 28:3106–20.

Cheng Shengquan 成胜权. 2012. "Ji yu RS he GIS de Bin Xian tudi liyong he turang qinshi de dingliang yanjiu" 基于 RS 和 GIS 的宾县土地利用和土壤侵蚀的定量研究 [Quantitative research on land use and erosion in Bin County on the basis of remote sensing and geographic information systems]. *Shuili Keji yu Jingji* 水利科技与经济 18 (9): 100, 105.

Cheng, Weixiao, Hong Chen, Chao Su, and Shuhai Yan. 2013. "Abundance and Persistence of Antibiotic Resistance Genes in Livestock Farms: A Comprehensive Investigation in Eastern China." *Environment International* 61 (1): 1–7.

Chengdurail. 2017–20. "Chengche zhinan" 乘车指南 [Passenger guide]. Accessed July 2020. www.chengdurail.com/index.html.

Cheung, Siu-woo. 1995. "Millenarianism and the Miao." In *Cultural Encounters on China's Ethnic Frontiers*, edited by Stevan Harrell, 217–47. Seattle: University of Washington Press.

Chi Wei. 1972. "Turning the Harmful into the Beneficial." *Peking Review* 1972 (4): 5–7.

Chiasson, Blaine. 2017. "Producing a Full-Fat Controversy: The Politicization of Dairy Production in Northern Manchuria, 1924–30." In *Empire and Environment in the Making of Manchuria*, edited by Norman Smith, 107–29. Vancouver: UBC Press.

Chik, Holly. 2021. "Yuan Longping: How a Career in Science Created a National Hero in China." *South China Morning Post*, 24 May 2021.

China Daily. 2005. "Steel Giant Makes Way for 2010 World Expo." 30 June 2005. www
.china.org.cn/english/2005/Jun/133540.htm.

———. 2011. "Prosecutors Looking at Dalian Oil Spill Case." 25 November 2011. www
.chinadaily.com.cn/china/2011–11/25/content_14158840.htm.

———. 2019. "China's Subway Network Reaches 4,600 Km." 10 June 2019. www
.chinadaily.com.cn/a/201906/10/WS5cfdf673a310176577230519.html.

China Dialogue. 2022. "Wind and Solar Projects Banned from Freshwater Bodies." 1
June 2022. https://chinadialogue.net/en/digest/wind-and-solar-projects-banned
-from-freshwater-bodies/.

China Maps. n.d. "China Climate Map of Annual Average Precipitation." Accessed July
2022. www.chinamaps.org/china/china-map-of-precipitation-annual.html.

China Power. 2020. "How Is China's Energy Footprint Changing?" Accessed January
2021. https://chinapower.csis.org/energy-footprint/.

China Three Gorges. 2002. *Sanxia Gongcheng Jianjie* 三峡工程简介 [Brief introduction
to the Sanxia Project]. https://web.archive.org/web/20100329113449/http://www
.ctgpc.com.cn/sxslsn/index.php.

Chiu, Hua-mei. 2020. "Environmental Activism in Kaohsiung, Taiwan." In *Greening
East Asia: The Rise of the Eco-Developmental State*, edited by Ashley Esarey, Mary
Alice Haddad, Joanna I. Lewis, and Stevan Harrell, 181–96. Seattle: University of
Washington Press.

Chotiner, Isaac. 2021. "Reconsidering the History of the Chinese Communist Party (Interview with Tony Saich)." *New Yorker*, 22 July 2021.

Chu, Junying, Jianhua Wang, and Can Wang. 2015. "A Structure-Efficiency Based Performance Evaluation of the Urban Water Cycle in Northern China and Its Policy
Implications." *Resources, Conservation, and Recycling* 104:1–11.

Clean Air Council. 2016. "Coke Ovens and Air Pollution: A Comprehensive Analysis—
Nature and Extent of Air Emissions." http://pacokeovens.org/what-is-coke/nature
-and-extent-of-air-emission/.

Clever, Jennifer. 2017. "China: Strong Domestic Demand and Declining Production Encourage Wood Imports in 2017." USDA Foreign Agricultural Service GAIN Report.
www.fas.usda.gov/data/china-strong-domestic-demand-and-declining-production
-encourage-wood-imports-2017.

Cliff, Thomas Matthew James. 2009. "Neo-Oasis: The Xinjiang *Bingtuan* in the Twenty-
First Century." *Asian Studies Review* 33:83–106.

Cliff, Tom. 2016. *Oil and Water: Being Han in Xinjiang*. Chicago: University of Chicago Press.

Collins, Leigh. 2021a. "Global Green-Hydrogen Pipeline Exceeds 250GW—Here's the 27
Largest Gigawatt-Scale Projects." *Recharge*, 2 August 2021. www.rechargenews.com
/energy-transition/global-green-hydrogen-pipeline-exceeds-250gw-heres-the-27
-largest-gigawatt-scale-projects/2-1-933755.

———. 2021b. "Hydrogen Now Firmly at the Heart of the Global Race to Net Zero—
for Better or for Worse." *Recharge*, 27 August 2021. www.rechargenews.com/energy
-transition/hydrogen-now-firmly-at-the-heart-of-the-global-race-to-net-zero-for
-better-or-worse/2-1-1058073.

Connell, Joseph H. 1978. "Diversity in Tropical Rain Forests and Coral Reefs."
Science 199 (4335): 1302–10.

Courtney, Chris. 2018a. "At War with Water: The Maoist State and the 1954 Yangzi Floods." *Modern Asian Studies* 52 (6): 1807–36.

———. 2018b. *The Nature of Disaster in China: The 1931 Yangzi River Flood*. Cambridge: Cambridge University Press.

CPPCC (Chinese People's Political Consultative Conference). 1949. "Zhonghua Renmin Gongheguo Zhengzhi Xieshanghui Gonggong Gangling" 中国人民政治协商会议共同纲领[The Common Program of the Chinese People's Political Consultative Conference]. www.cppcc.gov.cn/2011/12/16/ARTI1513309181327976.shtml.

Crosby, Alfred W. 1972. *The Columbian Exchange: Biological and Cultural Consequences of 1492*. Westport, CT: Greenwood.

Crow-Miller, Britt. 2015. "Discourses of Deflection: The Politics of Framing China's South-North Water Transfer Project." *Water Alternatives* 8 (2): 173–92.

Cui Zheng and Liu Zhiyi. 2013. "China's Urban Sludge Dilemma: Sinking in Stink." *Caixin* (English), 8 August 2013. www.caixinglobal.com/2013-08-08/chinas-urban-sludge-dilemma-sinking-in-stink-101014182.html.

Cumming, Graeme S., David H. M. Cumming, and Charles L. Redman. 2006. "Scale Mismatches in Social-Ecological Systems: Causes, Consequences, and Solutions." *Ecology and Society* 11 (1): article 14. www.ecologyandsociety.org/vol11/iss1/art14/.

Dai, Jingyun, and Anthony Spires. 2020. "Grassroots NGOs and Environmental Advocacy in China." In *Greening East Asia: The Rise of the Eco-Developmental State*, edited by Ashley Esarey, Mary Alice Haddad, Joanna I. Lewis, and Stevan Harrell, 225–38. Seattle: University of Washington Press.

Dai, Qili, Xiaohui Bi, Wenbin Song, Tingkun Li, Baoshuang Liu, Jing Ding, Jiao Xu, et al. 2019. "Residential Coal Combustion as a Source of Primary Sulfate in Xi'an, China." *Atmospheric Environment* 196:66–76.

Dai Qing. 1994. *Yangtze! Yangtze!* London: Earthscan.

d'Alpoim Guedes, Jade, Stevan Harrell, Keala Hagmann, Amanda H. Schmidt, and Thomas Hinckley. 2020. "Deep History in Western China Reveals How Humans Can Enhance Biodiversity." China Dialogue, 29 April 2020. www.chinadialogue.net/article/show/single/en/11984-Deep-history-in-western-China-reveals-how-humans-can-enhance-biodiversity.

Danjiangkou Online (Danjiangkou Zaixian 丹江口在线). 2021. "Yin Jiang bu Han gongcheng jiang shishi, Danjiangkou zaici zhandao lishi de qianyan" 引江补汉工程将实施，丹江口再次站到历史的前沿 [The Taking the Yangtze to Replenish the Han Project will happen, Danjiangou will once again stand at the forefront of history]. Sohu, 8 February 2021. www.sohu.com/a/449510218_2716141.

Dautcher, Jay. 2009. *Down a Narrow Road: Identity and Masculinity in a Uyghur Community in Xinjiang China*. Cambridge, MA: Harvard University Asia Center.

Davidson-Hunt, Iain. 2003. "Nature and Society through the Lens of Resilience: Toward a Human-in-Ecosystem Perspective." In *Navigating Social-Ecological Systems*, edited by Fikret Berkes, Johan Golding, and Carl Folke, 53–77. Cambridge: Cambridge University Press.

Day, Alexander. 2008. "The End of the Peasant? New Rural Reconstruction in China." *boundary 2* 35 (2): 49–73.

Day, Alexander S., and Mindi Schneider. 2018. "The End of Alternatives? Capitalist

Transformation, Rural Activism, and the Politics of Possibility in China." *Journal of Peasant Studies* 45 (7): 1221–46.

DeJesus, Erin. 2016. "Anthony Bourdain Parts Unknown in Sichuan: Just the One-Liners." Eater, 16 October 2016. www.eater.com/2016/10/16/13278532/anthony-bourdain -parts-unknown-sichuan-china-recap.

Demaria, Federico, François Schneider, Filka Sekulova, and Joan Martinez-Alier. 2013. "What Is Degrowth? From an Activist Slogan to a Social Movement." *Environmental Values* 22 (2): 191–215.

Deng, F. Frederic, and Youqin Huang. 2004. "Uneven Land Reform and Urban Sprawl in China: The Case of Beijing." *Progress in Planning* 61:211–36.

Deng, Kent G. 2004. "Unveiling China's True Population Statistics for the Pre-Modern Era with Official Census Data." *Population Review* 43 (2): 32–69.

Diamond, Jared. 2005. *Collapse: How Societies Choose to Fail or Succeed*. New York: Viking.

Diao Fanchao 刁凡超. 2021. "Jiangxi gongshi Jiangxi gongshi: Poyang hu jian zha yi naru guojia zhongda shuili gongcheng, jiang diaogu bu konghong" 江西公示：鄱阳湖建闸已纳入国家重大水利工程，将涸枯不控洪 [Jiangxi public notice: Building a sluice gate for Poyang Lake has been included in national large-scale water projects; It will solve dry-up but not control floods]. *Pengbai*, 8 January 2021.

Di Cosmo, Nicola. 1994. "Ancient Inner Asian Nomads: Their Economic Basis and Its Significance in Chinese History." *Journal of Asian Studies* 53 (4): 1092–1126.

———. 1998. "Qing Colonial Administration in Inner Asia." *International History Review* 20 (2): 287–309.

———. 1999. "State Formation and Periodization in Inner Asian History." *Journal of World History* 10 (1): 1–40.

Dikötter, Frank. 2011. *Mao's Great Famine: The History of China's Most Devastating Catastrophe, 1958–1962*. New York: Walker.

Dinda, Soumyananda. 2004. "Environmental Kuznets Curve Hypothesis: A Survey." *Ecological Economics* 49:431–55.

Ding Lüshu 丁履枢. 2003. "'Beidahuang' kaiken shi" "北大荒" 开垦史 [History of opening the "Great Northern Wilderness"]. *Yanhuang Chunqiu* 炎黄春秋2003 (4): 13–16.

Ding, Xueli, Xiaozeng Han, Yao Liang, Yunfa Qiao, Lujun Li, and Na Li. 2012. "Changes in Soil Organic Carbon Pools after 10 Years of Continuous Manuring Combined with Chemical Fertilizer in a Mollisol in China." *Soil and Tillage Research* 122:36–41.

Ding, Yanjun, Weijian Han, Qinhu Chai, Shuhong Yang, and Wei Shen. 2013. "Coal-Based Synthetic Natural Gas (SNG): A Solution to China's Energy Security and CO_2 Reduction?" *Energy Policy* 55:445–63.

Doll, Ross. 2020. "Place, Power, and Potential: Agricultural Modernization and the Remaking of China's Countryside." PhD diss., University of Washington, Seattle.

———. 2021. "Cultivating Decline: Agricultural Modernization Policy and Adaptive Resilience in the Yangtze Delta." *Human Ecology* 49 (1): 43–57.

———. 2022. "Agricultural Modernisation and Diabolic Landscapes of Dispossession in Rural China." *Antipode* 54 (6): 1738–59.

Dong, Feng, Jingyun Li, Xiaoyun Zhang, and Jiao Zhu. 2021. "Decoupling Relationship between Haze Pollution and Economic Growth: A New Decoupling Index." *Ecological Indicators* 129: article 107859.

Dong Ruzeng 董如增, Zhao Dingtao 赵定涛, and Yun Zhijie 尹志杰. 2002. "Chao Hu liuyu huanjing—jingji xietiao fazhan shishi zhanlüe" 巢湖流域环境—经济协调发展实施战略 [An implementary strategy of harmonious development on Chao Hu drainage basin environment-economy]. *Yuce* 预测21 (6): 73–77.

Dong, Xiao-yuan, and Lewis Putterman. 2003. "Soft Budget Constraints, Social Burdens, and Labor Redundancy in China's State Industry." *Journal of Comparative Economics* 31 (1): 110–33.

Dong, Xiao-yuan, Terrence S. Veeman, and Michele M. Veeman. 1995. "China's Grain Imports: An Empirical Study." *Food Policy* 20 (4): 323–38.

Dong, Yuan. 2014. "INTERACTIVE: Mapping China's 'Dam Rush.'" China Environment Forum, Wilson Center, 21 March 2014. www.wilsoncenter.org/publication/interactive-mapping-chinas-dam-rush.

Downs, Erica Strecker. 2000. *China's Quest for Energy Security*. Santa Monica, CA: RAND.

Du Jinping 杜金平, Li Mingda 李明达, and Li Jie 李杰. 1999. "Shilun Sanxia shuiku nongcun yimin feitudi anzhi tujing" 试论三峡水库农村移民非土地安置途径 [An attempt to discuss the route of relocating Sanxia Reservoir area farming migrants without land]. *Renmin Changjiang* 人民长江30 (11): 1–23.

Du, Ningrui, Henk Ottens, and Richard Sliuzas. 2010. "Spatial Impact of Urban Expansion on Surface Water Bodies: A Case Study of Wuhan, China." *Landscape and Urban Planning* 94:175–85.

Du, S. F., H. J. Wang, B. Zhang, F. Y. Zhai, and B. M. Popkin. 2014. "China in the Period of Transition from Scarcity and Extensive Undernutrition to Emerging Nutrition-Related Non-Communicable Diseases, 1949–1992." *Obesity Reviews* 15 (Suppl. 1): 8–15.

Du, Shanshan. 2002. *Chopsticks Only Work in Pairs: Gender Unity and Gender Equality among the Lahu of Southwest China*. New York: Columbia University Press.

Early, Catherine. 2017. "China Renews Clampdown on Waste Imports." China Dialogue, 31 July 2017. https://chinadialogue.net/en/pollution/9954-china-renews-clampdown-on-waste-imports/.

———. 2020. "What China's Waste Import Ban Has Meant for the West." China Dialogue, 28 January 2020. https://chinadialogue.net/en/cities/11816-what-china-s-waste-import-ban-has-meant-for-the-west/.

Eaton, Sarah, and Genia Kostka. 2014. "Authoritarian Environmentalism Undermined? Local Leaders' Time Horizons and Environmental Policy Implementation in China." *China Quarterly* 218:359–80.

Economy, Elizabeth. (2004) 2010. *The River Runs Black: The Environmental Challenge to China's Future*. 2nd ed. Ithaca: Cornell University Press.

EIA (US Energy Information Administration). 2020. "Country Analysis Executive Summary: China." Updated 30 September 2020. www.eia.gov/international/content/analysis/countries_long/China/china.pdf.

Ekvall, Robert B. 1968. *Fields on the Hoof: The Nexus of Tibetan Nomadic Pastoralism*. New York: Holt, Rinehart and Winston.

Ellis, Jeffrey C. 1996. "On the Search for a Root Cause: Essentialist Tendencies in Environmental Discourse." In *Uncommon Ground: Rethinking the Human Place in Nature*, edited by William Cronon, 256–68. New York: Norton.

Elvin, Mark. 1998. "The Environmental Legacy of Imperial China." *China Quarterly* 156:733–56.

———. 2004. *The Retreat of the Elephants: An Environmental History of China*. New Haven: Yale University Press.

Embassy of the People's Republic of China in the United States. n.d. "Chronology of the Three Gorges Project." Accessed July 2022. www.mfa.gov.cn/ce/ceus//eng/zt /sxgc/t36515.htm.

Engels, Friedrich. (1876) 1939. "The Part Played by Labour in the Ape's Evolution into Man." Translated by Morris Goldenberg. *Dialectics* 8:1–14.

———. (1880) 1970. *Socialism: Utopian and Scientific*. Moscow: Progress Publishers. Accessed at Marxists Internet Archive. www.marxists.org/archive/marx/works/1880 /soc-utop/index.htm.

EPA (US Environmental Protection Agency). 2017. Ambient Concentrations of Sulfur Dioxide. https://cfpub.epa.gov/roe/indicator.cfm?i=91.

———. n.d. Biosolids. Accessed July 2022. www.epa.gov/biosolids.

EPS China Data. n.d. Accessed July 2022. www.epschinadata.com/auth/platform.html ?sid=87559E30723ECBC355113380C8DD3ACC_ipv446394234.

Esarey, Ashley, Mary Alice Haddad, Joanna I. Lewis, and Stevan Harrell, eds. 2020. *Greening East Asia: The Rise of the Eco-Developmental State*. Seattle: University of Washington Press.

Escobar, Arturo. (1995) 2012. *Encountering Development: The Making and Unmaking of the Third World*. New ed. Princeton: Princeton University Press.

Esherick, Joseph W. 1995. "Ten Theses on the Chinese Revolution." *Modern China* 21 (1): 45–76.

Eyferth, Jacob. 2022. "State Socialism and the Rural Household: How Women's Handloom Weaving (and Pig-Raising, Firewood-Gathering, Food-Scavenging) Subsidized Chinese Accumulation." *International Review of Social History* (2022): 1–19.

Faber, Nathan. n.d. "Food Retail and Social Distinction in Chengdu." Unpublished BA honors thesis, University of Washington, Seattle.

Fabinyi, Michael, Kate Barclay, and Hampus Eriksson. 2017. "Chinese Trader Perceptions on Sourcing and Consumption of Endangered Seafood." *Frontiers in Marine Science* 4 (June): article 181. https://opus.lib.uts.edu.au/handle/10453/109308.

Fabinyi, Michael, and Neng Liu. 2014. "Seafood Banquets in Beijing: Consumer Perspectives and Implications for Environmental Sustainability." *Conservation and Society* 12 (2): 218–28.

Fabinyi, Michael, Neng Liu, Qingyu Song, and Ruyi Li. 2016. "Aquatic Product Consumption Patterns and Perceptions among the Chinese Middle Class." *Regional Studies in Marine Science* 7:1–9.

Fan, Fa-ti. 2017. "The People's War against Earthquakes: Cultures of Mass Science in Mao's China." In *Cultures without Culturalism: The Making of Scientific Knowledge*, edited by Karine Chemla and Evelyn Fox Keller, 296–323. Durham, NC: Duke University Press.

Fan, Liangxin, Lingtong Gai, Yan Tong, and Ruihua Li. 2017. "Urban Water Consumption and Its Influencing Factors in China: Evidence from 286 Cities." *Journal of Cleaner Production* 166:124–33.

Fan, Liangxin, Haipeng Niu, Xiaomei Yang, Wei Qin, Célia P. M. Bento, Coen J. Ritsema, and Violette Geissen. 2015. "Factors Affecting Farmers' Behaviour in Pesticide Use: Insights from a Field Study in Northern China." *Science of the Total Environment* 537:360–68.

Fan, Tinglu, B. A. Stewart, William A. Payne, Wang Yong, Junjie Luo, and Yufang Gao. 2005. "Long-Term Fertilizer and Water Availability Effects on Cereal Yield and Soil Chemical Properties in Northwest China." *Soil Science Society of America Journal* 69 (3): 842–55.

FAO (Food and Agriculture Organization). 1972. *Food Composition Table for Use in East Asia*. Rome: FAO. www.fao.org/3/X6878E/X6878E00.htm.

———. 2010. *State of World Fisheries and Aquaculture*. Rome: FAO.

———. 2022. *The State of World Fisheries and Aquaculture: Towards Blue Transformation*. Rome: FAO. www.fao.org/3/cc0461en/cc0461en.pdf.

FAS (USDA Foreign Agricultural Service) China Staff. 2019. "Industry Debates Forage Production and Imports at [2018] Conference." https://apps.fas.usda.gov/newgainapi /api/report/downloadreportbyfilename?filename=Industry%20Debates%20Forage %20Production%20and%20Imports%20at%20Conference_Beijing%20ATO_China%2 -%20Peoples%20Republic%20of_4-3-2019.pdf.

Fei, Hsiao-tung [Fei Xiaotong]. 1939. *Peasant Life in China: A Field Study of Country Life in the Yangtze Valley*. London: Kegan Paul, Trench and Trubner.

Feigon, Lee. 2000. "A Harbinger of the Problems Confronting China's Economy and Environment: The Great Chinese Shrimp Disaster of 1993." *Journal of Contemporary China* 9 (24): 323–32.

Feng, Chao, Jian-Bai Huang, and Miao Wang. 2019. "The Sustainability of China's Metal Industries: Features, Challenges, and Future Focuses." *Resources Policy* 60:215–24.

Feng, Chao, Jian-Bai Huang, Miao Wang, and Yi Song. 2018. "Energy Efficiency in China's Iron and Steel Industry: Evidence and Policy Implications." *Journal of Cleaner Production* 177:837–45.

Feng, Coco. 2021. "Sichuan Takes Lenient Stance on Bitcoin Mining amid National Crackdown to Deal with Rainy Season's Excess Hydropower." *South China Morning Post*, 7 June 2021.

Feng, Tong, Huibin Du, Zhongguo Lin, and Jian Zuo. 2020. "Spatial Spillover Effects of Environmental Regulations on Air Pollution: Evidence from Urban Agglomerations in China." *Journal of Environmental Management* 272: article 110998.

Feng Yandan 冯艳丹 and Wu Yiheng 伍奕衡. 2021. "Bihuan guanli yanjin 'digouyou' chong hui fanzhuo" 闭环管理严禁"地沟油" 重回饭桌 [Circular management strictly prohibits "gutter oil" returning to the dining table." *Nanfang Ribao*, 12 November 2021. www.xinhuanet.com/food/20211112/f8d9165e00214b089242980398 32f971/c.html.

Fernando, C. H. 1993. "Rice Field Ecology and Fish Culture: An Overview." *Hydrobiologia* 259:91–113.

Field, Robert Michael, and James A. Kilpatrick. 1978. "Chinese Grain Production: An Interpretation of the Data." *China Quarterly* 74:369–84.

Fish, Eric. 2013. "The Forgotten Legacy of the Banqiao Dam Collapse." *Jingji Guancha Bao* 经济观察报, 8 February 2013.

Fong, Vanessa L. 2004. *Only Hope: Coming of Age under China's One-Child Policy*. Stanford, CA: Stanford University Press.

FORHEAD (Forum on Health, Environment, and Development). 2018. *Air Pollution in China: An Interdisciplinary Perspective*. Summary report. www.forhead.org /upload/201811/15/201811151231287214.pdf.

Fortune. 2020. "Global 500." https://fortune.com/global500/.

Foster, John Bellamy. 1999. "Marx's Theory of Metabolic Rift: Classical Foundations of Environmental Sociology." *American Journal of Sociology* 105 (2): 366–405.

Foster, Stephen, Hector Garduno, Richard Evans, Doug Olson, Yuan Tian, Weizhen Zhang, and Zaisheng Han. 2004. "Quaternary Aquifer of the North China Plain: Assessing and Achieving Groundwater Resource Sustainability." *Hydrogeology Journal* 12:81–93.

Francis, Robert C. n.d. "EBFM [Ecosystem Based Fisheries Management]: What's It All About, Anyway?" Unpublished paper. Accessed July 2022. http://faculty.washington.edu/stevehar/Francis%20on%20EBFM.pdf.

Frank, Andre Gunder. 1969. *Capitalism and Underdevelopment in Latin America: Historical Studies of Chile and Brazil.* New York: Monthly Review Press.

Freedman, Maurice. 1958. *Lineage Organization in Southeastern China.* London: Athlone Press.

———. 1966. *Chinese Lineage and Society.* London: Athlone Press.

Fried, Morton H. 1967. *The Evolution of Political Society.* New York: Random House.

Friedman, Edward, Paul G. Pickowicz, and Mark Selden. 1991. *Chinese Village: Socialist State.* New Haven: Yale University Press.

———. 2005. *Revolution, Resistance, and Reform in Village China.* New Haven: Yale University Press.

Fu, Jia, Jun Zhong, Demin Chen, and Qiang Liu. 2020. "Urban Environmental Governance, Government Intervention, and Optimal Strategies: A Perspective on Electronic Waste Management in China." *Resources, Conservation & Recycling* 154: article 104547.

Fu, Jianru. 2020. "China's Sewage Treatment Industry Status Quo and Improvement Measures." *IOP Conference Series: Earth and Environmental Sciences* 514: article 030264. https://iopscience.iop.org/article/10.1088/1755-1315/514/3/032064.

Fubao 富宝废钢网. 2021. "2021 nian Zhongguo feigangtie shichang niandu baogao" 2021年中国废钢铁市场年度报告 [2021 annual report on China's scrap iron and steel market]. https://zhuanlan.zhihu.com/p/311654939.

Gale, Fred, and Dinghuan Hu. 2009. "Supply Chain Issues in China's Milk Adulteration Incident." Contributed paper prepared for presentation at the International Association of Agricultural Economists Conference, Beijing, China, 16–22 August 2009.

Gamble, Sidney. 1963. *North China Villages: Social, Political, and Economic Activities before 1933.* Berkeley: University of California Press.

Gao Baiyu. 2020a. "Vast River Diversion Plan Afoot in Western China." China Dialogue, 7 January 2020. www.chinadialogue.net/article/show/single/en/11762-Vast-river-diversion-plan-afoot-in-western-China.

———. 2020b. "China's List of Protected Animals to Be Updated after 32 Years." China Dialogue, 1 October 2020. https://chinadialogue.net/en/nature/chinas-list-of-protected-animals-to-be-updated-after-32-years/.

Gao, Jay, and Yansui Liu. 2011. "Climate Warming and Land Use Change in Heilongjiang Province, Northeast China." *Applied Geography* 31:476–82.

Gao, Jian Hua, Jianjun Jia, Albert J. Kettner, Fei Xing, Ya Ping Wang, Xia Nan Xu, Yang Yang, et al. 2014. "Changes in Water and Sediment Exchange between the Changjiang River and Poyang Lake under Natural and Anthropogenic Conditions, China." *Science of the Total Environment* 481:542–53.

Gao, Jie. 2015. "Pernicious Manipulation of Performance Measures in China's Cadre Evaluation System." *China Quarterly* 223:618–37.

Gao, Jinlong, Yehua Dennis Wei, Wen Chen, and Jianglong Chen. 2014. "Economic Transition and Urban Land Expansion in Provincial China." *Habitat International* 44:461–73.

Gao, Meiling, Huanfeng Shen, Xujun Han, Huifang Li, and Liangpei Zhang. 2017. "Multiple Timescale Analysis of the Urban Heat Island Effect Based on the Community Land Model: A Case Study of the City of Xi'an, China." *Environmental Monitoring and Assessment* 190 (8). https://doi.org/10.1007/s10661-017-6320-9.

Gao, Mobo C. F. 1999. *Gao Village: A Portrait of Rural Life in Modern China.* Honolulu: University of Hawai'i Press.

Gao, Yuan, Sina Shahab, and Negar Ahmadpoor. 2020. "Morphology of Urban Villages in China: A Case Study of Dayuan Village in Guangzhou." *Urban Science* 4 (23): 23 pages.

Gao Yuanyuan, Li Jia, Hao Qichen, Yu Chu, and Meng Suhua. 2018. "Analysis of the Effect of Groundwater Overexploitation Control in Water Receiving Region of the First Phase of the South-North Water Transfer Project." *MATEC Web of Conferences* 246: article 01069. www.researchgate.net/publication/329476078_Analysis_on_the _Effect_of_Groundwater_Overexploitation_Control_in_Water_receiving_region_of _the_First_Phase_of_the_South-North_Water_Transfer_Project.

Gao, Yuchen, Zehao Liu, Ruipeng Li, and Zhidan Shi. 2020. "Long-Term Impact of China's Returning Farmland to Forest Program on Rural Economic Development." *Sustainability* 12 (4): 1492–1508.

Gardner, Daniel K. 2015. "China's 'Silent Spring' Moment? Why 'Under the Dome' Found a Ready Audience in China." *New York Times*, 18 March 2015.

Garnaut, Anthony. 2014. "The Geography of the Great Leap Famine." *Modern China* 40 (3): 315–48.

Gaubatz, Piper. 1998. "Understanding Chinese Urban Form: Contexts for Interpreting Continuity and Change." *Built Environment* 24 (4): 251–70.

———. 1999. "China's Urban Transformation: Patterns and Processes of Morphological Change in Beijing, Shanghai, and Guangzhou." *Urban Studies* 36 (9): 1495–1521.

———. 2002. "Looking West towards Mecca: Muslim Enclaves in Chinese Frontier Cities." *Built Environment* 28 (3): 231–48.

Gaudreau, Matthew. 2019. "State Food Security and People's Food Sovereignty: Competing Visions of Agriculture in China." *Canadian Journal of Development Studies / Revue canadienne d'études du développement* 40 (1): 12–28.

Ge, Shengqiang, Jinming Li, Xiaoxu Fan, Fuxiao Liu, Lin Li, Qinghua Wang, Weijie Ren, et al. 2018. "Molecular Characterization of African Swine Fever Virus, China, 2018." *Emerging Infectious Diseases* 24 (11): 2131–33.

Ge, Shidong, and Shuqing Zhao. 2017. "Organic Carbon Storage in China's Urban Landfills from 1978–2014." *Environmental Research Letters* 12: article 104013.

Geheb, Kim, and Diana Suhardiman. 2019. "The Political Ecology of Hydropower in the Mekong River Basin." *Current Opinion in Environmental Sustainability* 37:8–13.

Geren Tushuguan 个人图书馆. 2018. *Zhongguo da he [8]: Lancang Jiang* 中国大河 (8): 澜沧江 [China's great rivers (8): Lancang River]. www.360doc.com/content/18 /0125/18/34003195_725057633.shtml.

Gibson, John, Chao Li, and Geua Boe-Gibson. 2014. "Economic Growth and Expansion of China's Urban Land Area: Evidence from Administrative Data and Night Lights, 1993–2012." *Sustainability* 2014 (6): 7850–7865.

Giersch, C. Patterson. 2006. *Asian Borderlands: The Transformation of Qing China's Yunnan Frontier*. Cambridge, MA: Harvard University Press.

Ginsburg, Norton S. 1951. "China's Railroad Network." *Geographical Review* 41 (3): 470–74.

Global Economy. n.d. Percent of world GDP, 2018—country rankings. Accessed July 2021. www.theglobaleconomy.com/rankings/gdp_share/.

Global IP News. 2018. "State Intellectual Property Office of China Receives University China Three Gorges CTGU's Patent Application for Water Surface Garbage Cleaning Device." *Environmental Patent News*, 9 April 2018. https://go-gale-com.offcampus.lib.washington.edu/ps/i.do?p=ITBC&u=wash_main&id=GALE%7CA534040710&v=2.1&it=r.

Glover, Denise M., Jack P. Hayes, and Stevan Harrell. 2021. "Resilience, Rationalism, and Response in Modern Chinese Social-Ecological Systems." *Human Ecology* 49 (1): 3–5.

Golany, Gideon. 1992. *Chinese Earth-Sheltered Dwellings: Indigenous Lessons for Modern Urban Design*. Honolulu: University of Hawai'i Press.

Gold, Thomas B. 1986. *State and Society in the Taiwan Miracle*. Armonk, NY: M. E. Sharpe.

Goldstein, Joshua. 2021. *Remains of the Everyday: A Century of Recycling in Beijing*. Oakland: University of California Press.

Goldstein, Melvyn C. 1971. "Stratification, Polyandry, and Family Structure in Central Tibet." *Southwestern Journal of Anthropology* 27 (1): 64–74.

———. 1989. *A History of Modern Tibet, 1913–1951: The Decline of the Lamaist State*. Berkeley: University of California Press.

———. 2014. *A History of Modern Tibet, Volume 3: The Storm Clouds Descend, 1955–57*. Berkeley: University of California Press.

Goldstein, Melvyn C., and Cynthia M. Beall. 1990. *Nomads of Western Tibet*. Berkeley: University of California Press.

———. 1994. *The Changing World of Mongolia's Nomads*. Berkeley: University of California Press.

Gong, Weigang, and Qian Forrest Zhang. 2016. "Betting on the Big: State-Brokered Land Transfers, Large-Scale Agricultural Producers, and Rural Policy Implementation." *China Journal* 77 (1): 1–26.

González, Roberto J. 2001. *Zapotec Science: Farming and Food in the Northern Sierra of Oaxaca*. Austin: University of Texas Press.

Gooch, Elizabeth, and Fred Gale. 2018. *China's Foreign Agriculture Investments*. US Department of Agriculture Economic Research Service Economic Information Bulletin No. 192. April. www.ers.usda.gov/webdocs/publications/88572/eib-192.pdf?v=43213.

Greaves, Wilfrid. 2018. "Damaging Environments: Land, Settler Colonialism, and Security for Indigenous Peoples." *Environment and Society: Advances in Research* 9:107–24.

Greenhalgh, Susan. 1990. "The Peasantization of the One-Child Policy in Shaanxi." In *Chinese Families in the Post-Mao Era*, edited by Deborah Davis and Stevan Harrell, 219–50. Berkeley: University of California Press.

———. 2003. "Science, Modernity, and the Making of China's One-Child Policy." *Population and Development Review* 29 (2): 163–96.

Greenpeace. 2012. "Thirsty Coal 2: Shenhua's Water Grab." www.banktrack.org/download/thirsty_coal_2/thirsty_coal_2.pdf.

———. 2018. "Analysis of Air Quality Trends in 2017." https://storage.googleapis.com/planet4-eastasia-stateless/2019/11/2aad5961-2aad5961-analysis-of-air-quality-trends-in-2017.pdf.

Griffin, Robert, Robert Ogelsby, Thomas Sever, and Udaysankar Nair. 2014. "Agricultural Landscapes, Deforestation, and Drought Severity." In *The Great Maya Droughts in Cultural Context: Case Studies in Resilience and Vulnerability*, edited by Giles Iannone, 71–86. Boulder: University Press of Colorado.

Grime, J. P. 1973. "Competitive Exclusion in Herbaceous Vegetation." *Nature* 242:344–47.

Grossman, Gene M., and Alan B. Krueger. 1994. "Environmental Impacts of a North American Free Trade Agreement." In *The Mexico-U.S. Free Trade Agreement*, edited by Peter Garber, 13–56. Cambridge, MA: MIT Press.

Gu, Hallie, and Dominique Patton. 2022. "Record Chinese Wheat Prices Raise Risk of Pricier Noodles." Reuters, 2 June 2022. www.reuters.com/markets/commodities/record-chinese-wheat-prices-raise-risk-pricier-noodles-2022-06-02/.

Gu Kangkang, Liu Jingshuang, and Wang Yang. 2009. "Relationship between Economic Growth and Water Environmental Quality of Anshan City in Northwest China." *China Geographical Science* 19 (1): 17–24.

Guan, Baohua, Shuqing An, and Binhe Gu. 2011. "Assessment of Ecosystem Health during the Past 40 Years for Lake Taihu in the Yangtze River Delta, China." *Limnology* 2011 (12): 47–53.

Guangdong (Guangdong Province People's Government) 广东省人民政府. 2019. "Guangdong zui da shuli gongcheng quanmian kaijian: Shushuixian chang 113.2 gongli, 'xi shui dong song' wei dawanqu "jieke" 广东最大水利工程全面开建 输水线长113.2公里 "西水东送"为大湾区"解渴" [Guangdong's largest hydraulic project fully begins constructing the 113.2-kilometer "sending western water east" water transfer route, the "thirst quencher" for the Greater Bay Area]. 7 May 2019. www.gd.gov.cn/gdywdt/bmdt/content/post_2382539.html.

Guanyan Baogao Wang 观研报告网. 2022. "2020 nian Zhongguo chengqu mianji, jianchengqu mianji ji shi renkou midu tongji qingkuang" 2020 年中国城区面积，建成区面积及市人口密度统计情况 [Statistical situation of China's urban area, built up area, and urban population density in 2020]. https://www.chinabaogao.com/detail/586704.html

Gunderson, Lance. 2003. "Adaptive Dancing: Interactions between Social Resilience and Ecological Crises." In *Navigating Social-Ecological Systems*, edited by Fikret Berkes, Johan Golding, and Carl Folke, 33–52. Cambridge: Cambridge University Press.

Gunderson, Lance, and C. S. Holling, eds. 2002. *Panarchy: Understanding Transformations in Human and Natural Systems*. Washington, DC: Island Press.

Gunderson, Lance H., C. S. Holling, Lowell Pritchard Jr., and Garry D. Peterson. 2002. "Resilience of Large-Scale Resource Systems." In *Resilience and the Behavior of Large-Scale Systems*, edited by Lance H. Gunderson and Lowell Pritchard Jr., 3–20. Washington, DC: Island Press.

Guo, Haipeng, Zhuchen Zhang, Guoming Cheng, Wenping Li, Tiefeng Li, and Jiu

Jimmy Jiao. 2015. "Groundwater-Derived Land Subsidence in the North China Plain." *Environmental Earth Science* 74:415–27.

Guo Shutian 郭书田. 2009. "1949–1979: 30 nian zhongguo nongcun de 12 jian da shi" 1949–1979: 30 年中国农村的12件大事 [1949–1979: Twelve big events over 30 years in Chinese agricultural villages]. *Nongcun Gongzuo Tongxun* 农村工作通讯8:44–47.

Guo, Z. G., L. F. Sheng, J. L. Feng, and Ming Fang. 2003. "Seasonal Variation of Solvent Extractable Organic Compounds in the Aerosols of Qingdao, China." *Atmospheric Environment* 37 (13): 1825–34.

Guoji Taiyangneng Guangyou Wang 国际太阳能光优网. 2020. "Gangyou boli channeng, jiage, chengben ji shichang geju fenxi" 光伏玻璃产能，价格，成本及市场格局分析 [Analysis of the structure of capital and markets in the production and pricing of plate glass]. https://solar.in-en.com/html/solar-2366416.shtml.

Guowuyuan 国务院 (State Council of the People's Republic of China). 2016. "Zhongguo jiaotong yunshu fazhan baipi shu" 中国交通运输发展白皮书 [White paper on the development of transportation in China]. www.scio.gov.cn/wz /Document/1537413/1537413.htm.

———. 2019. "Guowuyuan guanyu cujin xiangcun chanye zhenxing de zhidao yijian" 国务院关于促进乡村产业振兴的指导意见 [Directive opinions regarding the revitalization of rural production]. 28 June 2019. www.gov.cn/zhengce/content/2019-06/28 /content_5404170.htm.

———. 2022a. *Henan Zhengzhou "7.20" teda baoyu zaihai diaocha baogao* 河南郑州"7·20" 特大暴雨灾害 调查报告 [Report on an investigation of the huge rain disaster in Zhengzhou Henan on July 20]. www.mem.gov.cn/gk/sgcc/tbzdsgdcbg/202201 /P020220121639049697767.pdf.

———. 2022b. *Nanshui beidiao dongzhongxian yiqi gongcheng leiji diaoshui liang da 500 yi lifangmi* 南水北调东中线一期工程累计调水量达500亿立方米 [The cumulative total amounts of water delivered by the first stage projects of the eastern and middle routes of the South-North Water Transfer reach 50 billion cubic meters]. www .gov.cn/xinwen/2022-01/08/content_5667068.htm.

Hacking, Ian. 1999. *The Social Construction of What?* Cambridge, MA: Harvard University Press.

Haddad, Mary Alice, and Stevan Harrell. 2020. "Introduction: The Evolution of the Eco-Developmental State." In *Greening East Asia: The Rise of the Eco-Developmental State*, edited by Ashley Esarey, Mary Alice Haddad, Joanna Lewis, and Stevan Harrell, 5–31. Seattle: University of Washington Press.

Hale, Matthew A. 2013. "Reconstructing the Rural: Peasant Organizations in a Chinese Movement for Alternative Development." PhD diss., University of Washington, Seattle.

Han Shaogong, Huang Ping, Li Shaojun, Li Tuo, Wang Hongsheng, Wang Xiaoming, Chen Sihe, and Chen Yangu. 2004. "Why Must We Talk about the Environment? A Summary of Nanshan Seminar." Translated by Yan Hairong. *positions* 12 (1): 237–46.

Han, Zhiyong, Haining Ma, Guozhong Shi, Li He, Luoyu Wei, and Qingqing Shi. 2016. "A Review of Groundwater Contamination near Municipal Solid Waste Landfill Sites in China." *Science of the Total Environment* 569–70:1255–64.

Hansen, Mette Halskov. 1999. *Lessons in Being Chinese: Minority Education and Ethnic Identity in Southwest China*. Seattle: University of Washington Press.

Hanson, Jennifer. 2004. *Nations in Transition: Mongolia*. New York: Facts on File.

Hardin, Garrett. 1968. "The Tragedy of the Commons." *Science* 162:1243–48.

Harrell, Stevan. 1982. *Ploughshare Village: Culture and Context in Taiwan.* Seattle: University of Washington Press.

———. 1985. "Why the Chinese Work So Hard: Reflections on an Entrepreneurial Ethic." *Modern China* 11 (2): 203–26.

———. 1995a. "Introduction: Microdemography and the Modeling of Population Process in Late Imperial China." In *Chinese Historical Demography,* edited by Stevan Harrell, 1–20. Berkeley: University of California Press.

———. 1995b. "Introduction: Civilizing Projects and the Reaction to Them." In *Cultural Encounters on China's Ethnic Frontiers,* edited by Stevan Harrell, 1–36. Seattle: University of Washington Press.

———. 2001. *Ways of Being Ethnic in Southwest China.* Seattle: University of Washington Press.

———. 2020. "The Eco-Developmental State and the Environmental Kuznets Curve." In *Greening East Asia: The Rise of the Eco-Developmental State,* edited by Ashley Esarey, Mary Alice Haddad, Joanna Lewis, and Stevan Harrell, 241–65. Seattle: University of Washington Press.

———. 2021. "The Four Horsemen of the Ecopocalypse: The Agricultural Ecology of the Great Leap Forward." *Human Ecology* 49 (1): 7–18.

Harrell, Stevan, Y. Qingxia, S. J. Viraldo, R. K. Hagmann, A. H. Schmidt, and T. Hinckley. 2016. "Forest Is Forest and Meadows Are Meadows: Cultural Landscapes and Bureaucratic Landscapes in Jiuzhaigou County, Sichuan." *Archiv Orientální* 84 (3): 595–623.

Harrell, Stevan, Amanda H. Schmitt, Brian D. Collins, R. Keala Hagmann, and Thomas M. Hinckley. 2022. "Sunny Slopes Are Good for Grain; Shady Slopes Are Good for Trees: Nuosu Yi Agroforestry in Southwestern China." In *The Cultivated Forest: People and Woodlands in Asian History,* edited by Ian M. Miller, 161–84. Seattle: University of Washington Press.

Harrell, Stevan, Wang Yuesheng, Han Hua, Gonçalo D. Santos, and Zhou Yingying. 2011. "Fertility Decline in Rural China: A Comparative Analysis." *Journal of Family History* 36 (1): 15–36.

Harwood, Russell. 2013. *China's New Socialist Countryside: Modernity Arrives in the Nu River Valley.* Seattle: University of Washington Press.

Hassan, Jennifer. 2021. "Inside the Flooded China Subway, Which Trapped Commuters Up to Their Necks in Water." *Washington Post,* 21 July 2021.

Hathaway, Michael. 2013. *Environmental Winds: Making the Global in Southwest China.* Berkeley: University of California Press.

Hayashi, Daisuke, Joern Huentler, and Joanna I. Lewis. 2018. "Gone with the Wind: A Learning Curve Analysis of China's Wind Power Industry." *Energy Policy* 120:38–51.

Hayes, Jack Patrick. 2021. "Fire Suppression and the Wildfire Paradox in Contemporary China: Policies, Resilience, and Effects in Chinese Fire Regimes." *Human Ecology* 49 (1): 19–32.

He, Fanneng, Quansheng Ge, Junhu Dai, and Yujuan Rao. 2008. "Forest Change of China in Recent 300 Years." *Journal of Geographical Sciences* 18:59–72.

He, Fanneng, Shicheng Li, and Xuezhen Zhang. 2015. "A Spatially Explicit Reconstruction of Forest Cover in China, 1700–2000." *Global and Planetary Change* 131:73–81.

He, Feng, Qingzhi Zhang, Jiasu Lei, Weihui Fu, and Xiaoning Hu. 2013. "Energy Efficiency and Productivity Change of China's Iron and Steel Industry: Accounting for Undesirable Outputs." *Energy Policy* 54:204–13.

He, Guizhen, Yonglong Lu, Arthur P. J. Mol, and Theo Beckers. 2012. "Changes and Challenges: China's Environmental Management in Transition." *Environmental Development* 3:25–38.

He, Jia, and Hua Wang. 2012. "Economic Structure, Development Policy and Environmental Quality: An Empirical Analysis of Environmental Kuznets Curves with Chinese Municipal Data." *Ecological Economics* 76:49–59.

He Naiwei 何乃維. 1998. "Pohuai shengtai hui zhu hongshui weinüe" 破壞生態會助洪水為虐 [Ecological damage can aid the destructive effects of floods]. *Zhonghua Ernü haiwai ban* 中華兒女海外版1998 (9): 55–58.

He, Qingsong, Yan Song, Yaolin Liu, and Chaohui Yin. 2017. "Diffusion or Coalescence? Urban Growth Pattern and Change in 363 Chinese Cities from 1995 to 2015." *Sustainable Cities and Society* 35:729–39.

Heberer, Thomas, and René Trappel. 2013. "Evaluation Processes, Local Cadres' Behaviour, and Local Development Processes." *Journal of Contemporary China* 22 (84): 1048–66.

Heilmann, Sebastian. 2008. "From Local Experiments to National Policy: The Origins of China's Distinctive Policy Practice." *China Journal* 59:1–30.

Henck, Amanda C., Katharine W. Huntington, John O. Stone, David R. Montgomery, and Bernard Hallet. 2011. "Spatial Controls on Erosion in the Three Rivers Region, Southeastern Tibet and Southwestern China." *Earth and Planetary Science Letters* 303 (1): 71–83.

Henck, Amanda, James Taylor, Hongliang Lu, Yongxian Li, Qingxia Yang, Barbara Grub, Sara Jo Breslow, et al. 2010. "Anthropogenic Hillslope Terraces and Swidden Agriculture in Jiuzhaigou National Park, Northern Sichuan, China." *Quaternary Research* 73:201–7.

Herman, John E. 1993. "National Integration and Regional Hegemony: The Political and Cultural Dynamics of Qing State Expansion, 1650–1750." PhD diss., University of Washington, Seattle.

———. 2007. *Amid the Clouds and Mist: China's Colonization of Guizhou, 1200–1700.* Cambridge, MA: Harvard University Asia Center.

Hesketh, Therese, Li Lu, and Zhu Wei Xing. 2005. "The Effect of China's One-Child Family Policy after 25 Years." *New England Journal of Medicine* 353 (11): 1171–176.

Hessburg, Paul F., and James G. Agee. 2003. "An Environmental Narrative of Inland Northwest United States Forests, 1800–2000." *Forest Ecology and Management* 178:23–59

Hill, Ann Maxwell. 2001. "Captives, Kin, and Slaves in Xiao Liangshan." *Journal of Asian Studies* 60 (4): 1033–49.

Hinton, William. 1966. *Fanshen: A Documentary of Revolution in a Chinese Village.* New York: Vintage.

Ho, Joanne. 2004. "Pockets of Poverty in a Fast-Growing Economy: Quantifying Market Shares in Rural Southwest China." BA honors thesis, University of Washington, Seattle.

Ho, Samuel P. S. 1978. *Economic Development of Taiwan, 1860–1970.* New Haven: Yale University Press.

Hobor, George. 2015. "New Orleans' Remarkably (Un)Predictable Recovery: Developing a Theory of Urban Resilience." *American Behavioral Scientist* 59 (10): 1214–30.

Holeman, John N. 1968. "The Sediment Yield of Major Rivers of the World." *Water Resources Research* 4 (4): 737–47.

Holling, C. S., Lance H. Gunderson, and Garry D. Peterson. 2002. "Sustainability and Panarchies." In *Panarchy: Understanding Transformations in Human and Natural Systems*, edited by Lance H. Gunderson and C. S. Holling, 63–102. Washington, DC: Island Press.

Holling, C. S., and Gary K. Meffe. 1996. "Command and Control and the Pathology of Natural Resource Management." *Conservation Biology* 10 (2): 328–37.

Holst, Arne. 2020. "Refrigerators—Statistics and Facts." Statista, 11 June 2020. www .statista.com/topics/2182/refrigerators-and-freezers/.

Hong, Sheng, Yifan Jie, Xiaosong Li, and Nathan Lu. 2019. "China's Chemical Industry: New Strategies for a New Era." McKinsey & Company. www.mckinsey.com/industries /chemicals/our-insights/chinas-chemical-industry-new-strategies-for-a-new-era.

Hoshuyama, Tsutomu, Guowei Pan, Chieko Tanaka, Yiping Feng, Tiefu Liu, Liming Liu, Tomoyuki Hanaoka, and Ken Takahashi. 2006. "Mortality of Iron-Steel Workers in Anshan, China: A Retrospective Cohort Study." *International Journal of Occupational and Environmental Health* 11 (3): 193–202.

Hou Li. 2018. *Building for Oil: Daqing and the Formation of the Chinese Socialist State.* Cambridge, MA: Harvard University Asia Center.

Hsieh, Shih-chung. 1995. "On the Dynamics of Tai/Dai Lue Ethnicity: An Ethnohistorical Analysis." In *Cultural Encounters on China's Ethnic Frontiers*, edited by Stevan Harrell, 301–28. Seattle: University of Washington Press.

Hu, Danian. 2007. "The Reception of Relativity in China." *Isis* 98 (3): 539–57.

Hu, Dinghuan, Thomas Reardon, Scott Rozelle, Peter Timmer, and Honglin Wang. 2004. "The Emergence of Supermarkets with Chinese Characteristics: Challenges and Opportunities for China's Agricultural Development." *Development Policy Review* 22 (5): 557–86.

Hu, R. L., Z. Q. Yue, L. C. Wang, and S. J. Wang. 2004. "Review on Current Status and Challenging Issues of Land Subsidence in China." *Engineering Geology* 76:65–77.

Hu Yinan. 2011. "Life behind the Three Gorges Dam." *China Daily*, 3 June 2011.

Hu Yue 胡月, ed. 2020. "Mai shen 1800 mi diyi shijiao kan 'yin Han ji Wei' shushui suiddong chuanyue qinling zhuji" 埋深1800米第一视角看"引汉济渭"输水隧洞穿越秦岭主脊 [Buried 1800 meters deep: First look at the "Taking the Han to Fill the Wei" water transport tunnel that goes through the main spine of the Qinling Mountains]. CCTV.com, 1 December 2020. http://m.news.cctv.com/2020/12/01/ARTI40J0Ru FbYJd3TRQVnS4G201201.shtml.

Hua, Ye. 2015. "Influential Factors of Farmers' Demands for Agricultural Science and Technology in China." *Technological Forecasting and Social Change* 100:249–54.

Huang, Jikun, and Scott Rozelle. 1996. "Technological Change: Rediscovering the Engine of Productivity Growth in China's Rural Economy." *Journal of Development Economics* 49:337–69.

Huang, Jikun, Jun Yang, Xiangzheng Deng, Jinxia Wang, and Scott Rozelle. 2015. "Urbanization, Food Production and Food Security in China." AgEcon Search. https:// ageconsearch.umn.edu/record/189685/files/Huang_%20J.pdf.

Huang, Kristin. 2020. "Tibet Hydropower Plans Will Boost International Cooperation, Says Chinese State Company Boss Despite Risk of Indian Backlash." *South China Morning Post*, 29 November 2020.

Huang, Philip C. C. 1985. *The Peasant Economy and Social Change in North China*. Stanford, CA: Stanford University Press.

———. 1990. *The Peasant Economy and Rural Development in the Yangzi Delta, 1350–1988*. Stanford, CA: Stanford University Press.

Huang, Rong, Lupei Zhu, John Encarnacion, Yixian Xu, Chi-Chia Song Luo, and Xiaohuan Jiang. 2018. "Seismic and Geologic Evidence of Water-Induced Earthquakes in the Three Gorges Reservoir Region of China." *Geophysical Research Letters* 45 (12): 5929–36.

Huang Ronghan. 1988. "Development of Groundwater for Agriculture in the Lower Yellow River Alluvial Basin." In *Efficiency in Irrigation: The Conjunctive Use of Surface and Groundwater Resources*, edited by Gerald T. O'Mara, 80–84. Washington, DC: World Bank.

Huang, Shaoqing, and Weiguo Qiao. 2005. "Development of China's Electronics Industry: Causes and Constraints." *China and World Economy* 13 (3): 107–23.

Huang Shu-min. 1989. *The Spiral Road: Change in a Chinese Village through the Eyes of a Communist Party Leader*. Boulder, CO: Westview Press.

Huang Xiao-rong, Gao Lin-yun, Yang Peng-peng, and Xi Yuan-yuan. 2018. "Cumulative Impact of Dam Construction on Streamflow and Sediment Regime in Lower Reaches of the Jinsha River, China." *Journal of Mountain Science* 15 (12): 2752–65.

Huang, Yu. 2012. "Vibrant Risks: Scientific Aquaculture and Political Ecologies in China." PhD diss., University of Washington, Seattle.

———. 2015. "Can Capitalist Farms Defeat Family Farms? The Dynamics of Capitalist Accumulation in Shrimp Aquaculture in South China." *Journal of Agrarian Change* 15 (3): 392–412.

Hubei Daily 湖北日报. 2021. "Quan chang 194 gongli, touzi 700 duo yi yin Jiang bu Han kaigong zhi deng 'yi sheng ling xia'" 全长194公里, 投资700多亿引江补汉开工只等"一声令下" [The 194-kilometer long, 70 billion-plus *yuan* taking the Chang to Replenish the Han project just awaits "an order to come down"]. 8 February 2021. www.sohu.com/a/449375102_120207620.

Huisman, J., I. Botezatu, L Herreras, M. Liddane, J. Hintsa, V. Luda di Cortemiglia, P. Leroy, et al. 2015. *Countering WEEE Illegal Trade (CWIT) Summary Report, Market Assessment, Legal Analysis, Crime Analysis and Recommendations Roadmap*. Lyon, France: CWIT Consortium.

Huobao Nongye Zhaoshang Wang 火爆农业招商网. 2014. "Hongshu muchanliang: Hongshu muchan duoshao jin?" 红薯亩产量: 红薯亩产多少斤 [Per-*mu* yield of sweet potatoes: How many *jin* does a *mu* of sweet potatoes yield?]. http://m.3456.tv/bk/13526.html.

icity. 2015. "Chongqing Zhongguo Sanxia Bowuguan" 重庆中国三峡博物馆 [China Three Gorges Museum Chongqing]. https://art.icity.ly/museums/r77u7d6.

IEA (International Energy Agency). 2020. Global EV Outlook 2020. www.iea.org/reports/global-ev-outlook-2020.

———. 2021. Key World Energy Statistics 2021. https://iea.blob.core.windows.net/assets/52f66a88-0b63-4ad2-94a5-29d36e864b82/KeyWorldEnergyStatistics2021.pdf.

Index Mundi. 2022. Agriculture—China. www.indexmundi.com/agriculture/?country
=cn&graph=production.

Ingman, Mark Christian. 2012. "The Role of Plastic Mulch as a Water Conservation
Practice for Desert Oasis Communities of Northern China." MS thesis, Oregon State
University.

International Rivers. 2017. "Press Release: Disappointing and Lengthy Mediation
Leaves Impacts of Xayaburi Dam Unaddressed." 20 July 2017. https://archive.inter
nationalrivers.org/resources/press-release-disappointing-and-lengthy-mediation
-leaves-impacts-of-xayaburi-dam.

IPCC (Intergovernmental Panel on Climate Change). 2019. "Special Report on Climate
Change and Land: Summary for Policymakers." www.ipcc.ch/srccl/chapter/summary
-for-policymakers/.

IUCN (International Union for the Conservation of Nature). 2022a. "Chinese Sturgeon."
www.iucnredlist.org/species/236/146104213.

———. 2022b. "Yangtze Sturgeon." www.iucnredlist.org/species/231/61462199.

———. 2022c. "Chinese Paddlefish." www.iucnredlist.org/species/18428/146104283.

Ives, Mike. 2016. "China's Largest Freshwater Lake Shrinks, Solution Faces Criticism."
New York Times, 28 December 2016.

IWP (International Water Power and Dam Construction). 2018. "IFC Starts Pilot Assessment in Lao PDR River Basin." 4 September 2018. www.waterpowermagazine
.com/news/newsifc-starts-pilot-assessment-in-lao-pdr-river-basin-6734365.

———. 2019. "Nam Ngiep 1 Starts Commercial Operation." 13 September 2019. www
.waterpowermagazine.com/news/newsnam-ngiep-1-starts-commercial-operation
-7409287.

Jaffe, Daniel, Justin Putz, Greg Hof, Gordon Hof, Jonathan Hee, Dee Ann Lom-
mers-Johnson, Francisco Gabela, et al. 2015. "Diesel Particulate Matter and Coal
Dust from Trains in the Columbia River Gorge, Washington State, USA." *Atmospheric Pollution Research* 6 (6): 946–52.

Jagchid, Sechen, and Paul Hyer. 1979. *Mongolia's Culture and Society*. Boulder, CO:
Westview Press.

Jahiel, Abigail. 1998. "The Organization of Environmental Protection in China." *China
Quarterly* 156:757–87.

Jemio, Miriam Telma. 2019. "Will Exporting Beef to China Cause Deforestation in
Bolivia?" In *Tracking China's Soy and Beef Imprint on South America*, edited by
Isabel Hilton, 19–21. Diálogo Chino, 5 December 2019. https://dialogochino.net
/32059-tracking-chinas-soy-and-beef-imprint-on-latin-america/.

Jia Kehua 贾科华. 2020. "Yalu Zangbu Jiang xiayou shuidian kaifa juece qiaoding:
Guimo jin 6000 wan qianwa, xiangdang yu 'zai zao sange Sanxia'" 雅鲁藏布江下游
水电开发决策敲定：规模近6000万千瓦，相当于"再造3个三峡" [The decision to de-
velop hydroelectricity on the lower reaches of the Yarlung Tsangpo River is settled;
its extent will be close to 60 gigawatts, equivalent to "building three more Sanxias"]. Guanchazhe 观察者, 28 November 2020. www.guancha.cn/politics
/2020_11_28_572852.shtml.

Jia, Tianqi, Wei Guo, Wenbin Liu, Ying Xing, Rongrong Lei, Xiaolin Wu, and Shurui
Sun. 2021. "Spatial Distribution of Polycyclic Aromatic Hydrocarbons in the Water-

Sediment System near Chemical Industry Parks in the Yangtze River Delta, China. *Science of the Total Environment* 754: article 142176, 9 pages.

Jia, Xiaofang, Jiawu Liu, Bo Chen, Donghui Jin, Zhongxi Fu, Huilin Liu, Shufa Du, Barry M. Popkin, and Michelle A. Mendez. 2017. "Differences in Nutrient Energy Contents of Commonly Consumed Dishes Prepared in Restaurants v. at Home in Hunan Province, China." *Public Health Nutrition* 21 (7): 1307–1318.

Jiang, Chang. 2011. "A General Investigation of Shanghai Sewerage Treatment System." Master's thesis, Halmstad University. www.diva-portal.org/smash/get/diva2:427129 /FULLTEXT01.pdf.

Jiang Guanghui, Ma Wenqiu, Wang Deqi, Zhou Dingyang, Zhang Ruijian, and Zhou Tao. 2017. "Identifying the Internal Structure Evolution of Urban Built-Up Land Sprawl (UBLS) from a Composite Structure Perspective: A Case Study of the Beijing Metropolitan Area, China." *Land Use Policy* 62:258–67.

Jiang Hua 江华. 2012. "1975: Henan Banqiao kuiba canju" 1975: 河南板桥溃坝惨剧 [1975: The tragedy of the Banqiao Dam collapse in Henan]. *Gonchandang Yuan* 共产党员 September (2): 50.

Jiang, Qiang, Xinguang Cheng, Zhitao Liu, Deepthi S. Varma, Rong Wang, and Shiwen Zhao. 2017. "Diet Diversity and Nutritional Status among Adults in Southwest China." *PLoS One* 12 (2): 1–9.

Jiang, Xiaoyu, Yangfen Chen, and Lijian Wang. 2018. "Can China's Agricultural FDI in Developing Countries Achieve a Win-Win Goal? Enlightenment from the Literature." *Sustainability* 11 (1): 41–62. www.mdpi.com/2071-1050/11/1 /41/htm.

Jiang, Xinyuan, Sven G. Sommer, and Knud V. Christensen. 2011. "A Review of the Biogas Industry in China." *Energy Policy* 39:6073–81.

Jiang Yifan. 2019. "Waste Sorting, an Imposed 'Social Contract' with Potential." China Dialogue, 11 July 2019. https://chinadialogue.net/en/pollution/11373-waste-sorting -an-imposed-social-contract-with-potential/.

Jiang, Yilun, Ximei Hou, Guoshun Zhuang, Juan Li, Qiongzhen Wang, Rong Zhang, and Yanfen Lin. 2009. "The Sources and Seasonal Variations of Organic Compounds in $PM_{2.5}$ in Beijing and Shanghai." *Journal of Atmospheric Chemistry* 62:175–92.

Jiang, Yong, Chris Zevenbergen, and Dafang Fu. 2017. "Understanding the Challenges for the Governance of China's 'Sponge Cities' Initiative to Sustainably Manage Urban Stormwater and Flooding." *Natural Hazards* 89:521–29.

Jin Chengmei 靳承美 2009. "'Dayuejin' yu da lian gangtie jishi" "大跃进"与大炼钢铁 记事 [Record of the events of the "Great Leap Forward" and the Big Iron and Steel Smelting]. *Chunqiu* 春秋 2009 (3).

Jin, Jiyun. 2012. "Changes in the Efficiency of Fertiliser Use in China." *Journal of the Science of Food and Agriculture* 92:1006–9.

Jin, Wenting, Jianxia Chang, Yimin Wang, and Tao Bai. 2019. "Long-Term Water-Sediment Multi-Objectives Regulation of Cascade Reservoirs: A Case Study in the Upper Yellow River, China." *Journal of Hydrology* 577: article 123978. www.sciencedirect .com/science/article/pii/S0022169419306985?via% 3Dihub.

Jin Zhonghuan 金中桓. 2011. "Shuanglun shuanghua li qianshi jinsheng" 双轮双铧犁 前世今生 [The rebirth of the double-wheeled, double-shared plow]. *Wenzhou Ribao*

　　Ouwang 温州日报瓯网, 18 September 2011. Inaccessible July 2022. www.wzrb.com
　　.cn/mobile_show.aspx? id=299816.

Jing, Jun. 2000. *The Temple of Memories: History, Power, and Morality in a Chinese Village.* Stanford, CA: Stanford University Press.

Jing, Ran, Jack C. P. Cheng, Vincent J. L. Gan, Kok Sin Woon, and Irene M. C. Lo. 2014. "Comparison of Greenhouse Gas Emission Accounting Methods for Steel Production in China." *Journal of Cleaner Production* 83:165–72.

Jingchuhao 荆楚号. 2019. "Sanxia da ba bianxing yanzhong, you tu you zhenxiang? Jieguo bei Zhongguo hangtian kandaole" 三峡大坝变形严重，有图有真相？结果被中国航天看到了 [Serious shape-change in the Sanxia Dam—are the pictures genuine? A Chinese satellite looked at the results]. http://m.cnhubei.com/cmdetail /411624.

Jintian Toutiao 今天头条. 2019. "Beijing ditie zui xin guihua: 2020 nian Beijing jiang fen 40 xian" 北京地铁最新规划：2020年北京将分40线 [The newest plans for Beijing's subways: In 2020 Beijing will have 40 lines]. 12 December 2019. https://twgreatdaily .com/zh-cn/oqKvJGoBJleJMoPMLqs7.html.

Johnson, Chalmers A. 1962. *Peasant Nationalism and Communist Power: The Emergence of Revolutionary China, 1937–1945.* Stanford, CA: Stanford University Press.

Johnston, Barbara Rose, and Holly M. Barker. 2008. *Consequential Damages of Nuclear War: The Rongelap Report.* Walnut Creek, CA: Left Coast Press.

Johnston, R. F. 1910. *Lion and Dragon in Northern China.* New York: Dutton.

Ju, Hanyu, Sijia Li, Y. Jun Xu, Guangxin Zhang, and Jiquan Zhang. 2019. "Intensive Livestock Production Causing Antibiotic Pollution in the Yinma River of Northeast China." *Water* 11 (10): 2006–21.

Kahn, Joseph. 2007. "In China, a Lake's Champion Imperils Himself." *New York Times,* 14 October 2007.

Kaika, Dimitra, and Efthimios Zervas. 2013. "The Environmental Kuznets Curve (EKC) Theory—Part A: Concept, Causes and the CO_2 Emissions Case." *Energy Policy* 62:1392–1402.

Kallis, Giorgios, Vasilis Kostakis, Steffen Lange, Barbara Muraca, Susan Paulson, and Matthias Schmelzer. 2018. "Research on Degrowth." *Annual Review of Environment and Resources* 43:291–316.

Kambara, Tetsu. 1992. "The Energy Situation in China." *China Quarterly* 131:608–36.

Kander, Astrid, Paolo Malanima, and Paul Warde. 2013. *Power to the People: Energy in Europe over the Last Five Centuries.* Princeton: Princeton University Press.

Kang Jia 康佳. 2022. "Poyang Hu tiqian 'ruku' wei you jilu yilai zui zao" 鄱阳湖提前 "入枯" 为有记录以来最早 [Poyang Lake "enters dryness" the earliest since there have been records]. *Caixin* 财新, 8 August 2022. https://science.caixin.com/2022-08-08 /101923483.html.

Kang Ning. 2020. "Ten-Year Yangtze Fishing Ban Not Enough to Save Migratory Species." China Dialogue, 4 November 2020. https://chinadialogue.net/en/nature/ten -year-yangtze-fishing-ban-not-enough-to-save-migratory-species/.

Kepu. n.d. *Kepu jjylur shy-wa-te* ꋔꁌꐯꊏꄸ [Realigning the guardian spirits by means of horoscopes]. Nuosu-language ritual manuscript, catalog no. 1998-83/361, Burke Museum of Natural History and Culture, Seattle, WA.

Khan [Haan], Almaz. 1996. "Who Are the Mongols? State, Ethnicity, and the Politics of

Representation in the PRC." In *Negotiating Ethnicities in China and Taiwan*, edited by Melissa J. Brown, 125–59. Berkeley, CA: Institute of East Asian Studies.

Kharas, Homi, and Megan Dooley. 2020. "China's Influence on the Global Middle Class." Brookings Institution. www.brookings.edu/wp-content/uploads/2020/10/FP _20201012_china_middle_class_kharas_dooley.pdf.

Kibler, Kelly M., and Desiree D. Tullos. 2013. "Cumulative Biophysical Impact of Large and Small Hydropower Development in Nu River, China." *Water Resources Research* 49 (6): 3104–18.

Kim, Su Min "Alana." 2014. "Shunning the Authority: Symbolic Purity and Autonomy in Consuming Non-certified Organic Foods in Chengdu, China." BA honors thesis, University of Washington, Seattle.

Kinzley, Judd C. 2018. *Natural Resources and the New Frontier: Constructing Modern China's Borderlands*. Chicago: University of Chicago Press.

Kirin Holdings Company. 2019. "Kirin Beer University Report on Global Beer Production by Country in 2018." www.kirinholdings.co.jp/english/news/2019/1003_01.html.

Knapp, Ronald. 2008. "Beijing's Hutongs." *Guanxi: The China Letter* 2 (10).

Koitabashi Tsutomu 小板橋努. 2009. "Chūgoku nōgyaku kindaishi" 中国農薬近代史 [Modern history of agrochemicals in China] *Journal of Pesticide Science* 34 (4): 289–94.

Kong Dechen 孔德晨. 2022. "Dao 2025 nian, shenghuo laji fenlei shouyun nengli dadao 70 dun/ri zuoyou—"shisi wu," shenghuo laji zheyang guan" 到2025年, 生活垃圾分类收运能力达到70吨/日左右-"十四五", 生活垃圾这样管 [By 2025, capacity to sort, collect, and transport household waste will reach about 70 tons per day—during the "fourteenth five-year [plan]," household waste will be handled like this]. *People's Daily International Edition*, 23 February 2022.

Kong, Yang, Weijun He, Liang Yuan, Zhaofang Zhang, Xin Gao, Yu'e Zhao, and Dagmawi Mulugeta Degefu. 2021. "Decoupling Economic Growth from Water Consumption in the Yangtze River Economic Belt, China." *Ecological Indicators* 123: article 107344.

Kong, Zheng-Hong, Lindsay Stringer, Jouni Paavola, and Qi Lu. 2021. "Situating China in the Global Effort to Combat Desertification." *Land* 10 (7): 702.

Koop, Fermín. 2020. "Chinese Pork Investment Sparks Criticism in Argentina." *Diálogo Chino*, 9 October 2020. https://dialogochino.net/en/trade-investment/37359-chinese -pork-investment-sparks-criticism-in-argentina/.

Kornai, Janos. 1986. "The Soft Budget Constraint." *Kyklos* 39 (1): 3–30.

Kostka, Genia, and Chunman Zhang. 2018. "Tightening the Grip: Environmental Governance under Xi Jinping." *Environmental Politics* 27 (5): 769–81.

Kremen, Claire, Alistair Iles, and Christopher Bacon. 2012. "Diversified Farming Systems: An Agroecological, Systems-Based Alternative to Modern Industrial Agriculture." *Ecology and Society* 17 (4): 44. http://dx.doi.org/10.5751/ES-05103-170444.

Kremen, Claire, and Albie Miles. 2012. "Ecosystem Services in Biologically Diversified versus Conventional Farming Systems: Benefits, Externalities, and Trade-Offs." *Ecology and Society* 17 (4): article 40. http://dx.doi.org/10.5751/ES-05035-170440.

Kuaiyi Licai Wang 快易理财网. n.d. "Zhongguo linian renkou zongshu tongji" 中国历年人口总数统计 [Yearly statistics on China's total population]. Accessed June 2022. www.kylc.com/stats/global/yearly_per_country/g_population_total/chn.html/.

Kuang Yaohui 匡跃辉. 1999. "Hunan shuizai wenti yanjiu" 湖南水灾问题研究 [Research on the question of floods in Hunan]. *Hunan Administrative College Journal*, inaugural issue: 86–91.

Kueh, Y. Y. 1984. "A Weather Index for Analysing Grain Yield Instability in China, 1952–81." *China Quarterly* 97:68–83.

———. 2006. "Mao and Agriculture in China's Industrialization: Three Antitheses in a 50-Year Perspective." *China Quarterly* 187:700–723.

Kung, James Kai-sing, Xiaogang Wu, and Yuxiao Wu. 2012. "Inequality of Land Tenure and Revolutionary Outcome: An Economic Analysis of China's Land Reform of 1946–1952." *Explorations in Economic History* 49 (4): 482–97.

Laakkonen, Simo. 2020. "Urban Resilience and Warfare: How Did the Second World War Affect the Urban Environment?" *City and Environment Interactions* 5: article 100035.

Lai Nianyue 赖年悦, Hu Wanming 胡万明, and Fang Kai 方凯. 2016. "Shengtai wenming shijiao xia Chaohu yuye ziyuan yanghu yu liyong yanjiu" 生态文明视角下巢湖渔业资源养护与利用研究 [Research of Chaohu fishery resources conservation and utilization from the perspective of ecological civilization]. *Zhongguo Yuye Jingji* 中国渔业经济 2016 (3): 68–73.

Lai, Xijun, David Shankman, Claire Huber, Herve Yesou, Qun Huang, and Jiahu Jiang. 2014. "Sand Mining and Increasing Poyang Lake's Discharge Ability: A Reassessment of Causes for Lake Decline in China." *Journal of Hydrology* 519: 1698–1706.

Lamb, Vanessa. 2019. "Salween: What's in a Name?" In *Knowing the Salween River: Resource Politics of a Contested Transboundary River*, edited by Carl Middleton and Vanessa Lamb, 17–26. Cham, Switzerland: Springer Nature.

Land Matrix. n.d. "Global: By Investor Country." Accessed July 2022. https://landmatrix.org/data/by-investor-country/china/?order_by=&more=220.

Lao Sun 老孙. 2020. "Piaobo zai shuimian de taiyang neng, yuanlai liyong taiyang zuo zenme duo shih hai keyi xunzhao waixing ren" 漂浮在水面的太阳能, 原来利用太阳做怎么多事, 还可以寻找外星人 [Solar energy floating on the water, already using the sun to do so many things, we can even look for space aliens]. *Mei ri toutiao* 每日头条, 4 January 2020. https://kknews.cc/zh-sg/science/ljlayye.html.

Lary, Diana. 2007. Introduction to *The Chinese State at the Borders*. Vancouver: UBC Press.

———. 2017. "Manchuria: History and Environment." In *Empire and Environment in the Making of Manchuria*, edited by Norman Smith, 28–52. Vancouver: UBC Press.

Lattimore, Owen. 1929. "The Desert Road to Turkestan." *National Geographic Magazine* 55:661–702.

———. 1940. *Inner Asian Frontiers of China*. New York: Oxford University Press.

Latour, Bruno. 1998. "From the World of Science to the World of Research." *Science* 280 (5361): 208–9.

———. 2002. "The Science Wars: A Dialogue." *Common Knowledge* 8 (1): 71–79.

———. 2018. *Down to Earth: Politics in the New Climatic Regime*. Medford, MA: Polity Press.

Lau, Jack. 2021. "Probe into Deaths in Chinese Mine Accident after Claims of Cover-Up." *South China Morning Post*, 18 March 2021.

Lau, Mimi. 2021. "Xi Jinping Declares Extreme Poverty Has Been Wiped Out in China." *South China Morning Post*, 25 February 2021.

Lavely, William, and Ronald Freedman. 1990. "The Origins of the Chinese Fertility Decline." *Demography* 27 (3): 357–67.

Lavely, William R., and R. Bin Wong. 1998. "Revising the Malthusian Narrative: The Comparative Study of Population Dynamics in Late Imperial China." *Journal of Asian Studies* 57 (3): 714–48.

Lawrence, Martha, Richard Bullock, and Ziming Liu. 2019. *China's High-Speed Rail Development*. Washington, DC: World Bank. https://documents1.worldbank.org /curated/en/933411559841476316/pdf/Chinas-High-Speed-Rail-Development.pdf.

Lazzeri, Thais. 2019a. "China's Demand for Brazilian Beef Raises Deforestation Risk." In *Tracking China's Soy and Beef Imprint on South America*, edited by Isabel Hilton, 4–6. Diálogo Chino, 5 December 2019. https://dialogochino.net/32059-tracking -chinas-soy-and-beef-imprint-on-latin-america/.

———. 2019b. "How Soy and Beef Spark Amazon Fires." In *Tracking China's Soy and Beef Imprint on South America*, edited by Isabel Hilton, 7–9. Diálogo Chino, 5 December 2019. https://dialogochino.net/32059-tracking-chinas-soy-and-beef-imprint -on-latin-america/.

Leach, Edmund R. 1954. *Political Systems of Highland Burma*. Boston: Beacon.

Lee, Alexander Tse-Yan. 2005. "Faked Eggs: The World's Most Unbelievable Invention." *Internet Journal of Toxicology* 2 (1). http://web.archive.org/web/20060428034330 /http://www.ispub.com/ostia/index.php?xmlFilePath=journals/ijto/vol2n1/eggs.xml.

Lee, Amanda. 2021. "China Food Security: Soybean Imports 'Exceptionally Large' as Pig Population Nears Pre-African Swine Fever Level." *South China Morning Post*, 14 June 2021.

Lee, Ching Kwan. 1995. "Engendering the Worlds of Labor: Women Workers, Labor Markets, and Production Politics in the South China Economic Miracle." *American Sociological Review* 60 (3): 378–97.

———. 2007. *Against the Law: Labor Protests in China's Rustbelt and Sunbelt*. Berkeley: University of California Press.

Lee, James Z., and Wang Feng. 1999. *One Quarter of Humanity: Malthusian Mythology and Chinese Realities, 1700–2000*. Cambridge, MA: Harvard University Press.

Lee, Yen-Han, Timothy C. Chiang, Ching-Ti Liu, and Yan-Chang Chang. 2020. "Investigating Adolescents' Sweetened Beverage Consumption and Western Fast Food Restaurant Visits in China, 2006–2011." *International Journal of Adolescent Medicine and Health* 32 (5): article 20170209.

Leonard, Pamela. n.d. "Social Conflict and Its Spatial Dimension." In "Moral Landscape in a Sichuan Mountain Village: A Digital Ethnography of Place," edited by John Flower and Pamela Leonard. Accessed June 2021. www.sichuanvillage.org /chapters/xiakou/spaceandconflict.

Leonard, Pam, and John Flower. n.d. "New House: Livelihood, Debt and Imaginings of the Rural Landscape in Sichuan." Unpublished paper.

Lepawsky, Josh, Joshua Goldstein, and Yvan Schulz. 2015. "Criminal Negligence?" *Discard Studies*, 24 June 2015. https://discardstudies.com/2015/06/24/criminal-negligence/.

Levenson, Joseph R. 1962. "The Place of Confucius in Communist China." *China Quarterly* 12:1–18.

Levin, Simon A. 1992. "The Problem of Pattern and Scale in Ecology: The Robert H. MacArthur Award Lecture." *Ecology* 73 (6): 1943–67.

Levine, Mark D., Nan Zhou, and Lynn Price. 2009. *The Greening of the Middle Kingdom: The Story of Energy Efficiency in China*. Berkeley, CA: Lawrence Berkeley National Laboratory. www.nae.edu/14951/The-Greening-of-the-Middle-Kingdom-The-Story-of-Energy-Efficiency-in-China.

Levins, Richard. 1966. "The Strategy of Model Building in Population Biology." *American Scientist* 54:421–31.

Lewis, Joanna I. 2013. "China's Environmental Diplomacy: Climate Change, Domestic Politics and International Engagement." In *China across the Divide: The Domestic and Global in Politics and Society*, edited by Rosemary Foot, 200–226. Oxford: Oxford University Press.

———. 2017. "Getting More from Wind Farms." *Nature Energy* 1 (June): 1–2.

———. 2020. "China's Low-Carbon Energy Strategy." In *Greening East Asia: The Rise of the Eco-Developmental State*, edited by Ashley Esarey, Mary Alice Haddad, Joanna I. Lewis, and Stevan Harrell, 47–61. Seattle: University of Washington Press.

Leys, Colin. 1977. "Underdevelopment and Dependency: Critical Notes." *Journal of Contemporary Asia* 7 (1): 92–107.

Li Chunfeng 李春峰. 2010. "'Dayuejin' de qianzou: Xingxiu nongtian shuili jianshe yundong" "大跃进"的前奏: 兴修农田水利建设运动 [Prelude to the "Great Leap Forward": The campaign to build agricultural water conservancy]. *Cangsang* 2010 (1): 113–14.

Li, Dongfeng, Xi Xi Lu, Xiankun Yang, Li Chen, and Lin Lin. 2018. "Sediment Load Responses to Climate Variation and Cascade Reservoirs in the Yangtze River: A Case Study of the Jinsha River." *Geomorphology* 322:41–52.

Li Donghua 李冬华, Yu Dongmei 于冬梅, and Zhao Liyun 赵丽云. 2014. "Zhongguo jiu sheng chengren lintang yinliao xiaofei ji tianjia tang sheru liang de jushi fenxi" 中国九省成人含糖饮料消费及添加糖摄入量的趋势分析 [Analysis of the trends of Chinese adults' intake of sugar-sweetened beverages and of added sugar in nine provinces]. *Weisheng Yanjiu* 卫生研究43 (1): 70–72.

Li Heming, Paul Waley, and Phil Rees. 2001. "Reservoir Resettlement in China: Past Experience and the Three Gorges Dam." *Geographical Journal* 167 (3): 195–212.

Li Huazhong 李华忠. 1990. "Zuguo de gangtie yaolan: sishi nian de fazhan licheng" 祖国的钢铁摇篮，四十年的发展历程 [The cradle of the ancestral land's iron and steel: The process of forty years of development]. In *Anshan sishi nian* 鞍山四十年 [Forty years of Anshan], edited by Yu Liren 于利人, 24–30. Anshan: Anshan Difangzhi Bangongshi.

Li, Huizhen, Eddy Y. Zeng, and Jing You. 2014. "Mitigating Pesticide Pollution in China Requires Law Enforcement, Farmer Training, and Technological Innovation." *Environmental Toxicology and Chemistry* 33 (5): 963–71.

Li Jian. 2010. "The Decline of Household Pig Farming in Rural Southwest China: Socioeconomic Obstacles and Policy Implications." *Culture and Agriculture* 32 (2): 61–77.

Li Jing. 2004. "22 Detained for Fake Milk Products." *China Daily*, 26 April 2004.

Li Jingxu 李景旭 and Zheng Ruoyu 郑若愚. 2014. "2014 nian digou you huiliu canzhuo de chengyin yu zhili" 2014 年地沟油回流餐桌的成因与治理 [Origins and governance of gutter oil returning to dining tables]. *Ha'erbin Shifan Daxue Shehui Kexue Xuebao* 哈尔滨师范大学社会科学学报2014 (6): 76–79.

Li, Juan-juan, Xiang-rong Wang, Xin-jun Wang, Wei-chun Ma, and Hao Zhang. 2009.

"Remote Sensing Evaluation of Urban Heat Island and Its Spatial Pattern of the Shanghai Metropolitan Area, China." *Ecological Complexity* 6:413–20.

Li, Lillian M. 2007. *Fighting Famine in North China: State, Market, and Environmental Decline, 1690s–1990s*. Stanford, CA: Stanford University Press.

Li, Lyric. 2022. "'Monster' Fish Eludes Capture As Chinese City Drains Lake, Millions Watch." *Washington Post*, 25 August 2022.

Li, M. S. 2006. "Ecological Restoration of Mineland with Particular Reference to the Metalliferous Mineland of China." *Science of the Total Environment* 357:38–53.

Li, Nan, Ding Ma, and Wenying Chen. 2017. "Quantifying the Impacts of Decarbonization in China's Cement Sector: A Perspective from an Integrated Assessment Approach." *Applied Energy* 185:1840–48.

Li, Pei, Yi Lu, and Jin Wang. 2020. "The Effects of Fuel Standards on Air Pollution: Evidence from China." *Journal of Development Economics* 146 (September): article 102488.

Li Pengxiang 李鹏翔, Wu Zhi 吴植, Liang Jianqiang 梁建强, and Tan Yuanbin 谭元斌. 2016. "Jiemi|Sanxia 'da chuan pa louti, xiao chuan zuo dianti' zhuangjing sha yang?" 揭秘 | 三峡 "大船爬楼梯，小船坐电梯" 壮景啥样? [Secret revealed: What's the setup like at Sanxia's "Big Ships Climb the Stairs; Little Boats Ride the Elevator?"] *Huanqiu Wang* 环球网 , 19 September 2016. https://m.huanqiu.com/article/9CaKrnJXFF6.

Li Shaoming 李绍明. 2000. "Duli baiyi" 独立白彝 [Independent "White Yi"]. Paper presented at the Third International Yi Studies Conference, Shilin, Yunnan, August 2000.

Li, Wenbo, Dongyan Wang, Hong Li, and Shuhan Liu. 2017. "Urbanization-Induced Site Condition Changes of Peri-urban Cultivated Land in the Black Soil Region of Northeast China." *Ecological Indicators* 80:215–23.

Li, Xinyu, Hontao Chen, Xueyan Jiang, Zhigang Yu, and Qingzhen Yao. 2017. "Impacts of Human Activities on Nutrient Transport in the Yellow River: The Role of the Water-Sediment Regulation Scheme." *Science of the Total Environment* 592:161–70.

Li, Yifei, and Judith Shapiro. 2020. *China Goes Green: Coercive Environmentalism for a Troubled Planet*. Cambridge: Polity Press.

Li, Yinghui, Shuaijin Huang, and Xuexin Qu. 2017. "Water Pollution Prediction in the Three Gorges Reservoir Area and Countermeasures for Sustainable Development of the Water Environment." *International Journal of Environmental Research and Public Health* 14 (11): 1307–24.

Li, Ying-ying, Hao Zhang, and Wolfgang Kainz. 2012. "Monitoring Patterns of Urban Heat Islands of the Fast-Growing Shanghai Metropolis, China: Using Time-Series of Landsat TM/ETM+ Data." *International Journal of Applied Earth Observation and Geoinformation* 19:127–36.

Li Yongxiang. 2005. "State Power and Sustainable Development in Southwest China: A Case Study from Ailao Shan, Yunnan." PhD diss., University of Washington, Seattle.

Li, Yue, Xueliang Yuan, Yuzhou Tang, Qingsong Wang, Qiao Ma, Ruimin Mu, Junhua Fu, et al. 2020. "Integrated Assessment of the Environmental and Economic Effects of the 'Coal-to-Gas Conversion' Project in Rural Areas of Northern China." *Environmental Science and Pollution Research* 27:14503–514.

Li, Yuxuan, Weifeng Zhang, Lin Ma, Gaoqiang Huang, Oene Oenema, Fusuo Zhang, and Zhengxia Dou. 2014. "An Analysis of China's Fertilizer Policies: Impacts on the Industry, Food Security, and the Environment." *Journal of Environmental Quality* 42:972–81.

Lian, Jijian, Xiaozhong Sun, and Chao Ma. 2016. "Multi-year Optimal Operation Strategy of the Danjiangkou Reservoir after Dam Heightening for the Middle Route of the South-North Water Transfer Project." *Water Science & Technology: Water Supply* 16 (4): 961–70.

Liang Fuqing 梁福庆. 2009. "Sanxia kuqu wenwu baohu yi qi kaizhan" 三峡库区文物保护及其开发 [Cultural relic preservation and development in the Three Gorges Reservoir area]. *Chongqing shehui kexue* 重庆社会科学2009 (6): 85–88.

Liang Hao. 2015. "Toward a Resilient Landscape: The Eco-Cultural Redevelopment in Rural Chengdu Plain." Master's thesis, University of Washington, Seattle. http://hdl.handle.net/1773/33986.

Liang Huigang, Xiang Xiaowei, Huang Cui, Ma Haixia, and Yuan Zhiming. 2020. "A Brief History of the Development of Infectious Disease Prevention, Control, and Biosafety Programs in China." *Journal of Biosafety and Biosecurity* 2:23–26.

Liang Jing 梁京. 2019. "Liang Jing pinglun: Guanyu Sanxia kui ba fengxian sikao" 梁京评论：关于三峡溃坝风险的思考 [Liang Jing commentary: Thoughts on the danger of a Sanxia collapse]. RFA (Radio Free Asia), 16 July 2019. www.rfa.org/cantonese/commentaries/lj/com-07162019074826.html.

Liang, Zai, Yiu Por Chen, and Yanmin Gao. 2002. "Rural Industrialization and Internal Migration in China." *Urban Studies* 39 (12): 2175–87.

Liangshan Yizu Zizhizhou Gaikuang Bianxiezu 凉山彝族自治州概况编写组. 1985. *Liangshan Yizu Zizhizhou gaikuang* 凉山彝族自治州概况 [The general situation of the Liangshan Yi Autonomous Region]. Chengdu: Sichuan Minzu Chubanshe.

Lichtenberg, Erik, and Chengri Ding. 2008. "Assessing Farmland Protection Policy in China." *Land Use Policy* 25 (1): 59–68.

Lieberthal, Kenneth, and Michel Oksenberg. 1988. *Policy Making in China: Leaders, Structures, and Processes*. Princeton: Princeton University Press.

Lin, Chen, Sangwon Suh, and Stephan Pfister. 2012. "Does South-to North Water Transfer Reduce the Environmental Impact of Water Consumption in China?" *Journal of Industrial Ecology* 16 (4): 647–54.

Lin Chun. 2019. "群众路线 Mass Line." In *Afterlives of Chinese Communism: Political Concepts from Mao to Xi*, edited by Christian Sorace, Ivan Franceschini, and Nicholas Loubere, 121–26. Canberra: ANU Press.

Ling Daxie. 凌大燮. 1983. "Woguo senlin ziyuan de bianqian" 我国森林资源的变迁 [Changes in our country's forest resources]. *Zhongguo Nongshi* 1983 (2): 26–36.

Lin, George C. S. 2017. "Water, Technology, Society and the Environment: Interpreting the Technopolitics of China's South-North Water Transfer Project." *Regional Studies* 51 (3): 383–88.

Lin, Hu. 2011. "Perceptions of Liao Urban Landscapes: Political Practices and Nomadic Empires." *Archaeological Dialogues* 18 (2): 223–43.

Lin, Sihao, Xiaorong Wang, Ignatius Tak Sun Yu, Wenjuan Tang, Jianying Miao, Jin Li, Siying Wu, and Xing Lin. 2011. "Environmental Lead Pollution and Elevated Blood

Lead Levels among Children in a Rural Area of China." *American Journal of Public Health* 101 (5): 834–41.

Linli Wang 林立网. n.d. "Huanghe Yangqu shuidianzhan kaigong jianshe—Chengdu budui caigou zong zhuangji rongliang 120 wan qianwa" 黄河羊曲水电站开工建设—成都部队采购 总装机容量120万千瓦 [The Yangqu hydroelectric station on the Yellow River begins work on construction—Chengdu military units buy its total capacity of 1.2 gigawatts]. www.linlinshan.com/show/82226442.html.

Lippit, Victor D. 1978. "The Development of Underdevelopment in China." *Modern China* 4 (3): 251–328.

Liu, Binhui, Mark Henderson, Yandong Zhang, and Ming Xu. 2010. "Spatiotemporal Change in China's Climatic Growing Season: 1955–2000." *Climatic Change* 99:93–118.

Liu Chaohui 刘朝晖 and Jing Yan 景延. 2016. "Shuili sheshi nianjiu shixiu cheng yinhuan: shei zhi quan shei zhi ze?" 水利设施年久失修成隐患：谁之权谁之责 [Irrigation equipment, long neglected, becomes a hidden plague: Whose rights, whose responsibility?]. *Yangshi caijing "jingji ban xiaoshi"* 央视财经 «经济半小时. http://finance.sina.com.cn/china/gncj/2016-05-07/doc-ifxryhhh1739884.shtml.

Liu Dong 刘东, Feng Zhiming 封志明, Yang Yanzhao 杨艳昭, and You Zhen 游珍. 2011. "Zhongguo liangshi shengchan fazhan tezheng ji tudi ziyuan chengzaili kongjian geju xianzhuang" 中国粮食生产发展特征及土地资源承载力空间格局现状 [Characteristics of grain production and spatial pattern of land carrying capacity of China]. *Nongye gongcheng xuebao* 农业工程学报 27 (7): 1–4.

Liu Fei 刘飞, Lin Pengcheng 林鹏程, Li Mingzheng 黎明政, Wang Chunling 王春伶, and Liu Huanzhang 刘焕章. 2019. "Changjiang liuyu yulei ziyuan xianzhuang yu baohu duice" 长江流域鱼类资源现状与保护对策 [Situations and conservation strategies of fish resources in the Chang River basin]. *Shuishengwu xuebao* 水生生物学报 43 (S1). www.aquaticjournal.com/article/doi/10.7541/2019.177.

Liu, Gang. 2014. "Food Losses and Food Waste in China: A First Estimate." OECD Food and Fisheries paper no. 66. www.oecd-ilibrary.org/docserver/5jz5sq5173lq-en.pdf?expires=1672691632&id=id&accname=guest&checksum=7080EE32BECE935C2A143349297AF903.

Liu, Hai, Jie Yin, and Lian Feng. 2018. "The Dynamic Changes in the Storage of the Danjiangkou Reservoir and the Influence of the South-North Water Transfer Project." *Nature Scientific Reports* 2018:8710–21.

Liu Hongmao, Xu Zaifu, Xu Youkai, and Wang Jinxiu. 2002. "Practice of Conserving Plant Diversity through Traditional Beliefs: A Case Study in Xishuangbanna, Southwest China." *Biodiversity and Conservation* 11 (4): 705–13

Liu, Hongyan, Yi Yin, Shilong Piao, Fengjun Zhao, Mike Engels, and Philippe Ciais. 2013. "Disappearing Lakes in Semiarid Northern China: Drivers and Environmental Impact." *Environmental Science and Technology* 47 (21): 12107–114.

Liu, Huan, Yangwen Jia, and Cunwen Niu. 2017. "'Sponge City' Concept Helps Solve China's Urban Water Problems." *Environmental Earth Sciences* 76: article 473.

Liu Huixian, George W. Housner, Xie Lili, and Hu Dexin. 2002. *Overview to the English Version, Report on the Great Tangshan Earthquake of 1976*. Pasadena: Earthquake

Engineering Laboratory, California Institute of Technology. https://authors.library
.caltech.edu/26539/1/Tangshan/Overview.pdf.

Liu, Kang Ernest, Hung-hao Chang, and Wen S. Chern. 2011. "Examining Changes in
Fresh Fruit and Vegetable Consumption over Time and across Regions in Urban
China." *China Agricultural Economic Review* 3 (3): 276–96.

Liu Li 刘莉. 2019. "Daming dingding de Xizang Zangmu shuidianzhan xianzai zenme-
yang?" 大名鼎鼎的西藏藏木水电站现在怎么样? [How's the awesome Zangmu hydroelectric station doing now?]. *Zhongguo Zizang Wang* 中国西藏网, 22 August 2019.
Republished by CCTV.com. http://news.cctv.com/2019/08/24/ARTIA1AMNQ
moJBhh2LYKl9rh190824.shtml.

Liu, Shiwei, Pingyiu Zhang, and Kevin Lo. 2014. "Urbanization in Remote Areas: A
Case Study of the Heilongjiang Reclamation Area, Northeast China." *Habitat International* 42:103–10.

Liu, Wei, Hongye Yao, Wei Xu, Guangbing Liu, Xuebing Wang, Yong Tu, Peng Shi,
Nanyang Yu, Aimin Li, and Si Wei. 2020. "Suspect Screening and Risk Assessment
of Pollutants in the Wastewater from a Chemical Industry Park in China." *Environmental Pollution* 263: article 114493.

Liu Wenhua 刘文华 and Xu Bijiu 徐必久, eds. 2019. *Zhongguo huanjing tongji nianjian.*
中国环境统计年鉴—2018 [China Statistical Yearbook on Environment—2018]. Beijing: National Bureau of Statistics and Ministry of Ecology and Environment.

Liu, Xianbing, Masaru Tanaka, and Yasuhiro Matsui. 2006. "Electrical and Electronic
Waste Management in China: Progress and the Barriers to Overcome." *Waste Management and Research* 24 (1): 92–101.

Liu Xiaohong 刘晓弘 and Wang Yong 王永. 2013. "She digou you fanzui xiangguan
yinan wenti tanxi" 涉地沟油犯罪相关疑难问题探析 [Exploring some difficult questions relating to gutter oil crimes]. *Huanan Ligong Daxue Xuebao (Shehui Kexue Ban)*
华南理工大学学报(社会科学版) 16 (2): 67–72.

Liu Xiaowen 刘晓文. 2000. "Angang Jituan Anshan Kuangye Gongsi wushi nian jubian
shixian lishishing kuayue" 鞍钢集团鞍山矿业公司五十年巨变实现历史性跨
越 [Fifty years of momentous change in the Angang Mining Company: A real historical surpassing]. *Zhongguo Kuangye* 2000 (9): 3–9.

Liu, Xin. 2000. *In One's Own Shadow: An Ethnographic Account of the Condition of
Post-Reform China*. Berkeley: University of California Press.

Liu Xirui. 1958. "Ren you duo da dan, di you duo da chan" 人有多大胆，地有多大产
[The product of the earth is as great as the courage of the people]. *People's Daily*, 27
August 1958.

Liu, Xuejun, Ying Zhang, Wenxuan Han, Aohan Tang, Jianlin Shen, Zhenling Cui, Peter
Vitousek, et al. 2013. "Enhanced Nitrogen Deposition over China." *Nature* 494:459–63.

Liu, Xuewei, Zhengwei Yuan, Xin Liu, You Zhang, Hui Hua, and Songyan Jiang. 2020.
"Historic Trends and Future Prospects of Waste Generation and Recycling in China's Phosphorus Cycle." *Environmental Science and Technology* 54:5131–39.

Liu Yanqin 刘艳芹. 2019. "Shanxi yin Han ji Wei diaoshui gongcheng you wang 2020
nian tong shui jin Xi'an" 陕西引汉济渭调水工程有望2020年通水进西安 [Shaanxi's
"Taking the Han to Fill the Wei" Project hopes to bring water to Xi'an in 2020].
Shuili Bu Changjiang Shuili Weihuanhui 水利部长江水利委员会. http://sn.ifeng.com
/a/20190111/7158020_0.shtml.

Liu Yiman. 2021. "Poyang Lake: Caught Between a Dam and a Sluice Wall." China Dialogue, 11 May 2021. https://chinadialogue.net/en/nature/poyang-lake-caught-between-a-dam-and-a-sluice-wall/.

Liu Yinmei 刘银妹. 2018. "Xiangcun xin yimin: shidu guimo jingying yu nongmin liudong—yi Nacun wei li" 乡村新移民：适度规模经营与农民流动—以那村为例 [New rural migrants: Appropriate large-scale production and peasant mobility—Nacun as an example]. PhD diss., Guangxi Nationalities University.

Liu, Yonghong, Xiaoyi Fang, Yongming Xu, Shuo Zhang, and Qingzu Luan. 2018. "Assessment of Surface Urban Heat Island across China's Three Main Urban Agglomerations." *Theoretical and Applied Climatology* 133:473–88.

Liu, Zhengjia, Nan Xu, and Jieyong Wang. 2020. "Satellite-Observed Evolution Dynamics of the Yellow River Delta in 1984–2018." *IEEE Journal of Selected Topics in Applied Earth Observations and Remote Sensing* 13:6-44-6-50.

Lo, Kinling. 2021. "China-India Relations: Beijing Should Speed Up Hydropower Project, Tibetan Official Says." *South China Morning Post*, 10 March 2021.

Lo, Kinling, and Keegan Elmer. 2020. "Could China's New Dam Plans Unleash More Trouble with India?" *South China Morning Post*, 8 November 2020.

Lohmar, Bryan, Jinxia Wang, Scott Rozelle, Jikun Huang, and David Dawe. n.d. "Investment, Conflicts and Incentives: The Role of Institutions and Policies in China's Agricultural Water Management on the North China Plain." CCAP Working Paper 01-E7. Center for Chinese Agricultural Policy.

Looney, Kristen Elizabeth. 2012. "The Rural Developmental State: Modernization Campaigns and Peasant Politics in China, Taiwan, and South Korea." PhD diss., Harvard University.

————. 2015. "China's Campaign to Build a New Socialist Countryside: Village Modernization, Peasant Councils, and the Ganzhou Model of Rural Development." *China Quarterly* 224:909–32.

————. 2020. *Mobilizing for Development: The Modernization of Rural East Asia*. Ithaca: Cornell University Press.

Lora-Wainwright, Anna. 2017. *Resigned Activism: Living with Pollution in Rural China*. Cambridge, MA: MIT Press.

Low, Bobbie, Elinor Ostrom, Carl Simon, and James Wilson. 2003. "Redundancy and Diversity: Do They Influence Optimal Management?" In *Navigating Social-Ecological Systems*, edited by Fikret Berkes, Johan Golding, and Carl Folke, 83–114. Cambridge: Cambridge University Press.

Lu, Duanfang. 2006. "Travelling Urban Form: The Neighbourhood Unit in China." *Planning Perspectives* 21 (4): 369–92.

Lu, Jixia, and Anna Lora-Wainwright. 2014. "Historicizing Sustainable Livelihood: A Pathways Approach to Lead Mining in Central China." *World Development* 62:189–200.

Lu, Mingming, Yiying Jin, and Qingshi Tu. "The Gutter Oil Issue in China." *Waste and Resource Management* 166 (WR3): 142–49.

Lü, Xiaobo, and Elizabeth Perry. 1997. "Introduction: The Changing Chinese Workplace in Historical and Comparative Perspective." In *Danwei: The Changing Chinese Workplace in Historical and Comparative Perspective*, edited by Xiaobo Lü and Elizabeth Perry, 3–20. Armonk, NY: M. E. Sharpe.

Lu, Xinhai, Yan Li, and Shangan Ke. 2020. "Spatial Distribution Pattern and Its Optimization Strategy of China's Overseas Farmland Investments." *Land Use Policy* 91 (online only). www.sciencedirect.com/science/article/abs/pii/S0264837718319616.

Ludwig, Donald L., Brian H. Walker, and C. S. Holling. 2002. "Models and Metaphors of Sustainability, Stability and Resilience." In *Resilience and the Behavior of Large-Scale Ecosystems*, edited by L. H. Gunderson and L. Pritchard, 21–48. Washington, DC: Island Press.

Luo Guoqing 罗国清. 1998. "Huiyi Zhaojue xian Bi'er qu Pingxi nulizhu wuzhuang panluan douzheng" 回忆昭觉县比尔区平息奴隶主武装叛乱斗争 [Memoir of the struggle to pacify the slavelords' armed rebellion in Bi'er District, Zhaojue County]. In *Liangshan Yizu wenshi ziliao zhuanji* 凉山彝族文史资料专辑, edited by Liangshan Zhou Zhengxie. Chengdu: Sichuan Minzu Chubanshe 186–95.

Ma, Jianxiong. 2013. *The Lahu Minority in Southwest China: A Response to Ethnic Marginalization at the Frontier*. New York: Routledge.

Ma, Josephine, and Echo Xie. 2021. "China Opens More Research Centres Dedicated to Xi Jinping Thought." *South China Morning Post*, 9 July 2021.

Ma, Jun, Hao Chen, Xiang Gao, Jianhua Xiao, and Hongbin Wang. 2020. "African Swine Fever Emerging in China: Distribution Clusters and High-Risk Areas." *Preventive Veterinary Medicine* 175: article 104861.

Ma, Junwei, Huiming Tang, Xinli Hu, Antonio Bobet, Ming Zhang, Tingwei Zhu, Youjian Song, and Mutasim A. M. Ez Eldin. 2016. "Identification of Causal Factors for the Majiagou Landslide Using Modern Data Mining Methods." *Landslides* 14:311–22.

Ma, Yu-Jun, Xiao-yan Li, Maxwell Wilson, Xiu-Chen Wu, Andrew Smith, and Jianguo Wu. 2016. "Water Loss by Evaporation from China's South-North Water Transfer Project." *Ecological Engineering* 95:206–15.

MacFarquhar, Roderick. 1983. *The Origins of the Cultural Revolution 2: The Great Leap Forward, 1958–1960*. New York: Columbia University Press.

Maciel, Vinícius Gonçalves, Rafael Batista Zortea, Igor Barden Grillo, Cássia Maria Lie Ugaya, Sandra Einloft, and Marcus Seferin. 2016. "Greenhouse Gases Assessment of Soybean Cultivation Steps in Southern Brazil." *Journal of Cleaner Production* 131:747–53.

Magee, Darrin L. 2006a. "New Energy Geographies: Powershed Politics and Hydropower Decision Making in Yunnan, China." PhD diss., University of Washington, Seattle.

———. 2006b. "Powershed Politics: Yunnan Hydropower under Great Western Development." *China Quarterly* 185:23–41.

Mah, Alice, and Xinhong Wang. 2019. "Accumulated Injuries of Environmental Injustice: Living and Working with Petrochemical Pollution in Nanjing, China." *Annals of the American Association of Geographers* 109 (6): 1961–77.

Mallory, Tabitha Grace. 2016. "Fisheries Subsidies in China: Quantitative and Qualitative Assessment of Policy Coherence and Effectiveness." *Marine Policy* 68:74–82.

Maloney, Chris. 2019. "What Does the Future Hold for Biogas in China?" World Biogas Association, 11 March 2019. www.worldbiogasassociation.org/what-does-the-future-hold-for-biogas-in-china/.

Malthus, Thomas Robert. 1798. *Essay on the Principle of Population*. Available online at Econlib. www.econlib.org/library/Malthus/malPlong.html.

Manning, Kimberly Ens. 2005. "Marxist Maternalism, Memory, and the Mobilization of Women in the Great Leap Forward." *China Review* 5 (1): 83–110.

Mansur Sabit and Rahman Yusup. 2007. "Jianguo yilai Xinjiang renkou shikong dongtai bianhua tezheng ji qi chengyin fenxi" 建国以来新疆人口时空动态变化特征及其成因分析 [Analysis on the spatial and temporal changes of population and its influencing factors in Xinjiang in the last fifty years]. *Renwen dili* 人文地理 2007 (6): 114–19, 46.

Mao, Ruoxuan. 2018. "The Commercialization of Beijing Hutongs." *Journal of Geography and Geology* 10 (4): 39–49.

Mao Zedong. 1937. "Shixian lun" 实践论 [On practice]. Marxists Internet Archive. www.marxists.org/chinese/maozedong/marxist.org-chinese-mao-193707.htm.

———. 1940. "Zai Shanganning Bianqu ziran kexue yanjiu hui chengli da hui shang de jianghua" 在陕甘宁边区自然科学研究会成立大会上的讲话 [Speech at the inaugural meeting of the Natural Science Research Society of the Shaan-Gan-Ning border region]. Marxists Internet Archive. www.marxists.org/chinese/maozedong/1968/1-128.htm.

———. 1957. "Xiang Mosike de quanti zhongguo liuxuesheng, shixisheng, shiguan jiguan ganbu de jianghua (zhailu)" 向莫斯科的全体中国留学生、实习生、使馆机关干部的讲话（摘录） [Talk to Chinese exchange students, practice students, and diplomatic personnel in Moscow] (recording). Marxists Internet Archive. www.marxists.org/chinese/maozedong/1968/3-120.htm.

———. 1958a. "Zhongguo Gongchandang ba jie san zhong quan hui zongjie shi de jianghua" 中国共产党八届三中全会总结时的讲话 [Talk summarizing the third plenum of the Eighth Party Congress]. Marxists Internet Archive. www.marxists.org/chinese/maozedong/1968/3-117.htm.

———. 1958b. "Shicha Xushui shi de tanhua (zhailu)" 视察徐水时的谈话（摘录） [Conversations at the time of a visit to Xushui] (transcribed recording). Marxists Internet Archive. www.marxists.org/chinese/maozedong/1968/4-044.htm.

———. 1959. "Guanyu fazhan xumuye wenti" 关于发展畜牧业问题 [On the question of developing animal husbandry: Letter to Wu Lingxi], 31 October 1959. https://marxistphilosophy.org/maozedong/mx8/027.htm

———. 1963. "Ren de zhengque xixiang shi nali lai de?" 人的正确思想是从哪里来的? [Where do correct human thoughts come from?]. http://dangjian.people.com.cn/n/2015/0316/c117092-26697670.html

———. 1964a. "Zai yi ci huibao de chayu" 在一次汇报时的插话 [Remarks at a briefing]. Marxists Internet Archive. www.marxists.org/chinese/maozedong/1968/5-073.htm.

———. 1964b. "Bu gao Jiuquan he Panzhihua gangtie chang da qi zhang lai zenme ban?" 不搞酒泉和攀枝花钢铁厂打起仗来怎么办? [If we don't do Jiuquan and the Panzhihua steel mills, and then war comes, what should we do?]. In *Zhongguo Gongchandang yu san xian jianshe* 中国共产党与三线建设 [The Chinese Communist Party and the construction of the Third Front], edited by Chen Donglin 陈东林 et al., 40. Beijing: Zhongguo Dangshi Chubanshe.

Marks, Robert B. 1998. *Tigers, Rice, Silk, and Silt: Environment and History in Late Imperial South China*. Cambridge: Cambridge University Press.

———. 2012. *China: Its Environment and History*. Lanham, MD: Rowman and Littlefield.

Marx, Karl. 1845. *The German Ideology*. Marxists Internet Archive. www.marxists.org/archive/marx/works/1845/german-ideology/ch01a.htm.

———. 1875. *Critique of the Gotha Program*. Marxists Internet Archive. www.marxists.org/archive/marx/works/1875/gotha/ch01.htm.

Matalas, N. C., and C. F. Nordin Jr. 1980. "Water Resources in the People's Republic of China." *Eos* 61 (46): 891–901.

Mathews, John A., and Carol X. Huang. 2021. "The Global Green Shift in Electric Power: China in Comparative Perspective." *Asia Pacific Journal* 19 (8): article 5589. https://apjjf.org/2021/8/Mathews-Huang.html.

Matsushita, Kyohei, Fumihiro Yamane, and Kota Asano. 2016. "Linkage between Crop Diversity and Agro-ecosystem Resilience: Nonmonotonic Agricultural Response under Alternate Regimes." *Ecological Economics* 126:23–31.

mayor box. n.d. "top 10 países productores de Calzado." Accessed February 2021. https://mayorbox.com/en/blog-fabricantes-león-guanajuato/24_Top-10-shoe-producing-countries-in-the-world.html.

Meadows, Donella H. 2008. *Thinking in Systems*. White River Junction, VT: Chelsea Green.

MEE (Ministry of Ecology and Environment, PRC). n.d. a. "Environmental Quality Standards for Surface Water." Accessed May 2020. http://english.mee.gov.cn/SOE/soechina1997/water/standard.htm.

———. n.d. b. $PM_{2.5}$ Lishi shuju $PM_{2.5}$ 历史数据 [Historical statistics of $PM_{2.5}$]. www.aqistudy.cn/historydata/.

———. 2021. About MEE: Departments. http://english.mee.gov.cn/About_MEE/Internal_Departments/.

Mei, Xuefei, Zhijun Dai, Jinzhou Du, and Jiyu Chen. 2015. "Linkage between Three Gorges Dam Impacts and the Dramatic Recessions in China's Largest Freshwater Lake, Poyang Lake." *Nature Scientific Reports* 5: article 18197.

Mekong Dam Monitor. n.d. Stimson Center. Accessed July 2022. www.stimson.org/project/mekong-dam-monitor/.

Mele, Marco, and Cosimo Magazzino. 2020. "A Machine Learning Analysis of the Relationship among Iron and Steel Industries, Air Pollution, and Economic Growth in China." *Journal of Cleaner Production* 277: article 123293.

Meng, Xiaojie, Yan Zhang, Xiangyi Yu, Junhong Bai, Yingying Chai, and Yating Li. 2014. "Regional Environmental Risk Assessment for the Nanjing Chemical Industry Park: An Analysis Based on Information-Diffusion Theory." *Stochastic Environmental Research Risk Assessment* 28:2217–33.

MEP (Ministry of Environmental Protection 环境保护部) and MOA (Ministry of Agriculture 农业部) [China]. 2014. "Quango turang wuran zhuangkuang diaocha gongbao" 全国土壤污染状况调查公报 [Public report on the nationwide situation of contamination]. www.gov.cn/foot/site1/20140417/782bcb88840814ba158d01.pdf.

Mertha, Andrew C. 2008. *China's Water Warriors: Citizen Action and Policy Change*. Ithaca: Cornell University Press.

Meyer, Michael. 2015. *In Manchuria: A Village Called Wasteland and the Transformation of Rural China*. New York: Bloomsbury.

Meyskens, Covell. 2020. *Mao's Third Front: The Militarization of Cold War China*. Cambridge: Cambridge University Press.

———. 2021. "Dreaming of a Three Gorges Dam amid the Troubles of Republican China." *Journal of Modern Chinese History* 15 (2): 176–94.

Mgebbu Lunzy [Ma Erzi]. 2003. "Nuosu and Neighbouring Ethnic Groups: Ethnic Groups and Ethnic Relations in the Eyes and Ears of Three Generations of the Mgebbu Clan." Translated by Stevan Harrell. *Asian Ethnicity* 4 (1): 129–45.

Miao, Guangping, and R. A. West. 2004. "China Collective Forestlands: Contributions and Constraints." *International Forestry Review* 6 (3–4): 282–98.

Miao, Zhuang, Jichuan Sheng, Michael Webber, Tomas Baležensis, Yong Geng, and Weihai Zhou. 2018. "Measuring Water Use Performance in the Cities along China's South-North Water Transfer Project." *Applied Geography* 98:184–200.

Middleton, Carl, Alec Scott, and Vanessa Lamb. 2019. "Hydropower Politics and Conflict on the Salween River." In *Knowing the Salween River: Resource Politics of a Contested Transboundary River*, edited by Carl Middleton and Vanessa Lamb, 27–48. Cham, Switzerland: Springer Nature.

Milhorance, Flávia. 2020. "Major Brazilian Suppliers of Beef to China Linked to Deforestation." Diálogo Chino, 15 June 2020. https://dialogochino.net/en/agriculture /35891-major-brazilian-suppliers-of-beef-to-china-linked-to-deforestation-says-new -report/.

Millward, James, and Nabijan Tursun. 2004. "Political History and Strategies of Control, 1884–1978." In *Xinjiang: China's Muslim Borderland*, edited by S. Frederick Starr, 63–100. Armonk, NY: M. E. Sharpe.

Mingione, Enzo, and Martin Spence. 1993. "The Second Contradiction of Capitalism." *Capitalism, Nature, Socialism* 4:85–100.

Minwei (Guojia Minwei Zhengfu Wang) 国家民委政府网. 2017. "1959: Zhou Enlai wei Baogang di yi hao gaolu chutie qian cai" 1959: 周恩来为包钢第一号高炉出铁剪彩 [1959: Zhou Enlai cuts the ribbon for steel production from the #1 Blast Furnace at Baotou Steel]. www.neac.gov.cn/seac/c100466/201708/1075536.shtml.

MOA (Ministry of Agriculture and Villages). 2022. "Guojia nongzuowu pinzhong shending weihuanhui guanyu yinfa guojia ji zhuan jiyin dadou yumi pinzhong shending biaozhun de tongzhi" 国家农作物品种审定委员会关于印发国家级转基因大豆玉米品种审定标准的通知 [Notice from the National Committee for Authorization of Agricultural Varieties on the publication of national-level standards for genetically modified soybeans and corn]. www.moa.gov.cn/govpublic/nybzzj1/202206 /t20220608_6401924.htm.

Moko Technology. 2022. "A Brief Guide to Electronic Manufacturing in China." www .mokotechnology.com/a-brief-guide-to-electronic-manufacturing-in-china/.

MoR (Ministry of Railways) 铁道部. (1952) 2009. "Tiedao Bu laodong baohu gongzuo baogao tiyao" 铁道部劳动保护工作报告摘要 1952 年12 月 [Summary of Ministry of Railroads reports on worker protection work, December 1952]. Published in *Zhonghua Renmin Gongheguo jingji dang'an ziliao xuanbian 1949–1952 laodong gongzi he zhigong fuli juan* 中华人民共和国经济档案资料选编, 1949–1952 劳动工资和职工福利卷.

Mostern, Ruth. 2019. "Loess Is More: The Spatial and Ecological History of Erosion on China's Northwest Frontier." *Journal of the Economic and Social History of the Orient* 62:560–98.

MTA (Metropolitan Transit Authority of New York). 2020. Subway and Bus Facts. Up-

dated 14 April 2020. https://new.mta.info/agency/new-york-city-transit/subway
-bus-facts-2019.

Muir, John. 1911. *My First Summer in the Sierra*. Boston: Houghton Mifflin.

Muldavin, Joshua S. S. 1997. "Environmental Degradation in Heilongjiang: Policy Reform and Agrarian Dynamics in China's New Hybrid Economy." *Annals of the Association of American Geographers* 87 (4): 579–613.

MWR (Ministry of Water Resources) 水利部. 2021a. *Quanguo shuili fazhan tongji baogao* 全国水利发展统计公报 [2020 statistical bulletin on China water activities]. Beijing: Ministry of Water Resources. www.mwr.gov.cn/sj/tjgb/slfztjgb/202202 /P020220209630479640795.pdf.

———. 2021b. "Shuili Bu guanyu Huanghe liuyu shui ziyuan chaozai diqu zhanting xinzeng qu shui xuke de tongzhi" 水利部关于黄河流域水资源超载地区暂停新增取水许可的通知 [Notice from the Ministry of Water Resources regarding a temporary moratorium on issuing additional water withdrawal permits in overburdened water resource areas in the Yellow River watershed]. www.waizi.org.cn/doc/96108 .html.

Myers, Steven Lee, Keith Bradsher, and Chris Buckley. 2021. "As China Boomed, It Didn't Take Change into Account. Now It Must." *New York Times*, 26 July 2021.

Myrdal, Jan. 1965. *Report from a Chinese Village*. New York: New American Library.

Naughton, Barry. 1988. "The Third Front: Defence Industrialization in the Chinese Interior." *China Quarterly* 115:351–86.

———. 2007. *The Chinese Economy: Transitions and Growth*. Cambridge, MA: MIT Press.

NETL (National Energy Technology Laboratory). 2015. "China Gasification Database." https://netl.doe.gov/research/coal/energy-systems/gasification/gasification-plant -databases/china-gasification-database.

Newman, Leonard. 1981. "Environmental Concerns in China." *Environmental Science and Technology* 15 (7): 740–46.

Ng, Eric. 2021. "Success of Nuclear Reactor Hualong One Suggests It Can Compete with Wind and Solar to Drive China's Decarbonisation." *South China Morning Post*, 22 March 2021.

Niazi, Tarique. 2004. "Rural Poverty and the Green Revolution: The Lessons from Pakistan." *Journal of Peasant Studies* 31 (2): 242–60.

Nickum, James E. 1988. "All Is Not Wells in North China: Irrigation in Yucheng County." In *Efficiency in Irrigation: The Conjunctive Use of Surface and Groundwater Resources*, edited by Gerald T. O'Meara, 88–93. Washington, DC: World Bank.

———. 1998. "Is China Living on the Water Margin?" *China Quarterly* 156:880–98.

Ning, P., and Z. Zhang 2010. "On Formation and Features of Floating Garbage on Water Surface in Three Gorges Area." *Sanxia Daxue Xuebao* 三峡大学学报26:1–4.

Nongye Bu (Zhonghua Renmin Gongheguo Nonge Bu) 中华人民共和国农业部 (Ministry of Agriculture of the People's Republic of China). 2009. *Xin Zhongguo nongye 60 nian tongji ziliao* 新中国农业60年统计资料 [Statistics on sixty years of agriculture in the new China]. Beijing: Zhongguo Nongye Chubanshe.

Nujiang Net 怒江网. 2018. "Nu Jiang shuidian jidi" 怒江水电基地 [The Nu River hydroelectric base]. 11 November 2018. Inaccessible July 2022. http://nujiangw.com/index .php?con=cms_index&fun=details&id=1062&catid=172.

Nyabiage, Jevans. 2021. "Locals Fear Chinese Fishing Plant Threatens 'Environmental Catastrophe" in Sierra Leone." *South China Morning Post*, 31 May 2021.

Oksenberg, Michel. 1973. "On Learning from China." In *China's Developmental Experience*, edited by Michel Oksenberg, 1–16. New York: Praeger.

O'Neill, Aaron. 2021a. "Crude Birth Rate of China 1930–2020." Statista, 1 March 2021. www.statista.com/statistics/1037919/crude-birth-rate-china-1930-2020/.

———. 2021b. "Total Fertility Rate of China 1930–2020." Statista, 2 March 2021. www.statista.com/statistics/1033738/fertility-rate-china-1930-2020/.

Osgood, Cornelius. 1963. *Village Life in Old China: A Community Study of Kao Yao, Yünnan.* New York: Ronald Press.

Ostrom, Elinor. 1990. *Governing the Commons: The Evolution of Institutions for Collective Action.* Cambridge: Cambridge University Press.

Ostrom, Elinor, and Michael Cox. 2010. "Moving beyond Panaceas: A Multi-tiered Diagnostic Approach for Social-Ecological Analysis." *Environmental Conservation* 37 (4): 451–63.

Our World in Data. n.d. a. Animal Protein Consumption, 2019. Accessed December 2022. https://ourworldindata.org/grapher/animal-protein-consumption?tab=table&country=GBR~USA~ESP~BRA~JPN~China+%28FAO%29~LKA~IND~BGD.

———. n.d. b. Fish and Seafood Supply per Person, 2017. Accessed December 2022. https://ourworldindata.org/grapher/fish-and-seafood-consumption-per-capita?tab=table.

Ozlu, Ekrem, and Sandeep Kumar. 2018. "Response of Soil Organic Carbon, pH, Electrical Conductivity, and Water Stable Aggregates to Long-Term Animal Manure and Organic Fertilizer." *Soil Science Society of America Journal* 82:1243–51.

Palma, Dechen. 2019. "Tibet's Rivers Will Determine Asia's Future." *The Diplomat*, 19 November 2019.

Paltemaa, Lauri. 2011. "The Maoist Urban State and Crisis: Comparing Disaster Management in the Great Tianjin Flood in 1963 and the Great Leap Forward Famine." *China Journal* 66:25–51.

Pan Yue 潘岳. 2003. "Huanjing wenhua yu minzu fuxing" 环境文化与民族复兴 [Environmental culture and national revival]. *Lü Ye* 绿叶 [Green leaf] issue 6: 7–11.

———. 2006. "Lun shehuizhuyi shengtai wenming" 论社会主义生态文明 [On a socialist ecological civilization]. *Lü Ye* 绿叶 [Green leaf] issue 10: 10–18.

Panayotou, Theodore. 1993. "Empirical Tests and Policy Analysis of Environmental Degradation at Different Stages of Economic Development." World Employment Programme Research Working Paper, Geneva.

Parker, Joseph Lynn. 2013. "Beyond Sustainable Bounds: Changing Weather, Emigration, and Irrigation in a Farming Village of Sichuan, China, 1945–2012." MS thesis, Portland State University.

Pecht, Michael, Chung-Shing Lee, Zong Xiang Fu, Jiang Jun Lu, and Wang Yong Wen. 1999. *The Chinese Electronics Industry.* Boca Raton: CRC Press.

Pederson, Neil, Amy E. Hessl, Nachin Baaterbileg, Kevin J. Anchukaitis, and Nicola Di Cosmo. 2014. "Pluvials, Droughts, the Mongol Empire, and Modern Mongolia." *PNAS* 111 (12): 4375–79.

Pei, Xiaofang, Annuradha Tandon, Anton Alldrick, Liana Giorgi, Wei Huang, and Rui-jia Yang. 2011. "The China Melamine Scandal and Its Implications for Food Safety Regulation." *Food Policy* 36:412–20.

Peled, Micha, dir. 2005. *China Blue*. Documentary film, 86 minutes.

Peña, Devon G. 2005 *Mexican Americans and the Environment: Tierra y Vida*. Tucson: University of Arizona Press.

Peng, Wei, Junnan Yang, Xi Lu, and Denise L. Mauzerall. 2018. "Potential Co-Benefits of Electrification for Air Quality, Health, and CO_2 Mitigation in China." *Applied Energy* 218:511–19.

Peng, Xianzhi, Weihui Ou, Chunwei Wang, Zhifang Wang, Qiuxin Huang, Jiabin Jin, and Jinhua Tan. 2014. "Occurrence and Ecological Potential of Pharmaceuticals and Personal Care Products in Groundwater and Reservoirs in the Vicinity of Municipal Landfills in China." *Science of the Total Environment* 490:889–98.

People's Daily (*Renmin Ribao* 人民日报). 2005. "Haerbin shimin zhiyi zai gongshui shuizhi shengzhang cheng xian he di yi kou" 哈尔滨市民质疑再供水水质 省长称先喝第一口 [Citizens of Harbin doubt the quality of replacement water; governor says he will drink the first mouthful]. Sina News Center, 23 November 2005.

———. 2014. "Li Keqiang: Xiang dui pinkun xuan zhan yiyang, jianjue xiang wuran xuan zhan" 李克强：像对贫困宣战一样 坚决向污染宣战 [Li Keqiang: Like declaring war on poverty, resolve to declare war on pollution]. 5 March 2014.

———. 2018. "China's Per Capita Residential Space Surpasses 40 Square Meters." 13 December 2018.

People's Education Press (Renmin Jiaoyu Chubanshe 人民教育出版社). 2006. *Zhongguo jindai xiandai shi* 中国近代现代史 [Modern and contemporary history of China]. Beijing.

———. 2019. *Putong gaozhong jiaokeshu: Lishi bixiu zhongwai lishi gangyao* 普通高中教科书：历史必修中外历史纲要. [Ordinary high school textbook: History, required; Outline of Chinese and foreign history]. Beijing.

People's Government (Central People's Government of the People's Republic of China) (中华人民共和国中央人民政府). 1955. "Zhonghua Renmin Gongheguo fazhan guomin jingji de diyi wunian jihua" 中华人民共和国发展人民经济的第一五年计划 [The first five-year plan for developing the people's economy of the People's Republic of China]. www.gov.cn/test/2008-03/06/content_910770.htm.

Perdue, Peter C. 1982. "Official Goals and Local Interests: Water Control in the Dongting Lake Region during the Ming and Qing Periods." *Journal of Asian Studies* 41 (4): 747–65.

———. 2005. *China Marches West: The Qing Conquest of Central Eurasia*. Cambridge, MA: Belknap Press of Harvard University Press.

Pérez-Vicente, Luis, Miguel Dita, and Einar Martinez-de la Parte. 2014. *Technical Manual: Prevention and Diagnostic of* Fusarium *Wilt (Panama Disease) of Banana Caused by* Fusarium oxysporum f. sp. cubense *Tropical Race 4 (TR4)*. Rome: Food and Agriculture Organization.

Perkins, Dwight H. 1969. *Agricultural Development in China, 1368–1968*. Chicago: Aldine.

———. 1973. "Development of Agriculture." In *China's Developmental Experience*, edited by Michel Oksenberg, 55–67. New York: Praeger.

Pi, Chengdong, Zhang Rou, and Sarah Horowitz. 2014. "Fair or Fowl? Industrialization of Poultry Production in China." Institute for Agriculture and Trade Policy. www.iatp.org/sites/default/files/2017-05/2017_05_03_PoultryReport_f_web.pdf.

Pi, Ms. 皮小姐. 2019. "Ju bu fen lei hai bu naifan, Songjiang you ren bei fa 50 yuan! Laji fenlei shou ri, Shanghai kai chu 623 zhang zhenggai dan" 拒不分类还不耐烦，松江有人被罚50元！垃圾分类首日，上海开出623张整改单 [In accordance with not tolerating failure to sort garbage, people were fined 50 yuan! On the first day of garbage sorting, Shanghai issued 623 citations]. https://mp.weixin.qq.com/s?__biz
=MjM5ODI2NDMwMw==&mid=2652810646&idx=1&sn=d67a9452321552a22aaa408
8b227c711&chksm=bd2761168a50e8008a491784ad6ee68d7d0722a2a1767b0bb7c4c36
9d640497bf42c219877a4&mpshare=1&scene=1&srcid=07014PnC19RQNpn9U6Tqh50F
&pass_ticket=FoEBtSNwXXEsXX5w5sb0eybwUKkqsDZ7YRsq1FfD88VljibJLHs-
GdOjikBsC65CY#rd.

Pianporn Deetes. 2022. "Today Is a 'Day of Action for Rivers.'" *Bangkok Post*, 14 March 2022.

Pietz, David A. 2015. *The Yellow River: The Problem of Water in Modern China*. Cambridge, MA: Harvard University Press.

Pietz, David, and Mark Giordano. 2009. "Managing the Yellow River: Continuity and Change." In *River Basin Trajectories: Societies, Environments, and Development*, edited by François Molle and Philippus Wester, 99–122. Oxford: CAB International.

Pilkey, Orrin H., and Linda Pilkey-Jarvis. 2007. *Useless Arithmetic: Why Environmental Scientists Can't Predict the Future*. New York: Columbia University Press.

Pimentel, David, and Marcia Pimentel. "Sustainability of Meat-Based and Plant-Based Diets and the Environment." *American Journal of Clinical Nutrition* 78 (3 Suppl.): 660S–663S.

Plastics Europe. 2020. Plastics—the Facts 2020. https://issuu.com/plasticseuropeebook/docs/plastics_the_facts-web-dec2020.

Popkin, Barry M. 1994. "The Nutrition Transition in Low-Income Countries: An Emerging Crisis." *Nutrition Reviews* 52 (9): 285–98.

Popkin, Barry M., and Shufa Du. 2003. "Dynamics of the Nutrition Transition toward the Animal Foods Sector in China and Its Implications: A Worried Perspective." *Journal of Nutrition* 133:3898S–3906S.

Potter, Jack M. 1968. *Capitalism and the Chinese Peasant*. Berkeley: University of California Press.

Pritchard, Lowell, Jr., and Steven F. Sanderson. 2002. "The Dynamics of Political Discourse in Seeking Sustainability." In *Panarchy: Understanding Transformations in Human and Natural Systems*, edited by Lance H. Gunderson and C. S. Holling, 147–69. Washington, DC: Island Press.

Prybyla, Jan S. 1966. "Transportation in Communist China." *Land Economics* 42 (3): 268–81.

Ptáčková, Jarmila. 2020. *Exile from the Grasslands: Tibetan Herders and Chinese Development Projects*. Seattle: University of Washington Press.

Pun Ngai. 2005. *Made in China: Women Factory Workers in a Global Marketplace*. Durham, NC: Duke University Press.

Qi Ye, Song Qijiao, Zhao Xiaofan, Qiu Shiyong, Tom Lindsay, et al. 2020. *China's New Urbanisation Opportunity: A Vision for the 14th Five-Year Plan*. London: Coalition for Urban Transitions. https://urbantransitions.global/wp-content/uploads/2020/05/China's_New_Urbanisation_Opportunity_FINAL.pdf.

Qian Gang 钱钢. 2012. "'Kexue fazhan guan': qihao de mingyun" "科学发展观": 旗号 的命运 ["The scientific perspective on development": The fate of a banner slogan). *New York Times Chinese Language Network*, 27 September 2012. https://cn.nytimes .com/china/20120927/cc27qiangang09/.

Qian, Ying. 2020. "When Taylorism Met Revolutionary Romanticism: Documentary Cinema in China's Great Leap Forward." *Critical Inquiry* 46:578–604.

Qiao, Guanghua, Ting Guo, and K. K. Klein. 2012. "Melamine and Other Food Safety and Health Scares in China: Comparing Households with and without Young Children." *Food Control* 26:378–86.

Qiao Xuefeng 乔雪峰. 2019. "Zhujian Bu: 46 cheng 2020 niandi qian jiben jiancheng laji fenlei chuli xitong" 住建部： 46城2020年底前基本建成垃圾分类处理系统 [Ministry of Housing and Urban and Rural Development: 46 cities will basically implement systems of sorting and collecting garbage before the end of 2020]. *Renmin Wang* 人 民网, 28 June 2019.

Qiu Huanguang 仇焕广, Chen Ruijian 陈瑞剑, Liao Shaopan 廖绍攀, and Cai Yaqing 蔡亚庆. 2013. "Zhongguo nongye qiye 'zou chuqu' de xianzhuang, wenti yu duice" 中国农业企业"走出去"的现状、问题与对策 [Current situation, questions and responses regarding the "going out" of China's agricultural enterprises]. *Nongye Jingji Wenti* 农业经济问题 2013 (11): 44–49.

Räsänen, Timo A., Paridis Someth, Hannu Lauri, Jorma Koponen, Juha Sarkkula, and Matti Kummu. 2017. "Observed River Discharge Changes Due to Hydropower Operations in the Upper Mekong Basin." *Journal of Hydrology* 545:26–41.

Ren Huiru, Li Guosheng, Cui Linlin, and He Lei. 2015. "Multi-scale Variability of Water Discharge and Sediment Load into the Bohai Sea from 1950 to 2011." *Journal of Geographical Sciences* 25 (1): 85–100.

Ren Mei'e. 2015. "Sediment Discharge of the Yellow River, China: Past, Present, and Future—A Synthesis." *Acta Oceanologica Sinica* 34 (2): 1–8.

Ren Zhongping 任中评. 2013. "Shengtai wenming de Zhongguo juexing" 生态文明的中国觉醒 [The awakening of ecological civilization in China]. *Renmin Ribao* 人民日报, 22 July 2013.

Renmin Wang 人民网 (People's Daily Online). 2010. 昆明滇池暴发蓝藻 湖水如绿漆 "Kunming Dianchi baofa lanzao: Hushui ru lü qi" [Blue-green algae outbreak in Dian Lake, Kunming: The lake water is like green paint]. 30 November 2010. http:// gooootech.com/news/detail-10176194.html.

———. 2021. "Guotu 'san diao' shuju gongbu: woguo gengdi mianji 19.179 yi mu" 国土"三调"数据公布：我国耕地面积19.179亿亩 [Report of the statistics of the "Third Survey": Our country's cultivated area 1.979 billion *mu*]. 26 August 2021.

Reuters. 2020. "India Plans Brahmaputra Dam, after China Unveils Tibet Hydropower Project." *South China Morning Post*, 1 December 2020.

Reyes, Antonio R. 2009. "From SARS to Sanlu: The Mismanagement of Public Health Crises in China." Unpublished MA paper, University of Washington, Seattle.

RFA (Radio Free Asia). 2016. "Laos and Its Dams: Southeast Asia's Battery, Built by China." www.rfa.org/english/news/special/china-build-laos-dams/#seven.

Riskin, Carl. 1987. *China's Political Economy: The Quest for Development Since 1949.* Oxford: Oxford University Press.

Ritchie, Hannah, and Max Roser. 2022. "China: Energy Country Profile." Our World in

Data. https://ourworldindata.org/energy/country/china#what-sources-does-the
-country-get-its-energy-from.

Rithmire, Meg Elizabeth. 2017. "Land Institutions and Chinese Political Economy:
Institutional Complementarities and Macroeconomic Management." *Politics and
Society* 45 (1): 123–53.

Robbins, Alicia S. T., and Stevan Harrell. 2014. "Paradoxes and Challenges for China's
Forests in the Reform Era." *China Quarterly* 218:381–403.

Roberts, Sean R. 2020. *The War on the Uyghurs: China's Internal Campaign against a
Muslim Minority*. Princeton: Princeton University Press.

Roney, J. Matthew. 2013. "Wind Surpasses Nuclear in China." *Grist*, 19 February 2013.
https://grist.org/article/wind-surpasses-nuclear-in-china/.

Roney, Tyler, Kyaw Ye Lynn, Robert Bociaga, and Marc Jaffee. 2021. "China's Salween
Plans in Limbo in Post-coup Myanmar." Third Pole, 8 June 2021. www.thethirdpole
.net/en/energy/chinas-salween-plans-in-limbo-in-post-coup-myanmar/.

Rostow, W. W. 1960. "'Take-Off' into Economic Growth." *Challenge* 8 (8): 29–37.

Ruf, Gregory A. 1998. *Cadres and Kin: Making a Socialist Village in West China*. Stanford, CA: Stanford University Press.

Sahlins, Marshall. 1972. *Stone Age Economics*. Chicago: Aldine-Atherton.

Salaff, Janet. 1967. "The Urban Communes and Anti-City Experiment in Communist
China." *China Quarterly* 29:82–110.

Salimjan, Guldana 2019. "The Politics of Returning and Emplacement." Paper
prepared for the China Environment Symposium, Pacific Lutheran University,
April 2019.

———. 2021. "Naturalized Violence: Affective Politics of China's 'Ecological Civilization' in Xinjiang." *Human Ecology* 49 (1): 59–68.

Santos, Gonçalo D. 2004. "The Process of Kinship and Identity in a Common-Surname
Village among the Cantonese of Rural Southeastern China." PhD diss., Instituto Superior de Ciências do Trabalho e da Empresa, Lisbon.

Savadove, Bill. 2011. "China Activist Defies Officials in Fight to Save Lake." Rawstory,
2 October 2011. www.rawstory.com/2011/10/china-activist-defies-officials-in-fight-to
-save-lake/.

Scheffer, Marten, Stephen R. Carpenter, Timothy N. Lenton, Jordi Bascompte, William
Brock, Vasilis Dakos, Johan van de Koppel, et al. 2012. "Anticipating Critical Transitions." *Science* 338:344–48.

Scheffer, Marten, Frances Westley, William A. Brock, and Milena Holmgren. 2002.
"Dynamic Interaction of Societies and Ecosystems: Linking Theories from Ecology,
Economy, and Sociology." In *Panarchy: Understanding Transformations in Human
and Natural Systems*, edited by Lance H. Gunderson and C. S. Holling, 195–240.
Washington, DC: Island Press.

Schmalzer, Sigrid. 2014. "Self-Reliant Science: The Impact of the Cold War on Science
in Socialist China." In *Science and Technology in the Global Cold War*, edited by
Naomi Oreskes and John Krige, 75–106. Cambridge, MA: MIT Press.

———. 2016. *Red Revolution, Green Revolution: Scientific Farming in Socialist China*.
Chicago: University of Chicago Press.

Schmidt, Amanda H., David R. Montgomery, Katharine W. Huntington, and Chuan
Liang. 2011. "The Question of Communist Land Degradation: New Evidence from

Local Erosion and Basin-Wide Sediment Yield in Southwest China and Southeast Tibet." *Annals of the Association of American Geographers* 101 (3): 477–96.

Schmitt, Edwin A. 2016. "The Atmosphere of an Ecological Civilization: A Study of Ideology, Perception and Action in Chengdu, China." PhD diss., Chinese University of Hong Kong.

Schneider, Mindi. 2011. "Feeding China's Pigs: Implications for the Environment, China's Smallholder Farmers and Food Security." Institute for Agriculture and Trade Policy, Minneapolis. www.iatp.org/documents/feeding-china's-pigs-implications -for-the-environment-china's-smallholder-farmers-and-food.

———. 2015. "Wasting the Rural: Meat, Manure, and the Politics of Agro-industrialization in Contemporary China." *Geoforum* 78:89–97.

———. 2017. "Dragon Head Enterprises and the State of Agribusiness in China." *Journal of Agrarian Change* 17 (1): 3–21.

Schneider, Mindi, and Shefali Sharma. 2014. "China's Pork Miracle? Agribusiness and Development in China's Pork Industry." Institute for Agriculture and Trade Policy. www.iatp.org/documents/chinas-pork-miracle-agribusiness-and-development -chinas-pork-industry.

Schoppa, R. Keith. 1989. *Xiang Lake: Nine Centuries of Chinese Life*. New Haven: Yale University Press.

Schueneman, Jean J., M. D. High, and W. E. Biye. 1963. *Air Pollution Aspects of the Iron and Steel Industry*. Cincinnati: US Public Health Service.

Schwartz, Benjamin. 1973. "China's Developmental Experience, 1949–72." In *China's Developmental Experience*, edited by Michel Oksenberg, 17–26. New York: Praeger.

Science. 2013. Infographic: Pesticide Planet. *Science* 341 (6147): 730–31.

Scott, James C. 1976. *The Moral Economy of the Peasant: Rebellion and Subsistence in Southeast Asia*. New Haven: Yale University Press.

———. 1998. *Seeing Like a State: How Certain Schemes to Improve the Human Condition Have Failed*. New Haven: Yale University Press.

———. 2009. *The Art of Not Being Governed: An Anarchist History of Upland Southeast Asia*. New Haven: Yale University Press.

Scott, Steffanie, Zhenzhong Si, Theresa Schumilas, and Aijuan Chen. 2014. "Contradictions in State- and Civil Society-Driven Developments in China's Ecological Agriculture Sector." *Food Policy* 45:158–66.

SCSHKX (Sichuan Shehui Kexue Zai Xian 四川社会科学在线). 2006. "Sanxia gongcheng zhi zui" 三峡工程之最 [The mosts of the Sanxia Project]. https://web.archive.org /web/20140606203624/http://www.sss.net.cn/ReadNews.asp?NewsID=7271&BigClass ID=22&SmallClassID=67&SpecialID=0&belong=ts.

Seow, Victor. 2014. "Socialist Drive: The First Auto Works and the Contradictions of Connectivity in the Early People's Republic of China." *Journal of Transport History* 35 (2): 145–61.

———. 2021. *Carbon Technocracy: Energy Regimes in Modern East Asia*. Chicago: University of Chicago Press.

Shahbandeh, M. 2022. "Per Capita Meat Consumption in the United States in 2020 and 2030, by Type." Statista. www.statista.com/statistics/189222/average-meat -consumption-in-the-us-by-sort/.

Shakya, Tsering. 1999. *The Dragon in the Land of Snows: A History of Modern Tibet since 1947*. New York: Columbia University Press.

Shanchuan Wang 山川网. 2018. "31 ge shengfen chengshi jianshe da shuju: shiqu mianji, chengqu mianji, jianchengqu mianji paiming" 31 个省份城市建设大数据：市区面积，城区面积，建成区面积排名. [Total statistics for urban areas in 31 provinces: list of the urban area, built-up area, and building area.] https://mp.weixin.qq.com/s/8WI1z5zzqEY9tFqoHqJhnw?

Shandong Shuili 山西水利 (Shanxi Waterworks). 1993. "Hongqi Qu bei zha an de qianqian houhou" 红旗渠被炸案的前前后后 [Beginning to end of the incident of blowing up the Red Flag Canal]. *Shadong Shuili* 山东水利 1993 (3): 8–9.

Shang Guo'ao 尚国傲, ed. 2009. "Huanghe ruhaikou ganshou duanliu zhi tong" 黄河入海口感受断流之痛 [The pain felt when the Yellow River stopped running). Dahe Wang 大河网, posted 12 December 2009.

Shankman, David, and Qiaoli Liang. 2003. "Landscape Changes and Increasing Flood Frequency in China's Poyang Lake Region." *Professional Geographer* 55 (4): 434–45.

Shao, Chaofeng, Juan Yang, Xiaogang Tian, Meiting Ju, and Lei Huang. 2013. "Integrated Environmental Risk Assessment and Whole-Process Management System in Chemical Industry Parks." *International Journal of Environmental Research and Public Health* 2013 (10): 1609–30.

Shapiro, Judith. 2001. *Mao's War against Nature: Politics and the Environment in Revolutionary China*. New York: Cambridge University Press.

Shapiro-Bengtsen, Sara. 2020. "Is China Building More Waste Incinerators Than It Needs?" China Dialogue, 12 August 2020. https://chinadialogue.net/en/pollution/is-china-building-more-waste-incinerators-than-it-needs/.

Sharma, Shefali. 2014. *The Need for Feed: China's Demand for Industrialized Meat and Its Impacts*. Minneapolis: Institute for Agriculture and Trade Policy. www.iatp.org/sites/default/files/2017-05/2017_05_03_FeedReport_f_web_0.pdf.

Sharma, Shefali, and Zhang Rou. 2014. *China's Dairy Dilemma: The Evolution and Future Trends of China's Dairy Industry*. Minneapolis: Institute for Agriculture and Trade Policy. www.iatp.org/sites/default/files/2017-05/2017_05_03_DairyReport_f_web.pdf.

Shen Guozheng 沈果正. 1979. "Dujiangyan de bianqian" 都江堰的变迁 [Changes in the Dujiangyan]. *Sichuan Daxue Xuebao, Shehui Kexue Ban* 四川大学学报，社会科学版 1979 (4): 101–6.

Shen, Jianming, Xiaohong Zhang, Yanfeng Lv, Xiangdong Yang, Jun Wu, Lili Lin, and Yanzong Zhang. 2018. "An Improved Emergy Evaluation of the Environmental Sustainability of China's Steel Production from 2005 to 2015." *Ecological Indicators* 103:55–69.

Shen, Lei, Tao Dai, and Aaron James Gunson. 2009. "Small-Scale Mining in China: Assessing Recent Advances in the Policy and Regulatory Framework." *Resources Policy* 34:150–57.

Shen, Xiaowei, and Ping Xiao. 2014. "McDonald's and KFC in China: Competitors or Companions?" *Marketing Science* 33 (2): 287–307.

Sheng, Jichuan, and Michael Webber. 2019. "Governance Rescaling and Neoliberalization of China's Water Governance: The Case of China's South-North Water Transfer Project." *Environment and Planning* A 51 (8): 1644–64.

Shepherd, Christian, and Ian Livingston. 2022. "China's Summer Heat Wave Is Breaking All Records." *Washington Post*, 24 August 2022.

Shi Bonian 史柏年. 1990. "1958 nian da lian gang tie yundong shuping" 1958年大炼钢铁运动述评 [Narrative and evaluation of "big smelting of iron and steel" in 1958]. *Zhongguo jingji shi yanjiu* 中国经济史研究 1990 (2): 124–33.

Shi De-qing, Zhang Jian, Gui Zhao-long, Dong Jian, Wang Tian-li, Valentina Murygina, and Sergey Kalyuzhnyi. 2007. "Bioremediation of Oil Sludge in Shengli Oilfield." *Water, Air, and Soil Pollution* 185:177–84.

Shi, Y. 2004. "Optimum Strategy on Farmland Preservation: A Case Study in Jingzhou City, Hubei Province; Land Resource Bureau of Jianzhou City, Jingzhou." Cited in Lichtenberg and Ding 2008.

————. 2020. "Are China's New 2030 Climate Targets Ambitious Enough?" China Dialogue, 15 December 2020. https://chinadialogue.net/en/climate/are-chinas-new-2030-climate-targets-ambitious-enough/.

Shidai Zhoubao 時代週報. 2010. "Huanbao bu guanyuan zhiyi shuidian da ba cheng qi bi huodian wuran geng yanzhong" 环保部官员质疑水电大坝 称其比火电污染更严重 [Environmental official casts doubt on big hydroelectric dams, calls their pollution more serious than thermoelectric power]. 30 December 2010. http://green.sohu.com/20101230/n278593509.shtml.

Shih, Chuan-kang. 2010. *Quest for Harmony: Moso Traditions of Sexual Union and Family Life*. Stanford, CA: Stanford University Press.

Shou, Ming-Huan, Zheng-Xin Wang, Dan-Dan Li, and Yi Wang. 2020. "Assessment of the Air Pollution Emission Reduction Effect of the Coal Substitution Policy in China: An Improved Grey Modeling Approach." *Environmental Science and Pollution Research* 27:34357–368.

Shu Ti 舒俚. 1998. "Da lian gangtie de qiyi yu beiju" 大炼钢铁的奇迹与悲剧 [The miracle and the tragedy of big iron and steel smelting]. *Wenshi jinghua* 文史精华 1998 (6): 49–55.

Shue, Vivienne. 1980. *Peasant China in Transition: The Dynamics of Development toward Socialism*. Berkeley: University of California Press.

Si, Zhenzhong, Jenelle Regnier-Davies, and Steffanie Scott. 2018. "Food Safety in Urban China: Perceptions and Coping Strategies of Residents in Nanjing." *China Information* 32 (3): 377–99.

Si, Zhenzhong, Theresa Schumilas, and Steffanie Scott. 2015. "Characterizing Alternative Food Networks in China." *Agriculture and Human Values* 32 (2): 299–313.

Si, Zhenzhong, Steffanie Scott, and Cameron McCordic. 2019. "Wet Markets, Supermarkets, and Alternative Food Sources: Consumers' Food Access in Nanjing, China." *Canadian Journal of Development Studies / Revue canadienne d'études du développement* 40 (1): 78–96

Sichuan Sheng Dujiangyan Guanliju 四川省都江堰管理局 (Management Bureau of Dujiangyan, Sichuan Province). 1994. "Gu yan feng chun bei zenghui" 古堰逢春倍增辉 [The ancient weir finds a multitudinous new spring]. *Shuili Dianli Qiye Guanli* 水利电力企业管理 1994:86–87.

Silveira, Maria L., João M. B. Vendramini, Hiran M. da Silva, and Mariana Azenha. 2013. "Nutrient Cycling in Grazed Pastures." University of Florida Extension Service. http://edis.ifas.ufl.edu/pdffiles/SS/SS57800.pdf.

Simangunsong, Tonggo. 2021. "Indonesian Community Seeks World Bank Mediation against Chinese-Owned Zinc Mine." China Dialogue, 22 March 2021. https://china dialogue.net/en/business/indonesian-community-seeks-world-bank-mediation -against-chinese-owned-zinc-mine/.

Sina News Center. 2005. "Shipin: Zhongshiyou Jilin gongsi fasheng baozha" 视频：中石油吉林公司发生爆炸 [Video: Explosion happens at China petroleum Jilin Corporation]. http://news.sina.com.cn/c/2005-11-13/21178285891.shtml.

———. 2017. "Taxian Hu huoshi Xihu 100 bei: shijie zuida piaobo fadian zhan dansheng" 塌陷湖或是西湖100倍：世界最大漂浮发电站诞生 [Collapsed lake or 100 West Lakes? The world's largest floating power station is born]. 7 July 2017. http://news .sina.com.cn/c/nd/2017-07-13/doc-ifyiakwa3996245.shtml.

———. 2021. "Henan Zhengzhou Jialu He fasheng chao lishi hongshui" 河南郑州贾鲁河发生超历史洪水 [The Jialu River at Zhengzhou, Henan, experiences historic floods]. 21 July 2021. https://news.sina.com.cn/c/2021-07-21/doc-ikqciyzk6707701 .shtml.

Sinkule, Barbara J., and Leonard Ortulano. 1995. *Implementing Environmental Policy in China*. Westport, CT: Praeger.

Siu, Helen F. 1989. *Agents and Victims in South China: Accomplices in Rural Revolution*. New Haven: Yale University Press.

———. 2007. "Grounding Displacement: Uncivil Urban Spaces in Postreform South China." *American Ethnologist* 34 (2): 329–50.

Skinner, G. William. 1964–65. "Marketing and Social Structure in Rural China" (parts 1–3). *Journal of Asian Studies* 24 (1): 3–43; 24 (2): 195–228; 24 (3): 363–99.

———. 1977a. "Regional Urbanization in Nineteenth-Century China." In *The City in Late Imperial China*, 211–49. Stanford, CA: Stanford University Press.

———. 1977b. "Cities and the Hierarchy of Local Systems." In *The City in Late Imperial China*, 275–364. Stanford, CA: Stanford University Press.

———. 2017. *Rural China on the Eve of Revolution: Sichuan Field Notes, 1949–50*. Edited by Stevan Harrell and William Lavely. Seattle: University of Washington Press.

Skinner, G. William, Mark Henderson, and Jianhua Yuan. 2000. "China's Fertility Transition through Regional Space Using GIS and Census Data for a Spatial Analysis of Historical Demography." *Social Science History* 24 (3): 615–52.

Skinner, G. William, and Edwin A. Winckler. 1969. "Compliance Succession in Rural Communist China: A Cyclical Theory." In *A Sociological Reader on Complex Organizations*, edited by Amitai Etzioni, 410–38. 2nd ed. New York: Holt, Rinehart and Winston.

Smil, Vaclav. 1993. *China's Environmental Crisis*. Armonk, NY: M. E. Sharpe.

———. 1995. "Who Will Feed China?" *China Quarterly* 143:801–13.

———. 1999a. "China's Great Famine: 40 Years Later." *British Medical Journal* 319:1619–21.

———. 1999b. "China's Agricultural Land." *China Quarterly* 158:414–29.

———. 2002. "Nitrogen and Food Production: Proteins for Human Diets." *Ambio* 31 (2): 126–31.

Smith, Arthur H. (1899) 1970. *Village Life in China*. Introduction by Myron L. Cohen. Boston: Little, Brown.

Sniadecki, J. P. 2008. *Demolition/Chai Qian* 拆迁. Documentary film, 62 minutes.

———. 2012. *People's Park*. Documentary film, 78 minutes.

Solow, Robert. 1993. "Sustainability: An Economist's Perspective." In *Economics of the Environment: Selected Readings*, edited by Robert Dorfman and Nancy S. Dorfman, 179–87. 3rd ed. New York: Norton.

Song, Song, Christian Albert, and Martin Prominski. 2020. "Exploring Integrated Design Guidelines for Urban Wetland Parks in China." *Urban Forestry and Urban Greening* 53: article 126712.

Sönnichsen, N. 2020a. Primary Energy Consumption in China 2000 to 2019. Statista, 3 November 2020. www.statista.com/statistics/265580/primary-energy-consumption -in-china-since-1998/.

———. 2020b. Primary Energy Consumption in China in 2018 and 2019, by Fuel Type. Statista, 3 November 2020. www.statista.com/statistics/265612/primary-energy -consumption-in-china-by-fuel-type-in-oil-equivalent/.

Sorace, Christian P. 2017. *Shaken Authority: China's Communist Party and the 2008 Sichuan Earthquake*. Ithaca: Cornell University Press.

SSB (State Statistical Bureau) (Guojia Tongji Ju 国家统计局). 2014. *Zhongguo huanjing tongji nianjian* 中国环境统计年鉴 [China environmental statistical yearbook]. Beijing: Zhongguo Tongji Chubanshe.

———. 2018a. *Zhongguo tongji nianjian* 中国统计年鉴 [China statistical yearbook]. www.stats.gov.cn/tjsj/ndsj/2018/indexch.htm.

———. 2018b. *Zhongguo huanjing tongji nianjian* 中国环境统计年鉴 [China environmental statistical yearbook]. Beijing: Zhongguo Tongji Chubanshe.

———. 2021. *Di qi ci quanguo renkou pucha zhuyao shuju qingkuang* 第七次全国人口普查主要数据情况 [General statistical situation from the seventh national census]. www.stats.gov.cn/tjsj/zxfb/202105/t20210510_1817176.html.

———. n.d. *Zhongguo 2010 nian renkou pucha ziliao* 中国2010年人口普查资料 [Data from the 2010 population census of China]. www.stats.gov.cn/tjsj/pcsj/rkpc/6rp /indexch.htm.

Standaert, Michael. 2021. "Despite Pledges to Cut Emissions, China Goes on a Coal Spree." *Yale Environment 360*, 24 March 2021. https://e360.yale.edu/features/despite -pledges-to-cut-emissions-china-goes-on-a-coal-spree.

Standing Committee of the National People's Congress. 2008. "Zhonghua Renmin Gongheguo xunhuan jingji cujin fa" 中华人民共和国循环经济促进法 [Circular Economy Promotion Law of the People's Republic of China]. www.lawinfochina.com /display.aspx?id=7025&lib=law.

State Council (Guowuyuan 国务院). 2002. "Guowuyuan guanyu jin yi bu wanshan tui geng huan lin zhengce shishi ruogan yijian" 国务院关于进一步完善退耕还林政策实施的若干意见 [Some opinions of the State Council regarding taking a step toward successfully implementing the policy of returning farmland to forest]. www.china .com.cn/hinese/zhuanti/xbkf5/798409.htm.

State Council Information Office (Guowuyuan Xinwen Bangongshi 国务院新闻办公室). 2019. "'Zhongguo de liangshi anquan' baipi shu" 《中国的粮食安全》白皮书 [White paper on "food security in China"]. www.scio.gov.cn/ztk/dtzt/39912/41906/index.htm.

State Forestry Administration (Guojia Linye Ju 国家林业局). 2020. *Zhongguo linye he caoyuan tongji nianjian* 中国林业和草原统计年鉴 [China forestry and grassland statistical yearbook]. Beijing: China Forestry Publishing House.

———. n.d. a. *Changjiang shangyou, Huanghe shangzhongyou diqu tianran lin baohu gongcheng shishi fang'an* 长江上游，黄河上中游地区天然林保护工程实施方案 [Plan for implementing the Natural Forest Protection Program in the upper Chang and upper and middle Yellow River areas].

———. n.d. b. *Dongbei, Nei Menggu deng zhongdian guoyou linqu tianran lin ziyuan baohu gongcheng shishi fang'an* 东北，内蒙古等重点国有林区天然林资源保护工程实施方案 [Plan for implementing the Natural Forest Resource Protection Project in the key areas of the northeast and Inner Mongolia]. Accessed December 2022. https://wenku.baidu.com/view/c5ea7b5cbe23482fb4da4c3d.html?_wkts_=1672516703757.

Statista. 2019. Import Volume of Soybeans Worldwide in 2018/19, by Country. Accessed July 2022, behind paywall. www.statista.com/statistics/612422/soybeans-import-volume-worldwide-by-country/.

———. 2022a. Production Volume of Major Grain Crops in China from 2011 to 2021, by Type of Crop. www.statista.com/statistics/1198896/china-grain-crop-production-volume-by-main-crop/.

———. 2022b. Degree of Urbanization in China from 1980 to 2021. www.statista.com/statistics/270162/urbanization-in-china/.

———. 2022c. Number of Privately Owned Vehicles in China from 2009 to 2020. www.statista.com/statistics/278475/privately-owned-vehicles-in-china/.

———. 2022d. Total Volume of Passenger [*sic*] Transported by Highspeed Railways in China from 2008 to 2021. www.statista.com/statistics/1120071/china-passenger-transport-volume-of-highspeed-rail/.

Steenberg, Rune, and Alessandro Rippa. 2019. "Development for All? State Schemes, Security, and Marginalization in Kashgar, Xinjiang." *Critical Asian Studies* 51 (2): 274–95.

Stone, Bruce. 1988. "Developments in Agricultural Technology." *China Quarterly* 116:767–822.

Stone, Richard. 2011. "Mayhem on the Mekong." *Science* 222 (6044): 814–18.

Strassberg, Richard A. 1994. *Inscribed Landscapes: Travel Writing from Imperial China.* Berkeley: University of California Press.

Sturgeon, Janet C. 2005. *Border Landscapes: The Politics of Akha Land Use in China and Thailand.* Seattle: University of Washington Press.

Su, Chang, Hatef Madani, and Björn Palm. 2018. "Heating Solutions for Residential Buildings in China: Current Status and Future Outlook." *Energy Conversion and Management* 177: 493–510.

Su, Chen, Zhongshuang Cheng, Wen Wei, and Zongyu Chen. 2018. "Assessing Groundwater Availability and the Response of the Groundwater System to Intensive Exploitation in the North China Plain by Analysis of Long-Term Isotropic Tracer Data." *Hydrogeology Journal* 26:1401–15.

Sui, Yue, Dabang Jiang, and Zhiping Tian. 2012. "Latest Update of the Climatology and Changes in the Seasonal Distribution of Precipitation over China." *Theoretical and Applied Climatology* 113:599–610.

Sumberg, James, Dennis Keeney, and Benedict Dempsey. 2012. "Public Agronomy: Norman Borlaug as 'Brand Hero' for the Green Revolution." *Journal of Development Studies* 48 (11): 1587–1600.

Sun, Lian, Meng Xu, Junxiang Jia, and Chunhui Li. 2016. "Risk Identification of Water

Pollution Sources in Water Source Areas of Middle Route of the South-to-North Water Diversion Project." *International Journal of Environmental Science and Development* 7 (8): 576–80.

Sun, Xingsong, Xiaogang Wang, Lipeng Liu, and Ruizhi Fu. 2019. "Development and Present Situation of Hydropower in China." *Water Policy* 21:565–81.

Sun Zexue 孙泽学. 1992. "1978–1984 nian nongcun gaige zhi zhongyang, difang, nongmin de huzhu guanxi yanjiu—yi bao chan dao hu, bao gan dao hu wei zhongxin" 1978–1984 年农村改革之中央、地方、农民的互动关系研究—以包产到户、包干到户为中心 [Research on cooperative relationships among center, localities, and peasants in the 1978–1984 rural reforms—centered on Baochan Dao Hu and Baogan Dao Hu]. *Zhongguo Jingji Yanjiu* 中国经济史研究 2006 (1): 79–87.

Sun Zhiyu 孙志禹 and Hu Lianxing 胡连兴. 2019. "Zhongguo shuidian fazhan de xianzhung yu zhanwang" 中国水电发展的现状与展望 [The current situation and outlook for China's hydroelectric power]. https://zhuanlan.zhihu.com/p/160057178.

Svanberg, Ingvar. 1988. "The Nomadism of the Orta Guz Kazaks in Xinjiang, 1911–1949." In *The Kazaks of China: Essays on an Ethnic Minority*, edited by Linda Benson and Ingvar Svanberg, 107–40. Uppsala: University of Uppsala.

Swamy, Subramanian, and Shahid Javed Burki. 1970. "Foodgrains Output in the People's Republic of China, 1958–1965." *China Quarterly* 41:58–63.

Sze, Julie. 2015. *Fantasy Islands: Chinese Dreams and Ecological Fears in an Age of Climate Crisis*. Oakland: University of California Press.

Tan, Gillian G. 2016. *In the Circle of White Stones: Moving through Seasons with Nomads of Eastern Tibet*. Seattle: University of Washington Press.

Tan Shukui 谭树魁. 1999. "1998 nian Changjiang zhongxiayou teda hongzai de tudi liyong sikao" 1988 年长江中下游特大洪灾的土地利用思考 [Consideration of land use in the mega-floods of 1998 in the middle and lower Yangtze]. *Dili Kexue* 地理科学 18 (6): 493–500.

Tan, Su-lin. 2021. "China-Australia Relations: Is This the Start of 'Mining Boom 2.0' as Canberra Rakes in Cash from Iron Ore Rally?" *South China Morning Post*, 18 May 2021.

Tan Ye 谭野. 2003. "Sanmenxia da ba: Zengjing 'Zhongguo di yi ba' rujin ming xuan yi xian" 三门峡大坝：曾经"中国第一坝"如今命悬一线 [The big dam at Sanmenxia: The life of the former 'First Dam of China' is now hanging by a thread]. *Zhongguo Baodao Zhoukan* 中国报道周刊 [China report weekly], 30 November 2003. www.china-week.com/info/03669.htm.

Tang, Caihong, Yujun Yi, Zhifeng Yang, and Xi Cheng. 2014. "Water Pollution Risk Simulation and Prediction in the Main Canal of the South-to-North Water Transfer Project." *Journal of Hydrology* 519:2111–20.

Tang Hao. 2012. "Shifang: A Crisis of Local Rule." China Dialogue, 18 July 2012. https://chinadialogue.net/en/pollution/5049-shifang-a-crisis-of-local-rule/.

Tang, Ming, Jin Quan, Xiaohui Lei, Hezheng Zheng, and Tao Tang. 2018. "Risk Factors Analysis and Risk Assessment of Sudden Pollution from Traffic Accidents in the Middle Route of South-to-North Water Diversion Project." *MATEC Web of Conferences* 246 (02015). https://doi.org/10.1051/matecconf/201824602015.

Tang, Zhenwu, Qifei Huang, Yufei Yang, Xiaohua Zhu, and Haihui Fu. 2013. "Organochlorine Pesticides in the Lower Reaches of the Yangtze River: Occurrence, Ecological Risk and Temporal Trends." *Ecotoxicology and Environmental Safety* 87:89–97.

Tan-Mullins, May, Frauke Urban, and Grace Mang. 2017. "Evaluating the Behaviour of Chinese Stakeholders Engaged in Large Hydropower Projects in Asia and Africa." *China Quarterly* 230: 464–88.

Tao Yan 陶炎. 1983. "Beidahuang kaiken hou shengtai huanjing de bianqian" 北大荒开垦后生态环境的演变 [The evolution of ecological environment after reclamation in the Great Northern Wilderness]. *Shengtaixue Zazhi* 生态学杂志 1983 (1): 23–25, 41.

Tapp, Nicholas. 2001. *The Hmong of China: Context, Agency, and the Imaginary*. Leiden: Brill.

Taylor, Rodney. 1998. "Companionship with the World: Roots and Branches of a Confucian Ecology." In *Confucianism and Ecology: The Interrelation of Heaven, Earth, and Humans*, edited by Mary Evelyn Tucker and John H. Berthong, 37–58. Cambridge, MA: Harvard University Press.

The Guardian. 2005. "100 Tonnes of Pollutants Spilled into Chinese River." 25 November 2005.

Thorp, H. Holden. 2020. "Science Has Always Been Political." *Science* 369 (6501): 227.

Tian, Jinping, Han Shi, Ying Cheng, and Lujun Chen. 2012. "Assessment of Industrial Metabolisms of Sulfur in a Chinese Fine Chemical Industrial Park." *Journal of Cleaner Production* 32:262–72.

Tian, Li. 2015. "Land Use Dynamics Driven by Rural Industrialization and Land Finance in the Peri-urban Areas of China: The Examples of Jiangyin and Shunde." *Land Use Policy* 45:117–27.

Tian, Li, Yongfu Li, Yaqi Yan, and Boyi Wang. 2017. "Measuring Urban Sprawl and Exploring the Role Planning Plays: A Shanghai Case Study." *Land Use Policy* 67:426–35.

Tian, Li, and Zhihao Yao. 2018. "From State-Dominant to Bottom-Up Redevelopment: Can Institutional Change Facilitate Urban and Rural Redevelopment in China[?]." *Cities* 76:72–83.

Tian, Yiyu. 2019. "Trapped in Time: Bodily Experiences of Family Dependent Workers (*jiashu*) in Daqing, a Model Industrial City in High-Socialist China." MA thesis, University of Washington, Seattle.

Tilt, Bryan D. 2007. "The Political Ecology of Pollution Enforcement in China: A Case from Sichuan's Rural Industrial Sector." *China Quarterly* 192:915–32.

———. 2010. *The Struggle for Sustainability in Rural China: Environmental Values and Civil Society*. New York: Columbia University Press.

———. 2015. *Dams and Development in China: The Moral Economy of Water and Power*. New York: Columbia University Press.

Tilt, Bryan, and Zhuo Chen. 2021. "Population Resettlement for Hydropower Development in the Lancang River Basin: An Evolving Policy Framework and Its Implications for Local People." In *The Political Economy of Hydropower in Southwest China and Beyond*, edited by Jean-François Rousseau and Sabrina Habich-Sobiegalla, 89–106. London: Palgrave.

Tilt, Bryan, and Pichu Xiao. 2007. "Industry, Pollution, and Environmental Enforcement in Rural China: Implications for Sustainable Development." *Urban Anthropology and Studies of Cultural Systems and Economic Development* 36 (1–2): 115–43.

Tippins, Jennifer Laura. 2013. "Planning for Resilience: A Proposed Landscape Evaluation for Redevelopment Planning in the *Linpan* Landscape." Master's thesis, University of Washington, Seattle.

Tong, Songying, Lingdong Kong, Kejing Yang, Jiandong Shen, Lu Chen, Shengyan Jin, Cao Wang, et al. 2020. "Characteristics of Air Pollution Episodes Influenced by Biomass Burning Pollution in Shanghai, China." *Atmospheric Environment* 238: article 117756.

Toops, Stanley W. 2004. "The Ecology of Xinjiang: A Focus on Water." In *Xinjiang: China's Muslim Borderland*, edited by S. Frederick Starr, 264–75. Armonk, NY: M. E. Sharpe.

Trac, Christine Jane, Stevan Harrell, Thomas M. Hinckley, and Amanda C. Henck. 2007. "Reforestation Programs in Southwest China: Reported Success, Observed Failure, and the Reasons Why." *Journal of Mountain Science* 4 (4): 275–92.

Trac, Christine Jane, Amanda H. Schmidt, Stevan Harrell, and Thomas M. Hinckley. 2013. "Is the Returning Farmland to Forest Program a Success? Three Case Studies from Sichuan." *Environmental Practice* 15 (3): 350–66.

Trase. 2020. "China's Exposure to Environmental Risks from Brazilian Beef Imports." *Trase Issue Brief*, no. 3 (June). http://resources.trase.earth/documents/issuebriefs /IssueBrief3_EN.pdf.

Trotsky, Leon. 1924. "Revolutionary and Socialist Art." Marxists Internet Archive. www.marxists.org/archive/trotsky/1924/lit_revo/cho8.htm.

Turley, Samuel T., Robert L. Pitman, Barbara L. Taylor, Jay Barlow, Tomonari Akamatsu, Leigh Barrett, Xiujiang Zhao, Randall R. Reeves, et al. 2007. "First Human-Caused Extinction of a Cetacean Species?" *Biology Letters* 3 (5): 537–40.

Turner, Sarah, Christine Bonnin, and Jean Michaud. 2015. *Frontier Livelihoods: Hmong in the Sino-Vietnamese Borderlands*. Seattle: University of Washington Press.

Tverberg, Gail. 2011. "Is It Really Possible to Decouple GDP Growth from Energy Growth?" Our Finite World, 15 November 2011. https://ourfiniteworld.com /2011/11/15/is-it-really-possible-to-decouple-gdp-growth-from-energy-growth.

UCS (Union of Concerned Scientists). 2022. "Each Country's Share of CO_2 Emissions." www.ucsusa.org/resources/each-countrys-share-co2-emissions.

UNDAC (United Nations Disaster Assessment and Coordination Team). 1998. *Final Report on 1998 Floods in the People's Republic of China*. New York: UN Office for the Coordination of Human Affairs. https://reliefweb.int/report/china/final-report -1998-floods-peoples-republic-china.

UNPD (United Nations Department of Economic and Social Affairs, Population Division). 2022a. World Population Prospects 2022: Line Graphs. https://population .un.org/wpp/Graphs/DemographicProfiles/Line/156.

———. 2022b. World Population Prospects 2022: Population Pyramids. https:// population.un.org/wpp/Graphs/DemographicProfiles/Pyramid/156.

———. 2022c. World Population Prospects 2022: Probabilistic Projections. https:// population.un.org/wpp/Graphs/Probabilistic/POP/TOT/156.

Urban, Frauke, Sam Geall, and Yu Wang. 2016. "Solar PV and Solar Water Heaters in China: Different Pathways to Low Carbon Energy." *Renewable and Sustainable Energy Reviews* 64:531–42.

Urban, Frauke, Johan Nordensvärd, Deepika Khatri, and Yu Wang. 2012. "An Analysis of China's Investment in the Hydropower Sector in the Greater Mekong Sub-Region." *Environment, Development, and Sustainability* 15 (2): 301–24.

Urban, Frauke, Giuseppina Siciliano, and Johan Nordensvärd. 2018. "China's Dam

Builders: Their Role in Transboundary River Management in Southeast Asia." *International Journal of Water Resources Development* 34 (5): 747–70.

Urgenson, Lauren S., R. Keala Hagmann, Stevan Harrell, Amanda C. Henck, Thomas M. Hinckley, Sara Jo Shepler, Barbara L. Grub, and Phillip M. Chi. 2010. "Socio-Ecological Resilience of a Nuosu Community-Linked Watershed, Southwest Sichuan, China." *Ecology and Society* 15 (4): 2. www.ecologyandsociety.org/vol15/iss4/art2/.

Urgenson, Lauren S., Amanda H. Schmidt, Julie K. Combs, Stevan Harrell, Thomas M. Hinckley, Qiangxia Yang, Ziyu Ma, et al. 2014. "Traditional Livelihoods, Conservation, and Meadow Ecology in Jiuzhaigou National Park, China." *Human Ecology* 42 (3): 481–91.

Van Genderen, Eric, Maggie Wildnauer, Nick Santero, and Nadir Sidi. 2016. "A Global Life Cycle Assessment for Primary Zinc Production." *International Journal of Life Cycle Assessment* 21:1580–93.

van Schendel, Willem. 2002. "Geographies of Knowing, Geographies of Ignorance: Jumping Scale in Southeast Asia." *Environment and Planning D: Society & Space* 20 (6): 647–68.

Veeck, Ann, Hongyan Yu, and Alvin C. Burns. 2010. "Consumer Risks and New Food Systems in Urban China." *Journal of Macromarketing* 30 (3): 222–37.

Vermeer, Eduard. 1977. *Water Conservancy and Irrigation in China: Social, Economic, and Agrotechnical Aspects.* Leiden: Leiden University Press.

Viard, V. Brian, and Shihe Fu. 2015. "The Effect of Beijing's Driving Restrictions on Pollution and Economic Activity." *Journal of Public Economics* 125:98–115.

Visser, Robin. 2016. "The Chinese Eco-City and Suburbanization Planning: Case Studies of Tongzhou, Lingang, and Dujiangyan." In *Ghost Protocol: Development and Displacement in Global China*, edited by Carlos Rojas and Ralph A. Litzinger, 36–61. Durham, NC: Duke University Press.

Vitousek, P. M., R. Naylor, T. Crews, M. B. David, L. E. Drinkwater, E. Holland, P. J. Johnes, et al. 2009. "Nutrient Imbalances in Agricultural Development." *Science* 324 (5934): 1519–20.

Wagner, Donald B. 2011a. "The Great Leap Forward in Iron and Steel, 1958–1960." http://donwagner.dk/MS-English/GreatLeap.html.

———. 2011b. "Background to the Great Leap Forward in Iron and Steel: The Traditional Chinese Iron Industry and Its Modern Fate." http://donwagner.dk/MS-English/MS-English.html.

Wakeman, Frederic, Jr. 1985. *The Great Enterprise: The Manchu Reconstruction of Imperial Order in Seventeenth-Century China.* Berkeley: University of California Press.

Walder, Andrew. 2014. "Rebellion and Repression in China, 1966–1971." *Social Science History* 38 (3–4): 513–39.

Walker, Brian, Lance H. Gunderson, Ann Kinzig, Carl Folke, Steve Carpenter, and Lisen Schultz. 2006. "A Handful of Heuristics and Some Propositions for Understanding Resilience in Social-Ecological Systems." *Ecology and Society* 11 (1): article 13. www.ecologyandsociety.org/vol11/iss1/art13/.

Walker, Brian, and David Salt. 2006. *Resilience Thinking: Sustaining Ecosystems and People in a Changing World.* Washington, DC: Island Press.

Walker, Kenneth R. 1966. "Collectivization in Retrospect: The 'Socialist High Tide' of Autumn 1955–Spring 1956." *China Quarterly* 27:1–43.

Wallerstein, Immanuel. 2010. "A World-System Perspective on the Social Sciences." *British Journal of Sociology* 61 (Suppl. 1): 167–76.

Walling, Des E. 2008. "The Changing Sediment Loads of the World's Rivers." *Annals of Warsaw University of Life Sciences—SGGW Land Reclamation* 39:3–20.

Wan, Guanghua, Dongqing Zhu, Chen Wang, and Xun Zhang. 2020. "The Size Distribution of Cities in China: Evolution of Urban System and Deviations from Zipf's Law. *Ecological Indicators* 111: article 106003.

Wan, Rongrong, Xue Dai, and David Shankman. 2018. "Vegetation Response to Hydrological Changes in Poyang Lake, China." *Wetlands* 39 (S1): 99–112.

Wang Chen. 2020. "Can China's Waste Incinerators Appease Local Opposition?" China Dialogue, 13 January 2020. https://chinadialogue.net/en/cities/11777-can -china-s-waste-incinerators-appease-local-opposition/.

Wang, Dan, and Hong Li. 2010. "On the Subject of China's Agricultural Science and Technology Optimization." *Asian Social Science* 6 (2): 117–21

Wang, Feng, Yong Cai, and Baocheng Gu. 2012. "Population, Policy, and Politics: How Will History Judge China's One-Child Policy?" *Population and Development Review* 38 (Suppl.): 115–29.

Wang Fusheng 王福生. 2020. *Nan shui bei diao xi xian gongcheng de xin silu yu xin fang'an—xi xian diaoshui ying cong Nujiang, Palong Jiang huo Yalu Zangbu Jiang xuan dian de diaoyan* 南水北调西线工程的新思路与新方案—西线调水应从怒江、帕龙江或雅鲁藏布江选点的调研 [New ideas and new possibilities for the western route of the North-South Water Transfer Project—Researching transferring water on the Western Route from the Nu, Parlung Tsangpo, or Yarlung Tsangpo Rivers]. www.gsass.net.cn/zhongdaxiangmunew/zsrgnew/ktzcg/2020-05-07/1680.html.

Wang Guangbin 王光彬. 2019. "Wunonglong shuidian zhan shou tai jizu touru shangye yunxing" 乌弄龙水电站首台机组投入商业运行 [The first group of generators at the Wunonglong Hydroelectric Station enters commercial production]. *Zhongguo Gezhouba Jituan Jidian Jianshe Youxian Gongsi* 中国葛洲坝集团机电建设有限公司, 3 January 2019. www.gzbjd.ceec.net.cn/art/2019/1/3/art_14961_1813971.html.

Wang Hanjuan 王菡娟. 2021. "Nan shui bei diao xi xian gongcheng weihe chichi meiyou kaigong jianshe?" 南水北调西线工程为何迟迟没有开工建设? [Why is the western route of the South-North Water Transfer dilly-dallying and not starting construction?]. *Shangguan Xinwen* 上观新闻, 20 May 2021.

Wang, Houjie, Naishuang Bi, Yoshiki Saito, Yan Wang, Xiaoxia Sun, Jia Zhang, and Zuosheng Yang. 2010. "Recent Changes in Sediment Delivery by the Huanghe (Yellow River) to the Sea: Causes and Environmental Implications in Its Estuary." *Journal of Hydrology* 391:302–13.

Wang, Hua, Jie He, Yoonhee Kim, and Takuya Kamata. 2011. "Municipal Solid Waste Management in Small Towns: An Economic Analysis Conducted in Yunnan, China." World Bank Policy Research Working Paper 5767. https://papers.ssrn.com/sol3 /papers.cfm?abstract_id=1914978.

Wang Jing 王京. 2005. "Huanbao zongju tongbao Songhua jiang wuran an wuran dai chang yue 80 gongli" 环保总局通报松花江污染案 污染带长约80公里 [The Environmental Protection Office reports that the band of pollution in the Sungari River has stretched out about 80 kilometers]. Sina News Center, 23 November 2005. http:// news.sina.com.cn/c/2005-11-23/14487519345s.shtml.

Wang, Jinxia, Jikun Huang, Qiuqiong Huang, and Scott Rozelle. 2006. "Privatization of Tubewells in North China: Determinants and Impacts on Irrigated Area, Productivity and the Water Table." *Hydroecology Journal* 14:275–85.

Wang Lina and Wang Yongchen. 2011. "Yellow River Decade (5): Silt at the Sanmenxia Power Station." China Green News. http://eng.greensos.cn/ShowArticle.aspx ?articleId=579.

Wang, Luo, Hongliang Jia, Xianjie Liu, Yeqing Sun, Meng Yang, Wenjun Hong, Hong Qi, and Yi-Fan Li. 2013. "Historical Contamination and Ecological Risk of Organochlorine Pesticides in Sediment Core in Northeastern Chinese River." *Ecotoxicology and Environmental Safety* 93:112–20.

Wang, Mark. 2022. "Adapting China to Extreme Weather." China Dialogue, 19 October. https://chinadialogue.net/en/climate/adapting-china-to-extreme-weather/.

Wang, Mark, Michael Webber, Brian Finlayson, and Jon Barnett. 2007. "Rural Industries and Water Pollution in China." *Journal of Environmental Management* 86:648–59.

Wang Ningsheng. 1985. "Rock Paintings in Yunnan, China: Some New Light on the Old Shan Kingdom." *Explorations* 27 (1): 25–33.

Wang, Orange. 2020. "China Food Security: Beijing Tells Farmers to Stick to Grain Amid Global Uncertainty." *South China Morning Post*, 18 November 2020.

———. 2021. "African Swine Fever Still 'Major Risk Factor' for China as It Bans Malaysian Pig Imports over Outbreak Fears." *South China Morning Post*, 12 March 2021.

Wang, Peng, Xiuxiu Zhang, and Shuhua Qi. 2019. "Was the Trend of Net Sediment Flux in Poyang Lake, China, Altered by the Three Gorges Dam?" *Environmental Earth Sciences* 78 (3).

Wang, Pu, James P. Lassoie, Shikui Deng, and Stephen J. Morreale. 2013. "A Framework for Social Impact Analysis of Large Dams: A Case Study of Cascading Dams on the Upper-Mekong River, China." *Journal of Environmental Management* 117:131–40.

Wang, Q., M. Shao, Y. Zhang, Y. Wei, M. Hu, and S. Guo. 2009. "Source Apportionment of Fine Organic Aerosols in Beijing." *Atmospheric Chemistry and Physics* 9:8573–85.

Wang Ruifang 王瑞芳. 2008. "Dayuejin shiqi nongtian shuili jianshe deshi wenti yanjiu pingshu" 大跃进时期农田水利建设得失问题研究评述 [Critique of research on the question of gains and losses from construction of waterworks during the Great Leap Forward]. *Beijing Keji Daxue Xuebao: Shehui Kexue Ban* 24 (4): 122–30.

Wang Weiluo. 2019. "Sanxia da ba xingbian shishi yidan kui ba minzhong ruhe zijiu?" 王维洛：三峡大坝形变是事实 一旦溃坝民众如何自救 [The changes in the shape of the Sanxia Dam are real; if the dam fails, how will people save themselves?] (video interview). *Boxun*, 12 July 2019. www.boxun.com/news/gb/pubvp /2019/07/201907121108.shtml.

Wang, Xi. 2019. "People's Republic of China." In *The Oxford Handbook of Comparative Environmental Law*, edited by Emma Lees and Jorge E. Viñuales, 128–48. Oxford: Oxford University Press.

Wang, Ya Ping. 2011. "Recent Housing Reform Practice in Chinese Cities: Social and Spatial Implications." In *China's Housing Reform and Outcomes*, edited by Joyce Y. Man, 19–46. Cambridge, MA: Lincoln Institute of Land Policy.

Wang, Yanping, Xiangzhen Kong, Zhaoliang Peng, Hui Zhang, Gang Liu, Weiping Hu, and Xiangqian Zhou. 2020. "Retention of Nitrogen and Phosphorus in Lake Chaohu,

China: Implications for Eutrophication Management." *Environmental Science and Pollution Research* 27:41488–502.

Wang, Yanxin, Chunmiao Zheng, and Rui Ma. 2018. "Review: Safe and Sustainable Groundwater Supply in China." *Hydroecology Journal* 26:1301–24.

Wang Yongchen. 2011. "Yellow River Decade (4): The Case of the Xiaolangdi Water Control Project." China Green News. http://eng.greensos.cn/ShowArticle.aspx ? articleId=606.

Wang, Z. H., F. Y. Zhai, J. G. Zhang, W. W. Du, C. Su, J. Zhang, H. R. Jiang, and B. Zhang. 2015. "Secular Trends in Meat and Seafood Consumption Patterns among Chinese Adults, 1991–2011." *European Journal of Clinical Nutrition* 69:227–33.

Wang, Zuoyue. 2014. "The Cold War and the Reshaping of Transnational Science in China." In *Science and Technology in the Global Cold War*, edited by Naomi Oreskes and John Krige, 343–69. Cambridge, MA: MIT Press.

Warren, Kayanna. 2005. "To Market: China's Changing Market Participation in Remote Rural Areas." BA honors thesis, University of Washington, Seattle.

Wasserstrom, Jeffrey. 1984. "Resistance to the One-Child Family." *Modern China* 10 (3): 345–74.

Watson, James L. 1974. *Emigration and the Chinese Lineage*. Berkeley: University of California Press.

Watson, Rubie S. (1981) 2004. "Class Differences and Affinal Relations in South China." In *Village Life in Hong Kong*, edited by James L. Watson and Rubie S. Watson, 73–104. Hong Kong: Chinese University of Hong Kong Press.

Webb, Warren, Stan Szarek, William Lauenroth, Russell Kinerson, and Milton Smith. 1978. "Primary Productivity and Water Use in Native Forest, Grassland, and Desert Ecosystems." *Ecology* 59 (6): 1239–47.

Webber, Michael, M. T. Li, J. Chen, B. Finlayson, D. Chen, Z. Y. Chen, M. Wang, and J. Barnett. 2015. "Impact of the Three Gorges Dam, the South-North Water Transfer Project and Water Abstractions on the Duration and Intensity of Salt Water Intrusions in the Yangtze River Estuary." *Hydrology and Earth System Sciences* 19 (11): 4411–25.

Webber, Michael, Britt Crow-Miller, and Sarah Rogers. 2017. "The South-North Water Transfer Project: Remaking the Geography of China." *Regional Studies* 51 (3): 370–82.

Wee, Sui-Lee, and Elsie Chen. 2018. "China's Tech Farms Are Mapping Pig Faces." *New York Times*, 24 February 2018.

Wei, Jin-Bao, Jan-Dirk Herbell, and Shao Zhang. 1997. "Solid Waste Disposal in China: Situation, Problems and Suggestions." *Waste Management and Research* 15 (6): 573–83.

Wei, Liangliang, Fengyi Zhu, Qiaoyang Li, Chonghua Xue, Xinhui Xia, Hang Yu, Qingliang Zhao, et al. 2020. "Development, Current State and Future Trends in Sludge Management in China: Based on Exploratory Data and CO_2-Equivalent Emissions Analysis." *Environment International* 144 (November): article 106093.

Wei, Meng, Guoqing Hu, Hui Wang, Edith Bai, Yanhong Lou, Aijun Zhang, and Yuping Zhuge. 2017. "35 Years of Manure and Chemical Fertilizer Application Alters Soil Microbial Community Composition in a Fluvo-aquic Soil in Northern China." *European Journal of Soil Biology* 82:27–34.

Wei Yaqiang 魏亚强, He Yuanyuan 何媛媛, Zhang Xiaokun 张晓坤, and Jin Xu 金旭.

2010. "Qianyi Chao Hu shuizhi bianhua yu yuye fazhan de xianghu yingxiang jili" 浅议巢湖水质变化与渔业发展的相互影响机理 [Superficially considering the mutual influences between changes in the water quality and the development of fishing in Chao Lake]. *Zhongguo Gaoxin Jishu Qiye* 中国高新技术企业2010 (32): 102–9.

Weiji Baike 维基百科 (Wikipedia Chinese). 2017. "Shouzhang xian" 寿张县 [Shouzhang County]. https://zh.wikipedia.org/wiki/寿张县.

Weinstein, Jodi L. 2013. *Empire and Identity in Guizhou: Local Resistance to Qing Expansion*. Seattle: University of Washington Press.

Weisheng Yingji Bangongshi 卫生应急办公室 (Office of Health Emergency Response). 2022. "Henan Sheng faxian yi li ren ganran H3N8 qinliugan bingli" 河南省发现一例人感染H3N8禽流感病例 [A human case of H3N8 avian influenza discovered in Henan Province]. www.nhc.gov.cn/yjb/s3578/202204/8dbeadf0efed45b0b2ea22 928523e289.shtml.

Wen Jianwu 文兼武, ed. 2017. *Zhongguo gongye tongji nianjian* 中国工业统计年鉴—2017 [China industry statistical yearbook—2017]. Beijing: China Statistics Press.

Weng, Naiqun. 1993. "The Mother House: The Symbolism and Practice of Gender among the Naze in Southwest China." PhD diss., University of Rochester.

White, Leslie A. 1943. "Energy and the Evolution of Culture." *American Anthropologist* 45 (3, part 1): 335–56.

White, Tyrene. 1994. "Two Kinds of Production: The Evolution of China's Family Planning Policy in the 1980s." *Population and Development Review* 20 (Suppl.): 137–58.

Whiting, Susan H. 2001. *Power and Wealth in Rural China: The Political Economy of Institutional Change*. Cambridge: Cambridge University Press.

Whiting, Susan H., Daniel Abramson, Yuan Shang, and Stevan Harrell. 2019. "A Long View of Resilience on the Chengdu Plain, China." *Journal of Asian Studies* 78 (2): 257–84.

Whitney, J. B. R. 1973. "Ecology and Environmental Control." In *China's Developmental Experience*, edited by Michel Oksenberg, 95–109. New York: Praeger.

———. 1980. "East Asia." In *World Systems of Traditional Resource Management*, edited by Gary A. Klee, 101–29. New York: Wiley.

Whittaker, Jacob Tyler. 2008. "Yi Identity and Confucian Empire: Indigenous Local Elites, Cultural Brokerage, and the Colonization of the Lu-ho Tribal Polity of Yunnan, 1174–745." PhD diss., University of California, Davis.

WHO (World Health Organization). 2006. "WHO Air Quality Guidelines for Particulate Matter, Ozone, Nitrogen Dioxide and Sulfur Dioxide: Global Update 2005." https://apps.who.int/iris/bitstream/handle/10665/69477/WHO_SDE_PHE_OEH_06.02_eng.pdf?sequence=1.

Wiemer, Calla. 2004. "The Economy of Xinjiang." In *Xinjiang: China's Muslim Borderland*, edited by S. Frederick Starr, 163–89. Armonk, NY: M. E. Sharpe.

Wiens, Herold J. 1954. *China's March toward the Tropics*. Hamden, CT: Shoe String Press.

Wikipedia. 2019. "List of Largest Hydroelectric Power Stations." https://en.wikipedia.org/wiki/List_of_largest_hydroelectric_power_stations.

Wikipedia 维基百科. 2013. "Fenghuang Shan gu jianzhu qun" 凤凰山古建筑群 [Phoenix Mountain old structures group]. https://zh.wikipedia.org/wiki/凤凰山古建筑群.

———. 2021a. "Xin Shu he" 新沭河 [New Shu River]. https://zh.wikipedia.org/wiki/新沭河.

———. 2021b. "Yingxia tielu" 鹰厦铁路 [Yingtan-Xiamen railway]. https://zh.wikipedia .org/wiki/鹰厦铁路.

———. n.d. "Chengdu ditie" 成都地铁 [Chengdu subway]. Accessed October 2020. https://zh.wikipedia.org/wiki/成都地铁 .

Will, Pierre-Etienne. 1990. *Bureaucracy and Famine in Eighteenth-Century China*. Stanford, CA: Stanford University Press.

Williams, Dee Mack. 1996. "The Barbed Walls of China: A Contemporary Grassland Drama." *Journal of Asian Studies* 55 (3): 665–91.

———. 2002. *Beyond Great Walls: Environment, Identity, and Development on the Chinese Grasslands of Inner Mongolia*. Stanford, CA: Stanford University Press.

Wilmsen, Brooke. 2015. "After the Deluge: A Longitudinal Study of Resettlement at the Three Gorges Dam, China." *World Development* 84:41–54.

Winckler, Edwin A. 2018. "Powers in Times: Policies, Regimes, Worlds." Paper presented at the Conference on China in Time and Space: G. William Skinner's Ideas Going Forward, Hong Kong University of Science and Technology, 20–22 June 2018.

Wittfogel, Karl A. 1957. *Oriental Despotism: A Comparative Study of Total Power*. New Haven: Yale University Press.

WNN (World Nuclear News). 2019. "Long-Term Operation of China's Oldest Reactor Assessed." 17 May 2019. www.world-nuclear-news.org/Articles/Long-term-operation -of-China-s-oldest-reactor-asse.

———. 2021. "Nuclear Power in China." www.world-nuclear.org/information-library /country-profiles/countries-a-f/china-nuclear-power.aspx.

Wolf, Audra. 2020. "All Together Now: Science Is Political." Never Just Science, 27 October 2020. https://neverjustscience.substack.com/p/all-together-now-science-is -political.

Wolman, Abel, and W. H. Lyles. 1978. "John Lucian Savage, 1879–1967: A Biographical Memoir." National Academy of Sciences. www.nasonline.org/publications /biographical-memoirs/memoir-pdfs/savage-john-l.pdf.

Wong, Edward. 2014. "China's Energy Plans Will Worsen Climate Change, Greenpeace Says." *New York Times*, 23 July 2014.

———. 2016. "China's Last Wild River Carries Conflicting Environmental Hopes." *New York Times*, 18 June 2016.

Wong, Samantha. 2020. "China's Water Use in 2019, by Type." Statista, 11 December 2020. www.statista.com/statistics/281679/water-use-in-china-by-type/.

Wood, James W. 1998. "A Theory of Preindustrial Population Dynamics: Demography, Economy, and Well-Being in Malthusian Systems." *Current Anthropology* 39(1): 99–135.

World Bank. n.d. a. Air Transport, Passengers Carried—China. Accessed November 2020. https://data.worldbank.org/indicator/IS.AIR.PSGR?locations=CN.

———. n.d. b. Cereal Yield (kg per hectare). Accessed February 2021. https://data. worldbank.org/indicator/AG.YLD.CREL.KG?end=2017&start=1961&view=map &year=1980.

World Footwear. 2022. "All the Facts and Numbers of the Footwear Industry in 2021." www.worldfootwear.com/news/all-the-facts-and-numbers-of-the-footwear-industry -in-2021/8152.html.

Wright, Tim. 2012. *The Political Economy of the Chinese Coal Industry: Black Gold and Blood-Stained Coal*. London: Routledge.

Wu, C., Y. Ye, and B. Fang. 2004. "Optimum Strategy on Farmland Preservation: A Case Study in Pinghu City, Zhejiang Province." College of Southeastern Land Management, Zhejiang University, Hangzhou. Cited in Lichtenberg and Ding 2008.

Wu, Changhua, Crescencia Maurer, Yi Wang, Shouzheng Xue, and Devra Lee Davis. 1999. "Water Pollution and Human Health in China." *Environmental Health Perspectives* 107 (4): 251–56.

Wu Fengsheng 吴凤声. 1999. "Qianxi '1998 nian te da hongzai' de chengyin ji qi fanghong duice" 浅析 "1998年特大洪灾"的成因及其防洪对策 [Superficial analysis of the contributing causes and anti-flood policies of the "mega-flood of 1998"). *Tianjin Shida Xuebao (ziran kexue ban)* 天津师大学报 (自然科学版) 19 (4): 64–68.

Wu, Jichun, Xiaoqing Shi, Yuqun Xue, Yun Zhang, Zixin Wei, and Jun Yu. 2007. "The Development and Control of Land Subsidence in the Yangtze Delta, China." *Environmental Geology* 55 (8): 1725–35.

Wu, Mei. 1993. "Expressing Dissent: Politics, Media, and Environmental Issues in China: The Case of the Media Debate on the Three Gorges Project." MA thesis, Concordia University, Montreal.

Wu Minliang 吴敏良. (1983) 2010. "Dujiangyan lao gongcheng de kexue jiazhi ji 'gu wei jin yong' de juda xiaoyi" 都江堰古老工程的科学价值及 '古为今用' 的巨大效益 [Scientific value of the ancient Dujiangyan project and its great benefit as 'the past serving the present']. *Nongye Kaogu* 农业考古 2:59–65. Reprinted in privately published collection of writings, September 2010.

Wu Qiyue 吴其乐. 1998. "Minbei 'dayuejin' zhi fansi" 闽北"大跃进"之反思 [Reflecting on the "Great Leap Forward" in northern Fujian]. *Fujian Dangshi* 福建党史 1998 (7): 137–39.

Wu, Xiao, Naishuang Bi, Jingping Xua, Jeffrey A. Nittrouer, Zuosheng Yang, Yoshiki Saito, and Houjie Wang. 2017. "Stepwise Morphological Evolution of the Active Yellow River (Huanghe) Delta Lobe (1976–2013): Dominant Roles of Riverine Discharge and Sediment Grain Size." *Geomorphology* 292:115–27.

Wu Xinhua 吴新华. 1998. "Shui wuran ji qi fangzhi dangyi" 水污染及其防治刍议 [Humble opinions on water pollution and ways of controlling it]. *Suzhou Jiaoyu Xueyuan Xuebao (zonghe ban)* 苏州教育学院学报 (综合版) 1998 (12): 109–12.

Wu Yixiu. 2019. "Shanghai's Compulsory Waste Sorting Begins." China Dialogue, 2 July 2019. https://chinadialogue.net/en/cities/11349-shanghai-s-compulsory-waste-sorting -begins/.

Wu You 吴悠. 2017. "Huanjing baohu 'yi piao foujue' zai hezhang zhi de yunyong" 环境保护 "一票否决" 在河长制的运用 [The use of environmental protection "one ticket denial" in the river master system]. *Zhongguo Xiangcun Qiye Kuaiji* 中国乡镇企业会计 2017 (7): 33–35.

Wu Zhijun 吴志军. 2006. "Shilun 1957 nian dong, 1958 nian chun nongcun shuili jianshe yundong" 试论1957 年冬, 1958 年春农田水利建设运动 [Provisional discussion of the agricultural waterworks construction campaign of winter 1957 and spring 1958]. *Beijing Dangshi* 北京党史 2006 (1): 12–15.

Wyatt, Jessa, and Maggie Kristian. 2021. "The True Land Footprint of Solar Energy." Great Plans Institute, 14 September 2021. https://betterenergy.org/blog/the-true -land-footprint-of-solar-energy/.

Xia, Haoming, Jinyu Zhao, Yaochen Qin, Jia Yang, Yaoping Cui, Hongquan Song, Liqun

Ma, et al. 2019. "Changes in Water Surface Area during 1989–2017 in the Huai River Basin using Landsat Data and Google Earth Engine." *Remote Sensing* 11 (15): article 1824.

Xia, Min, Guang Ming Ren, and Xin Lei Ma. 2013. "Deformation and Mechanism of Landslide Influenced by the Effects of Reservoir Water and Rainfall, Three Gorges, China." *Natural Hazards* 68 (2): 467–82.

Xia Zenglu 夏增禄 and Li Senzhao 李森照. 1985. "Wo guo wushui guangai de fenbu tezheng ji qi chengyin" 我国污水灌溉的分布特征及其成因 [Distribution and aspects of wastewater irrigation in our country]. *Dili Yanjiu* 地理研究 4 (3): 40–46.

Xiao Donglian 箫冬连. 2018. "Zhongguo nongcun gaige shi ruhe lüxian tupo de?" 中国农村改革是如何率先突破的? [How did China's rural reforms initially break through?] *Zhonggong Dangshi Yanjiu* 中共党史研究 2018 (8): 18–30.

Xie, Echo. 2021a. "China Puts Nuclear Power, Waste Disposal on the Front Burner in Bid to Meet Climate Targets." *South China Morning Post*, 9 March 2021.

———. 2021b. "How a Pristine Chinese Mountain Fell Prey to the Country's Property Boom." *South China Morning Post*, 20 June 2021.

Xinhua. 2019a. "Poyang Hu tui chu jingxie shuiwei: Jiangxi jiesu fangxun si ji yingji xiangying" 鄱阳湖退出警戒水位 江西结束防汛四级应急响应 [Poyang Lake retreats from the warning water level; Jiangxi ends the fourth-stage emergency alert]. 29 July 2019. http://m.xinhuanet.com/2019-07/28/c_1124808539.htm.

———. 2019b. "Qiandao Hu pei gong shui gongcheng zhengshi tongshui" 千岛湖配供水工程正式通水 [Water is officially flowing in the Qiandao Lake Water Supply Project]. 29 September 2019. http://m.xinhuanet.com/2019-09/29/c_1125057624.htm.

———. 2020a. "Dian zhong yinshui gongcheng jianshe quanmian tisu" 滇中引水工程建设全面提速 [All facets of construction on the Central Yunnan Water Project pick up speed]. 27 April 2020. www.xinhuanet.com/local/2020-04/27/c_1125913738.htm.

———. 2020b. "Quebao liangshi anquan, Xi Jinping ba zheji da wenti tantou le" 确保粮食安全，习近平把这几大问题谈透了 [Assure grain security, Xi Jinping thoroughly discusses these questions]. 16 October 2020. www.xinhuanet.com/politics/xxjxs/2020-10/16/c_1126617636.htm.

———. 2022. "Shuili Bu: Dao 2025 nian jianshe yipi guojia shuiwang gugan gongcheng" 水利部：到2025年建设一批国家水网骨干工程 [Ministry of Water Resources: By 2025, build a set of backbone projects for the national water grid]. 2 January 2022. https://politics.gmw.cn/2022-01/02/content_35423156.htm.

Xinhuanet. 2019. "China's Major Milk-Producing Region to Promote Large-Scale Dairy Farming." 25 August 2019. www.xinhuanet.com/english/2019-08/25/c_138337546.htm.

Xiong, Wu, Xiao-Wei Jiang, Yu-Fu Chen, Hong Tian, and Neng-Xiong Wu. 2009. "The Influences of Mining Subsidence on the Ecological Environment and Public Infrastructure: A Case Study at the Haolaigou Iron Ore Mine in Baotou, China." *Environmental Earth Sciences* 59:803–10.

Xiu, Changbai, and K. K. Klein. 2010. "Melamine in Milk Products in China: Examining the Factors that Led to Deliberate Use of the Contaminant." *Food Policy* 35:463–70.

Xu, Bochao, Disong Yang, William C. Burnett, Xiangbin Ran, Zhigang Yu, Maosheng Gao, Shaobo Diao, and Xueyan Jiang. 2016. "Artificial Water Sediment Regulation Scheme Influences Morphology, Hydrodynamics, and Nutrient Behavior in the Yellow River Estuary." *Journal of Hydrology* 539:102–12.

Xu, Jianchu, Erzi T. Ma, Duojie Tashi, Yongshou Fu, Zhi Lu, and David Melick. 2005.

"Integrating Sacred Knowledge for Conservation: Cultures and Landscapes in Southwest China." *Ecology and Society* 10 (2): article 7. www.ecologyandsociety.org /vol10/iss2/art7/.

Xu, Jin-Hua, Tobias Fleiter, Ying Fan, and Wolfgang Eichhammer. 2014. "CO_2 Emissions Reduction Potential in China's Cement Industry Compared to IEA's Cement Technology Roadmap up to 2050." *Applied Energy* 130:592–602.

Xu, Jin-Hua, Bo-wen Yi, and Ying Fan. 2016. "A Bottom-Up Optimization Model for Long-Term CO_2 Emissions Reduction Pathway in the Cement Industry: A Case Study in China." *International Journal of Greenhouse Gas Control* 44:199–216.

Xu, Min, Chunyang He, Zifeng Liu, and Yinyin Dou. 2016. "How Did Urban Land Expand in China between 1992 and 2015? A Multi-Scale Landscape Analysis." *PLoS One* 11 (5). www.ncbi.nlm.nih.gov/pmc/articles/PMC4856333/pdf/pone.0154839.pdf.

Xu, Muyu, and Chen Aizhu. 2020. "China's Primary Energy Use to Peak in 2035—CNPC Research." *Reuters*, 16 December 2020. www.reuters.com/article/us-china-energy -cnpc-forecast/chinas-primary-energy-use-to-peak-in-2035-cnpc-research-idUSKB N28R0SE.

Xu, Xibao, Yan Tan, and Guishan Yang. 2013. "Environmental Impact Assessments of the Three Gorges Project in China: Issues and Interventions." *Earth-Science Reviews* 124:115–25.

Xu, Zhun. 2017. "Decollectivization, Collective Legacy, and Uneven Agricultural Development in China." *World Development* 98:290–99.

Xun, Sean. 2020. *The Mineral Industry of China*. Reston, VA: US Geological Survey. https://d9-wret.s3.us-west-2.amazonaws.com/assets/palladium/production/s3fs -public/atoms/files/myb3-2017-18-ch.pdf.

Xunyang (浔阳晚报 Xunyang Evening News). 2017. "Mianhuai! 1998 nian Jiujiang liu lei de kanghong jiangjun Dong Wanrui zoule" 缅怀！1998年为九江流泪的抗洪将军董万瑞走了 [Cherish the memory! Dong Wanrui, the flood control general who cried in Jiujiang, has left us]. https://baike.baidu.com/tashuo/browse/content? id=9a703 5925c1f2f3e50cdea69.

Yan, D. H., H. Wang, H. H. Li, G. Wang, T. L. Qin, D. Y. Wang, and L. H. Wang. 2012. "Quantitative Analysis of the Environmental Impact of Large-Scale Water Transfer Project on Water Resource Area in a Changing Environment." *Hydrology and Earth System Sciences* 16:2685–2702.

Yan Dingfei. 2014. "Shifang Deadlock Is a "National Problem" for China." *China Dialogue*, 10 November 2014. https://chinadialogue.net/en/pollution/7470-shifang -deadlock-is-a-national-problem-for-china/. Originally published in *Southern Weekend* 南方周末.

Yan, Hairong. 2008. *New Masters, New Servants: Migration, Development, and Women Workers in China*. Durham, NC: Duke University Press.

Yan, Yunxiang. 1996. *The Flow of Gifts: Reciprocity and Social Networks in a Chinese Village*. Stanford, CA: Stanford University Press.

———. (1997) 2006. "McDonald's in Beijing: The Localization of Americana." In *Golden Arches East: McDonald's in East Asia*, edited by James B. Watson, 39–76. Stanford, CA: Stanford University Press.

———. 2003. *Private Life under Socialism: Love, Intimacy, and Family Change in a Chinese Village, 1949–1999*. Stanford, CA: Stanford University Press.

Yang, Bao, Achim Braeuning, Kathleen R. Johnson, and Shi Yafeng. 2002. "General Characteristics of Temperature Variation in China during the Last Two Millennia." *Geophysical Research Letters* 29 (9): 38-1–38-4.

Yang, C. K. 1959. *A Chinese Village in the Early Communist Transition.* Bound as part of *Chinese Communist Society: The Family and the Village.* Cambridge, MA: MIT Press.

Yang, Dali L. 1996. *Calamity and Reform in China: State, Rural Society, and Institutional Change since the Great Leap Famine.* Stanford, CA: Stanford University Press.

Yang, Dan, Shihua Qi, Jiaquan Zhang, Chenxi Wu, and Xinli Xing. 2013. "Organochlorine Pesticides in Soil, Water and Sediment along the Jinjiang River Mainstream to Quanzhou Bay, Southeast China." *Ecotoxicology and Environmental Safety* 89:59–65.

Yang, Fenglin, and K-M Lau. 2004. "Trend and Variability of China Precipitation in Spring and Summer: Linkage to Sea-Surface Temperatures." *International Journal of Climatology* 24:1625–44.

Yang Jian and Song Yiyang. 2020. "City's Iron and Steel Center to Be Reborn as a Smart City." *Shine,* 28 June 2020.

Yang Jisheng 杨继绳. 2008. *Mubei: Zhongguo liushi niandai da jihuang jishi* 墓碑: 中国六十年代大饥荒纪实 [Tombstone: A factual record of China's great famine of the sixties]. Hong Kong: Cosmos.

Yang Long 杨龙, Qiang Jiashan 强佳杉, and Li Xiangning 李湘宁. 2017. "Zhi Huai yu Zhonggong de difang zhengquan jianshe, 1950–1952" 治淮与中共的地方政权建设, 1950–1952 [Huai conservancy and the establishment of CCP local power, 1950–1952]. *Wenhua Congheng* 文化从横 (March): 118–27.

Yang, Martin C. 1945. *A Chinese Village: Taitou, Shantung Province.* New York: Columbia University Press.

Yang, Mingchuan. 1994. "Reshaping Peasant Culture and Community: Rural Industrialization in a Chinese Village." *Modern China* 20 (2): 157–79.

Yang, Qingshan, Jie Liu, and Yu Zhang. 2017. "Decoupling Agricultural Nonpoint Source Pollution from Crop Production: A Case Study of Heilongjiang Land Reclamation Area, China." *Sustainability* 9 (6): 1024–34.

Yang, S. L, J. D. Milliman, K. H. Xu, B. Deng, X. Y. Zhang, and X. X. Luo. 2014. "Downstream Sedimentary and Geomorphic Impacts of the Three Gorges Dam on the Yangtze River." *Earth-Science Reviews* 138:469–86.

Yang, Tianliang, Xuexin Yan, Xinlei Huang, and Jianzhong Wu. 2020. "Integrated Management of Groundwater Exploitation and Recharge in Shanghai Based on Land Subsidence Control." *Proceedings of the International Association of Hydrological Sciences* 382: 831–36.

Yang Xiaobao 杨小豹. 2005. "Jilin Shihua Gongsi yi queren zaocheng wu ren siwang yi ren shizong" 吉林石化公司已确认造成五人死亡一人失踪 [Jilin petrochemical company confirms five dead, one missing]. Sina News Center, 14 November 2005. http://news.sina.com.cn/c/2005-11-14/18258295142.shtml.

Yang Yaliang 杨亚良. 1996. "Zhongguo Changzhou kechixu fazhan de gongshui, wushui chuli he laji chuzhi celüe" 中国常州可持续发展的供水，污水处理和垃圾处置策略 [Strategies for water supply, sewage treatment, and solid waste treatment in the sustainable development of Changzhou, China]. *Ambio* 25 (2): 85–88.

Yang, Yanqun, Said M. Easa, Xinyi Zheng, Aixiu Hu, Fashui Liu, and Meifeng Chen. 2019. "Evaluation Effects of Two Types of Freeway Deceleration Markings in

China." *PLoS One* 14 (8). https://journals.plos.org/plosone/article?id=10.1371/journal
.pone.0220811.

Yang Yong. 2014. "World's Largest Hydropower Project Planned for Tibetan Plateau."
China Dialogue, 3 May 2014. https://chinadialogue.net/en/energy/6781-world-s
-largest-hydropower-project-planned-for-tibetan-plateau/.

Yang Zi 杨子, Liu Xiaoguang 刘晓光, Ning Jing 宁静, Dong Fangchen 董芳辰, Yu Jie
于杰, Zhang Peng 张鹏, and Wang Sai 王赛. 2017. "Dianxing hei tu longzuo gengdi
goushi dui turang yangfen de xingxiang jianjiu" 典型黑土垄作耕地沟蚀对土壤养分
的形象研究 [Effects of gully erosion on soil nutrients in ridge area of typical black
soil]. *Turang* 土壤 49 (2): 379–85.

Yao, Shujie, and Xiuyun Yang. 2008. "Airport Development and Regional Economic
Growth in China." Research paper of the Leverhulme Centre, Cambridge, UK.

Yao, Yun-sheng, Qiu-Liang Wang, Wu-Lin Liao, Li-Fen Zhang, Jun-Hua Chen, Jing-
Gang Li, Li Yuan, and Yan-Nan Zhao. 2017. "Influences of the Three Gorges Project
on Seismic Activities in the Reservoir Area." *Science Bulletin* 62 (15): 1089–98.

Ye, Jingzhong. 2015. "Land Transfer and the Pursuit of Agricultural Modernization in
China." *Journal of Agrarian Change* 15 (3): 314–37.

Ye Ruolin and Yuan Ye. 2021. "Untangling the Crossed Wires of China's 'Super Grid.'"
Sixth Tone, 9 March 2021. www.sixthtone.com/news/1006932/untangling-the-crossed
-wires-of-chinas-super-grid.

Ye, Zhiguo. 2015. "Cities under Siege: The Great Flood of 1931 and the Downsides of
Urban Modernization in China." Paper presented at the Pacific Northwest China
Environment Conference, Kwantlen Polytechnic University, September 2015.

Yeh, Emily T. 2013. *Taming Tibet: Landscape Transformation and the Gift of Chinese Development*. Ithaca: Cornell University Press.

Yin, Jie, Zhane Yin, Haidong Zhong, Shiyuan Xu, Xiaomeng Hu, Jun Wang, and Jinping
Wu. 2011. "Monitoring Urban Expansion and Land Use/Cover Changes of Shanghai
Metropolitan Area during the Transitional Economy (1979–2009) in China." *Envi-
ronmental Monitoring and Assessment* 177:609–21.

Yu, Fengling, Zhongyuan Chen, Xianyou Ren, and Guifang Yang. 2009. "Analysis of
Historical Floods on the Yangtze River, China: Characteristics and Explanations."
Geomorphology 113:210–16.

Yu, Hongyuan, Weixin Deng, Jiafa Luo, Ruilin Geng, and Zucong Cai. 2012. "Long-Term
Application of Organic Manure and Mineral Fertilizers on Aggregation and Aggre-
gate-Associated Carbon in a Sandy Loam Soil." *Soil and Tillage Research* 124:170–77.

Yu, Huan-Yun, Fang-Bai Li, Wei-Min Yu, Yong-Tao Li, Guo-Yi Yang, Shun-Gui Zhou,
Tan-Bin Zhang, Yuan-Xue Gao, and Hong-Fu Wan. 2013. "Assessment of Organo-
chlorine Pesticide Contamination in Relation to Soil Properties in the Pearl River
Delta, China." *Science of the Total Environment* 447:160–68.

Yu Jianrong 于建嵘. 2012. "Dangqian yali weiwen de kunjing yu chulu—zai lun Zhongguo shehui de gangxing wending" 当前压力维稳的困境与出路—再论中国社会的刚
性稳定 [The current predicament of pressure for stability and the way out—another
discussion of Chinese society's rigid stability]. *Aisixiang* 爱思想. www.aisixiang
.com/data/58296.html.

Yu Jintao 于津涛, Zhang Yue 张悦, and Shen Liang 沈亮. 2005. "Songhua Jiang wuran
weiji jiantao" 松花江污染危机检讨 [Discussion of the Sungari River pollution

crisis"). Sina News Center, from *Liaowang Dongfang Zhoukan* 瞭望东方周刊, 28 November 2005. http://news.sina.com.cn/c/2005-11-28/11548426150.shtml.

Yu Liren 于利人, ed. 1990. *Anshan sishi nian* 鞍山四十年 [40 years of Anshan]. Anshan: Anshan Difangzhi Bangongshi.

Yu Xiaogang, Chen Xiangxue, and Carl Middleton. 2019. "From Hydropower Construction to National Park Creation: Changing Pathways of the Nu River." In *Knowing the Salween River: Resource Politics of a Contested Transboundary River*, edited by Carl Middleton and Vanessa Lamb, 49–67. Cham, Switzerland: Springer Nature.

Yuan Yin 袁因, Wei Dezhong 魏德忠, Shi Baoxiu 石宝琇, and Yang Huan 杨寰. 2007. "Rengong tianhe Hongqi Qu" 人工天河红旗渠 [Artificial heavenly river: Red Flag Canal]. *Zhongguo Yichan* 中国遗产 [China heritage] 2007 (7): 110–28.

Yue, Wenze, Yong Liu, and Peilei Fan. 2013. "Measuring Urban Sprawl and Its Drivers in Large Chinese Cities: The Case of Hangzhou." *Land Use Policy* 31:358–70.

Yun, Xiao, Guofeng Shen, Huizhong Shen, Wenjun Meng, Yilin Chen, Haoran Xu, Yuang Ren, et al. 2020. "Residential Solid Fuel Emissions Contribute Significantly to Air Pollution and Associated Health Effects in China." *Science Advances* 6 (44).

Yunnan (Yunnan Dianzhong Yinshui 云南滇中引水 [Yunnan Central Yunnan Water Delivery]). 2018. "Dianzhong yinshui gongcheng chubu sheji huo Shuili Bu xingzheng xuke" 滇中引水工程初步设计获水利部行政许可 [Preliminary plans for the Central Yunnan Water Project receive administrative approval from the Ministry of Water Resources]. http://dzys.yn.gov.cn/dzysdt/201803/t20180321_859746.html.

Zacharias, John, Yue Hu, and Quan Le Huang. 2013. "Morphology and Spatial Dynamics of Urban Villages in Guangzhou's CBD." *Urban Studies Research* 2013. https://doi.org/10.1155/2013/958738.

Zao, Mandy. 2019. "Desperate Beijing Motorists Marrying Somebody Just So They Can Secure a License Plate for Their Car." *South China Morning Post*, 25 November 2109.

Zeng Yuesong 曾月松. 1988. "Yanyuan Xian yanye fazhan jianshi" 盐源县盐业发展简史 [Short history of the development of the salt industry in Yanyuan]. In *Yanyuan Wenshi* 盐源文史, edited by Feng Zonglu 冯宗禄 et al., 2:1–16.

Zeng Zhiyang 曾志杨. 2012. "Qiye pingxuan xianjin yao shixing 'huanjing yi piao foujue'" 企业评选先进要实行"环境一票否决" [Evaluating enterprises as "advanced" should employ "environmental one ticket denial"]. *Ziyuan yu Renju Huanjing* 资源与人居环境2012 (6): 58.

Zenz, Adrian. 2019. "'Thoroughly Reforming Them towards a Healthy Heart Attitude': China's Political Re-education Campaign in Xinjiang." *Central Asian Survey* 38 (1): 102–28.

Zhai Yang 翟杨. 2020. "Huai He zhili jianshi: Cong xuxie jianchou dao yi xie wei zhu, Huai He he ri zai anxian?" 淮河治理简史:从蓄泄兼筹到以泄为主，淮河何日再安澜? [Short history of regulating the Huai River: From storage and drainage together to mainly drainage, when will the Huai River be calmed again?]. 搜狐城市 *Souhu Chengshi*, 8 August 2020. www.sohu.com/a/411267690_120179484.

Zhan, Shaohua. 2015. "From Local State Corporatism to Land Revenue Regime: Urbanization and the Recent Transition of Rural Industry in China." *Journal of Agrarian Change* 15 (3): 413–32.

Zhan, Shaohua, Hongzhou Zhang, and Dongying He. 2018. "China's Flexible Overseas

Food Strategy: Food Trade and Agricultural Investment between Southeast Asia and China in 1990–2015." *Globalizations* 15 (5): 702–21.

Zhang, Amy. 2017. "China 2." In *Fueling Culture: 101 Words for Energy Development*, edited by Imre Szeman, Jennifer Wenze, and Patricia Yaeger, 80–82. New York: Fordham University Press.

———. 2019. "Invisible Labouring Bodies: Waste Work as Infrastructure in China." *Made in China Journal*, 23 July 2019. https://madeinchinajournal.com/2019/07/23 /invisible-labouring-bodies%EF%BB%BF-waste-work-as-infrastructure-in-china/.

Zhang, Chao, Ruifa Hu, Guanming Shi, Yanhong Jin, Mark G. Robson, and Xusheng Huang. 2015. "Overuse or Underuse? An Observation of Pesticide Use in China." *Science of the Total Environment* 538:1–6.

Zhang Shengyi 张晟义, Song Mingzhen 宋明珍, and Zhao Tong 赵彤. 2018. "Digou you dixia jingji qinru shenong (shipin) gongying lian: duo zhong sunhai ji qi guangsan zhuanyi jili" 地沟油地下经济侵入涉农 (食品) 供应链：多种损害及其广散转移机理 [The underground economy of gutter oil invades the agricultural (food) supply chain: Multiple kinds of harm and the mechanism of their diffusion and transfer]. *Xinjiang Caijing* 新疆财经 2018 (2): 14–21.

Zhang Di 张迪. 2021. Xiaolangdi Shuili Shuniu Gongcheng 小浪底水利枢纽工程 [The Xiaolangdi Water Diversion Project]. 黄河水利职业技术学院 [Yellow River Conservancy Technical Institute] https://www.yrcti.edu.cn/slgcxy/info/1057/5285.htm

Zhang, Hong. 2005. "Bracing for an Uncertain Future: A Case Study of New Coping Strategies of Rural Parents under China's Birth Control Policy." *China Journal* 54:53–76.

———. 2007. "From Resisting to 'Embracing?' the One-Child Rule: Understanding New Fertility Trends in a Central China Village." *China Quarterly* 192:855–75.

Zhang, Hongzhou. 2014. "Global Food Security: Debunking the 'China Threat' Narrative." RSIS Commentaries, S. Rajaratnam School of International Studies, 21 April 2014. www.rsis.edu.sg/rsis-publication/rsis/2198-global-food-security-debunkin/# .YuWq9S1h24I.

———. 2016. "Chinese Fishermen in Disputed Waters: Not Quite a 'People's War.'" *Marine Policy* 68:65–73.

Zhang, Hongzhou, and Fenshi Wu. 2017. "China's Marine Fishery and Global Ocean Governance." *Global Policy* 8 (2): 216–26.

Zhang, Jane. 2021. "Huawei Turns to AI Pig Farming as the Chinese Tech Giant Explores New Growth in Areas Outside Smartphones." *South China Morning Post*, 18 February 2021.

Zhang Jianlong 张建龙, ed. 2018. *Zhongguo linye he caoyuan tongji nianjian, 2018* 中国林业和草原统计年鉴, 2018 [China yearbook of forestry and grasslands, 2018]. Beijing: Zhongguo Linye Chubanshe.

Zhang Jiaran 张嘉然, Yang Jie 杨洁, and Yi Jiaxuan矣佳璇. 2019. "Zhongguo yuanmu jinkou yicundu fenxi" 中国原木进口依存度分析 [Analysis of China's dependence on log imports]. *World Journal of Forestry* 8 (4): 129–36.

Zhang, Jiayan. 2014. *Coping with Calamity: Environmental Change and Peasant Response in Central China, 1736–1949*. Vancouver: UBC Press.

Zhang Jing 张静. 2013. "Qinlizhe de Henan Banqiao shuiku kui ba jiyi" 亲历者的河南

板桥水库溃坝记忆 [Memories of people who personally survived the Banqiao Dam collapse in Henan]. 百姓生活月刊 *Baixing Shenghuo Yuekan* 2013 (1): 56–57.

Zhang Juan 张隽 and Guan Xiyan 关喜艳, eds. 2020. "Nan shui bei diao zhong xian gongshui dadao sheji mubiao" 南水北调中线供水达到设计目标 [Water delivery from the central route of the South-North Water Transfer reaches the planned quota]. *Hubei Ribao* 湖北日报, 11 December 2020.

Zhang, Kai, Wen Gong, Jizhong Lv, Xiong Xiong, and Chengxi Wu. 2015. "Accumulation of Floating Microplastics behind the Three Gorges Dam." *Environmental Pollution* 204:117–23.

Zhang, Kai, Jerald L. Schnoor, and Eddy Y. Zeng. 2012. "E-Waste Recycling: Where Does It Go from Here?" *Environmental Science and Technology* 46 (20): 10861–67.

Zhang, Kate. 2022. "New Railway Completes 2,700 km Loop of Taklamakan Desert in Move to Integrate Xinjiang with Rest of China." *South China Morning Post*, 17 June 2022.

Zhang, Li. 2002. *Strangers in the City: Reconfiguration of Space, Power, and Social Networks within China's Floating Population*. Stanford, CA: Stanford University Press.

———. 2010. *In Search of Paradise: Middle-Class Living in a Chinese Metropolis*. Ithaca: Cornell University Press.

Zhang, Li, and Gubo Qi. 2019. "Bottom-Up Self-Protection Responses to China's Food Safety Crisis." *Canadian Journal of Development Studies / Revue canadienne d'études du développement* 40 (1): 113–30.

Zhang, Lifei, Liang Dong, Wenlong Yang, Li Zhou, Shuangxin Shi, Xiulan Zhang, Shan Niu, Linglng Li, Zhongxiang Wu, and Yeru Huang. 2013. "Passive Air Sampling of Organochlorine Pesticides and Polychlorinated Biphenyls in the Yangtze River Delta, China, and Cancer Risk Assessment." *Environmental Pollution* 181:159–66.

Zhang, Ling. 2016. *The River, the Plain, and the State: An Environmental Drama in Northern Song China, 1048–1128*. Cambridge: Cambridge University Press.

Zhang Pingyu. 2008. "Revitalizing Old Industrial Base of Northeast China: Process, Policy, and Challenge." *China Geographical Sciences* 18 (2): 109–18.

Zhang, Q. H., W. N. Yang, H. H. Ngo, W. S. Guo, P. K. Jin, Mawuli Dzakpasu, S. J. Yang, Q. Wang, X. C. Wang, and D. Ao. 2016. "Current Status of Wastewater Treatment Plants in China." *Environment International* 92–93:11–22.

Zhang, Qian Forrest, Carlos Oya, and Jingzhong Ye. 2015. "Bringing Agriculture Back In: The Central Place of Agrarian Change in Rural China Studies." *Journal of Agrarian Change* 15 (3): 299–313.

Zhang, Qian Forrest, and Zi Pan. 2013. "The Transformation of Urban Vegetable Retail in China: Wet Markets, Supermarkets, and Informal Markets in Shanghai." *Journal of Contemporary Asia* 43 (3): 497–518.

Zhang Ruoting 张若婷. 2019. "Zongli bushu nan shui bei diao houxu gongcheng gongzuo: ba shuili gongcheng zuo wei guangda touzi zhongdian" 总理部署南水北调后续工程工作：把水利工程作为扩大投资重点 [Premier's office's work on the follow-up phase of the South-to-North Water Transfer Project: Make waterworks into a focus of expanded investment]. *Pengbai Xinwen* 澎湃新闻, 19 November 2019.

Zhang Shuwen, Li Fei, Li Tianqi, Yang Jiuchun, Bu Kun, Chang Liping, Wang Wenjuan, and Yan Yechao. 2015. "Remote Sensing Monitoring of Gullies on a Regional Scale:

A Case Study of Kebai Region in Heilongjiang Province, China." *China Geographical Sciences* 25 (5): 602–11.

Zhang Tongle 张同乐and Guo Qi 郭琪. 2008. "'Dayuejin' shiqi shengtai huanjing wenti lunxi: Yi Hebei Sheng wei li" "大跃进"时期生态环境问题论析: 以河北省为例 [Discussion and analysis of environmental and ecological questions from the time of the "Great Leap Forward": The example of Hebei Province]. *Hebei Shifan Daxue Xuebao, zhexue shehui kexue ban* 河北师范大学学报/哲学社会科学版 31 (2): 143–49.

Zhang Wenjiang, Jiang Hong, and Sarah Rogers. 2022. "The Next Phase of China's Water Infrastructure: A National Water Grid." China Dialogue, 16 March 2022. https://chinadialogue.net/en/cities/the-next-phase-of-chinas-water-infrastructure -a-national-water-grid/.

Zhang, XiaoHong, Jun Cao, JinRong Li, ShiHai Deng, YanZong Zhang, and Jun Wu. 2015. "Influence of Sewage Treatment on China's Energy Consumption and Economy and Its Performances." *Renewable and Sustainable Energy Reviews* 49:1009–18.

Zhang Xiaosong 张晓松 and Lin Hui 林晖. 2022. "Xi Jinping tan liangshi anquan: youyou wan shi, chifan wei da" 习近平谈粮食安全: 悠悠万事, 吃饭为大 [Xi Jinping discusses grain security: Among ten thousand worries, eating is the biggest]. www.gov.cn/xinwen/2022-03/07/content_5677598.htm.

Zhang, Xiaowei. 2020. *Blockchain Chicken Farm and Other Stories of Tech in China's Countryside*. New York: FSGO/Logic.

Zhang, Xuelian, Yanxia Li, Bei Liu, Jing Wang, Chenghong Feng, et al. 2014. "Prevalence of Veterinary Antibiotics and Antibiotic-Resistant *Escherichia coli* in Surface Water of a Livestock Production Region in Northern China." *PLoS One* 9 (11).

Zhang Yue 张悦 and Shen Liang 沈亮. 2005. "Zaihai you maidanzhe ma?" 灾害有埋单者吗? [Is someone going to pay for the disaster?]. *Liaowang Dongfang Zhoukan* 瞭望東方週刊, 28 November 2005.

Zhang Yulin 张玉林. 2013. "Zaihai de zaishengchan yu zhili weiji—Zhongguo jingyan de Shanxi yangben" 灾害的再生产与治理危机—中国经验的山西样本 [Disaster reproduction and the crisis of governance: The Shanxi example in China's experience]. *Rural China* 10:83–100.

Zhang Yuxing 张煜星. 2008. "Zhongguo senlin ziyuan 1950–2003 nian jingying zhuangkuang ji wenti" 中国森林资源1950–2003年经营状况及问题 [Forest management conditions and problems from 1950 to 2003 in China]. *Beijing Linye Daxue Xuebao* 北京林业大学学报 2008 (5): 91–96.

Zhang Zhihong. 1999. "Rural Industrialization in China: From Backyard Furnaces to Township and Village Enterprises." *East Asia* 17 (3): 61–87.

Zhao Feipeng 赵飞鹏. 2010. "Dongbei heitu shushi nian hou keneng xiaoshi: yi chang hongshui hui 8 wan miao di" 东北黑土数十年后可能消失 一场洪水毁8万亩地 [In a few decades, the black soil of Heilongjiang may disappear: One flood wipes out 80,000 *mu*]. *Zhongguo qingnian bao* 中国青年报, 3 June 2010.

Zhao, Simon X. B., and Kenneth K. K. Wong. 2002. "The Sustainability Dilemma of China's Township and Village Enterprises: An Analysis from Spatial and Functional Perspectives." *Journal of Rural Studies* 18:257–73.

Zhao Zheng 赵峥. 2014. "Dangqian Beijing 'digou you' wenti de chengyin yu duice fenxi"

当前北京"地沟油"问题的成因与对策分析 [Analysis of the causes and countermeasures of the 'gutter oil' problem in Beijing]. *Jingji Jie* 经济界 2014 (6): 63–67.

Zheng, Lan, and Ming Na. 2020. "A Pollution Paradox? The Political Economy of Environmental Inspection and Air Pollution in China." *Energy Research and Social Science* 70: article 101773.

Zheng, Mei, Lynn G. Salmon, James H. Schauer, Limin Zeng, C. S. Kiang, Yuanhang Zhang, and Glenn R. Cass. 2005. "Seasonal Trends in PM2.5 Source Contributions in Beijing, China." *Atmospheric Environment* 39:3967–76.

Zheng, William. 2020. "China Aims to Upgrade Farmland the Size of the Republic of Ireland: Here's Why." *South China Morning Post*, 18 December 2020.

Zheng, Zheng, Junqi Gao, Zhen Ma, Zhengfang Wang, Xiaoying Wang, Xingzhang Luo, Thierry Jacquet, and Guangtao Fu. 2016. "Urban Flooding in China: Main Causes and Policy Recommendations." *Hydrological Processes* 30:1149–2.

Zhong Linxiang 钟霖湘. 2011. "Lüelun 'Dayuejin' shiqi de shenfan tudi yundong" 略论 "大跃进"时期的深翻土地运动 [General discussion of the deep tilling campaign at the time of the "Great Leap Forward"]. *Dangshi Yanjiu yu Jiaoxue* 党史研究与教学 2011 (4): 14–20.

Zhong, Shuru, Mike Crang, and Guojun Zhang. 2020. "Constructing Freshness: The Vitality of Wet Markets in Urban China." *Agriculture and Human Values* 37:175–85.

Zhongyang (Zhongyang Xinwen Jilu Dianying Zhipianchang 中央新闻纪录电影片厂). 1970. *Hongqi Qu* 红旗渠 [Red Flag Canal]. www.youtube.com/watch?v=EjHLt8yogyE

Zhou, Cissy. 2020. "China Reports Outbreak of Deadly Bird Flu among Chickens in Hunan Province." *South China Morning Post*, 2 February 2020.

Zhou Jun 周军. 2015. "Sanxia wenwu fujian qu lüyou jiazhi pinggu tixi yanjiu" 三峡文物复建区旅游价值评估体系研究 [Investigation of the system for evaluating the tourist value of (feature) reconstruction districts in Sanxia]. *Sanxia Wenwu Yanjiu* 三峡文物研究11:321–30.

Zhou, Kate Xiao. 1996. *How the Farmers Changed China*. Boulder, CO: Westview Press.

Zhou, Mengling, Bing Wang, and Zhongfei Chen. 2020. "Has the Anti-Corruption Campaign Decreased Air Pollution in China?" *Energy Economics* 91: article 104878.

Zhou, Shuxuan. Forthcoming. *From Forest Farm to Sawmill: Stories of Labor, Gender, and the Chinese State*. Seattle: University of Washington Press.

Zhou, Yijing, Shufa Du, Chang Su, Bing Zhang, Huijun Wang, and Barry M. Popkin. 2015. "The Food Retail Revolution in China and Its Association with Diet and Health." *Food Policy* 55 (August): 92–100.

Zhou, Ying, Teng Zi, Lianlei Lang, Dawei Huang, Peng Wei, Dongsheng Chen, and Shuiyuan Cheng. 2020. "Impact of Rural Residential Coal Combustion on Air Pollution in Shandong, China." *Chemosphere* 260: article 127517.

Zhou, Yong, Zhongxu Chen, Mengping Cheng, Jian Chen, Tingting Zhu, Rui Wang, Yaxi Liu, et al. 2018. "Uncovering the Dispersion History, Adaptive Evolution, and Selection of Wheat in China." *Plant Biotechnology Journal* 16:280–91.

Zhu Guosen 朱国森, Qiu Dongying 邱冬英, and Ma Qingshen 麻庆申. 2019. "Bai nian Shougang, bai lian cheng gang" 百年首钢, 百炼成钢 [One hundred years of Capital Steel, one hundred firings made into steel]. *Shijie Jinshu Daobao* 世界金属导报, 15 August 2019. www.worldmetals.com.cn/viscms/tupianxinwen3693/20190815/248754.html.

Zhu Xianling 朱显灵 and Hu Huakai 胡化凯. 2009. "Shuang lun shuang hua li yu
    Zhongguo xinshi nongju tuiguang yundong" 双轮双铧犁与中国新式农具推广运
    动 [The double-wheeled, double-shared plow and China's campaign to disseminate
    new-style agricultural implements]. *Dangdai Zhongguo Shi Yanjiu* 当代中国史研究
    16 (3): 65–64.

Zhu Yan 朱妍. 2020. "Jizhe diaocha: qi da tegaoya zhiliu yunneng xiuzhi chao san
    cheng" 记者调查：七大特高压直流运能闲置超三成 [Reporter's investigation: Un-
    used capacity on the seven major ultra-high voltage direct current lines exceeds
    30%]. *Zhongguo Nengyuan Bao* 中国能源报, 16 December 2020. http://energy.people
    .com.cn/n1/2020/1216/c71661-31968207.html.

Zhu, Z. L., and D. L. Chen. 2002. "Nitrogen Fertilizer Use in China—Contributions to
    Food Production, Impacts on the Environment and Best Management Strategies."
    Nutrient Cycling in Agroecosystems 63:117–27.

Zhuang, Wen, Samantha C. Ying, Alexander L. Frie, Qian Wang, Jinming Song,
    Yongxia Liu, Qing Chen, and Xiaoying Lai. 2019. "Distribution, Pollution Status, and
    Source Apportionment of Trace Metals in Lake Sediments under the Influence of
    the South-to-North Water Transfer Project, China." *Science of the Total Environment*
    671:108–18.

Zinda, John Aloysius, Christine J. Trac, Deli Zhai, and Stevan Harrell. 2017. "Dual-
    Function Forests in the Returning Farmland to Forest Program and the Flexibility
    of Environmental Policy in China." *Geoforum* 78:119–32.

Zito, Angela. 2011. *Writing in Water*. Documentary film, 48 minutes.

Zong, Yongqiang, and Xiqing Chen. 2000. "The 1998 Flood on the Yangtze, China." *Nat-
    ural Hazards* 22:165–84.

Zou Huabin 邹华斌. 2010. "Mao Zedong 'yi liang wei gang' fangzhen de tichu ji qi
    zuoyong" 毛泽东"以粮为纲"方针的提出及其作用 [Mao Zedong's raising of the pol-
    icy of "emphasize staple production" and its effects]. *Dangshi Yanjiu yu Jiaoxue* 党史
    研究与教学 2010 (6): 46–52.

Zou Jinchun 邹今春 and Ma Li 马力. 2018. "Yunnan Lancang Jiang Lidi shuidian zhan
    shou tai ji fadian mubiao yuanman shixian" 云南澜沧江里底水电站首台机发电目标
    圆满实现 [The goal of generating electricity from the first group of generators at the
    Lidi Hydroelectric Station on the Lancang River in Yunnan is completely realized].
    Nengyuan Jie 能源界, 31 October 2018. Inaccessible July 2022. www.nengyuanjie.net
    /article/20481.html.

Zuo, Mandy. 2022. "More 'Monster Fish' on the Loose: Sightings of Alligator Gar
    Emerge in 8 Chinese Provinces after a Pair Caught Over Weekend." *South China
    Morning Post*, 29 August 2022.

Zweig, David. 1989. *Agrarian Radicalism in China, 1968–1981*. Cambridge, MA: Harvard
    University Press.

INDEX

The letters *f*, *m*, and *t* following a page number denote, respectively, a figure, map, or table.

accidents: industrial, 346–49; mining, 295, 335

acid rain, 415, 417

Adams, Donald, 296–97

adaptive cycle, 18f, 19–21; applied to PRC history, 79–89, 438; future possibilities, 438; in Qing period, 94–95; retarded by imperialism, 67

administrative accomplishments (*zhengji*), 184–85, 338, 348; fix-fixing as, 382–83; in waste management, 400

aesthetics of landscape, 39–40, 51–52, 61

Africa: Chinese farms in, 268; wood imports from, 389

African swine fever, 233, 262–63; and pork imports, 267

agribusiness, 250–53; purchases of foreign, 268

agricultural producers' cooperatives, 109–11

agriculture: air pollution from, 412; among Akha, 46; animal, 165–66, 247–48, 251 (*see also* CAFO agriculture); antibiotics in, 262; "big farms," 249; capital inputs, 142–55, 266f; CCCM (consolidated, chemicalized, commercialized, mechanized), 231, 246–50, 256–64, 272–73, 276, 431–32, 439; chemicals, 1–2, 256–62; in Common Program, 80–81; cultivated area, 100–102, 243; *da hu*, 249; declining percentage of rural income, 329–30; dragon head, 150; ecological, 246; epizootics and, 262–64; export of pollution from, 270; food safety standards, 277; fossil fuels in, 256–57; fruit, 163–65; "going out" strategy, 265–70; Great Leap Forward, 82–83; hired labor, 249, 250; increasing scale, 246; inputs, 100–104; intensification of, 17, 135–36, 142–55; international investment, 246, 247; investment in overseas, 251, 267–68; investment lacking in early years, 82, 98–100; market reform, 246; mechanization, 1–2, 152, 246; mechanization lacking in early years, 98, 101f; monocrop industrial, 18–19; Nuosu, 45, 135; "Opinions" directive, 246–47; organic, 271, 277, 429, 431–32; peasant, end of, 246–47, 253–55, 277; pollution beyond the farm, 256–62; productivity, 237, 337; regenerative, 246, 431–32; resilience, 254–55, 261–62, 266f, 431–32; rice, 12–13, 19, 27–29; "scientific," 246; shifting (swidden), 11–12, 96, 135; specialization, 163–66, 251–53; supply chain integration, 246; vertical integration, 250–51, 273; water usage, 386–87; yields, 103. *See also* rice; farmland

Ahlers, Anna, 298, 411

air conditioning, 379

air-pocalypses, 413, 421

air pollution: from agriculture, 412; benzene in, 296, 297; from biomass aerosols, 412; from cement, 417; from chemical industry, 345–46; from cigarette smoke, 412; from coal, 296–97, 332, 412; from coking, 332–33; compared to Los Angeles, 415; compared to Taipei, 416; compared to US, 414; compared to water pollution, 421; complaints about, 425; control measures, 297, 412, 413–18; from cooking, 412, 416; from copper smelting, 297; diseases from, 297–98; dust as, 296–97, 412, 428; environmental justice and, 420; and environmental Kuznets curve, air pollution (*continued*)

421, 428, 436; from exhaust, 412; export from cities to rural areas, 332, 378–79; fuel standards and, 414, 417; from glass manufacture, 417; growth and, 344; from incineration, 392–93, 399–400; indices, 413; from iron smelting, 296; lack of regulation, 332; mining and, 295; from motor vehicles, 366, 412, 415, 417; as necessary evil, 298; ozone as, 416–17; PAH in, 296, 297; particulate matter, 296, 412–13, 414, 417; from pesticides, 260; Prevention and Control Law, 415; reduction in 2010s, 411–17, 421; Reform and Opening period, 411–12; regulation and, 417, 428; renewable energy reduces, 428; retrospective on high socialist period, 296–97; rural, 418; seasonal variations, 412, 414; from sintering, 296; from sludge incineration, 399; from steelmaking, 296, 417; sulfur dioxide in, 412; urbanization and, 351, 366, 378–79; WHO guidelines, 415–16; Zhu Rongji on, 412–13; from zinc smelting, 297, 332–33

air quality index, 414

air travel, 376–77

Akha (Hani) people, 42; agriculture, 46, 48; ancestral beliefs, 50; deforestation among, 142; ecological buffers, 48; forestry, 46; in Great Leap Forward, 136; house construction, 46; household ecology of, 46; hunting, 46; landscape plasticity, 48; property rights, 46; sacred precincts, 50; villages, 46, 47m

alfalfa, 242

algal blooms: fertilizer and, 258; Lake Tai, 173–75; Sanxia reservoir, 215

alternative food networks, 271, 278–82; Buddhism and, 281; as critique of Chinese consumerism, 282; mistrust and, 278, 280–81; motivations for, 280–81; significance of, 282–83

aluminum, 317; urbanization and, 387

Amazon Basin: deforestation in, 267

American crops, 10

Angang iron and steel complex, 287; air pollution from, 297; early history, 289–90; under Japanese control, 289, 290; production in 1950s, 290–91; worker health at, 297

animal foods, 233–34

Anlong ecological farm, 264, 278, 281–82

Anshan, Liaoning, 289–90, 291m. *See also* Angang iron and steel complex

antibiotics: in agriculture, 262–64; fed to chickens, 273; resilience and, 261f, 262; resistance to, 262

Anti-Rightist Campaign, 120

apples, 164–65

appliances, household, 387

aquaculture: increases in 21st century, 235; organic, 271; in Reform and Opening, 166; replaces farmland, 244; replaces inland capture fisheries, 219

aquifers, 145, 170–71, 190, 194

Argentina: agricultural imports from, 234, 242, 245–67; agricultural investment in, 250; deforestation in, 267; pork imports from, 267; soy production in, 267

Australia: agricultural investment in, 251; beef imports from, 265–67; greenhouse gas emissions, 403; iron ore imports from, 344–45; wood imports from, 389

autarky: decline of, 378

authoritarianism: and environmental outcomes, 426–27, 434–35

automobiles, private: air pollution and, 366, 412, 417; electric, 404; growth in ownership, 355, 374, 417; traffic jams and, 366

avian influenza, 263

Baiwu Valley, Liangshan: apple cultivation in, 164–65; collectivization in, 135–36; deforestation, 142; diets, 43–45, 238; reforestation, 184; rice cultivation, 142. *See also* Yanyuan County bamboo, 45

bananas, 252, 265

Banqiao dam collapse, 107, 147–48, 438

Baotou, Inner Mongolia: air pollution in, 414; designated heavy industrial city, 289; steelmaking at, 290, 291

barley, 26

beef: imports, 265–67; increase in consumption, 165t, 166, 233, 234t; not favored in China Proper, 32. *See also* cattle

beer: and autarky, 378; production, 342t

Beijing: air pollution in, 296–97, 412, 413, 414–16; airports, 376–77; coal stoves in, 296–97; cold islands, 385; designated heavy industrial city, 289; expressways, 360; heat islands, 384; land subsidence at, 383–84; ring roads, 360; scrap markets around, 394; sludge dumping in, 399; steel production in, 289, 291; subway, 366, 367m

Belt-and-Road Initiative, 222–23

benzene, 296, 297, 347, 348

Bessemer converters, 300; air pollution from, 296; small-scale, 301

bicycles, 310

Bingtuan, 101, 171

biodiversity, 2; diminished in Great Leap Forward, 136; environmental Kuznets curve and, 431; loss of, 430–31; and management, 14–15; sacred precincts as sites of, 51; Sanxia dam and decline of, 218–19

biofuels: gutter oil reused as, 274–75

black holes (illegal mines), 334

black workshops, 274–75

Blaikie, Piers, 16

blast furnaces: and air pollution, 296; small-scale, 300–304, 303f; in Tangshan earthquake, 325

blue-green algae, 173–75

blue skies (for special events), 413, 420

Bolivia: deforestation in, 267

Borlaug, Norman, 153, 154

Boserup, Ester, 10–11

boulevards, 361–62, 366

Brazil: agricultural imports from, 234, 242, 245; agricultural investment in, 251, 265–67; deforestation in, 245, 265–67, 389; soy cultivation in, 267; wood imports from, 389

bribery, 340

brickmaking: and air pollution, 332; and water pollution, 331

Brookfield, Harold, 16

Brown, Lester, 230, 245, 265

buckwheat, 42, 45, 71, 238

Buddhism: and alternative food networks, 281; sacred precincts in, 50; Tibetan, 56

buffers, 16–17; in Central Asia, 58–62; in China Proper, 36–41; decline in Qing period, 94–95; industrialization eliminates, 323; intensification and, 41; in Zomia, 46–52. *See also* cultural buffers; ecological buffers; infrastructural buffers; institutional buffers

building safety, neglect of, 323, 324–25

"bulldozer state," 371–72

bureaucracy: alternations with campaigns, 77–78; as conservative force, 76; contradictions with ideology, 74–77; as institutional buffer, 76; internal conflicts, 75; perquisites of, 75; and resilience, 76; as rigidity trap, 76; and stability, 20. *See also* governance

cabbage, 237

cadmium: as soil contaminant, 430; as water pollutant, 332

cadre evaluation system, 338, 420–21, 426

CAFO agriculture, 247; antibiotics in, 262–64; epizootics and, 263; manure from, 257; soil damage from, 432; water pollution from, 432

California: water shortages in, 242

Cambodia: dams in, 220–22; land purchases in, 268

camels, 53, 59

campaigns: alternation with bureaucratic rule, 77–78; decline after 1980s, 77

Canada: environmental governance in, 435; greenhouse gas emissions, 403; renewable energy in, 409; and Sanxia dam, 204

canteens (cafeterias), 82, 135–36; abolition of, 139

capitalism: as cause of pollution, 69–70; developmental ideologies of, 64–66. *See also* development: capitalist

carbon dioxide (CO_2), 296, 405, 417

carbon intensity, 404

carbon neutrality, 404, 411, 432, 436. *See also* peak carbon

carrying capacity, 10; overshoot of, in Qing, 93

cashmere, 53

cattle: in Central Asia, 53, 59; feed for, 242; greenhouse gases and, 378; increase in 21st century, 242, 272–73; ranching replaces herding, 248; stall-feeding, 242. *See also* beef; milk products

CCCM (consolidated, chemicalized, commercialized, mechanized) agriculture. *See under* agriculture

cellularization, 306–7; decline of, 378

cement: and air pollution, 332, 378; and energy consumption, 344; growth, 343t; small rural factories, 322, 328; urbanization and, 387–88; and water pollution, 331, 378

Central Asia, 7m, 8, 52–62; agriculture in, 52; buffers in, 17, 58–62; climate, 52; collectivization in, 119; cultural buffers in, 61–62; ecological buffers in, 59–60; ecological divisions of, 52–54; ecological limitations of, 53–54, 65; extractive colonialism in, 293; farming in, 56; grasslands of, 53–54; household ecology, 58–62; institutional buffers in, 60; livestock in, 52–54; milk production in, 272–73; oases in, 52–53; political repression in, 434; political systems of, 6; population density in, 54; Qing rule in, 93, 96; renewable energy in, 410

ceramics: and air pollution, 332

Chai Jing, 413

Chan, Kam Wing, 353, 354

Chang (Yangtze) River: channel capacity, 180–81; diversions from, 194–95; erosion after Sanxia dam, 210–11; fisheries, 218–19; floods, 108, 116, 176–85; flow rates, 178, 197; as hanging river, 181; hydraulic generation capacity, 216–17; level changes, 180–81; pollution in, 219; Poyang Lake connections, 210, 211–13; sediment regime, 208–11; sewage discharge into, 398; Three Gorges, 201–4 (*see also* Sanxia dam); water quality, 430

Chang River delta: erosion, 211, 217; farmland, 244; land subsidence in, 383; saltwater intrusion in, 197

Changchun, Jilin, 293

Chao Lake, 174–75

charcoal, 304–5

charismatic megafauna, 431

chemical industry, 345–49; growth, 343t; and water pollution, 331

chemical oxygen demand, 258; near landfills, 391; and urbanization, 398

Chengdu: air pollution in, 414–16; designated heavy industrial city, 289; expressways and ring roads, 360; replacement of traditional housing, 359; subway, 366; university campuses, 364

Chengdu Plain: farmland loss, 244; household pig prohibition, 235; housing patterns, 244; irrigation on, 114–15; resilience of communities, 264–65; rice imports, 244; village consolidation, 264

chestnuts, 252

chickens, 32; antibiotics fed to, 262, 273; decline in household, 247–48; large scale farms, 247–48; in 21st century, 234–35, 234t, 247–48, 273

China dream, 424

China Proper, 6, 7m, 26m; buffers in, 17, 36–41; climate zones, 25–27; cultural buffers in, 38–40; ecological buffers in, 36; ecological limits of, 25; houses, 30–33; infrastructural buffers in, 37; institutional buffers in, 37–38; irrigation in, 26; as morally superior civilization, 65; scale in ecology of, 29–35; topography, 25–27; villages, 33–35

Chinese Communist Party. *See* Communist Party

Chinese People's Political Consultative Conference (CPPCC), 63, 204

Chongqing: aluminum at, 317; Sanxia museum, 206; and Sanxia reservoir, 203; steel at, 300

circular economy, 390

cities: building materials for, 387–88; in Central Asia, 370–73; consumer, 355; *danwei*-dominated, 306–13, 327; ghost, 365; growth patterns, 355–59; heavy industrial, 289; housing in, 294–95, 310–11, 315, 324, 325; master plans for, 356; minority-themed, 368–69; preindustrial, 389; producer, 294, 327; rebuilding, 359–62; resilience of, 355; resource consumption, 387–88; satellite, 356, 364–65; small and medium, 366–67; sponge, 382; in Tibet, 370; and water usage, 386–87; in Zomia, 367–69. *See also* housing; urbanization

climate: of Central Asia, 52, 55–56; of China Proper, 25–28

climate change: China as largest contributor to, 403, 411; China as leader in mitigation, 411; commitment to reduce, 406; as long-term problem, 436; possibilities of future, 438. *See also* carbon neutrality; greenhouse gases; peak carbon

clothing: manufacture, 341, 342t; Wenzhou merchants, 357

coal: and air pollution, 296–97, 332, 412; at Anshan, 289; artisinal mining of, 333; ban on high-sulfur, 417; in Beijing, 296–97; and desertification, 295; in early industrialization, 289, 291–92; as electric power source, 404–5; in energy mix, 401–3, 411; expansion

of output, 340, 403–4; Fushun mines, 287, 292; gasification, 419; in Great Leap Forward, 304–5; growth, 343t; inefficient labor, 332–34; and land subsidence, 334; mining accidents, 295, 335; at Panzhihua, 320; pollution controls, 340; replacement of, 417–18; and sediment deposition, 334; share of world consumption, 403–4; and soil pollution, 334; in Third Front, 317; and water supply, 295, 334–35, 419

coke, 291; and air pollution, 296, 304, 305, 332–33; in Great Leap Forward, 304

collapse, 14, 17, 19; future possibility, 438; Great Leap as, 13, 20; of Soviet Union, 20, 438–39

collectivization, 109–16; among Akha, 136; beginning-level cooperatives, 109–10; and double-cropped rice, 115; higher-level cooperatives, 110–11; large-scale, in Great Leap Forward, 123–24; of livestock in Central Asia, 134; among Nuosu, 135–36; in Zomia, 135–38

college campuses, 360, 364

colonialism: as cause of underdevelopment, 66; extractive, 293, 322, 324; Japanese, 287, 289; Russian, 102; semi-, 67, 74

Common Program of CPPCC, 63; developmental goals of, 80–81

common property systems, 17; among Akha, 46, 48; in Chinese villages, 34–35; design principles for, 48–49; as institutional buffer, 60; in Mongolia, 58; among Nuosu, 45, 49

commune and brigade enterprises, 322–23, 328–29; rechristened as TVEs, 329

communes. *See* people's communes

Communist Party: congresses, 10; historiography by, 3–4; 100th anniversary, 438–39; totalizing visions, 62

communist society, as Great Leap Forward goal, 124; retreat from, 139

community supported agriculture (CSA), 264, 271, 279–80

concrete: and floods, 381; urbanization and, 354–55, 387–88. *See also* cement

cones of subsidence, 170, 190

construction work: housing and, 357; peasant migrants and, 353–54

consumer goods: manufacturing, 340–43, 351, 355–56; migration and, 351; steel and durable, 341

consumerism, 276, 278–83; and NRR movement, 279–80; and trust, 277–78

consumption: growth of urban, 385–86; urban-rural differences, 385

cooperatives, 109–11. *See also* collectivization

copper: and air pollution, 297, 348; recycling, 396; urbanization and, 387

corn, 28, 93, 149; as animal feed, 232, 233–34, 242; genetically modified, 250; hybrid, 153, 155; imports, 231, 233; increased production in 21st century, 231; introduction of, 27; among Nuosu, 45; as staple in China Proper, 34

corruption, 418, 420

cotton: in Great Leap Forward, 129; pesticide use on, 260

courtyard, 30, 31f

COVID-19 pandemic, 401, 427–28; China's death rates, 434

crisis, environmental, 84–85

cross-scale mismatches, 5–6, 124; disrupt eco-feedbacks, 129–33; in Great Leap Forward, 131–32

crowding in animal agriculture, 262–64

cryptocurrency mining, 433

cultural buffers, 16–17; aesthetic appreciation as, 39, 51–52; in Central Asia, 61–62; in China Proper, 38–40; frugality as, 39, 61; generational continuity as, 38–39; reciprocity as, 39; respect for land as, 60; social-ecological parallels as, 51; supernatural beliefs as, 49–51; suspicion of work ethic as, 61; temporal prohibitions as, 51; in Zomia, 49–52

Cultural Revolution, 4, 52, 76; blamed for urban housing neglect, 325; in conventional periodization, 79, 155; effects on ecosystems, 84

curtailment: and power shortages, 410; of wind power, 408–9

cutting with one knife. *See* panacea solutions

cycles: adaptive, 18f, 19–21; climatic, 8–10; in ecosystems, 8–10; in social systems, 8–10

da hu (big [farm] firms), 249; and dragon heads, 250–51

Dai Qing, 205

dairy products. *See* milk products

Dalai Lama, 119

dams: builders of, 220, 222–23, 224–25, 227; China as world leader in, 201–3; contradictions among purposes, 205; ecological tradeoffs, 405–6; as escape from poverty, 225, 226; eutrophication and, 406; evolution in function and methods, 229; failures, 104, 107, 147–48, 214, 229, 325; hydrological changes from, 130–31, 208–11, 220; as industrialization stage, 201, 229; international effects, 219–28; opposition to, 204–6, 225–27, 425; Nu River, dispute over, 224–28, 406; as replacement for fossil fuels, 405–6; resettlement from, 187, 204, 206–8, 220, 229; as rigidity traps, 219; Sanmenxia, 130–31, 168, 186–87, 201; in Sipsongpanna, 220; Three Gorges (*see* Sanxia Dam)

Danjiangkou Reservoir: pollution of, 197; raising level for South–North Water Transfer, 193; resettlement from, 194, 207; variable volumes in, 194–95

danwei urban organization, 309–13; conversion of abandoned, 357, 359, 360; and environmental injustice, 306–7; as living space, 311–13; origins of, 306–7; at Panzhihua, 319–20; replaced in Reform and Opening, 313

Daqing oilfield, Heilongjiang, 313–16; discovery of, 313; housing at, 314–15;

as model for Third Front, 317; 1998 floods, 177

DDT, 149–50, 259–60

decollectivization, 161–62; pigs and, 165; in Reform and Opening, 161–62, 329

decoupling, 437

deep tilling and dense planting, 83, 124, 125–27; Mao's support for, 125; of rice, 126, 127; and soil fertility, 129; and wheat yields, 125

deforestation: and beef imports, 265–67; environmental Kuznets curve and, 428, 436; export of, 270; and floods, 178–79, 211; in Great Leap Forward, 136–37; in Heilongjiang Reclamation Area, 141; for orchards, 252; in Qing period, 95; resettlement and, 207; sedimentation and, 141; in South America, 265–67; and staple food production, 140–43; in Zomia, 135, 142, 178–79

degraded wastelands, 295

degrowth, 437

democracy: and environmental governance, 435

Democratic Reforms (1956–57), 105, 118–19

demography: as anti-Marxist, 97; Mao on, 69. *See also* population

Deng Xiaoping, 84; assumes power, 157; begins Reform and Opening, 157, 327; on development, 89, 336–37; and "seek truth from facts," 423

dependency theory, 66

desertification, 431

developers, real estate, 365, 380

development, 1–3; capitalist, 64–66, 323–24; and CCP ideology and legitimacy, 63–66, 85, 88–89, 423–24; and Cold War, 63–64; contradiction with environment, 88–89, 323–24; critique of, 64, 66, 348; "ecological," 365; and ecological civilization, 88–89; and environmental injustice, 323–24; Land Reform as, 105–7; "low-carbon," 365; as "only solid truth," 89; and reduced resilience, 178–82, 323–24; revolution

as means to, 66; science as signifier of, 89; scientific perspective on, 424; social and political constraints on, 68–69

developmentalism: capitalist and Communist contrasted, 65–66, 67f, 323–24; as common 20th century ideology, 63–64, 73; replaced by eco-developmentalism, 423–26

Dian Lake (Dianchi), 141–42, 175

diets: breadth in Reform and Opening, 162–66; breadth in 21st century, 230, 232–37; and economic class, 35; energy restrictions on, 35; expecation of rich, 245; farmers', 35, 97–98; grain-based, 140; Nuosu traditional, 43; and receding famine pattern, 240; regional variation in, 238–39; restaurant, 239–41; rural, 238; urban-rural differences, 235, 238; variety, 232–37; Xinjiang, change in, 238–39; Zomia, change in, 238

dikes: breaches, 177, 210; illegal, 38; as infrastructure buffers, 37, 104; and reduced resilience, 37, 104, 180

dimo (plastic mulch), 150–51

disease: artificial intelligence monitoring of, 263; diet-related, 240; infectious, 323; pesticide-resistant agricultural, 260; pig, 20, 262–63; shrimp, 171–73; viral animal, 262–63

disposable income, urban, 355

dissent, suppression of, 434

disturbance: and resilience, 12–15

diversity: in Central Asia, 61; as ecological buffer, 36, 48, 59, 61; ethnic, in Zomia, 41–42; patch, 48, 49; reduced, and shrimp disaster, 172–73; and resilience, 14, 264–65

dogmatic uniformity, 128

Doll, Ross, 249

Dongting Lake, 36, 180–81, 211, 212m

donkeys, 32, 45

dormitories, collective, 124

dragon fruit, 253

dragon head agriculture, 250–51, 252

drainage, urban, 382–83

droughts, 14, 103, 120; crop diversity as buffer against, 36; in Heilongjiang Reclamation District, 141; and hydropower, 229; on North China Plain, 94, 145; in Qing period, 38, 94; and Red Flag Canal, 169; and Xayaburi Dam, 222

drugs, manufacture, 341; pollution from manufacture, 346

ducks, 29, 32, 235; industrial raising of, 248

Dujiangyan irrigation system, expansion of, 114–15, 131, 147

dumping: floods and, 390; random, 390; of sludge, 399

dust: desertification and, 431; as souce of air pollution, 412, 431

dyes and dyeing, 331, 345

earthquakes: and death of leaders, 325–26; Horinger, Inner Mongolia, 235–26; Longling, Yunnan, 325–26; and Sanxia dam, 214; Tangshan (1976), 324–25, 438; Wenchuan, 368–69, 438

East Turkestan. *See* Xinjiang

eco-developmentalism, 226; and economic growth, 435–36; floods and, 182–85, 424; origins of, 85, 182–85, 349–50, 379; results of, 426–27; turn to, 423–26; urbanization and, 379, 382–83

ecological buffers, 16–17, 36–37, 48, 60, 323; in Central Asia, 59–60; in China Proper, 36; diversity as, 36, 48; fallowing as, 36; forests as, 36, 323; herd size as, 60; industrialization eliminates, 323; landscape plasticity as, 48; replaced by institutional and infrastructural buffers, 25, 36, 185–86, 189, 199, 253, 400; restoration of, 400; surplus as, 36; temporal diversity as, 60; tradeoffs with productivity, 25; urbanization eliminates, 377, 381–83; wetlands as, 36–37; in Zomia, 48. *See also* ecosystem services

ecological civilization, 4, 85–89; circular economy and, 390; contradictions

within, 88–89, 263–64; and Marxist-Leninist teleology, 437; as ordinary language, 420; urban pollution and, 400; Xi Jinping adoption of, 349, 424

"ecological food," distinguished from organic food, 280, 282

ecological history: defined, 3–4; lessons for, 433–39

economic growth, 1–2; acceleration in Reform period, 84; and eco-developmentalism, 435–36

ecosystem services: as buffers, 16; provided by forests, 45; among Nuosu, 45; urbanization and loss of, 381. *See also* ecological buffers

eggs, 32; centralization of production, 248; China as exporter of, 245; fake, 273; increase in consumption, Reform and Opening, 165t, 166; in 21st century, 234t, 235–36

electric power, 404–5; coal-generated, 404–5, 411; curtailment, 410; governance problems, 410; grid, 410; hydroelectric, 405–7; mix of sources, 404–5; nuclear, 407–8; renewable, 408–11; replaces coal, 417–18; residential, 404; solar, 409–10; transmission, 410; in transportation, 404; wind, 408–9, 428

electronics: manufacturing, 340, 342t; recycling, 396–97

electrostatic precipitators, 296

elevators: replace stairways, 362

"Eliminate the Four Pests" campaign, 125, 133

energy: biogas, 409; coal share, 401–3, 411; conservation, 341–44; consumption growth, 342–43, 401–11; efficiency growth, 341, 343, 401–3; electrification of, 404–5; forecast peak use, 401; fossil fuel share, 401–4; green, as ordinary language, 421; increasing inputs in Reform period, 84; intensity, 310, 321, 327–28, 332–33, 338, 341–44, 378, 402; mix, 401–5; oil share, 403–4; and poverty alleviation, 411; renewable, 408–11, 428; solar, 409–10, 428; wind,

408–9, 428. *See also* hydroelectric
power generation
Engels, Friedrich, 71
entrepreneurs, 354
environmental crisis, 84–85
environmental culture, 84; as implicit critique of Marxism-Leninism, 86
environmental degradation: export of, 344–45, 389, 432
Environmental Impact Assessment Law, 225
environmental injustice, 432; air pollution and, 420; coal gasification and, 419; cryptocurrency and, 433; *danwei* model and, 306–7, 308; governance problems and, 418, 420, 432; incinerators and, 392–93, 432; industrialization and, 298, 323, 346; insensitivity of leaders to, 432; minority peoples and, 324, 432; oil and, 316; petrochemical industry and, 346, 432; Third Front and, 322; waste treatment and, 392–93
environmental Kuznets curve (EKC): air pollution and, 421; deceleration vs. reversal, 436; differential applicability of, 427–33, 436; and green growth, 437; inapplicable cases, 429–33; limitations of, 435–36; and Marxist-Leninist teleology, 437; organic agriculture and, 429; reforestation and, 428; soil contamination and, 430; support for, 427–29
environmental protection: bureaus, 339, 426; law, 415, 426; as ordinary language, 421
epizootics, 233–34, 262–63; as impetus to farm consolidation, 263
erosion: coal mining and, 334; forests as protection against, 36, 45, 122; in Heilongjiang Reclamation District, 141; land clearing and, 179; in Qing period, 94; resettlement and, 204, 207; and Sanxia dam, 208–11
Escobar, Arturo, 64
eutrophication, 430; dams and, 406; Lake Tai, 173–75; Sanxia reservoir, 215; sewage and, 398

exploitation: in adaptive cycle, 18f, 19, 80f, 94f, 95f; buffers against, 16; cause of backwardness, 66–69; by landlord class, 67, 69; of peasant migrants, 434; Revolution as rescue from, 4; in urban-rural relationship, 308. *See also* Land Reform; *and specific peoples and resources*
exports: of consumer goods, 328; of photovoltaic cells and panels, 409; from TVEs, 330. *See also* deforestation: export of; environmental degradation: export of
expressways, 360, 373–74
extinction: and biodiversity, 430–31; of fish and aquatic animals, 218–19
extractive colonialism, 293, 322, 324
extreme weather events, 381, 431, 438. *See also* floods

factory workers: dormitory housing, 357; peasant migrants as, 354
fallowing: as ecological buffer, 36
false consciousness: science as, 69
"family farms" model: in grain farming, 249–50; in shrimp farming, 251
famine: absent after 1962, 146; avoided in 1998 floods, 178; after Great Leap Forward, 83, 128–29, 133–34; in 19th and 20th centuries, 93; in Qing, 94, 95
farmers: as housing renters, 357; as "ignorant" or "primitive," 257, 262, 272
farmers' markets, ecological, 280
farmland: Basic Protection Regulations, 243; consolidation in 21st century, 249, 264–65; contamination by heavy metals, 430; conversion of, 358–59, 365; expropriation of, 358–59; fragmentation in Reform and Opening, 249; loss to aquaculture, 244; loss to industry, 380; loss to urbanization, 242–45, 355, 379–81; New Administration Law, 243; preservation policies flouted, 380; "red lines," 244; regularization of, 246; returning to forest, 183–85, 243; solar farms built on, 410; statistics on area of, 243
fast food restaurants, 240

feedbacks, disruption of, in ecosystems, 125, 129–33

Fei Xiaotong (Fei Hsiao-tung), 39

fencing (of pastures), 185

fertility: decline, 159–61, 427–29; high in 1950s, 81, 97; rural vs. urban, 159–60

fertilizer: comparative use rates, 258; limits to uptake, 258; organic, 28, 32; sewage sludge used as, 399. *See also* manure

fertilizer, chemical, 148–49, 254, 257–59; diminishing returns to, 258, 261–62; lack of, in 1950s, 98, 101f; and lake eutrophication, 174; overuse of, 174, 258–59; small rural factories, 322, 328; and soil quality, 258–59

feudalism, 42, 66, 87f; semi-, 67, 74

First Auto Works (Changchun), 293; as model, 317; urban design at, 294

fish, 218–19; consumption, in 21st century, 233, 234t, 235–36; as luxury foods, 235–36, 269–70; in rice fields, 29

fisheries, capture: in Cambodia, 221–22; dams and, 218–19, 221–22; decline of, 218–19, 269; expansion in Reform and Opening, 166, 269; expansion in 21st century, 269–70; marine/ocean, 235, 269; overseas processing, 269; pressure on world, 245, 269–70, 432; Tonle Sap, 221–22; unsustainable practices, 269. *See also* aquaculture

five-year plans: 1st, 81–82; 11th, 426; 12th, 406; 13th, 409; 14th, 404, 408; steel and coal in, 289. *See also* Great Leap Forward

fix to fix the fix (fix-fixing), 21, 199; as administrative accomplishments, 382–83; at Poyang Lake, 213; South–North water transfer as, 190, 194; in urban flood remediation, 382–83; wastewater treatment and, 399; waterworks as, 185–86, 190, 194, 195–96; Xiaolangdi dam as, 187; on Yellow River, 189

floating population. *See* peasant migrants

floods, 20, 176–79; 1998 artificial, 187; from Banqiao dam collapse, 148; Chang and Huai Basins 1954, 108, 438; Chang Basin, 107–8, 116, 176–85, 424, 431, 438; coal mining and, 334; concrete and, 381; damage in 1957, 120; diking and, 211; dumping and, 390–91; and eco-developmentalism, 182–85, 424; greenhouse gases and, 432; Huai Basin, 107–8, 431; impervious surfaces and, 381; increasing freaquency of urban, 381–82; intractability of, 189; mining and, 335; as political tool, 103–4; Poyang Lake, 211; in Qing period, 94; remediation of urban, 382; Sanmenxia Dam and, 130–31; Sanxia Dam and, 205, 210; sponge cities as defense against, 382–83; urban vulnerability to, 181; urbanization and, 211, 381–83, 400; as wake-up call, 85, 178, 424; Wei River (Shaanxi), 186; Wuhan, 210, 381; Yellow River, 169, 189, 431

food buying clubs, 280

food diversity: in Reform and Opening, 158, 159t; in 21st century, 231–37

food safety, 271–78; alternative food networks as reaction to, 281; CCCM agriculture as reaction to, 275–78; certification standards, 277; consumer strategies for, 276–77; law, 273; official strategies to ensure, 277–78; pesticides and, 275–78; replaces food security as public concern, 276; and unsafe handling, 276–77

food scandals: gutter oil, 273–75; milk adulteration, 271–73

food security: achieved in Reform and Opening, 157–66; for North China, 146; and overseas investment, 267–68; as precarious in 1970s, 155; urbanization and, 379–80; worries over, in 21st century, 243–44, 245, 267–68, 379–81, 400

food web, disruption of, 173

foreign investment: in TVEs, 330

forests: as commons, 35; conversion to agriculture, 140–41; coverage, 184; as ecological buffer, 36, 45; in Great Leap, 136–37; at Jiuzhaigou, 48; in Nuosu livelihood, 45; quality, 184; reforestation, 143, 183–85, 207, 388–89; sawmills and, 311–12. *See also* deforestation

fossil fuels: China as world's largest consumer, 411; and development, 10, 11, 19, 253; share in electric power generation, 404–5; share in energy mix, 401–4

Four Modernizations, 204

frogs, 28

"from point to field" experimentation, 77

frugality: abandonment of, 215–16; as cultural buffer, 39, 61; in high-socialist era, 394

fruit: increase in 21st century, 237; in Leizhou peninsula, 238; among Nuosu, 238; pesticide use on, 260; in Reform and Opening, 163–65; in Xinjiang, 239

Fu Zuoyi, 121–22

fuel standards, 415, 417, 436

Fukushima triple disaster, 407

furniture: export, 388; manufacture, 340, 388

Fusarium wilt, 252, 265

Fushun collieries, 287, 292

gandalei house, 314–15

Gao, Mobo, 167

garbage: crisis, 390–91; dumping, 390–91; "harmless treatment," 391–92; incineration, 392–93; landfilling, 391–93; Sanxia reservoir, 215–17

gasoline, 387, 388t; exhaust, 412; sulfur in, 415, 417

Gaubatz, Piper, 309

geese, 29

generational continuity: as cultural buffer, 38–39

genetically modified organisms (GMOs), 250, 276

ger (Mongolian mobile home), 54

Germany: environmental governance in, 435; renewable energy in, 409

Gezhou dam, 201, 203; and decline of fish(eries), 218

ghost cities, 365

glass: energy consumption, 344; production, 343t; reprocessing, 396; urbanization and, 354–55, 378, 387–88

goats: in Central Asia, 53, 59; in China Proper, 33; in Zomia, 45

golden monkey, 431

González, Roberto, 72–74

governance, environmental, 12–13, 15; and air pollution, 418; autoritarianism and, 435; citizen pressure and, 418; compliance with, 420; coordination of, 418–19; corruption in, 418, 420; democracy and, 435; enforcement, 425–26; and environmental injustice, 418, 420; inspections in, 420–21; and political ideology, 435; problems in, 184, 189, 382–83, 400, 418–21; of sponge city program, 382–83; of waste treatment, 400; of water pollution, 197–98; of water use inefficiency, 197–98. *See also* local government; management

grain. *See* staple foods

Grain-for-Green. *See* Returning Farmland to Forest Program

Great Green Wall, 431

Great Leap Forward (1958–61): agriculture, 82–83, 125–29; among Akha people, 136; blamed on disasters, 137–38, 139; canteens, 82, 135–36; Central Asia, 134–35, 307; coal mining, 304–5; as collapse, 13, 20, 83; in conventional historical periodization, 79; cotton, 129; decline of staple harvests, 128–29, 133–34; deforestation, 131, 136–37, 305; destruction of infrastructural and institutional buffers, 83; disruption of system feedbacks, 129–33, 134, 138; as ecological overreaching, 129; as ecosystem transformation, 83–84; evaluation of specific projects, 131–32; famine after, 83, 128–29, 133–34, 137–38; Great Iron and Steel Smelting, 300–304, 307; historical evaluation, 4, 137–38; *hukou* system and, 352; as hyperrationalism, 15, 134, 137; ideology, 82; in industry, 298–308; irrigation,

Great Leap Forward (*continued*)
121–23, 129–32, 144, 398; labor mobilization, 82, 123; large-scale collectivization, 123–24; local and regional differences, 129; logic of, 124, 137, 298–99; among Nuosu, 135–36; panacea solutions, 122, 138; among pastoral peoples, 134; preludes to, 118–23; psychology of speed, 122; recovery from, 83–84; reservoirs, 130; resilience diminished by, 122–23, 124, 135, 138; salinization, 122; science in, 126–27; sedentarization, 134; sewage for irrigation, 398; shoddy projects, 130; slogans, 299, 300–301; steel, 299–305; successful projects, 131; unidirectional thinking, 132, 137–38; unsustainability, 124–25; urbanization, 305–8, 351–52; voluntarism, 82–83, 123, 299, 300; waste, 302–4; waterlogging, 122; waterworks, 121–23, 129–32, 144; weather during, 137–38; women's labor, 122, 123–24, 126; Zomia, 135–37

Great Northern Wilderness. *See* Heilongjiang Reclamation Area

Great Western Development Program, 220, 224

Green Earth Volunteers, 225

green growth, 437

Green Revolution crops, 98, 153–55; CCP critique of, 153; contribution to food security, 154–55; and environmental degradation, 153, 173–75; and lake eutrophication, 173–75; negative socioeconomic effects of, 153; science and, 153–54

Green Watershed, 225

Greenhalgh, Susan, 160

greenhouse gases: China share of world, 401; coal gasification and, 419; commitment to reducing, 403, 404, 409, 428; and dust storms, 432; and floods, 432; growth and, 344; increase in emissions continues, 432; from iron and steel production, 296, 344; lack of resilience against effects, 432; from landfills, 399–400; per capita emissions, 403; recency of discourse prominence, 406; renewable energy and reduction of, 428

groundwater: land subsidence and, 383–84; landfills and pollution of, 391–92; nutrient leaching into, 258; pharmaceutical pollution of, 392. *See also* irrigation; water table

Guangzhou: air pollution in, 414–16; solid waste, 392; urbanization patterns, 358

Guanzhong Plain: flooding on, 186

Gunderson, Lance, 15, 20, 433; diagrams by, 8, 9, 18

Guo Laixi, 204

gutter oil, 271, 273–75; biofuel from, 274; contaminants in, 274; illness caused by, 274; prosecutions, 274–75

habitat decline, 431

Hacking, Ian, 264

Hale, Matthew, 281

Han River, 193, 194; diversion from, 195; pollution in, 197; reduced volume from South–North Water Transfer, 194; replenishment of, 195, 196–97

Hangzhou: farmland loss, 380; landfills, 391; water transfers to, 198

Hani people. *See* Akha (Hani) people

Hansen, Mette Halskov, 298, 411

He Gegao, 205

heat islands, urban, 379, 384–85

heavy metals: near landfills, 391; and soil pollution, 260; in water pollution, 332

herders. *See* pastoral peoples

Heilongjiang Reclamation Area, 102, 103m; drought in, 141; erosion in, 141; expansion of, 140–41; fertilizer use in, 258; shift to irrigated rice, 140–41; wheat in, 140

high modernism, 70–71; and waterworks, 111, 114

historic preservation, 360–62; in Kashgar, 371–73

history: ecocentric periodization of, 4; official PRC narrative of, 3–4. *See also* ecological history hoarding, 347

Holling, C. S., 8, 16, 20; diagrams by, 9f, 18f

horses: in Central Asia, 52–53, 59; in China Proper, 32; in Zomia, 45

hotels, 264

Hou Xueyu, 204

household: abolished as economic unit in Great Leap Forward, 123–24; ecology of Chinese, 29–33; ecology of Zomian, 43–46; and marketing, 34; water usage, 386–87

household responsibility system, 162; decline of buffers in, 167. *See also* decollectivization

houses: cave, 30; in China Proper, 30–33; construction materials, 30, 31f, 43; energy conservation in, 30; floor plans, 30–32, 43; *gandalei*, 314–15; ventilation, 30, 31f

housing, urban: amenities in, 361; "big yard" *dayuan*, 357; compared to rural, 324; construction worker sheds, 357; at Daqing, 314; and earthquakes, 325; *gandalei*, 314–15; high-rise, 354–55; lack of resilience, 313–15, 324; neglect of, in high socialist period, 294–95, 310–11, 315, 324, 325, 354; peasant migrants and, 357–58; per capita area, 354–55, 385; privatization, 356, 360; rented from farmers, 357; replacement of traditional, 359; repurposing, 294, 311; residential parks, 359; resource consumption in, 354–55; spatial hierarchy of desirability, 356, 362–63; suburban "villas," 363; total area, 354–55; university faculty, 360; Wenzhou merchants', 357. *See also* cities; urbanization

Hu Jintao, 4, 226, 257, 264

Hua Guofeng, 157

Huadian, 224–25

Huai River Basin, 106m; conservancy project, 107–8; expansion of reservoirs, 147–48; floods, 107–8, 189; turns black, 169–70; water pollution in, 169–70

Huanan Seafood Market, 362

Huang, Yu, 172–73

hukou system: as caste system, 354; Household Registration Law, 307; origins of, 307, 352; rationing and, 353

human rights, 425

humanism: Confucian, 69; Marxism-Leninism as, 68–71

hutong urban lanes: replaced, 359

hydraulic engineering. *See* waterworks

hydrological changes, 2, 36; in Qing period, 95; as result of dams, 130–31

hydroelectric power generation: as backup for wind and solar, 228–29; boom in, 405–7; compared to thermoelectric power, 406; and cryptocurrency mining, 433; as decarbonization strategy, 226–27, 228–29, 405–7; ecological tradeoffs of, 405–6; "five giants," 224–25; future plans, 407; and greenhouse gases, 428; growth of, 406–7; hydropower nationalism, 226; installed capacity, 203, 228, 406–7; pumped storage, 407; as revenue source, 220–21, 225; share of world's, 407, 428; Yellow River, 189. *See also* dams

hyperrationalism, 15–16, 70

hysteresis, 174

imitation ancient streets, 360–62

imperialism, 14, 67, 74

impervious surfaces: and floods, 381–82; and urban heat islands, 384

imports: beef, 265–67; corn, 231–33; edible oils, 245; iron ore, 344–45; milk products, 245; oil (petroleum), 293, 313; pork, 234, 267; rice, 232; soybeans, 230, 232, 233–34, 242, 245, 267; staple foods, 157, 163, 230–31, 249; sugar, 245; trucks, 292–93; wood, 388, 432; wheat, 231; zinc, 345

incinerators, 389, 392–93; protests, 392–93; replace landfills, 392–93, 394; sludge burned in, 399

India, dam disputes with, 228

industrialization: and agriculture, 288–
89, 329; chemical, 345–49; coal and,
289, 291–92; in Common Program, 81;
commune and brigade, 322–23, 328–29;
and consumer goods, 289, 340–43,
342t; defense, 317, 321; development
zones, 363–64; dynamics of early, 288f;
and environmental injustice, 322–23,
346; environmental legacy of early,
348; and factory location, 294, 318,
345–46, 363–64; five small industries,
322–23, 328; industrial parks, 345–46,
364; and pollution, 81, 295–96, 331–40;
priority of heavy, 81, 287–89; return of
heavy, in 21st century, 341, 343t; rural,
328–40, 364; Stalinist strategy, 287–98,
323, 327; steel, 289–91, 299–305; transport and, 288, 365–66; and urbanization, 293–95; water usage, 386–87;
worker health, 297. See also Third
Front industrialization
infant mortality, decline of, 97
infrastructural buffers, 16–17; buildup
in first five-year plan, 81; dikes as,
37; disrepair at the Founding, 97; increased reliance on, 189, 423, 429, 433;
lacking in Central Asia, 58–59; lacking
in mountains, 95; limits in Zomia,
46–48, 52; need for upkeep, 25; neglect
with decollectivization, 167; replacement for ecological buffers, 25, 199,
253, 400; reservoirs as, 37; terraces as,
37; on Yellow River, 189
Inner Mongolia, 119, 248
inspections: and pollution, 339
institutional buffers, 16–17; buildup
in first five-year plan, 81; in China
Proper, 37–38; credit as, 38; decline
with Reform and Opening, 167; disrepair at the Founding, 97; generalized
reciprocity as, 49; granaries as, 38;
labor exchange as, 37; lineages as, 38;
markets as, 38; reliance on, 423, 429,
433; replace ecological buffers, 25; upkeep, 25; in Zomia, 48–49

intensification: and buffer replacement,
36–38; and floods, 178–82; limits on, in
Zomia, 46–48; and loss of resilience,
178–82; and reduced lake capacity,
179–81
intermediate disturbance hypothesis,
14–15; and adaptive cycle, 20
internal combustion engine/oil block,
288
iron: impurities in, 302; mining, 290,
291m, 318–19, 335, 343t; ore imports,
344–45; smelting, 296, 300–304. See
also steel
irrigation, 103; on Chengdu Plain, 115; as
common property, 38; disrepair with
Reform and Opening, 167; expansion,
11, 121–23, 129–32, 143–48; and institutional buffers, 38, 104; on North China
Plain, 114, 121–22; and rigidity traps,
104; with sewage, 398; surface water,
111, 121–22, 144, 147–48; by tubewells,
84, 145–46; Yellow River withdrawals
for, 168–69, 189. See also waterworks
Italy: greenhouse gas emissions, 403;
solar energy in, 409

Japan: environmental trajectory of, 435;
hostility from China, 434; ozone pollution in, 416–17; sea cucumbers from,
270; solar energy in, 409; as transportation model, 374
Jilin City, chemical spill in, 346–48
Jinsha River: dams on, 201–3, 209–10,
216–17; deforestation in watershed,
217; downstream effects of dams,
216–17; hydaulic generation, 216; at
Panzhihua, 318; sediment load, 217
Jiuzhaigou, 48; airport, 376; reforestation
in, 184; sacred precincts in, 50
Johnston, R. F., 40
Junggar Basin, 56; Qing conquest of, 96;
water pollution in, 171

Kailuan (Kaiping) coal mines, 292, 325
Kander, Astrid, 287–88

kang (heated living platform), 30

Karamay, Xinjiang: city of, 371; oilfield, 293, 307, 316

karst, 335

Kashgar, Xinjiang: Han migration to, 371; historic preservation, 371–73; urban renewal, 371–73

Kazakh people: collectivization, 119; flight to USSR, 134; migration patterns, 57; sedentarization, 134

Khrushchev, Nikita, 299

kidney stones, 271–72

Kim, Su Min, 282

Koktokay mining complex, Xinjiang, 295, 307

Korea, South: environmental trajectory of, 435; ozone pollution in, 416–17

labor: agricultural, 28–29, 34, 115, 327–28; "bare" power, 157; diminishing incentives, 157, 327–28; exchange, 34; and food production, 116; of livestock, 32; migrant, 340–41; migratory, 265, 278–79; mobilization for waterworks, 82, 107–8, 120–22; mobilization in Great Leap Forward, 82, 120–22; reorganization in 1950s, 100, 101f, 104–16; and resilience, 116; rural surplus of, 278–79, 327–28, 329, 338; women's, in Great Leap Forward, 123–24, 306. *See also* peasant migrants

Lahu people, 48

lakes: artificial, 382; elimination of, 179–80, 381–82; eutrophication of, 173–75, 215, 398, 406, 430; overnutrification, 173; pollution in, 331; reduced area, 179–81; restoration of urban, 382; sediment deposition in, 179; sewage discharge into, 398; water quality, 430. *See also specific lakes*

Lancang River, 219–23

land: increases in agricultural, 100–102; ownership in China Proper, 34; redistribution, 104–9; transfers, 330; in Zomia, 105. *See also* farmland

Land Reform, 104–9; classification of households in, 105; evaluation of, 108–9; in Inner Mongolia, 119; local differences in, 109; motivations for, 105–6

landfills (sanitary), 389, 391–92; greenhouse gases and, 399–400; leaching from, 399–40; pollution from, 391–92; replaced by incinerators, 392–93, 394; sludge in, 399–400

landscape plasticity, 48

landscapes: aesthetics of, 39–40, 51–52, 61; village, in China Proper, 34–35

landslides: from Lancang dams, 220; in Qing period, 95; resettlement and, 204, 207; from Sanxia dam, 204, 207, 210, 214

Laos: dams in, 222–23; as hydropower producer, 221–22; land purchases in, 268

Latour, Bruno, 434

Lattimore, Owen, 58, 61

laundry detergent, 174

lavender, 264

laws, environmental, 225, 243, 273, 307, 390, 408, 415, 426

laws of nature: as false consciousness, 69–71

lead (metal): and disease, 336; mining, 335–36; urbanization and, 387

"leapfrog growth," 356

legitimacy, political, 2; and development, 85, 88–89; ecological civilization and, 85–88; and economic growth, 423; and environmental governance, 423–24; waterworks and, 111, 114

Leizhou peninsula: diets in, 238; shrimp farming, 166, 171–73, 251

Leonard, Pamela, 40

Lhasa, 370

Li Chunfeng, 122

Li Jingxu, 275

Li Keqiang, 191; declares war on pollution, 349, 409

Li, Lillian, 96

Liang Shuming, 279

liberation: in Land Reform, 106–7; market reform and, 162; of productive forces, 67–69, 81

"license to pollute," 339, 418

linear thinking, 261, 400

Ling Jiang, 406

linpan dispersed settlement pattern, 264–65

Lisu people, 48, 223

Liu Huixian, 324–25

Liu Yinmei, 252–53

livestock: among Akha, 46; antibiotics fed to, 262; in Central Asia, 52–54, 59; in China Proper, 31–33; collectivization, 134; crowding, 257, 262–64; diversity, as ecological buffer, 59; epizootics, 262–64; increasing farm size, 247–48; among Nuosu, 43–45; in Reform and Opening, 165–66; staples as feed for, 241–42

local government: autonomy of, 418–19; environmental inspections and, 339; relation to TVEs, 330, 338–40. *See also* governance

locomotives, manufacture of, 292

logging, 311; banned in southwest, 179, 183

logs: imports, 388–89; milling of, 311, 388. *See also* sawmill

Lora-Wainwright, Anna, 335–36

Los Angeles: air pollution compared with Chinese cities, 416

Lu, Jixia, 335–36

lumber: imports, 388–89. *See also* sawmill

lychee, 252

Lysenko, Trofim, 71

Ma Yinchu, 68

maintenance, neglect of, 122

Malthus-and-Boserup ratchet, 10–11; in early PRC, 97, 423; Qing period turn of, 96

Malthusianism: criticized as anti-Marxist, 97

management: agricultural, 250, 275; environmental, 15–16; of forests, 182–85; golden-rule, 16; in Great Leap Forward, 123–24; of herds, 134; hyperrational, 15–16; of pastures, 185; pathologies of, 20, 76; of resources, 20; top-down, 400; of waste, 396; of water systems, 186, 211. *See also* governance

Manchuria, 100, 102. *See also* Heilongjiang Reclamation District

manufacturing: consumer goods, 340–43; dominates TVEs, 330; energy and, 344; outsourcing of, 270; pollution and, 330, 336–37, 378; as stage in industrial development, 378, 437; steel, 289–92. *See also* industrialization

manure, 28, 32, 98, 125, 149, 276; decline in use, 247, 254, 257–58; as energy source, 409; as pollution, 258; replaced by chemical fertilizer, 257–58; and soil quality, 258–59; urban, 380. *See also* fertilizer

Mao Zedong: campaign against Ma Yinchu, 69; on collectivization, 110; on communist society, 82–83; death of, 157, 326; on deep tilling and close planting, 125–26; on demography, 69; and dissent, 120; and double-wheeled, double-bladed plow, 128; on Great Leap Forward, 82–83; on Huai River, 107; in "Hundred Flowers" campaign, 120; on pigs, 32; and Sanxia dam, 204; on science, 72; "seek truth from facts," 423; sides with leftists, 4, 162; on staple food production, 140; on Third Front, 317; and urban communes, 306; visits Zhengzhou, 306; on waterworks, 190; on women's liberation, 82

markets: absence in Zomia, 45; dependency on, 254–55, 270; and food sufficiency, 237; as institutional buffers, 38, 60; open, 277, 362; periodic, 34, 35; in Reform and Opening, 162–63; and resilience, 38, 60, 264–65; scrap, 394; standard area, 35; super-, 277–78; urban consumer preferences for, 277–78

Marx, Karl: on commmunist society, 66, 124

Marxism-Leninism: as constraint on environmental thinking, 86; contradictions within, 73–74; critique of, in environmental culture, 86; as humanist ideology, 68–70; lack of ecological awareness, 68–71, 85–89; as modernist ideology, 15, 64–66; and pollution, 298; and population, 97; teleology in, 68, 73–74, 85–87, 437; and women's liberation, 124

mass line, 77

materialism: antimaterial, 70; dialectical, 71–72; labor, 83

meat: in Nuosu diet, 43–45, 238; purchasing preferences, 277–78; as restaurant food, 240; in 21st century diet, 232–35

megaproject mentality, 191

Mekong River, 219–23; differences among countries along, 220–23; interaction with Tonle Sap, 221–22; seasonal flows of, 221–22. *See also* Lancang River

melamine, adulteration of milk with, 271–73

melons, sludge-fertilized, 399

Miao people, 42

microdistrict (*mikrorayon*), 294, 305; rejection of, 306–7

migration: Han to Tibet, 370; in Qing period, 95; among pastoralists, 57; restrictions on, 329, 337, 352; and urbanization, 351, 353–54. *See also* peasant migrants

milk products: absence of quality control, 272–73; adulteration, 271–73; baby formula, 236, 271–72; in CCCM agriculture, 272–73; in Central Asia, 52–54, 236, 272; consolidation in 21st century, 273; consumption, Reform and Opening, 165t, 166; excess iodine in, 273; imports, 245; increase in 21st century, 234t, 236, 272; melamine in, 271–73; milk powder, 236, 271–73; not in traditional Chinese diet, 33, 236, 248; as protein source, 241; as sign of modernity, 236; supply chain, 248, 272; watered down, 272–73. *See also* cattle

millet, 26; decline of, 232

Minamata disease, 348

Ming Tombs Reservoir, 121f

mining, 19, 341; accidents, 296, 335; and air pollution, 295; closure of small, 340; coal, 287, 291–92, 295–96, 304–5, 333–35; cryptocurrency, 433; and farmland loss, 336; in Great Leap Forward, 304–5; iron, 290, 291m, 318–19, 335; lead, 335–36; mountaintop, 320f; nonferrous metal, 293; in Qing period Zomia, 96; and renewable energy, 410; and resettlement, 295, 335; sand, 210, 213; small-scale (artisinal), 302, 303f, 333–35; wastes from, 295; water pollution, 295, 335; zinc, 335. *See also* coal

Ministries: Ecology and Environment, 412, 413, 425–26; Environmental Protection, 406, 413, 420, 425, 430; Soil Resources, 430; Water Resources, 189, 198, 410, 426; Waterworks and Electeric Power, 204

minority peoples: as backward, 74; and dams, 229; incorporation into Qing polity, 96. *See also specific peoples and regions*

misery (condition), 10–11, 16, 67; possibility of future, 423; in Qing period, 93

mobility: lack of, in high socialist period, 351–53; of pastoral peoples, 54, 57. *See also* peasant migrants; urbanization

modernism, scientific, 15–16

molybdenum, 348

money: and business ethics, 275; in collectives, 110; in early Reform era, 329; as institutional buffer, 38; lack of, in high socialist era, 162, 353; and pollution, 339; for relocation, 207; in 21st century, 253

Mongol people: communes among, 135–36; in Great Leap Forward, 134–35

Mongolia: climate, 55; ecology of, 55–56; migratory patterns in, 57; relations with China Proper, 55–56. *See also* Inner Mongolia monoculture: lack of resilience in, 252–53; and pesticides, 259–61. *See also* biodiversity

moral economy, 19, 36

moral hazards, of ecosystem intensification, 191, 199–200

mortality: decline with public health, 81, 97

motor vehicles: air pollution and, 366, 412, 417; electric, 404; growth in private ownership of, 355, 374, 417; lack of, in high socialist period, 310, 327; private, 355, 417; traffic jams and, 366. *See also* automobiles, private; trucks

mules, 32

Muslims, 56; diets, 239; urban living patterns, 370

mutton: increase in consumption, Reform and Opening, 165t, 166, 233, 234t

mutual-aid teams, 109

Myanmar: dams in, 227; effects of Lancang/Mekong dams on, 221; as hydropwer exporter, 227; land purchases in, 268; military coup of 2021, 228

Na (Mosuo) people, 42

Nanjing: air pollution, 297, 414–16; lakes and flooding in, 381–82; petrochemicals, 346

Nasu Yi people, 42

National People's Congress, 205–6

national water network, 198; as fix-fixing, 195–98

Natural Forest Protection Program, 183, 207

natural gas, 403–4; replaces coal, 405, 417–18

Naughton, Barry, 287

Naxi people, 49–50

negative externalities, 132

Nen River, 102; 1998 floods, 177

New Rural Reconstruction (NRR), 279–80

New Socialist Countryside campaign, 264–65, 278–79, 424

New World crops, 10; and Qing expansion, 93–95

nitrogen: deficiency, 149, 257; diminishing returns to, 258; fertilizers, 149; in industrial discharge, 332; lake eutrophication and, 173–75; leaching into groundwater, 258; recycling, 257; volatilization, 258; wastewater treatment and, 398; in Yellow River, 188

nomads. *See* pastoral peoples

non-ferrous metals, 343t

non-governmental organizations (NGOs): and alternative food networks, 281; oppose Nu River dams, 225, 226; oppose shark fin consumption, 269–70; proliferation of, 425

North China Aquifer, 145, 170–71, 190, 194, 383–84

North China Plain, 26–27; falling water table, 144–46, 167–68, 170–71, 190; land subsidence on, 383–84; rising water table, 144–45; waterworks on, 82, 107–8, 111–14

Northeast China. *See* Manchuria

Nuosu Yi people: agriculture, 43–45; common property among, 45, 49; Democratic Reforms among, 118–19; diets, 43–45, 238; ecological niches, 43; forestry, 45; in Great Leap, 135–36; history of, 43; house construction of, 43; household ecology of, 43–45; hunting, 45; lacked markets, 45; natural philosophy, 51–52; opium, 45; rebellion, 119; trade with Han, 45

Nuozhadu dam, 220; effect on downstream fisheries, 221–22

Nu/Salween River, 193, 223–28; dam disputes on, 224–28; plans for dams in Myanmar, 227; small hydro projects on, 223, 227; as "wild river," 223–24

nuclear power, 318, 403, 404; caution about, 405–6; environmental tradeoffs, 407–8; growth of, 407–8

nuclear weapons, 156, 316, 318

oases, 53m, 56; intensification of cultivation in, 96

obesity, 240

oil (petroleum): and air pollution, 316, 417; Daqing oilfield, 313–16; dependency on imports in 1950s, 293, 313; as developent bottleneck, 313; discoveries in 1950s, 293; exports in 1970s and '80s, 313; increased consumption in 21st century, 403–4; Karamay oilfield, 293, 307, 316; regulation of sulfur content, 415, 417; Shengli oilfield, 315; and soil contamination, 316; in Xinjiang, 293, 316

oils, edible: fixed-price sales to state, 110; imports of, 245; increase in production of, 158, 159t

Olympic Games, 2008: blue skies for, 413

"one family, two systems" agriculture, 276

one-child policy. See planned birth policy

open hearth furnaces, air pollution from, 296

opium, 45, 96

oranges, 163–64

orchards, 163–65; counted as forests, 184; deforestation for, 252; increase in 21st century, 237

organic chemical pollution, 391

organic foods, 277; certification, 277, 279; as different from "ecological foods," 279; NRR and, 278

ornamental plants, 264

Ostrom, Elinor, 48–49, 128

overfishing: inland, 219, 235; ocean, 269

overnutrification, 173, 174. See also eutrophication

overshoot, population, 10, 11f; in Great Leap, 129; in Qing period, 93–96

oxen, 32–33

ozone, 416–17

Pan Yue, 86–88, 349

panacea solutions, 76, 125–26, 127–28, 184; disrupt ecosystem feedbacks, 129–33, 138; in reforestation, 184–85

"pancake growth," 356

panda, giant, 431

Panzhihua, Sichuan, 316, 318–22; airport, 376; as danwei city, 319–20; iron mining in, 318–19; pollution in, 321; steelmaking in, 319–20

paper: import of recycled, 396; production, 331; recycling, 394

Papua New Guinea, 270

particulate matter, small ($PM_{2.5}$), 296, 412, 415–16; inspections and, 420

pastoral peoples, 52–62; as adaptive to environment, 52–53, 57, 61–62; blamed for pasture degradation, 185; collectivization of, 119; in Mongolia, 57; relations with agricultural peoples, 55–56; sedentarization of, 185, 248; in Tibet, 57, 119, 185; in Xinjiang, 56–57, 119

pasture: degradation, 54; fencing, 185; impossibility of intensification, 54; lack of, in China Proper, 33; property rights to, 58; restoration programs, 185; rotational grazing in, 54

pathologies of natural resource management, 20, 76

payment for ecosystem services, 183, 198

peak carbon, 403, 404, 432

Pearl River basin: water quality in, 430

Pearl River Delta, 127; loss of farmland in, 244; urbanization of, 357–58; "villages within the city" in, 358; water transfer to, 198–99

peasant agriculture: end of, 253–55

peasant migrants, 353–54, 434; and factory work, 353–54; housing of, 357–58; resettled from dam sites, 188; in plastic reprocessing, 396. See also mobility

people's communes, 82; de-communizing of, 139; as labor mobilization strategy, 123–24; renamed as townships, 329; in Zomia, 135–36

People's Daily, 83, 88, 125, 126, 128, 153, 301, 349, 354

People's Victory Canal, 111; partial shutdown of, 144

"people's war on terror," 373

periodization, 78–89; conventional and ecosystemic, 78–80

permeable urban surfaces, 382

pesticides: adverse effects of, 260–61;
contamination by, 150; dangers of, 150;
food safety fears over, 275–78; lack of,
in 1950s, 101f; monocrops and, 259–60,
260–61; overuse of, reasons for, 260–
61, 276; pollution from manufacture,
346; in 21st century agriculture, 259–
62; use of, beginning in 1960s, 149–50
petrochemical industry: and environmental injustice, 346; pollution from,
346–47; protests against, 348–49, 425.
See also chemical industry
petroleum. *See* oil (petroleum)
pharmaceuticals: manufacture of, 341,
346; pollution from, 346, 392
Philippines: agricultural investment in, 251
phosphorus: content in iron, 302; deficiency in rice agriculture, 149, 257; in
industrial discharge, 332; lake eutrophication and, 173–75; leaching into
groundwater, 258; overapplication of,
174; wastewater treatment and, 398
photovoltaic cells, 409–10
Pietz, David, 111, 130
pigs: antibiotics and hormones fed to,
242, 276; breeds, 276; in China Proper
ecology, 32; and farm size, 247; feed
for, 241–42; in Great Leap Forward,
129; international investment in, 247;
among Nuosu, 45; prohibitions on
raising, 235; in Reform and Opening,
165–66; in 21st century, 233–34, 247; in
Zomia, 241. *See also* pork
Planned Birth Policy, 159–61, 354
planned economy: as inhibition to innovation, 328–29; as means to development, 63–64
plastic mulch, 150–51
plastics: health hazards from reprocessing, 396; production of, 340, 342t; recycling of, 394, 396; as reservoir garbage,
215–16
plenty (condition), 10–11, 423
plow, double-wheeled, double-bladed, 125,
128
polders, 37, 38

pollution; abatement expenses, 338;
as capitalist problem, 69–70, 298;
chemical industry, 345–49; cleanup,
resistance to, 336–40; coal, 296–97,
332, 334, 412; coal gasification, 419;
compensation for, 339; controls, 339,
412–18, 424; decrease in TVEs, 339–40;
and development, 336–37; dilution,
347–48, 398; discourses, 349–50; and
energy intensity, 338; export to poor
areas within China, 418–19; export to
poor countries, 2; havens, 418; heavy
metal, 260, 332, 391, 397; increase in
Reform period, 84, 331–40; inspections
and, 420; intensity in Third Front, 321;
landfills and, 391–92; Li Keqiang and,
349; "license to pollute," 339, 418; organic chemicals as, 391, 397; penalties
for, 339; pesticide manufacture and,
346; preindustrial, 389; ratio to output,
308; recycling and, 394–97; shelters,
418; tolerance, 330, 337, 339; "war"
against, 349. *See also* air pollution;
water pollution
policyclic aromatic hydrocarbons (PAH),
296, 297; from petrochemical industry,
346
polyculture, 28–29
Popkin, Barry M., 240
population: Qing period, 93; urban, 351,
385–86
population control: and food production,
160–61; as institutional buffer in Tibet,
60; "late, spaced, and few" policy,
157, 160; planned birth program, 158,
159–61
population density: in Central Asia, 54,
60; in China Proper, 25, 29; and urban
core decline, 359; in Zomia, 46
population growth: decline after 1990,
161; in early PRC, 97; encroachment
on buffers, 41; end of, 380, 427–28; and
food supply, 81, 97; general trends in
PRC, 10–11; in 1950s, 81, 97; in 1960s
and '70s, 143–44; in Qing dynasty, 10,
94; and reduced lake capacity, 179–80;

and resources, 10; as slow variable, 14; urban, in Great Leap Forward, 307; urban, in Reform and Opening, 351, 354

pork: adulterants in, 273; imports of, 234, 267; increase in consumption, Reform and Opening, 165–66; increase in consumption, 21st century, 233, 234t. *See also* pigs

potatoes, 93, 238; decline of, 232; improved varieties of, 155; spread of, 10, 45

poultry, 234t, 234–35

poverty alleviation, energy and, 411

Poyang Lake, 36, 175, 180, 210, 211–13

precipitation. *See* rainfall

prepared foods, 239–41

primitive accumulation, 111, 117

private plots, 110–11; abolition of, 123–24

production teams, 139

productive forces: in Land Reform, 106–7; liberation of, 68–69, 81; market reform and, 162

productivity: of agricultural labor, 327–28; in China Proper households, 36; as low in Central Asia, 53–54; and resilience, 17–19, 36–37, 104, 256–62

pronatalism, 97

propaganda, 73, 87–88, 98, 139, 349; associated with specific programs, 301, 393; local, 127. *See also* slogans

protein, animal, 234, 240–41; conversion efficiencies, 241; in pig feed, 242

protest, environmental, 78, 348–49, 392–93; ubiquity in 21st century, 425

public health, 81, 97

pulp and paper: production, 331, 342t; and water pollution, 332

pumped storage, 407

punishments, supernatural, 49–50

Qiang people, 368–69

Qing empire: boundaries of, 6; ecological zones of, 52, 53m; environmental degradation in, 94–96; famine relief in, 38; governance in, 93–94; overshoot of carrying capacity, 93; population growth in, 10–11; rule over Central Asia, 51, 62, 93, 96; steelmaking in, 291

railroads: Beijing-Guangzhou, 375; Chengdu-Kunming, 317; high-speed, 2, 374–75; magnetic levitation, 376; restoration in 1950s, 292; South Manchurian, 314; as support for industrialization, 292; world's longest network, 375; Yingtan-Xiamen, 311

railway cars, manufacture of, 292

rainfall: buffers for, 36, 37; in Central Asia, 52, 53, 55, 56; in Han River Basin, 194–95; in Huai basin, 108; in North China, 26; in South China, 27. *See also* dams; floods

rainwater recovery, 382

rapeseed, 103

rationing, 353

real estate market, 264, 365

reciprocity: balanced, 49; as cultural buffer, 39; generalized, 49

reclamation: of Dian Lake, 141–42; limits on, 102–3; in Manchuria, 102

recycling: ban on imports for, 396; decline with urbanization, 378, 389; facilities in residential compounds, 362, 363f; health hazards from, 396, 397; in high socialist era, 394; modern, 394–96; of plastic, 394, 396; pollution from, 394–97; prohibition of informal, 397; slow start to, 215–16; specialized neighborhoods and, 357; of wastewater, 387. *See also* reprocessing

Red Flag Canal, 147, 169

red tides, 174

reforestation: environmental Kuznets curve and, 428; floods as impetus to, 182–85; Great Green Wall, 431; inappropriate, 184, 431; programs, 143, 183–85, 207, 388–89, 428; and sediment yield, 179; Three North Shelterbelt, 431; and wood imports, 432; in Zomia, 179

Reform and Opening, 4; animal agriculture in, 165–66; aquaculture in, 166; in conventional periodization, 79; de-collectivization in, 161–62; ecological continuity in, 84, 157–58; as economic break, 158; economic growth in, 84; economic rationale for, 327–28; fisheries in, 166; food security achieved in, 157–66; fruit cultivation in, 163–65; neglect of buffers with, 167; origins of, 327; revival of marketing in, 162; rural crisis and, 278–79

refrigerator manufacture, 342t

regulations: on air pollution, 418; enforcement of, 425–26; on water use, 188

relocation: of industry, 363–64; of population (*see* resettlement)

renewable energy, 408–11; in Central Asia, 410; China compared to other countries, 409, 411, 425; law, 408; reduces air pollution, 428; transmission problems, 410. *See also* solar power; wind power

reprocessing, 394–97. *See also* recycling

Republic of China: borders of, 8; and dams, 203–4; decrease of buffers and development, 96, 203–4; during, 81; famine during, 93

reservoirs: algal blooms and eutrophication, 215; floating garbage, 215–16; in Great Leap, 130; Huai Basin, 106m, 107, 147–48; as infrastructural buffers, 37, 104; and landslides, 204, 207, 210, 214; Ming Tombs, 121f; North China Plain, 114; and pollution, 215, 392; and population resettlment, 188, 204; and resilience, 37, 104, 117; Sanxia dam, 203; siltation, 130–31

resettlement: compensation for, 207–8, 229; conflicts with original populations, 207; from dams, 187, 204, 206–8, 229; and deforestation, erosion, and siltation, 204, 207, 208–9; in high socialist period, 207; and impoverish-

ment, 207–8; land for, 207; and landslides, 207; mining and, 295, 335; from South–North Water Transfer, 194; and urbanization, 207

residential compounds/parks, 359, 360, 362–63: governments and development of, 364; recycling in, 362, 363f; resemblances to socialist *danwei*, 362

resilience: and abundance, 255; and antibiotics, 261f, 262–64; buffers and, 39–40; of CCCM agriculture, 254–55, 261–62, 266f; of Chengdu Plain communities, 264–65; of cities, 355; decline in Qing period, 94–95; defined, 12–14; diminished in Great Leap forward, 123, 124, 135, 138; diversity and, 264–65; and efficiency, 19; famine and, 438; and fertilizer, 260–61; of food system, late 20th century, 178; general and specific, 433; intensification and loss of, 178–82; and interconnectedness, 254; knowable only in retrospect, 21, 438–39; and labor mobilization, 116; lack of, against floods, 431; lack of, in monocultures, 252–53, 260–61; of landscapes, 40; and markets, 38, 254, 264–65; and pesticides, 260–61; population and, 41; and productivity, 16–17, 36–37, 104, 256–62, 264–65; reservoirs and, 37, 104, 117; and scale, 254–55; and specialization, 265

resource depletion: urban consumption and, 385–86, 387–88

restaurants, 239–41; fast food, 240; gutter oil in, 273–74; and nutrition, 240; proliferation in 21st century, 239–40; in Reform and Opening, 354; scarcity in high socialist period, 239, 327

"Returning" Programs: Farmland to Forest, 143, 183–85, 424; Pasture to Grassland, 185, 424

revolution: in first five-year plan, 81–82; as means to development, 66–71

rice: cultivation of, 27–29; deep tilling and close planting of, 126; double-cropped,

103, 115, 152; hybrid, 154; imports, 232; increased production in 21st century, 231; labor demands of, 115; new varieties in Qing period, 94; replaced by specialized crops, 252; replaces corn in Nuosu diets, 238; replaces sweet potatoes, 232; as staple, 34

rich peasant economy, 105

rigidity traps, 20, 429; bureaucracy and, 76; dams as, 199; and food supply, 171; irrigation and, 104, 167–68, 171; South–North Water Transfer as, 199–200; and vulnerability, 21; and Yellow River, 189

ring roads, 359–60

Ripert, Eric, 239–40

rivers: pollution in, 331, 398, 430; restoration of urban, 382; sewage discharge into, 398; water quality, 331, 398, 430. *See also specific rivers*

road network, 292–93

Rostow, W. W., 64–65

Ruf, Gregory, quoted, 163–64

runoff, urban, 382–83

rural crisis, 278–79

rural-urban inequality, 278–79

Russia: agricultural investment in, 251; chemical spill and, 348; invasion of Ukraine, 230, 245, 433; wood imports from, 389

sacred precincts, 50–51

salinization, 130, 144–45; alleviated by tubewells, 146; and South–North Water Transfer, 190, 194; in Xinjiang, 171

Salt, David, 323

salt marshes, 210–11

Salween River. *See* Nu/Salween River

sand mining, 210, 213

sandstorms, 431

Sanlu Corporation, 271–72, 275

Sanmenxia Dam, 130–31, 168, 186–87, 201, 229; floods caused by, 186; problems corrected by Xiaolangdi, 186–89; re-

settlement from, 207; as rigidity trap, 199; siltation behind, 186

Sanxia dam, 203–19; algal blooms and eutrophication in reservoir, 215; and biodiversity decline, 218–19; Dai Qing's critique, 205–6; delayed in 1980s, 204; dimensions, 203; earthquakes and, 214; erosion and, 208–11; expert evaluation, 204–5; fear of collapse, 214–15; and grasslands, 212; interaction with other dams, 209–10; landslides and, 214; limited flood control function, 210; locks, 203; Mao on, 204; opposition to, 204–6; origins of, 203–6; People's Congress approval, 205–6; and Poyang Lake, 211–13; relocation from, 206–8; reservoir pollution and garbage, 215, 216, 217f; salvage archaeology, 206; sediment regime and, 208–11; socioeconomic effects, 206–8; storage capacity, 203, 209–10; and wetlands, 212; Zhou Enlai proposes, 204

Savage, John "Jack Dam," 204

sawmill, 311–12

scale: in agriculture, 246–50; in China Proper ecology, 29–35; mismatch, 5–6; spatial, 3–7; temporal, 8–12

Schmalzer, Sigrid, 63–64, 72, 74, 153–54

Schneider, Mindi, 247

science: agricultural, 250, 257; in CCP ideology, 71–75; definitions of, 72–73; and dissent, 120; and ecosystem concerns, 424; folk and professional, 73–74, 77, 301; in Great Leap Forward, 126; and Green Revolution crops, 153–54; in high socialism, 156; linear, 172–73, 185, 257, 261, 263–64, 271, 400; Mao on, 72; and Marxism-Leninism, 71–72; and Nu River dam disputes, 225; perspectives on development, 257, 424; political nature of, 434–35; as politically suspect, 74–75; pure and applied, 72; and Sanxia dam dispute, 205; as signifier of development, 89; as social activity, 71–72

Scott, James, 6–8, 70
scrap markets, 394
sea cucumber, 236, 270
seawalls, 188
seawater incursion, 169
Second Auto Works, 317, 322
sedentarization, 134
sediment deposition, 130–31; Chang River regime, 208–11; coal mining and, 334; in Dongting Lake, 211; Jinsha River, 217; and lake capacity, 179–80; precipitation and, 209; reforestation and, 209; regimes of, 186–88, 217, 208–11; resettlement and, 207; at Sanmenxia, 186; and Sanxia dam, 208–11; in Sungari River, 141; at Xiaolangdi, 187; at Yellow River mouth, 188–89
self-sufficiency: cities as decreasingly, 378; vs. entrepreneurship, 252; food, 158, 245; household and village, 254, 279; national, 230, 248–49; pastoral economy as lacking, 55; precariousness of, 243
semi-feudal, semi-colonial society, 67, 74
service industries, peasant migrants and, 354
sewage, 397–400
sewers: in Changzhou, 398; gutter oil clogging, 274; lack of, in premodern cities, 397–98; in Shanghai, 398
Shanghai, 171; air pollution in, 412; airports in, 376; land subsidence in, 383; sewers in, 398; urban heat islands in, 384; waste separation in, 393–94
Shapiro, Judith, 128, 323
shark fin, 236; and conservation, 269–70
sheep, 33, 45, 48; in central Asia, 53, 59
shoe manufacturing, 341, 342t
shrimp, 166, 171–73, 251
Sichuan peppercorns, 238
Sichuan-Tibet highway, 293
Sierra Leone, fish processing in, 269
silage, 242
Silk Road, 370
siltation. See sediment deposition
single variable maximization, 15, 124–27;

of agricultural chemicals, 261–62; and Banqiao dam collapse, 148; disrupts ecosystem feedbacks, 129–33; in Great Leap Forward, 131–32, 137; in shrimp disaster, 172–73; in socialist industrialization, 323; staple production, 140
sintering: and air pollution, 296
Sipsongpanna, 42; Akha people in, 42, 46–50; dams in, 220; Tai people in, 50
666 (pesticide), 149–50, 259–60
slogans, 424, 441–46. See also propaganda
slow variables, 11–12; in Qing period, 94–95
sludge, 399–400; as fertilizer, 399; "land application," 399
small industries, five, 322–23, 328
snowmelt, 176
social-ecological systems, 3–4; asymmetry of, 5, 35; China as, 6–8; core-periphery strucutre, 5, 35; emic and etic scales, 6; marketing hierarchy as, 35; overlap of componenent parts, 5; spatial scale in, 3–7; temporal scale in, 8–12
socialism: Engels on, 71; as nonpolluting, 69–70
socialist construction, 79
socialist high tide, 110–11
soil: CAFO agriculture and, 432; loss, 95, 141; permanency of pollution of, 421, 430, 436; pollution of, 260, 316, 321, 336, 430; quality of, 257, 258–59. See also erosion
solar energy, 409–10
Solomon Islands: wood imports from, 389
sorghum, 27; decline of, 232; hybrid varieties, 155
South-to-North Water Transfer, 190–96, 386–87; debates over, 190–91; fix-fixing in, 194, 199; and highways, 193; losses to evaporation, 191; moral hazards, 191, 199–200; pollution concerns, 191, 194, 197; and population resettlement, 194; and resilience, 194; as rigidity trap, 199; routes, 191, 193–94; and salinization, 190, 194; unused water, 198; volumes, 193–94; water pricing, 194
Soviet Union: collapse of, 20; collective

farms, 110; and dam construction, 130, 204; fear of war with, 316; as model for PRC, 75, 287; oil imports from, 313; payments to, in Great Leap Forward, 129; science in, 71; theft of Angang equipment, 289; urbanization models adopted from, 294; and waterworks, 111; withdrawal of aid from, 139, 315

soybeans: deforestation in South America and, 267, 432; genetically modified, 250; imports of, 230, 232, 233–24, 242, 245, 267; as pig feed, 234, 242

Spain: renewable energy in, 409

sparrow-bashing, 133

specialization, agricultural, 163–66, 247–48; and resilience, 265

sprawl, urban, 358–59

stability: contrasted with resilience, 20–21; rigid, 21

Stalin, Josef, 71, 287

staple foods: as animal feed, 241–42; and deforestation, 141, 243; emphasis on, 98, 139–42; enlargement of farms, 248–49; and erosion, 243; in Great Leap Forward, 128–29; imports, 157, 163, 230–31, 249; increases in, 97, 155, 158, 231–32, 248; lack of surplus, 241–42; losses to floods, 120; Mao on, 140; on marginal land, 243; and maximization of cropland, 140; record harvests, 245, 380–81; retreat from emphasis on, 163–66, 243; self-sufficiency in, 230, 248–49; trends in, 99f; upgrading, 232; world prices, 230

State Environmental Protection Administration/Office, 225, 332, 425

state farms: in Manchuria, 100, 102

steam-coal-steel development block, 288

steel: air pollution from, 296, 378; at Angang, 289–91, 299–300; Baoshan, 363–64; at Baotou, 290, 291, 300; as basis for industrialization, 287; at Beijing, 289, 291, 300, 364; China as world leader, 2, 344; and consumer durables, 341; electric-arc furnaces, 394–96; energy consumption, 344; in

Great Leap Forward, 82, 299–304; and greenhouse gas emissions, 344, 378; growth after Reform and Opening, 343t; at Jiuquan, 318; at Panzhihua, 316, 318–19; relocation of mills, 363–64; reproccesing, 394–96; at Shanghai, 363–64; small-scale, 300–304; unusable, 302–4; urbanization and, 354–55; and water pollution, 378; at Wuhan, 291, 300. See also iron

stochastic processes, 11; buffers against, 49

subsidence, land, 170, 190, 334, 383–84

subways, 366, 367m

sugar: cane farming, 252; imports of, 245

sulfur: content in coal, 307, 417; content in iron, 296, 302; in industrial waste, 346

sulfur dioxide (SO_2): as air pollutant, 412, 417; from chemical industry, 346; content in oil, 415, 417; from copper and zinc smelting, 297; from steelmaking, 296

Sun Yat-sen, 203–4

Sungari (Songhua) River, 102; chemical spill in, 346–48; floods, 177; sedimentation in, 141

supermarkets, 277–78

supply chains, agricultural, 246

surveillance, 263, 373

sustainable development, 424

Svarverud, Rune, 298, 411

sweet potatoes, 27; decline of, 232; improved varieties of, 155; as staple, 34; yields, 125–26

synthetic fibers, 340

Tai, Lake, 173–75

Tai people, 50, 51

Taiwan: environmental trajectory of, 435; ozone pollution in, 416–17

Taklimakan desert, 56

tangerines, 163

Tangshan, Hebei: air pollution at, 297–98, 413–14; coal mines at, 292; earthquake, 310, 324–25; steel production at, 394–96; worker health in, 297–98

Tarim Basin, 56, 171

tea: trade in, 55

technological lock-in, 108, 423, 429

tenancy, 34

terraces, 27; as infrastructural buffers, 37

Thailand: as hydropower importer, 221, 227

thermoelectric power: compared to hydropower, 406

Third Front industrialization, 84, 316–23; energy intensity of, 321; environmental injustice in, 322; evaluation of, 322; exacerbates environmental degradation, 321–22; Mao on, 317; regional projects, 316; transportation, 317

Third Plenum, 1978, 327, 329

"three *nong* questions," 278–79

Three North Shelterbelt Project, 431

"three parallel rivers" (Yunnan), 223, 225

three Rs (reduce, reuse, recycle), 390

throwaway ethic, 390, 394

Tianjin: earthquakes, 325; floods, 182; land subsidence, 170, 383–84; water transfers, 146, 193

Tibet: cities, 370; collectivization, 119; dams, 228; ecology, 56; Han migration to, 370; rebellions in, 119; rivers of, 193, 228; urbanization in, 370; waterworks in, 191–93

Tibetan antelope, 431

Tibetan Plateau, 220

Tilt, Bryan, 208, 224–25, 332–33, 418

Tippins, Jennifer, 282–83

tires, 388t

titanium, 318

toilets, 32, 380, 398

Tonle Sap, 221–22

towns, satellite, 364–65

township and village enterprises (TVEs), 329–40; decline of, 330–31; decreasing pollution in 1990s, 339–40; foreign investments in, 330; local government dependency on, 330, 336–40; loss of competitiveness, 339–40; manufacturing among, 329–30; numbers of, 329; and pollution, 330, 331–40; privatiza-tion of, 330–31, 338; profitability, 338; and rural income, 329–30

tractors: lack of, in early years, 99; small, 152, 253

trade-offs: admitting or ignoring a problem, 347; in dam building, 405–6; defense and environment, 321; development and environment, 194, 436; diet and ecosystem health, 436; dietary, 240–41; ecological and infrastructural buffers, 189; economic growth and decarbonization, 411; economy and environment, 318, 322, 324, 339, 345, 349–50, 424; energy efficiency and growth, 344; environment and greenhouse gases, 405–8, 419; growth and environmental justice, 432; housing and production, 310; industrialization and pollution, 298, 336–37; nuclear power, 408; pollution and wealth, 335–37; productivity and resilience, 36, 104, 241, 256–62, 438; renewable energy and environment, 410; resource efficiency and growth, 389; spatial, 143–44; temporal, 143–44, 257, 323; "warm and full" and "environmental protection" (*wenbao-huanbao*), 337

traditional knowledge: seen as "ignorance," 263

traffic, 366

transportation: air, 376–77; of coal, 419; intercity, 373–77; Japanese model, 374; public, 310, 365–66; railroad, 374–75; road, 373–74; traffic jams, 366; urbanization and, 365–66; US model, 374

trucks: coal hauled by, 419; import and manufacture of, 292–93. *See also* motor vehicles

tubewells, 84, 144–46, 170

TVEs. *See* township and village enterprises

Ukraine: Russian invasion of, 230, 245, 433

Ulanhu, 134–35

uniformity, dogmatic, 128

United Kingdom: agricultural yields, 258; renewable energy, 409; steel production, 299

United States: agricultural yields, 258; air pollution, 415; embassy of, 413; environmental governance in, 435; fear of war with, 316; greenhouse gas emissions, 403; hostility to China, 434; as model for agriculture, 246–50; as model for highway network, 373–74; nuclear weapons testing, 321; renewable energy in, 409; solar energy in, 409; as source of food imports, 234, 240–41, 242; steel production, 299

university campuses, 360, 364

urban communes, 305–6

"urban peasants" movement, 280

urban renewal, 359–62; in Xinjiang, 371–73

urban-rural inequality, 278–79; in consumption, 379, 385; Great Leap Forward exacerbates, 308

urbanization: and air pollution, 378–79; autarky declines with, 378; boulevards, 359–60, 361f; cement and, 387–88; in Central Asia, 307, 370–73; concrete and, 354–55, 387–88; consumer goods manufacturing and, 355–56; consumption and, 379, 385–86; controls on, 307; conversion of abandoned *danwei*, 357; copper and, 387; *danwei*-based model, 306–13, 327; Daqing model, 313–16; density, 356–57; and farmland loss, 242–45, 351, 355, 379–81; fiscal reform and, 358–59; floods and, 211, 381–83, 400; and food security, 379–80; glass and, 354–55; in Great Leap, 305–8, 351–52; growth patterns, 355–59; and heat islands, 379, 384–85; historic preservation, 360–62; housing conversion, 294; and industrialization, 293–95; "infill growth," 356; iron ore and, 387–88; and land subsidence, 170, 351, 382–83; "leapfrog growth," 356; and microclimate changes, 351, 384–85; migrant "villages," 357; migration and,

351, 353–54; neglect of consumer amenities, 294; in 1950s, 293–95; "pancake growth," 356; parks (public), 360; in Pearl River Delta, 357–58; "producer cities of socialism," 294, 327; property law reform, 356; rationing and, 353; recyclers' neighborhoods, 357; and reduced resilience, 377; and resettlement, 207–8; residential parks/compounds, 359; and resource consumption, 354–55, 385–89; ring roads, 359–60; and rural crisis of Reform and Opening, 278–79; sand and, 387–88; satellite cities, 364–65; Soviet models, 294; sponge cities, 382–83; sprawl, 244, 358–59; steel and, 354–55, 387–88; in Tibet, 370; traffic grid, 359–60; university campuses, 360; unplanned, 358–59; "villages within the city," 357–58; and waste, 351, 378; and water demand, 190, 351, 379, 383–84; and water pollution, 351, 378–79; wealth and, 354–55; wood products and, 388–89; in Xinjiang, 370–73; in Zomia, 367–69; zoning, 356. *See also* cities; housing

Uruguay: beef imports from, 265–67

Ürümchi, Xinjiang: air pollution in, 414–15; *mehelle* (neighborhoods), 371; urban renewal in, 371

utopianism, 70–71, 82–83

Uyghur people: Land Reform among, 119; surveillance of, 263

vaccination, 97

vanadium, 318

vegetables: consumption in 21st century, 237; exports, 245; gardening, 27; pesticide use on, 260, 276–77; urban gardens, 277

Vietnam: effects of Lancang/Mekong dams on, 221

villages: Akha, 46; in China Proper, 33–35; consolidation, 264, 278–79

Visser, Robin, 365

voluntarism: in policy, 76; and waterworks, 114

vulnerability: and development, 3, 423; increases in Reform period, 84–85; intensification and, 171–73; of irrigation works, 2; and rigidity traps, 20–21; of scientific modernism, 15; and slow variables, 12; in 21st century, 423. *See also* resilience

Walker, Brian, 323
walls: *danwei*, 309–10; dismantling of city, 309
Wang Fusheng, 191–93
Wang Yongchen, 225
"war against nature," 323
"warm and full" problem, 157, 240, 241; consequences of solving, 276, 336–37; and environmental degradation, 336–37
washing machines, 361
waste: bins, 363f, 393; chemical, 345–46; crisis, 389–90; dry, 393; dumping, 390–91; electronic, 396–97; import ban, 396; imports of, 394, 396; incineration of, 389; increases in urban, 351, 378; industrial, 321–22; infrastructure for, inadequate, 393; lake eutrophication and municipal, 174; landfilling, 389; mining, 295, 321; prioritization, 378; recycling, 357, 394–97; reprocessing, 394–96; separation, 393–94; solid, 390–94; wet, 393
wastelands, 295
wastewater: discharge, 398; fix-fixing for, 399; pollution from treatment, 346; recycling, 387; treatment, 346, 398–99. *See also* sewage
water allocation, 190
water buffaloes, 33
water heaters, solar, 409
water pollution: and algae, 392; antibiotics as, 392; CAFO agriculture and, 432; chemical industry and, 345–49; compared to air pollution, 421, 429–30; dilution as solution to, 347–48, 398; and disease, 335; in drinking water, 332; and fish, 219, 335, 392; growth and, 344; heavy metals in, 332, 391; and

industrial discharge, 169–70, 294, 331–32; international repercussions, 347–48; lack of controls in high socialist period, 294–95; landfills and, 391–92; mining and, 295, 335; organic chemicals, 391; at Panzhihua, 321; persistence of, 429–30; pharmaceuticals, 392; and rural industry, 331–32; in Sanxia Reservoir, 215; of South–North Water Transfer, 191; and urban discharge, 331, 351; from wastewater treatment, 346; in Xinjiang, 171. *See also* groundwater: pollution
water pricing: and South–North Water Transfer, 190, 194, 198
water quality, 331, 398, 429–30
water sediment regulation, 187–88, 189
water table: on North China Plain, 114, 167–68
water usage, 386–87
waterlogging: in Huai Basin, 107–8; on North China Plain, 121, 122, 144–45
waterworks: campaign, 1957–58, 82, 120–23; collectivization and, 111–16; as fix-fixing, 21, 194; high modernism and, 111; improvement in Qing, 94; and intensification, 37; legitimacy and, 111, 114; Mao on, 190; moratorium on Yellow River, 189; planned, 198–99; poorly planned, 122; repair of, 111; and resilience, 37; and salinization, 130; small-scale, 114; South–North Transfer, 190–95; Soviet influence, 111; transfers, 190–99, 386–87; and water pollution, 169–70; Xinjiang, 193; Yellow River, 111–14, 186–89. *See also* dams; irrigation
weather: buffers against, 94f, 381; extreme, 431, 438; and Great Leap Forward, 127, 137–38
Wei River (Henan), 111
Wei River (Shaanxi), 130; flooding, 186; as hanging river, 186; replenishing, 195
wells: and coal mining, 334–35; on North China Plain, 114. *See also* tubewells

Wen Jiabao, 225–26, 236, 264
Wen Tiejun, 280
"Western-style" democracy: avoiding, 79; as forbidden topic, 425; as scapegoat for environmental degradation, 86–87
wetlands: artificial, 382; Poyang Lake, 212; reclamation of, 180, 190; restoration of urban, 382
wheat, 26–27; deep tilling and close planting of, 125; hybrid, 153, 155; imports, 231; increased production in 21st century, 231; replaces coarse grains in diet, 232; as staple, 34
Wilmsen, Brooke, 208
Winckler, Edwin, 89
wind energy, 408–9
Wittfogel, Karl, 439
women's labor: devaluation of, 124, 306, 313–14, 449n7; in factories, 340–41; in Great Leap Forward, 123–24, 126; as jiashu (dependents), 311, 313; migrant, 340–41
women's liberation, 124; Mao on, 82
Wood, James, 10–11
wood products: urbanization and, 388–89
work ethic, 161
work points, 110–11, 157, 449n7
world-systems models, 66
wormholes: as sign of safe food, 276
Wuhan, Hubei: diminished resilience of, 381; floods in, 178, 204, 210, 381; lake eutrophication in, 175; lakes filled in, 381; landfills in, 391; steelmaking at, 291, 300

Xi Jinping, 79, 376; and China dream, 424; on climate change, 404; and dams, 201–3; and ecological civilization, 349, 424; on food security, 245, 380–81; institutes for the study of, 89; on peak carbon, 403; on poverty alleviation, 411, 421; on sponge cities, 382
Xichang, Sichuan: expressway to, 373, 374f; as medium-sized city, 366–68; railroad to, 373; steelmill at, 321
Xinjiang: "bulldozer state," 371–72;

cities, 370–73; cotton, 101; dietary change, 238–39; ecology, 56; as extractive colony, 293, 322; farming, 56; in Great Leap Forward, 134; and Han migration, 134, 307, 370; increases in agricultural land, 101–2; mining, 295, 307, 316; oases, 56, 96, 370; oil, 293, 307, 316; people's communes, 134; "people's war on terror," 373; Qing rule, 93, 96, 370–71; surveillance, 263, 373; urbanization, 307, 370–73; water pollution, 171; water transfers, 193, 199
Xishuangbanna. See Sipsongpanna
Xushui County, Hebei, 82–83, 126

yaks, 48, 53, 59; tents made from wool of, 54
Yan Yangchu (James Yen), 279
Yang, Martin, 27, 39, 40
Yangtze River. See Chang River
Yanyuan County, Liangshan, 164–65, 367–68. See also Baiwu Valley
Yarlung Tsangpo River, 193; dams on, 228
Yellow River: course changes, 113; delta morphology changes, 188–89; dikes, 113; discharge to ocean, 188; diversion of, 146; dry-ups, 144–45, 168–69, 185, 188, 190; floods, 113, 169, 189; General Plan for, 111–14; governance conflicts, 189; as hanging river, 113, 186; ice control, 189; nitrogen in, 188; regulation in 21st century, 186–89; replenishing, 195; rigidity traps, 189; sediment/siltation, 113, 169, 186–89; and South–North Water Transfer, 191; tributaries lacking, 113–14; water quality, 430; withdrawals, 111–14, 144–45, 168–69
yields, agricultural: increasing, 103
Yi people. See Nuosu people
Yu Jianrong, 21
Yu Xiaogang, 225–26
Yuan Longping, 154
Yunnan: as hydropower exporter, 220–21, 226
yurts, 54, 55

Zhang, Amy, 394

Zhang Boting, 406

Zhang Guangdou, 186

Zhang Zuoji, 348

Zhanjiang County, Guangdong, 166. *See also* Leizhou peninsula

Zheng Ruoyu, 275

Zhou, Kate Xiao, 161

Zhou Enlai, 204, 291; death of, 325–26

Zhu De: death of, 325–26

Zhu Rongji, 412–13

Zhungaria. *See* Junggar Basin

zinc: and air pollution, 297, 332–33; energy intensity of, 332–33; imports of, 345; mining, 335; urbanization and, 387

Zomia, 7m, 8, 41–42, 42m, 223; buffers in, 17, 46–52; cities in, 367–69; climate, 41; dams in, 219–29; Democratic Reforms in, 105, 118–19; ecological buffers in, 48; ethnic diversity in, 41–42; household ecology in, 43–46; industry in, 316; institutional buffers in, 48–49; Land Reform in, 105; Qing environmental changes in, 96; rebellions in, 119; rivers of, 219–28; solar farms in, 410; solar water heaters in, 409

zoning, urban, 356

zooplankton, 173